CIRCUMSTELLAR MATTER 1994

CIRCUMSTELLAR MATTER 1994

Proceedings of an International Conference to Celebrate
the Centenary of the Royal Observatory, Edinburgh,
held at the Edinburgh Conference Centre,
Heriot-Watt University,
Riccarton, Edinburgh, Scotland,
29 August – 2 September, 1994

Edited by

G. D. WATT and P. M. WILLIAMS

Reprinted from Astrophysics and Space Science
Volume 224, Nos. 1–2, 1995

SPRINGER-SCIENCE+BUSINESS MEDIA, B.V.

A. C.I.P. Catalogue record for this book is available from the Library of Congress.

ISBN 978-94-010-4066-2   ISBN 978-94-011-0147-9 (eBook)
DOI 10.1007/978-94-011-0147-9

Printed on acid-free paper

TABLE OF CONTENTS

SESSION C:
HIGH MASS STARS & ULTRACOMPACT HII REGIONS

SESSION D:
DUST & GAS; COMPOSITION & CHEMISTRY

CONCLUDING REMARKS

INDICES

CIRCUMSTELLAR MATTER

1994

PREFACE

The conference recorded in this volume was one of the events organised to celebrate the centenary of the (re)establishment of the Royal Observatory, Edinburgh, on Blackford Hill in 1884. Circumstellar Matter was selected as the topic because of important contributions toward research in the field by recent observations in the infrared and submillimetre, particularly with the two telescopes which the Observatory has both operated and built instrumentation for — the United Kingdom Infrared Telescope (UKIRT) and the James Clerk Maxwell Telescope (JCMT). The programme aimed to cover as many aspects of circumstellar matter as could fit into a one-week meeting, omitting only planetary nebulae, which had been well served by meetings in the previous two years. We thank the international scientific advisory committee (overleaf) for their help in selecting the Invited Reviewers around which the programme was built.

The Invited Reviews and oral contributions are included in the order and sections in which they were presented, even where re-ordering might have been more logical. We did not attempt to categorise the poster contributions but have included them in alphabetical order. An evening session for viewing and discussing posters in an unhurried atmosphere was very successful. A competition for the best poster was held and the prize was awarded for that by Lindqvist, Lucas, Olofsson, Omont, Eriksson & Gustafsson.

On another evening, the Royal Observatory was able to show participants some of the facilities on Blackford Hill, including the Schmidt plate library, the Visitor Centre, the Crawford Collection of historical astronomical books and the laboratory where the SCUBA bolometer array for the JCMT is being built. We are grateful to our colleagues Sue Tritton, Mike Read, Mark McAuley, Julie Mitchell, Angus Macdonald, Colin Cunningham and Walter Gear for their assistance. Other help from the Royal Observatory is gratefully acknowledged, particularly the time of and facilities for the local organising committee (overleaf). Nothing would ever have happened at all without the expertise of Anne Bryans, Dorothy Skedd and the pc!

It is a pleasure to acknowledge the receipt of generous sponsorship, without which the meeting could not have taken place. Grants were awarded by the Particle Physics & Astronomy Research Council (PPARC), Lothian and Edinburgh Enterprise Limited (LEEL), Herzberg Institute of Astrophysics, Kluwer Academic Publishers Limited, Cambridge University Press, John Wiley & Sons Limited and the Royal Astronomical Society. The excursions

Astrophysics and Space Science **224**: xiii–xiv, 1995.
© 1995 *Kluwer Academic Publishers.*

and buffet were partially funded by American Express Europe Limited. The International Science Foundation provided travel and lodging funds for several of the former-Soviet Union scientists. Inglis Allen subsidised the printing of the Book of Abstracts. We are grateful to the Lord Provost and Edinburgh District Council for the Civic Reception, and to Cllr Douglas Mackenzie for receiving the participants.
We thank the staff of the Heriot-Watt Conference Centre for efficient organisation, the Society of Writers to the Signet for allowing us to hold the Conference Dinner in their Library and the staff of The Peppermill for an superb dinner.
And finally, a brief thank you to all those who participated, especially if your contribution to this volume was camera-ready or on-time. We hope that you enjoyed the conference, the weather, the surroundings, the whisky, the firework display, etc, etc.

Graeme D Watt
Peredur M Williams
ROE, 29 November 1994

International Scientific Advisory Committee
John Dyson, Ian Howarth, Göran Sandell,
Rens Waters, Hans Olofsson & John MacLeod

Local Organising Committee
Isabelle Cherchneff, Suzanne Ramsay-Howat, Derek Ward-Thompson,
Mark Casali, Ray Wolstencroft, Sally McIntosh, Anne Bryans,
Dorothy Skedd, Graeme Watt & Peredur Williams

INTRODUCTION:
CIRCUMSTELLAR MATTER 1994

A. BOKSENBERG
Director, Royal Observatories

It gives me great pleasure to welcome you here in Edinburgh, both personally and on behalf of the Local Organising Committee for this, the Circumstellar Matter 1994 Conference.

Last year I had the task of opening a conference at the Royal Greenwich Observatory in Cambridge entitled Circumstellar Media in the late stages of stellar evolution and I am delighted to perform a similar task here for this conference for the Royal Observatory Edinburgh. Although the conferences bear a resemblance, this one is far broader in its scope and the contents of the two hardly overlap. The former concentrated on planetary nebulae, WR nebulae, supernovae ejecta and other mass loss manifestations, whereas this one has a high emphasis on star formation and the early stages of stellar/circumstellar evolution – protostellar and pre-main sequence shells and disks, pre-main sequence outflows, regions around newly formed stars - as well as progress to late evolutionary stages, and dealing with a host of related phenomena. I can see that some of the most exciting and interesting studies arise in the boundary regions or interface layers between stars and the interstellar medium. So, in all, it is perhaps not surprising that there is so much interest in the topic of Circumstellar Matter. The subject clearly has immense vitality and much progress is being made rapidly.

This meeting is intended as a centenary celebration for the Royal Observatory on Blackford Hill, so I will now say a few words about its history.

Originally the Royal Observatory was situated on Calton Hill overlooking the centre of the city, but as Edinburgh grew, the site became unsuitable due to light and smoke pollution. In 1888 Lord Crawford donated his priceless collection of telescopes and astronomy books to the nation, on the sole condition that a new Royal Observatory be built on Blackford Hill and be maintained by the government. Work commenced on the construction of the Blackford Hill site in the 1890s with a major feature of the Royal Observatory, the Library, being completed in 1894, as identified by the plaque which you can see on the end face of the building. The Crawford collection was installed within the site, where it remains today. For many years the Royal Observatory was also the home of the Astronomer Royal for Scotland. His house is now used for the business of the Royal Observatory and I have an office in there. Over the years the number of buildings has increased, the architectural styles have changed and the number of research activities on the site have expanded greatly. Today, 100 years later, the Royal Obser-

Astrophysics and Space Science **224**: xv-xvi, 1995.
© 1995 *Kluwer Academic Publishers.*

vatory still plays a key role in the world of astronomy. But of course, its site has been superseded. The major facilities which the Royal Observatory has established – the UK Schmidt Telescope with the powerful measuring machines, the UK Infra-Red Telescope and the James Clerk Maxwell Telescope – are particularly appropriate for the subject of this conference.

Last year the UK Observatories' structure was strengthened by a harmonisation of the Royal Observatory Edinburgh with the Royal Greenwich Observatory, which together with the telescope facilities on the islands of Hawaii and La Palma, now form a single organisation known as the Royal Observatories.

The people of Edinburgh take great pride in their Royal Observatory and this is reflected in the number of local public events in which the Observatory and its staff are involved. In addition to being an important venue for the Edinburgh International Science Festival, the Observatory incorporates an excellently developed and well-attended Visitor Centre. Members of staff regularly feature in television and radio interviews and frequently give public and schools lectures. Earlier this year, together with the University of Edinburgh, the Observatory hosted the European and National Astronomy Meeting which had over 400 participants.

It is appropriate and pleasing that this international gathering of astronomers and astrophysicists from all over the world should be here to celebrate the 100 years of astronomical history and achievement from this important institution.

Although the scientific agenda for this conference takes place entirely on this campus, there are several opportunities built into the programme to allow you to explore the city of Edinburgh and its surrounds, expose yourself to the Festival, and of course, to visit Blackford Hill. Tomorrow evening all conference participants are invited to attend a buffet on Blackford Hill when you will have a chance to tour the Visitor Centre to see the exhibits and to visit sections of the Royal Observatory to see some of the current activities.

The last major reviews in the broad field of Circumstellar Matter are contained in the Proceedings of the IAU Symposium 122 published in 1986. The intervening years have seen significant advances in angular resolution, vast improvements in sensitivity of instrumentation and increasingly more complex and sophisticated computer modelling. The agenda for the conference has been devised so that these features will be reviewed, summarised and developed, hopefully leading to improved theories for the origin and evolution of the circumstellar and interstellar material.

I wish all of you an enjoyable, informative, interesting and memorable week here in Edinburgh.

I now hand over to the Chairman of this morning's session. Let the fun begin!

SESSION A:

PROTOSTELLAR
&
PRE-MAIN SEQUENCE
SHELLS AND DISKS

CIRCUMSTELLAR ENVELOPES AND DISKS
OF PRE-MAIN SEQUENCE STARS

L. HARTMANN

Center for Astrophysics, 60 Garden St.,
Cambridge, MA 02138 USA

Abstract. The current state of knowledge about circumstellar matter of young stellar objects is briefly reviewed. It appears that some very young stars yet to accrete substantial amounts of mass may be seen through their dusty infalling envelopes even at optical wavelengths, because of the presence of holes or large departures from spherical symmetry in the envelopes. The evidence for this picture is summarized in the context of one well-studied young star, HL Tau, indicating that much of the large-scale structure originally identified as a rotating disk is probably a flattened infalling envelope. Departures from spherical symmetry in protostellar clouds are likely to lead to quite flattened structures once collapse gets under way, further suggesting that infall in large-scale toroids may be a general feature of low-mass star formation. The best kinematic evidence for Keplerian disk rotation comes from optical and near-infrared high-resolution spectroscopy of the innermost regions of circumstellar disks. Disk masses are uncertain but are likely to be at least the order of "minimum mass" solar nebula models, if not much larger.

Key words: Pre-main sequence stars · infall   circumstellar disks

1.   Introduction

Stars are thought to be formed by the rapid collapse of a cold molecular cloud core, leaving behind a remnant disk. Studies of the circumstellar gas and dust around young stars are needed to test this hypothesis. Over the last decade, dusty material near young stellar objects has been inferred from a variety of techniques, leading to the paradigm that low-mass pre-main sequence stars with strong emission lines (the so-called "classical" T Tauri stars) are surrounded by accretion disks, the latter plausibly representing precursors to planet formation.

In the last few years, observations of a few objects have provided tantalizing evidence for the rapid molecular envelope collapse predicted by the standard model. These developments have had the side effect of casting doubts upon some previous identifications of Keplerian disks. It now appears that many of the large scale dusty structures around young stellar objects are collapsing, flattened envelopes. The most general situation is probably one in which a flattened, toroidal envelope falls in to a much smaller, rotationally-supported disk. Detections of infall are likely to multiply in the next couple of years, as we struggle to understand where the envelope leaves off and the disk begins.

Astrophysics and Space Science **224**: 3–12, 1995.
© 1995 *Kluwer Academic Publishers.*

2.   Disks vs. Envelopes

2.1 GENERAL CONSIDERATIONS

To see why confusion could arise between dusty envelopes and disks, we start with some simple considerations. The parent cloud cores of low-mass stars are thought to be relatively quiescent, and mostly supported by thermal pressure against gravity, with modest contributions from magnetic fields and non-thermal motions (*e.g.* Shu *et al.* 1987). For a simple estimate, assume only thermal support; then the initial radius R of a protostellar cloud is approximately

$$R \approx GM/c_s^2.\tag{1}$$

If this cloud is rotating, then it will collapse to a disk of initial radius given by the angular momentum of the cloud. If we assume uniform rotation at angular velocity Ω, then the initial size of the Keplerian disk at the end of collapse will be given by the "centrifugal radius" r_c,

$$r_c = \Omega_o^2 R^4/(GM).\tag{2}$$

Now, if we assume a typical angular velocity seen in molecular cloud cores $\Omega = 3 \times 10^{-14}\,rad\,s^{-1} \approx 1\,km\,s^{-1}\,pc^{-1}$ (Goodman *et al.* 1993), we have the following results for a $M = 0.5M_\odot$ cloud for two representative temperatures of molecular clouds.

TABLE I

Cloud and centrifugal radii

T	R	r_c
10 K	5000 AU	29 AU
30 K	1666 AU	0.35 AU

One immediately sees that, even in the nearest star-forming regions ≈ 150 pc distant, initial cloud cores will be tens of arcseconds in size, and thus relatively easy to detect, while Keplerian disks are likely to be less than an arcsecond in size and thus difficult to detect. In other words, observed velocity gradients of cloud cores suggest that they are far from being rotationally supported. This argument is not definitive, since the centrifugal radius is quite sensitive to the assumed parameters, and one could easily imagine producing a much larger disk for individual objects. In addition, if any angular momentum transfer occurs in the disk during accretion, the outer edge of the disk can spread considerably (*e.g.* Lynden-Bell & Pringle 1974; Cassen

& Moosman 1981). Nevertheless, it is quite clear from these general considerations that caution must be employed in interpreting spatially-resolved (> 1 arcsec) structures as disks.

2.2 FLATTENED COLLAPSING ENVELOPES AND HOLES

The disk interpretation was favored for many optically-visible young stellar objects because of the argument that a star surrounded by its dusty collapsing envelope would be optically invisible for any reasonable parameters. On the other hand, a disk is an attractive and stable configuration to place relatively large amounts of dust for which the central star remains visible for almost any viewing angle (*e.g.* Beckwith *et al.* 1990).

Unfortunately, a disk is not the only configuration which satisfies this criterion, especially if continued infall eliminates the need for an equilibrium state. It has become apparent that circumstellar envelopes around young stellar objects often have large "holes" which allow short-wavelength light to escape readily. This was recognized several years ago by Adams, Lada & Shu (1987), who attempted to explain the emission from the so-called "Class I" sources, which are faint or invisible at optical wavelengths, and whose spectral energy distributions are dominated by mid- and far-infrared emission. Models of radiative equilibrium emission from dusty infalling envelopes were generally successful in explaining the observations, but the models fell far short in explaining the observed near-infrared emission. Only by positing enhanced scattering at short wavelengths could the observations be reconciled with infall models. This has been confirmed by from the work of Kenyon *et al.* (1993a,b), who similarly modelled a large fraction of the known Class I sources in the Taurus molecular cloud.

Powerful jets provide one possible mechanism to produce a hole in the circumstellar, possibly infalling envelopes (*e.g.* Shu *et al.* 1987; Whitney & Hartmann 1993). The momentum fluxes in these jets and molecular outflows, which appear to be common in the earliest stages of stellar evolution (cf. Terebey *et al.* 1989), are easily sufficient to punch holes in the surrounding envelope. Even if an object is not viewed directly down the outflow hole, it is quite possible to detect short-wavelength radiation scattered along the sides of the outflow hole (Whitney & Hartmann 1993). It appears that this actually occurs in some YSOs, whose optical emission is highly polarized because only scattered light is observed – the direct light of the central star is completely absorbed in the line of sight (Whitney & Hartmann, 1993).

There may be a more general and typical mechanism to produce holes in envelopes: an asymmetry in the protostellar cloud. Galli & Shu (1993a,b) have shown that, even with an initially spherical cloud, a modest internal magnetic field can deflect a portion of the collapsing material into a "pseudodisk", a flattened structure which is not rotationally supported, helping to reduce the amount of extincting material in the envelope.

Fig. 1. Evolution of a collapsing sheet. The grayscale images show the density distribution in a meridional plane for an axisymmetric calculation of collapse for an isothermal, non-rotating, self-gravitating sheet. The sheet is initially in hydrostatic equilibrium, but the mass enclosed in the spherical computational volume is larger than the Jeans mass, so instability and collapse eventually result. The image in the upper left shows that the density distribution after 2.7 free-fall times is still close to the original state. The upper right image shows the initial development of a massive central concentration at 6 free-fall times (roughly 3×10^5 yr). The lower left and right panels present the envelope evolution after 6.5 and 7 free-fall times, respectively, showing how the envelope becomes even more flattened as collapse proceeds, with an evacuated hole along the axis of symmetry. From Hartmann *et al.* (1994).

More generally, there is really no reason to assume an initially spherical cloud. Models of magnetically supported clouds in hydrostatic equilibrium necessarily show departures from spherical symmetry (Mouschovias 1976). Moreover, the molecular cloud cores thought to be the precursors of low-mass stars are elongated in approximately a 2:1 ratio (Myers *et al.* 1991). The fragmentation required to make binary or multiple star systems may be most naturally accomplished in sheet-like or filamentary cloud geometries (Larson 1985; Bonnell & Bastien 1992; Bonnell *et al.* 1992), for which there is ample observational evidence (Schneider & Elmegreen 1979). To investigate the potential effects of initial conditions, Hartmann *et al.* (1994) considered the axisymmetric collapse of a non-rotating, isothermal, self-gravitating sheet.

As shown in Figure 1, the initial response of the flattened cloud is to contract to a nearly spherical core. However, after a central mass has developed, the external material goes into nearly free-fall. The net result, as shown by the sequence of models in Figure 1, is that the cloud becomes much more flattened with time; the polar material collapses first, while material is still raining in from distant equatorial regions. The result is a toroidal distribution of dusty infalling material, *achieved with the complete absence of rotation*. Moreover, the relatively rapid collapse in the shortest dimension results in an evacuated, roughly conical cavity.

Qualitatively similar results have been obtained by Nakamura *et al.* (1994), who considered the collapse of filamentary molecular clouds, extending along relatively strong magnetic fields. In this case the filament initially collapses along its *longest* axis; nevertheless, the result is a dense, thin, non-rotationally supported disk-like structure, formed perpendicular to the magnetic field.

These simple simulations suggest that in general, infalling clouds may be quite flattened and quasi-toroidal. Obviously this weakens any case for Keplerian disks made mostly on the basis of flattened large-scale structure.

3.   HL Tau and Infall Models

To see how these ideas might be applied in detail, consider the classical T Tauri star HL Tau. Sargent & Beckwith (1987, 1991) showed that the ^{13}CO emission, observed with mm interferometry, was concentrated in a flattened distribution of extent ≈ 2000 AU, elongated perpendicular to the optical jet (Mundt *et al.* 1988). Sargent & Beckwith found that most of the dust mass was concentrated in an unresolved component 200 AU or less in extent, which they interpreted as the inner disk. However, they also interpreted the outer flattened structure as a disk as well, and presented evidence for rotation from the velocity channel maps.

Recently, Hayashi *et al.* (1993) also mapped the ^{13}CO emission of HL Tau. While their maps were reasonably consistent with those of Sargent & Beckwith, Hayashi *et al.* pointed out that the largest velocity gradient was in the direction of the *minor* axis, not the major axis as would be expected if the dominant motion of the flattened gas cloud were rotation. This observation implied that the largest motion is in the radial direction. Assuming that the flattened gas cloud is perpendicular to the optical jet, so that the southwest portion of the cloud is closest to us, Hayashi *et al.* concluded that the gaseous "disk" of HL Tau is in fact falling in toward the central star, with some component of rotation. Additional recent observations also suggest infall, with a rotation suggesting a current centrifugal radius of ≈ 200 AU (Koerner, personal communication).

These results are quite consistent with the general considerations leading to the table and with the flattened collapse models discussed in the previous section. Moreover, infall was suggested by two much earlier studies. Beckwith *et al.* (1989) explained the near-infrared scattered light nebula of HL Tau in terms of infalling material. Grasdalen *et al.* (1989) observed C_2 absorption redshifted by 20 kms^{-1} in the red optical spectrum of HL Tau, and also suggested infall. In both cases the inferred mass infall rate of a few times $10^{-6} M_\odot yr^{-1}$ is consistent with the infall rate inferred by Hayashi *et al.* (1993).

Another, independent argument for infall in HL Tau was developed by Calvet etal (1994), who showed that the "flat spectral energy distribution" of HL Tau was much more easily modelled with radiation from a dusty infalling envelope than from a disk. Simple disk models have difficulty in producing flat spectra; optically-thick, physically-thin disks, whether powered mostly by steady accretion, or heated by light from the central star impinging on the disk, have temperature distributions $T_{disk} \propto R^{-3/4}$, and therefore exhibit asymptotic spectral energy distributions with $\lambda L_\lambda \propto \lambda^{-4/3}$. This is far from the flat-spectrum $\lambda L_\lambda \propto \lambda^0$. Such a spectrum requires $T_{disk} \propto R^{-1/2}$, which implies that much of the luminosity is radiated at large distances, in contrast with simple disk models in which most of the energy is released deep in the gravitational potential well of the system. Several theoretical efforts have been made to develop mechanisms which might explain flat spectrum disks (Adams *et al.* 1988, 1989; Shu *et al.* 1990), but no convincing predictions of disk heating have yet been made.

Figure 2 shows that the SED of a typical Class I source in the Taurus molecular cloud, IRAS 04016+2610. It is immediately evident that the flat spectrum source HL Tau has a SED much closer to that of the Class I source than to the spectrum of the typical T Tauri star. This comparison suggests that the far-infrared emission of HL Tau also arises from an infalling dusty envelope rather than a disk. The dashed and solid lines in the middle panel of Figure 2 represent the thermal and scattered light emission, respectively, for a dusty envelope model with an infall rate of $4 \times 10^{-6} M_\odot yr^{-1}$, showing that infall models can indeed explain flat spectrum sources (Calvet *et al.* 1994).

Thus a variety of arguments – near-infrared scattered light, redshifted C_2 lines, mm-wave interferometry of ^{13}CO emission, and mid- to far-infrared emission models – all suggest that HL Tau is surrounded by an infalling, flattened, dusty envelope on $\approx 10^3$ AU scales.

More generally, several studies (*e.g.* Butner *et al.* 1991; Zhou *et al.* 1993; Kenyon *et al.* 1993a,b) have provided increasing evidence for infall around an increasing number of pre-main sequence objects. Disks and envelopes should be more easily discernable as resolution increases (*e.g.* the study of Keene & Masson (1990) who showed that the 2.7 mm interferometry of L1551 IRS 5 in Taurus is best explained by a "core-halo" structure, suggesting the presence of a disk of radius ≈ 45 AU situated within a much larger envelope). This

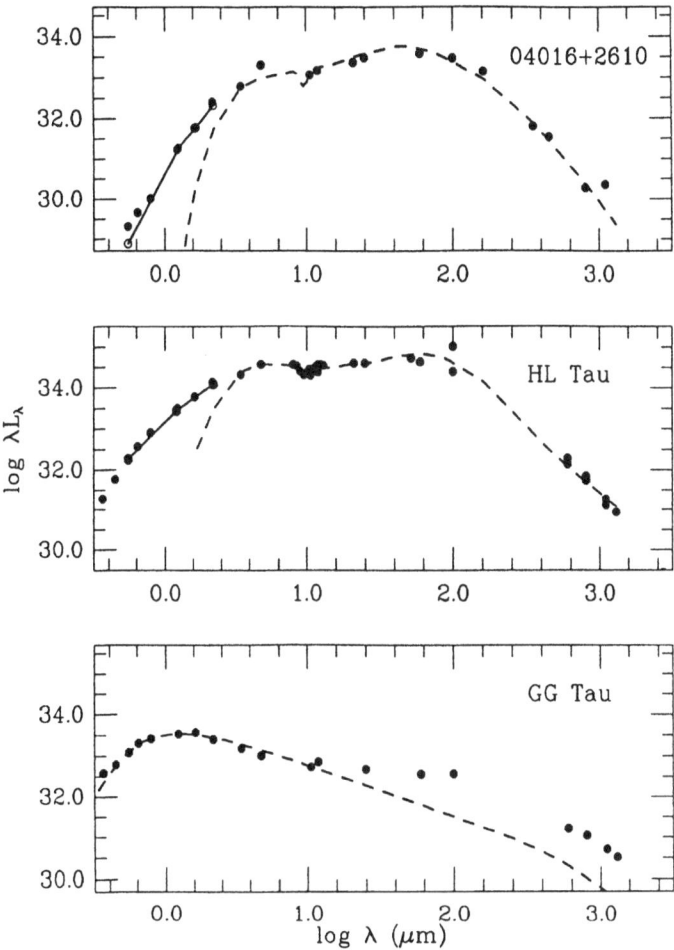

Fig. 2. Spectral energy distributions of three representative young stellar objects in Taurus. The uppermost object is a "Class I" source, whose spectral energy distribution is consistent with radiative equilibrium emission from a dusty infalling envelope model (dashed line), modified by extra light scattered out a hole in the envelope (solid line). The lowest panel represents a typical classical T Tauri star or Class II source, with the dashed line indicating the spectral energy distribution for a simple disk model. The middle panel shows the spectral energy distribution of HL Tau. The SED of HL Tau is much closer to that of the Class I source than to that of the T Tauri star. A dusty radiative equilibrium envelope model, with the same mass infall rate as the model in the uppermost panel but with more scattered light, accounts quite satisfactorily for the entire SED. The infall rate of $4 \times 10^{-6} M_\odot yr^{-1}$ is consistent with several independent estimates of mass infall rates for HL Tau (see text). From Calvet *et al.* (1994).

evidence should continue to accumulate even more rapidly in the next few years as radio interferometric maps become more sensitive and complete.

Infall may also play a role in dust and gas distributions even beyond the main protostellar collapse phase. Kenyon & Hartmann (1994) have found that the infrared colors of pre-main sequence stars in Taurus exhibit a continuous distribution from likely infall sources to colors consistent with standard disk models. This may suggest that infall does not terminate abruptly, but instead decays away over a finite time. Even low infall rates one or two orders of magnitude below the rates of the main infall phase could be important in explaining the excess infrared emission of many young stars, which in general is larger than that predicted by the standard disk models (Natta 1993).

4. Pre-Main Sequence Disks

4.1 KINEMATIC EVIDENCE

At present, the only reliable evidence for Keplerian motion in disks comes from optical and near-infrared high-resolution spectroscopy. These observations are sensitive to only the innermost regions of circumstellar disks, and therefore unfortunately are not spatially resolved. Nevertheless, the detection of double-peaked line profiles, indicative of a rotating flat object, strongly suggests of disk rotation, and this has been observed in the near-infrared spectra of several young stellar objects (see Carr's contribution, this volume).

In the case of a few FU Orionis objects, which are accreting so rapidly that the hot disks radiate at very short wavelengths, it is possible to measure the rotational broadening both at optical and near-infrared wavelengths (cf. Hartmann, Kenyon, & Hartigan 1993). Since observations at longer wavelengths sample larger disk regions, it is possible to use the wavelength-dependence of the disk emission to investigate the radial dependence of disk properties. The FU Ori objects show a differential rotation with wavelength, such that the cooler outer regions rotate more slowly than the hot inner regions, in a manner that is quantitatively consistent with Keplerian rotation (see also Welty et al. 1992).

4.2 DISK MASSES AND SIZES

From the discussion of the previous sections, it seems clear that we can be more confident of detecting disks on small spatial scales. Infalling or protostellar envelopes have most of their mass on large scales. For example, the mass of an envelope falling into a $0.5 M_\odot$ star at the standard rate for Taurus $\approx 4 \times 10^{-6} M_\odot yr^{-1}$ will have a mass of at most $\approx 4 \times 10^{-4} M_\odot$ within a radius of 100 AU. This mass scales as $r^{3/2}$, so that the enclosed mass is $\approx 10^{-2} M_\odot$ within 1000 AU. Thus, it would appear that a few hundredths of a solar mass contained within regions of a few hundred AU or less are most probably referring to disk material.

The landmark study of circumstellar material by Beckwith *et al.* (1990) for young stars in the Taurus molecular cloud provided the first systematic estimates of disk masses for low-mass pre-main sequence stars. In principle, the measurement of disk masses is best done from long-wavelength (*e.g.* mm) fluxes, where the disk grains should be generally optically thin, both from theoretical considerations and from observations of spectral indices *steeper* than Rayleigh-Jeans slopes. There is still substantial uncertainty in the calibrations, however, because dust opacities are poorly understood in this region. The observed spectral indices are inconsistent with the λ^{-2} dependence predicted for absorption of light with wavelengths much larger than the grain size, and with estimates from the diffuse interstellar medium. Moreover, there is the distinct possibility that grain accumulation/growth within the protoplanetary nebula quickly increases the population of larger grains. Several authors argue for an uncertainty of at most a factor of five in mm dust opacities, but this is based on flimsy arguments. My own guess is that the current estimates probably underestimate the amount of disk material.

In any event, the current estimates of $10^{-2} - 10^{-1} M_\odot$ for disk masses in Taurus are still suggestive, indicating that protoplanetary disks with masses greater than that of the so-called "minimum mass solar nebula" are reasonably common. The survey of Beckwith *et al.* (1990) indicated that roughly half of all young stars have substantial circumstellar disks at ages of a million years or so, consistent with estimates from near-infrared emission (Strom *et al.* 1993). Some objects without near-infrared excesses, and thus without inner disks, show evidence for more distant material from the mm fluxes; others with near-infrared excesses were not detectable in the mm survey. This suggests that the majority of low-mass stars once had circumstellar disks. Close binaries do not seem to have inner disks, unsurprisingly, but wide pairs may still have disks. Perhaps the most famous example of this is T Tau, which appears to have infrared disk emission despite the presence of an infrared companion at a projected distance of $\approx$ 100 AU. André (this volume) and collaborators have surveyed the dense star-forming region in the ρ Oph molecular cloud and have come to similar conclusions. Much work remains to be done on dense clusters which may be the sites of most star formation. Environmental effects, such as photoionization ablation of circumstellar disks by nearby hot, massive stars (*e.g.* Churchwell *et al.* 1987) could produce important differences in disk properties.

Acknowledgements

I wish to thank Scott Kenyon, Nuria Calvet, and Barbara Whitney for many useful conversations, and David Koerner for providing results in advance of publication. This work was supported in part by NASA grants NAGW 2306 and NAGW 2909.

References

Adams, F.C., Lada, C.J. & Shu, F.H.: 1987, Astrophys. J. **312**, 788.
Adams, F.C., Lada, C.J. & Shu, F.H.: 1988, Astrophys. J. **326**, 865.
Adams, F.C., Ruden, S.P. & Shu, F.H.: 1989, Astrophys. J. **358**, 495.
Beckwith, S.V.W., Sargent, A.I., Koresko, C.D. & Weintraub, D.A.: 1989, Astrophys. J. **343**, 393.
Beckwith, S., Sargent, A., Chini, R. & Gusten, R.: 1990, Astron. J. **99**, 1024.
Bonnell, I., Arcoragi, J.P., Martel, H. & Bastien, P.: 1992, Astrophys. J. **400**, 579.
Bonnell, I. & Bastien, P.: 1992, Astrophys. J. **401**, 654.
Butner, H.M., Evans, N.J., II, Lester, D.F., Levreault, R.M. & Strom, S.E.: 1991, Astrophys. J. **376**, 676.
Calvet, N., Hartmann, L., Kenyon, S. & Whitney, B.: 1994, Astrophys. J. in press.
Cassen, P. & Moosman, A.: 1981, Icarus **48**, 353.
Churchwell, E.B., Felli, M., Wood, D.O.S. & Massi, M.: 1987, Astrophys. J. **321**, 516.
Galli, D. & Shu, F.H.: 1993a, Astrophys. J. **417**, 220.
Galli, D. & Shu, F.H.: 1993b, Astrophys. J. **417**, 243.
Goodman, A.A., Benson, P.J., Fuller, G.A. & Myers, P.C.: 1993, Astrophys. J. **406**, 528.
Grasdalen, G.L., Sloan, G., Stout, M., Strom, S.E. & Welty, A.D.: 1989, Astrophys. J. **339**, L37.
Hartmann, L., Boss, A.P., Calvet, N. & Whitney, B.: 1994, Astrophys. J. **430**, L49.
Hartmann, L., Kenyon, S.J. & Hartigan, P.: 1993, *Young Stars: Episodic Phenomena, Activity, and Variability* in *Protostars and Planets III*, eds: E.H. Levy and J. Lunine, University of Arizona Press: Tucson, p497.
Hayashi, M., Ohashi, N. & Miyama, S.: 1993, Astrophys. J. **418**, L71.
Keene, J. & Masson, C.: 1990, Astrophys. J. **355**, 635.
Kenyon, S.J., Calvet, N. & Hartmann, L.: 1993a, Astrophys. J. **414**, 676.
Kenyon, S.J., Whitney, B., Gomez, M. & Hartmann, L.: 1993b, Astrophys. J. **414**, 773.
Larson, R.B.: 1985, Mon. Not. of the RAS **214**, 379.
Lynden-Bell, D. & Pringle, J. E.: 1974, Mon. Not. of the RAS **168**, 603.
Mouschovias, T.Ch.: 1976, Astrophys. J. **207**, 141.
Mundt, R., Ray, T.P. & Bührke, T.: 1988, Astrophys. J. **333**, L69.
Myers, P.C., Fuller, G.A., Goodman, A.A. & Benson, P.J.: 1991, Astrophys. J. **376**, 561.
Nakamura, F., Hanawa, T. & Nakano, T.: 1994, Astrophys. J. in press.
Natta, A.: 1993, Astrophys. J. **412**, 767.
Sargent, A.I. & Beckwith, S.: 1987, Astrophys. J. **323**, 294.
Sargent, A.I. & Beckwith, S.: 1991, Astrophys. J. **382**, L31.
Schneider, S. & Elmegreen, B.G.: 1979, Astrophys. J., Suppl. **41**, 87.
Shu, F.H., Adams, F.C. & Lizano, S.: 1987, Ann. Rev. of Astron. & Astrophys. **25**, 23.
Shu, F.H., Tremaine, S., Adams, F.C. & Ruden, S.P.: 1990, Astrophys. J. **358**, 495.
Strom, S.E., Edwards, S. & Skrutski, M.F.: 1993, *Evolutionary Time Scales for Circumstellar Disks Associated with Intermediate- and Solar-Type Stars* in *Protostars and Planets III*, eds: E.H. Levy and J. Lunine, University of Arizona Press: Tucson, p497.
Terebey, S., Vogel, S.N. & Myers, P.C.: 1989, Astrophys. J. **340**, 472.
Welty, A.D., Strom, S.E., Edwards, S., Kenyon, S.J. & Hartmann, L.W.: 1992, Astrophys. J. **397**, 260.
Whitney, B.A. & Hartmann, L.: 1993, Astrophys. J. **402**, 605.
Zhou, S., Evans, N.J., Kömpke, C. & Walmsley, C.M.: 1993, Astrophys. J. **404**, 232.

OBSERVATIONS OF DISKS AROUND PROTOSTELLAR SOURCES

WITH NOBEYAMA MILLIMETER ARRAY

NAGAYOSHI OHASHI *

Nobeyama Radio Observatory, National Astronomical Observatory, Japan

and

MASAHIKO HAYASHI

SUBARU project office, National Astronomical Observatory, Japan

Abstract. We present results from a survey observation of circumstellar disks around protostellar sources associated with the Taurus molecular cloud. Our result shows that the 98 GHz continuum emission tends to be weaker for embedded sources than for visible T Tauri stars, which is consistent with our previous interpretation of disk formation. Direct observations of the formation of a centrifugally supported viscous accretion disk around HL Tau is discussed.

Key words: circumstellar disks – disk formation – dynamical accretion – HL Tau

1. Nobeyama Millimeter Array Survey of Protostellar Sources

Recent observations of T Tauri stars have shown that $\sim 50\%$ of T Tauri stars are associated with compact circumstellar disks (*e.g.* Strom *et al.* 1989; Beckwith *et al.* 1990). The fact that T Tauri stars have compact circumstellar disks at high rate means that the disks should form in the prior stage to T Tauri phase, such as in the embedded phase when central sources are still optically invisible. Hence the observations of circumstellar structure around embedded sources are important for us to understand the formation processes of compact circumstellar disks. After reporting observations of 13 sources (Ohashi *et al.* 1991), we finished the survey of 19 young stellar objects associated with the Taurus molecular cloud.

Table 1 summarizes the result. The most striking result in Table 1 is the strong contrast of the continuum and line detectabilities between the embedded sources and T Tauri stars; embedded sources generally have detectable CS ($J = 2 - 1$) emission with upper limits to the 98 GHz continuum emission except for the two sources, whereas the opposite trend was observed for T Tauri stars. The upper limit to the disk mass around the observed embedded sources without continuum detection is ~ 0.03 M$_\odot$. This is consistent with the recent results that the disk mass around embedded sources in Taurus are at most ~ 0.01 M$_\odot$ (Moriarty-Schieven *et al.* 1994). Moreover,

Present address:Harvard-Smithsonian Center for Astrophysics, 60 Garden St., Cambridge, MA02138, USA

Table 1

Observed Sources	Optical Appearance	CS (J=2–1)			98GHz Continuum		
		I_{CS} (Jy kms^{-1})^a	Size (AU)^b	M (M$_\odot$)^c	F_ν(mJy)^a	Size (AU)^b	M (M$_\odot$)
L1489	invisible	5.6±0.32	2000×1700	1.1×10^{-1}	<9.6		<2.2×10^{-2}
04108+2803	invisible	1.4±0.32	2700×1200	2.6×10^{-2}	<21		<4.7×10^{-2}
04154+2823	invisible	<0.44			<19		<4.4×10^{-2}
04169+2702	invisible	2.6±0.32	1600×1500^d 2000×1200^d	4.9×10^{-2}	<14		<3.2×10^{-2}
04191+1523	invisible	0.80±0.085	1400×1400	1.5×10^{-2}	<9.0		<2.1×10^{-2}
04239+2436	invisible	1.4±0.066	2400×2100	2.6×10^{-2}	<12		<3.2×10^{-2}
04248+2612	invisible	1.7±0.17	2400×2100	4.3×10^{-2}	<12		<3.4×10^{-2}
04295+2251	invisible	<0.56			<21		<2.8×10^{-2}
L1551-IRS5	invisible	15±0.32	2300×1600	2.8×10^{-1}	131±4.0	<430×360	7.0×10^{-2}
04325+2402	invisible	0.85±0.080	1800×1400	1.6×10^{-1}	<9.2		<2.1×10^{-2}
04361+2547	invisible	1.4±0.32	2100×1500^d 2100×1500^d	2.6×10^{-2}	<14		<3.2×10^{-2}
04365+2535	invisible	2.5±0.095	1600×1600	4.6×10^{-2}	30±3.8	<940×700	6.8×10^{-2}
L1527	invisible	11±0.32	2600×2000	2.1×10^{-1}	<11		<2.5×10^{-2}
FS Tau	visible	<0.34			<11		<4.1×10^{-2}
T Tau	visible	<0.60			56±10	<550×420	1.0×10^{-1}
DG Tau	visible	1.4±0.32	1600×730	2.6×10^{-2}	57±4.0	<800×770	1.0×10^{-1}
HL Tau	visible	1.5±0.32	1200×850	2.8×10^{-2}	74±4.0	<530×420	1.5×10^{-1}
GG Tau	visible	<0.44			41±4.0	<780×670	1.6×10^{-1}
DL Tau	visible	<0.37			23±4.0	<1470×880	1.0×10^{-1}

^{a} Upper limit values are 3σ and errors are 1σ
^{b} FWHM size
^{c} Gas mass was derived under the assumption of optically thin CS emission
^{d} Two spatially separate components are identified

recent observations have shown that ∼ 40% of classical T Tauri stars in Taurus have disk mass larger than 0.01 M$_\odot$ (Osterloh and Beckwith 1994), which is also consistent with the high continuum detectability for T Tauri stars in the present study. Such a contrast between the embedded and visible objects means that disk mass around embedded sources is generally smaller than that around T Tauri stars. This suggests that the disk mass must grow rapidly in the course of evolution from embedded to visible sources, as has been discussed on the basis of a disk formation scenario in the previous paper (Ohashi *et al.* 1991)

2. Disk Formation around HL Tau

Recent observations of HL Tau with Nobeyama Millimeter Array have revealed direct evidence of dynamical infall in the gas disk around it (Hayashi, Ohashi, & Miyama 1993). This result gives evidence for the formation of a centrifugally supported compact disk at the very central part of the infalling disk as discussed below (Lin *et al.* 1994).

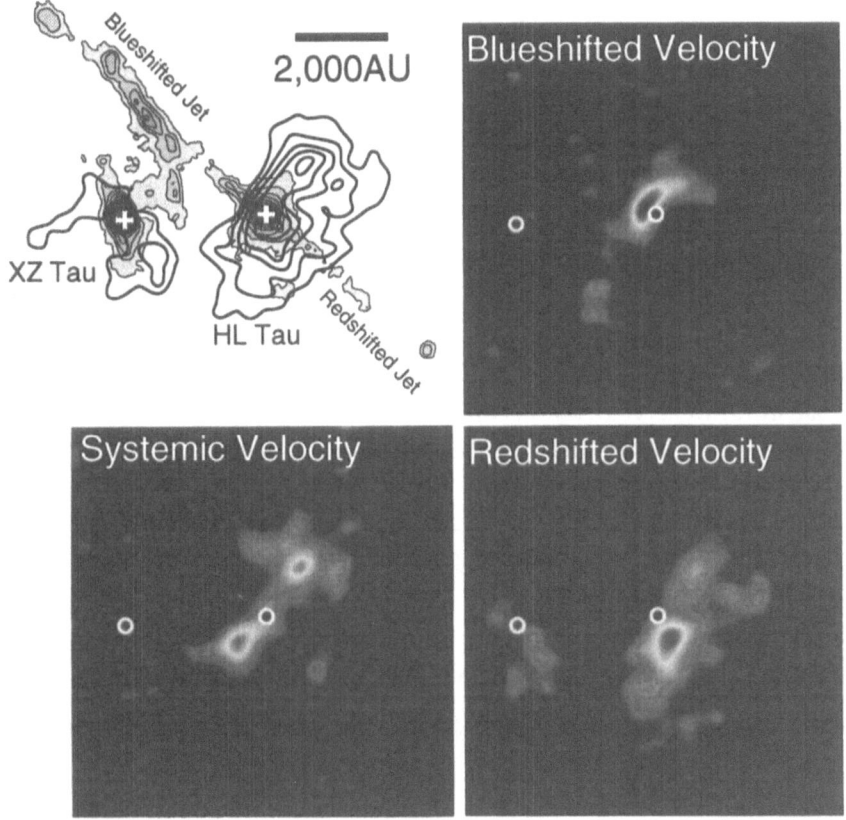

Fig. 1. ^{13}CO (J = 1 - 0) maps of HL Tau

Figure 1 summarizes the result. The upper left panel shows an integrated intensity map of ^{13}CO($J = 1 - 0$) superimposed on the optical jet (Mundt *et al.* 1988). The other three panels are velocity channel maps at blueshifted (upper right), systemic (lower left), and redshifted (lower right) velocities. The jet is blueshifted at its northeastern part and redshifted at its southwestern part. Because the disk is perpendicular to the jet, the northeastern part of the disk is, in turn, the far side, and its southwestern side is the near side. The three velocity channel maps show that the disk is approaching us at its far side, and is receding away at its near side, suggesting that the disk is shrinking toward the central star.

The radius and mass of the infalling disk are $\sim 1,400$ AU and 0.03 $M_\odot$, respectively. The measured infalling velocity, ~ 1 km s^{-1} at a radius of 700 AU, is consistent with that acquired when matter is freely accelerated by the central star with mass of 0.55 $M_\odot$. It is hence clear that the infalling disk is a dynamically accreting protostellar envelope. The inferred mass accretion rate of 5×10^{-6} $M_\odot yr^{-1}$ agrees well with theories of star formation (see *e.g.* Shu, Adams & Lizano 1987). Such a disk may form through dynamical collapse under the influence of magnetic fields (Galli & Shu 1993a, b).

The disk shows a velocity gradient also along its major axis, which implies that the infalling disk also has rotating motion. The rotation velocity is estimated as ~ 0.2 km s^{-1} at 700 AU in radius. Because the free-fall velocity is inversely proportional to the square root of radius and the rotation velocity is inversely proportional to radius, the rotation velocity increases much faster than the infalling velocity as matter infalls. The rotating velocity eventually dominates the kinematics of the disk, and a viscous accreting disk forms. The radius of the centrifugally supported disk is easily estimated as the radius where the rotational velocity becomes equal to the infalling velocity, and is ~ 30 AU. Recent interferometric observations have actually resolved the compact disk structure around HL Tau with its radius of ~ 60 AU (Lay *et al.* 1994). The case of HL Tau is, therefore, an excellent example of the formation of centrifugally supported viscous accretion disk through dynamical accretion from a protostellar envelope.

References

Beckwith, S.V.W., Sargent, A.I., Chini, R. & Güsten,R.: 1990, Astron. J. **99**, 924.
Galli, D. & Shu, F.H.: 1993a, Astrophys. J. **417**, 220.
Galli, D. & Shu, F.H.: 1993b, Astrophys. J. **417**, 243.
Hayashi, M., Ohashi, N. & Miyama, S.M.: 1993, Astrophys. J. **418**, L71.
Lay, O.P., Carlstrom, J.E., Hills, R.E. & Phillips, T.G.: 1994, Astrophys. J. in press.
Lin, D.N.C., Hayashi, M. Bell, K.R. & Ohashi, N.: 1994, Astrophys. J. in press.
Moriarty-Schieven, G.H., Wannier, P.G., Keene, J. & Tamura, M.: 1994, Astrophys. J. submitted.
Mundt, R., Ray, T.P. & Burke, T.: 1988, Astrophys. J. **333**, L69.
Ohashi, N., Kawabe, R., Hayashi, M. & Ishiguro, M.: 1991, Astron. J. **102**, 2054.
Osterloh, M. & Beckwith, S.V.W.: 1994, Astrophys. J. in press.
Shu, F.H., Adams, F.C. & Lizano, S.: 1987, Ann. Rev. of Astron. & Astrophys. **25**, 23.
Strom, K.M., Strom, S.E., Edwards, S., Cabrit, S. & Skrutskie, M.F.: 1989, Astron. J. **97**, 1451.

NEAR-IR SPECTROSCOPY AND THE PHOTOSPHERES OF EMBEDDED YSOS

MARK M. CASALI

Royal Observatory, Blackford Hill, Edinburgh EH9 3HJ

and

CARLOS EIROA

Dept. de Fisica Teorica, Un. Autonoma de Madrid

Abstract. We present a study of weak near-IR absorption lines in 44 low luminosity YSOs. Using a spectral resolution of ∼ 1000 most Class II sources show CO overtone absorption bands of varying strength in the K window, whether they have optical counterparts or not. Class I sources tend to show featureless 2 μm continua even though a S/N > 100 was achieved. High resolution (R=17000) echelle spectra were also obtained for a sub-sample of YSOs. Most show an unresolved ^{12}CO(2-0) bandhead, which when combined with inferred CO excitation temperatures and optical depths clearly points to a photospheric rather than a disk origin for the bands. They also show that embedded Class IIs are not rapidly rotating.

We find an excellent correlation between increasing near-IR colour excess and decreasing band strength and interpret this in a straightforward way as due to veiling of the stellar photosphere by circumstellar dust emission at 1000-1200 K, probably from a disk. A veiling correction was applied and intrinsic indices obtained for many YSOs. The results provide confirmation that Class II sources are equivalent to T Tauri stars.

Key words: Star Formation – Spectroscopy

1. Introduction

Since young embedded YSOs are often heavily obscured by dust extinction, photometric and spectroscopic studies require observations to be made at IR wavelengths. However, until recently IR spectra taken mainly with CVFs or single detector grating spectrometers at low resolution have tended to show that embedded YSOs generally have smooth spectra in the near-IR with few obvious features despite the fact that strong near-IR bands are observable in normal cool stars, especially giants and supergiants, due to CO and H_2O. But with the new generation of sensitive IR array detectors and spectrometers it has finally become possible to make high resolution observations of the near-IR radiation from YSOs. Casali and Matthews (1992) showed that absorption lines, particularly those of the ^{12}CO overtone bands in the K window, are common in young stars and Hodapp and Deane (1993) attempted the derivation of spctral types for an embedded cluster using IR spectroscopy.

In order to extend this early work we have carried out near-IR spectroscopy in the K band using the UK Infrared Telescope and the CGS4 spectrometer of 44 low luminosity YSOs of different types from Taurus,

Astrophysics and Space Science **224**: 17–20, 1995.
© 1995 *Kluwer Academic Publishers.* .

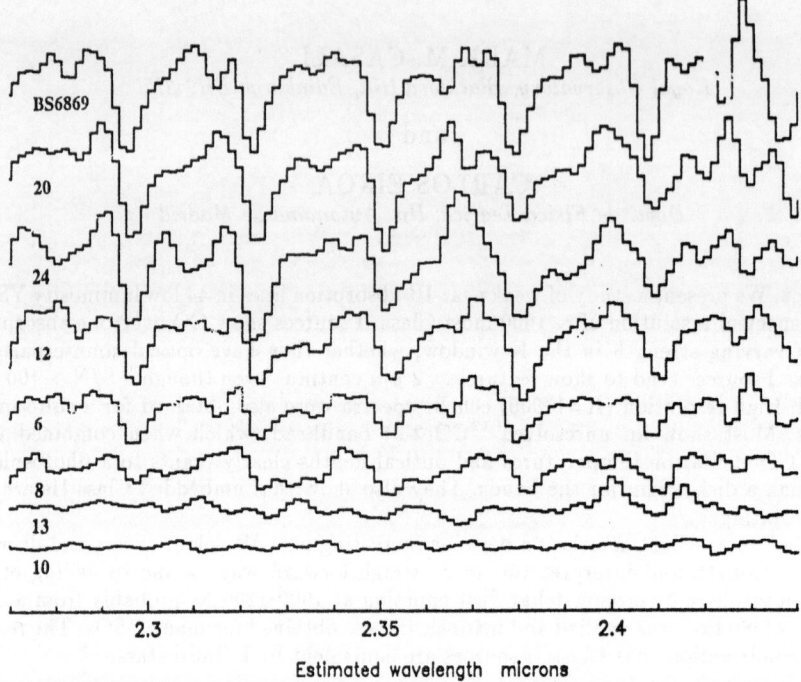

Fig. 1.   Sample spectra showing the observed range of CO band strengths in the YSOs.
BS6869 is a K2 giant for comparison.

Ophiuchus, Serpens, Perseus and L1641. Most of the YSOs have IRAS identifications and therefore well defined spectral energy distributions and luminosities.

2.   The Observed Spectra

^{12}CO bands are clearly visible in the absorption in the final spectra of many
of the YSOs studied and no obvious emission was observed in any of the
objects. A representative sample of spectra is shown in Fig. 1, illustrating
the range in band strengths observed. ^{13}CO bands are also weakly visible
in some of the spectra. Other weak lines such as MgI at 2.281 μm are
undetected in most of the spectra. Note that the ^{12}CO overtone bands are
visible from V=2-0 at 2.293 μm up to the V=7-5 transition at the long
wavelength edge of the K window.

Of the 44 sources observed 14/17 (80%) of the Class II sources show CO
absorption bands, while only 4/11 (36%) of Class I sources show absorption
features. Further, all the Class I sources which do show absorption bands
tend to have spectral indices close to zero, suggesting that these are in
effect in a transition region between the two classes. Since the spectral index

is normally calculated between 2.2 and 25 μm and since any foreground extinction can shift any Class II sources near zero index into the Class I region, the nature of these near flat spectrum sources must be considered ill-defined. With this in mind we conclude that the presence of CO absorption features in low luminosity YSOs is a good tracer of SED class.

3. The CO Bands are Photospheric

There are two likely physical origins for the absorption bands seen in the YSOs. The first and most obvious is that the CO arises in the YSO photosphere just as it does in normal stars. A second possibility is that the CO bands arise from a disk surrounding the YSOs. Such a mechanism would not be extraordinary since a large fraction of the IR radiation from YSOs can originate in a disk. To choose between the two cases we observed a sub-sample of the YSOs at high spectral resolution (17 km/s) with the echelle grating. The bandhead shows an unresolved drop in 7/9 cases, indicating a broadening of less that 17 km/s. This point is very important and shows that the CO bands do not usually arise in the inner part of a Keplerian disk since they should then show broadening of >100 km/s. Rather they are consistent with the normal microturbulent broadening of less than 10 km/s in the stellar atmosphere, as seen in the study of giants for example (Tsuji 1986). Further this also shows that the majority of YSOs are not rapidly rotating, a point of considerable importance in the study of angular momentum evolution in YSOs.

A photospheric origin for the CO bands also helps to explain an interesting correlation. In trying to find correlations between CO strength and various stellar parameters such as spectral index, luminosity etc., we found that the best correlation occurred between CO strength and near-IR excess as measured from the photometric indices J-H and H-K. It is the H-K excess beyond the normal reddening line from a stellar photosphere which is often taken to be an indicator of the amount of thermal dust emission from young stars. The correlation with CO is in the sense that sources with larger excesses tend to have the weakest CO strength. This can be quite simply understood as due to veiling of the photosphere and its CO bands by the dust emission. It also explains why in the steepest spectrum sources (Class I) the CO bands vanish altogether – they are very heavily veiled.

4. Conclusions

We conclude that (i) CO bands are photospheric in origin for most YSOs. (ii) The bands are usually broadened less than 17 km/s indicating that most obscured YSOs are also slow rotators. (iii) The bands are veiled by excess thermal dust emission. Steep spectrum sources show no CO bands. (iv) The

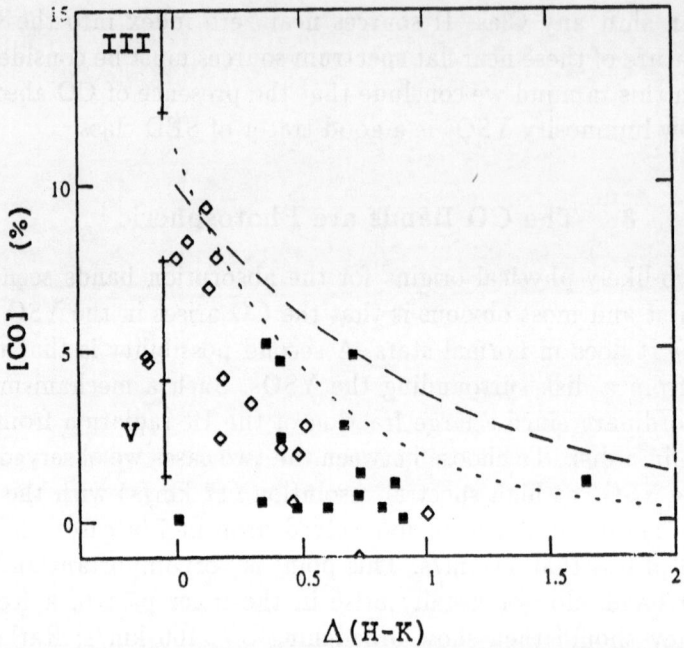

Fig. 2.   The Observed CO band strength vs H-K excess. Filled squares are sources more luminous and diamonds less luminous than 1 L⊙. Lines show the predicted behaviour of band strength given veiling at 1200 K (dotted line) and <900 K (dashed line)

CO bands are a good tracer of Class II sources.

References

Casali, M. & Matthews, H.: 1992, Mon. Not. of the RAS **258**, 399.
Hodapp, K. & Deane, J.: 1993, Astrophys. J., Suppl. **88**, 119.
Tsuji, T.: 1986, Astron. & Astrophys. **156**, 8.

MILLIMETER POLARIMETRY OF
STAR-FORMING REGIONS:
MAGNETIC FIELD STRUCTURE AROUND
LOW-MASS STARS

M. TAMURA and S. HAYASHI

National Astronomical Observatory, Mitaka, Japan

and

J. H. HOUGH

University of Hertfordshire

Abstract. We have measured polarization of the 1.1 mm and 0.8 mm continuum emission for 3 pre-T Tauri stars and 2 T Tauri stars. Positive detections were made for NGC 1333 IRAS 4 and IRAS 16293-2422, while L1551 IRS 5 and HL Tau were only marginally detected. For GG Tau we measured a $2\,\sigma$ upper limit of 3%. The polarization is interpreted in terms of thermal emission by magnetically aligned dust grains in circumstellar disks or envelopes. We have found a definite geometrical relation between the polarization and other circumstellar structure.

Key words: infrared: interstellar: continuum – polarization – stars: formation – stars

1. Introduction

Star-formation is believed to take place in the disk-like structure of gas and dust in the dense core region of molecular clouds. The existence of such disks around young stellar objects (YSOs) has been confirmed by radio and infrared observations. Recent radio observations have directly delineated a large-scale disk or envelope (more than several hundred AU) around low-mass YSOs in nearby dark clouds of Taurus and Ophiuchus (*e.g.* Sargent *et al.* 1988)

Evidence, although indirectly, has also been obtained for much smaller-scale disks around YSOs (less than 100 AU). The lack of red-shifted forbidden lines in most of the optical T Tauri stars (TTSs) is most likely due to the presence of obscuring disks of size 100 AU (Edwards *et al.* 1987). Large linear polarizations have been observed towards a number of TTSs at both optical (Bastien 1982) and near-infrared (Tamura and Sato 1989) wavelengths, which are probably due to the asymmetry of the dust disk structure. Near-IR (Strom *et al.* 1989) and millimeter data (Beckwith *et al.* 1990) have shown that roughly 50% of the TTSs have excess emission at these wavelengths, which is due to dust emission in the different parts of the small-scale disk.

The disks have a definite relationship with other circumstellar structure:

Astrophysics and Space Science **224**: 21–24, 1995.
© 1995 *Kluwer Academic Publishers.*

the major axis of the flattened disk is known to be perpendicular to the axis of the mass outflows observed as molecular outflows or optical/radio jets. This relationship suggests that the disk might play a vital role in the formation and/or collimation of outflows. Clearly it is important to understand how these flattened, axisymmetric structures are formed. Two of the most likely candidates affecting the shape of the cloud cores in the process of star-formation are rotation and magnetic fields. Although information on rotation can be obtained with mm and submm observations, it is very difficult to measure the magnetic field in the disk/envelope region.

Polarized thermal emission from magnetically aligned dust is of potential use in revealing the magnetic field structure in such dense regions like disks/envelopes (Hildebrand 1988). Successful observations of high mass stars have been made in several star-forming regions (*e.g.* Flett & Murray 1991; Kane *et al.* 1993). However, the observations toward the low-mass stars have been very limited (Tamura *et al.* 1993) because of their relative faintness at 1 mm.

2.   Observations and Results

The observations were made on the JCMT at Mauna Kea, Hawaii. The polarimetry was performed using the Aberdeen/QMC polarimeter and UKT 14 with a 1.1 mm or a 0.8 mm filter. The results for the five low-mass YSOs are summarized in Table 1.

IRAS 16293-2422, L1551 IRS5, and NGC 1333 IRAS4 are classified as pre-T Tauri stars, while GG Tau is a classical TTS. HL Tau is regarded as an transitional object from pre-TTS to classical TTS.

In Figure 1, we plot our 1.1 mm polarization vectors on the contour delineating the large envelope structure. Also shown are the directions of mass outflows and ambient magnetic fields.

3.   Discussion

The observed 1.1 mm continuum emission for pre-T Tauri stars is of thermal origin from dust in a disk-like structure of size several 1000 AU or an envelope, while that for T Tauri stars is of thermal origin from dust disk of size 100 AU. In the case of polarization due to thermal emission, the electric vector is perpendicular to the magnetic field direction projected on the sky.

In the three invisible IRAS sources, we found a common relationship between the envelope magnetic field inferred from our 1 mm polarimetry and the envelope geometry: the envelope magnetic field is perpendicular to the major axis of the envelope plane. In addition, excluding the ambiguous case of IRAS 4, we found that the envelope magnetic field in IRAS 16293-2422 and L1551 IRS 5 is parallel to the parent cloud magnetic field.

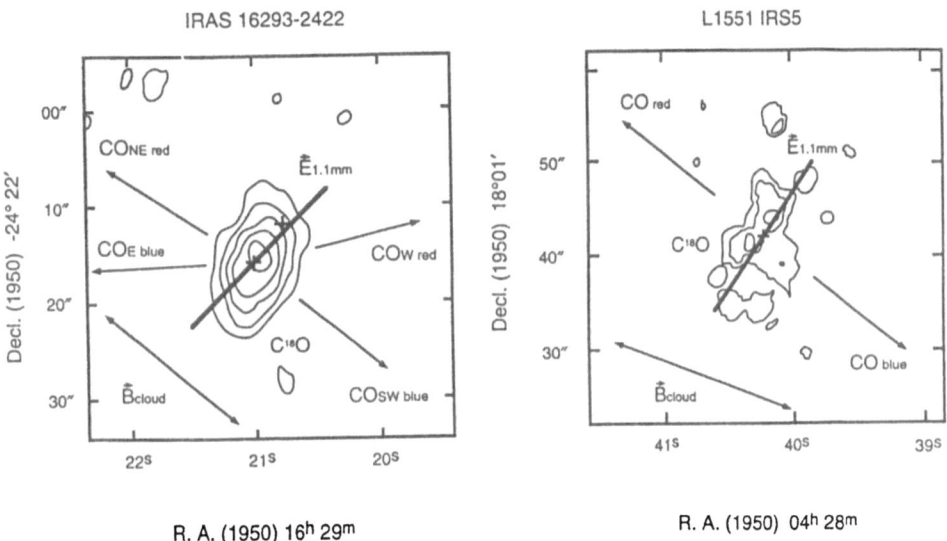

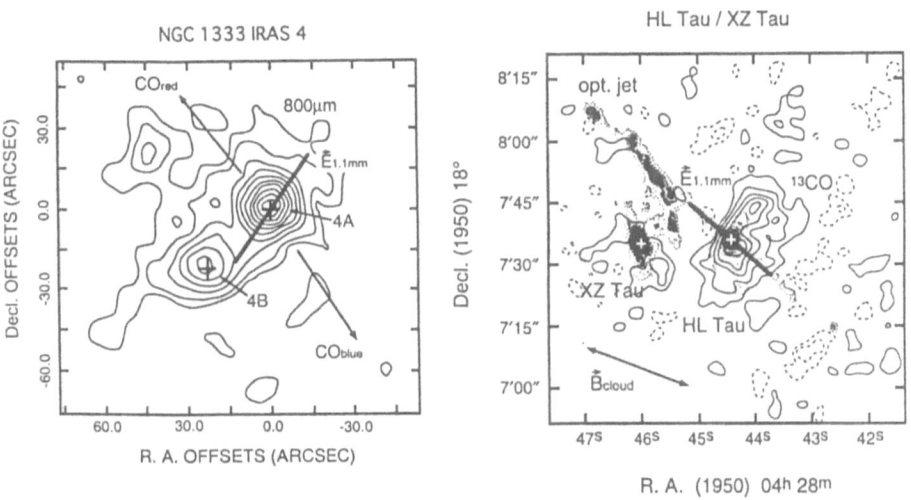

Fig. 1. Relationship between 1 mm polarization vector (this work) and other circumstellar structure (large-scale disks, CO outflows, and magnetic fields of embedding cloud).

TABLE I

1.1 mm polarimetry of low-mass YSOs

Object	λ (mm)	P (%)	P.A. (°)	Date	dis. (pc)	L (L⊙)
pre-T Tauri Stars						
IRAS16293-2422	1.1	2.2±0.4	135±5	91/06/25	160	29
N1333 IRAS4	1.1	4.6±0.8	145±5	92/02/12	350	28
	0.8	4.3±0.6	155±4	93/08/29		
L1551 IRS5	1.1	3.3±0.6	145±5	92/02/07	140	32
	0.8	0.2±0.5	—	93/08/30		
T Tauri Stars						
HL Tau	1.1	7.3±2.2	37±9	92/09/05	140	8.5
	1.1	3.4±1.8	89±20	93/08/31		
(average)	1.1	3.6±2.4	50±20			
	0.8	0.6±0.8	(65)	93/08/30		
GG Tau	0.8	0.8±1.1	—	93/08/31	140	2.8

In contrast, the 1 mm polarization of HL Tau, a T Tauri star, most likely represents the magnetic field in a transition zone between the disappearing envelope and disk. In this region, the magnetic field direction is perpendicular to the surrounding field direction. Thus, the field in the disk is detached from the parent cloud and might be dragged or twisted near the disk plane, suggesting that accretion or rotation dominates over the magnetic field.

We have not detected large mm polarization in GG Tau that clearly has a small-scale disk. This might suggest that the disk magnetic field is weak or that the dust grains in the disk region are not well aligned. Several theoretical arguments in fact suggest that magnetic fields would be decoupled from the gas in the disk because of very low ionization fraction in the disk.

References

Bastien, P.: 1982, Astron. & Astrophys. Suppl. **48**, 153.
Beckwith, S. V. W.: 1990, Astron. J. **99**, 924.
Edwards, S. *et al.* : 1987, Astrophys. J. **321**, 473.
Flett, A. M., & Murray, A. G.: 1991, Mon. Not. of the RAS **249**, 4P.
Hildebrand, R. H.: 1988, Quart. J. of the RAS **29**, 327.
Kane, B. D. *et al.* : 1993, Astrophys. J. **708**, 411.
Sargent, A. I. *et al.* : 1988, Astrophys. J. **333**, 936.
Strom, S. *et al.* : 1989, Astron. J. **97**, 1451.
Tamura, M. *et al.* : 1993, Astrophys. J. **404**, L21.
Tamura, M., & Sato, S.: 1989, Astron. J. **93**, 1368.

INFRARED CO EMISSION AND DISKS
AROUND YOUNG STARS

JOHN S. CARR

The Ohio State University, Astronomy Dept., Columbus, Ohio 43210-1106, USA

Abstract. High-resolution near-infrared spectroscopy of CO overtone emission bands has provided some of the best kinematic evidence for the existence of circumstellar disks around young stars. The CO emission flux and the detailed shape of the overtone bands are well matched by simple Keplerian disk models. A brief overview of the use of infrared CO emission as a diagnostic of the kinematics and conditions of gas in the inner disks of young stars is presented.

1. Introduction

The overtone bands of CO near 2.3 μm can be a useful probe of dense neutral circumstellar gas in the temperature range of 1000 to 6000 K. This was first demonstrated by Scoville *et al.* (1983) for the BN object. Modeling their high-resolution FTS spectrum of the CO bandhead emission, they deduced a temperature of $\approx$ 3500 K, an emitting region < 1 AU in size, and line widths for the individual CO transitions of 50 to 100 km s^{-1}. Since their discovery, emission in the overtone CO bands has been observed in a number of other young stellar objects (Thompson 1985; Geballe and Persson 1987; Carr 1989), but the low spectral resolution of these subsequent observations could not provide any information on the gas kinematics. Fabry-Perot observations of CO emission at a resolution of 85 km s^{-1} by Dent and Geballe (1991) revealed very large velocity dispersions in WL 16 ($\approx$ 400 km s^{-1}) and S106IR ($\approx$ 250 km s^{-1}). Further progress in studying the kinematics of the CO gas, and hence the origin of the emission, had to await a new generation of sensitive high-resolution echelle spectrographs.

2. Recent Work

CO overtone emission has now been observed in several young stellar objects using near-infrared echelle spectrographs with velocity resolutions of $13 - 38$ km s^{-1} (Carr and Tokunaga 1992; Carr *et al.* 1993; Chandler *et al.* 1993; Najita *et al.* 1994). The striking result from these new data is that in a large fraction of the objects the CO bandhead shows the characteristic shape expected for emission from a rotating disk, as was first shown clearly for the case of WL 16 (Carr *et al.* 1993).

A high-resolution spectrum of the $v = 2 - 0$ bandhead emission from gas with an intrinsically small velocity dispersion would resolve the individual CO transitions away from the bandhead, as shown in Figure 1a. A *single*

Astrophysics and Space Science **224**: 25–28, 1995.

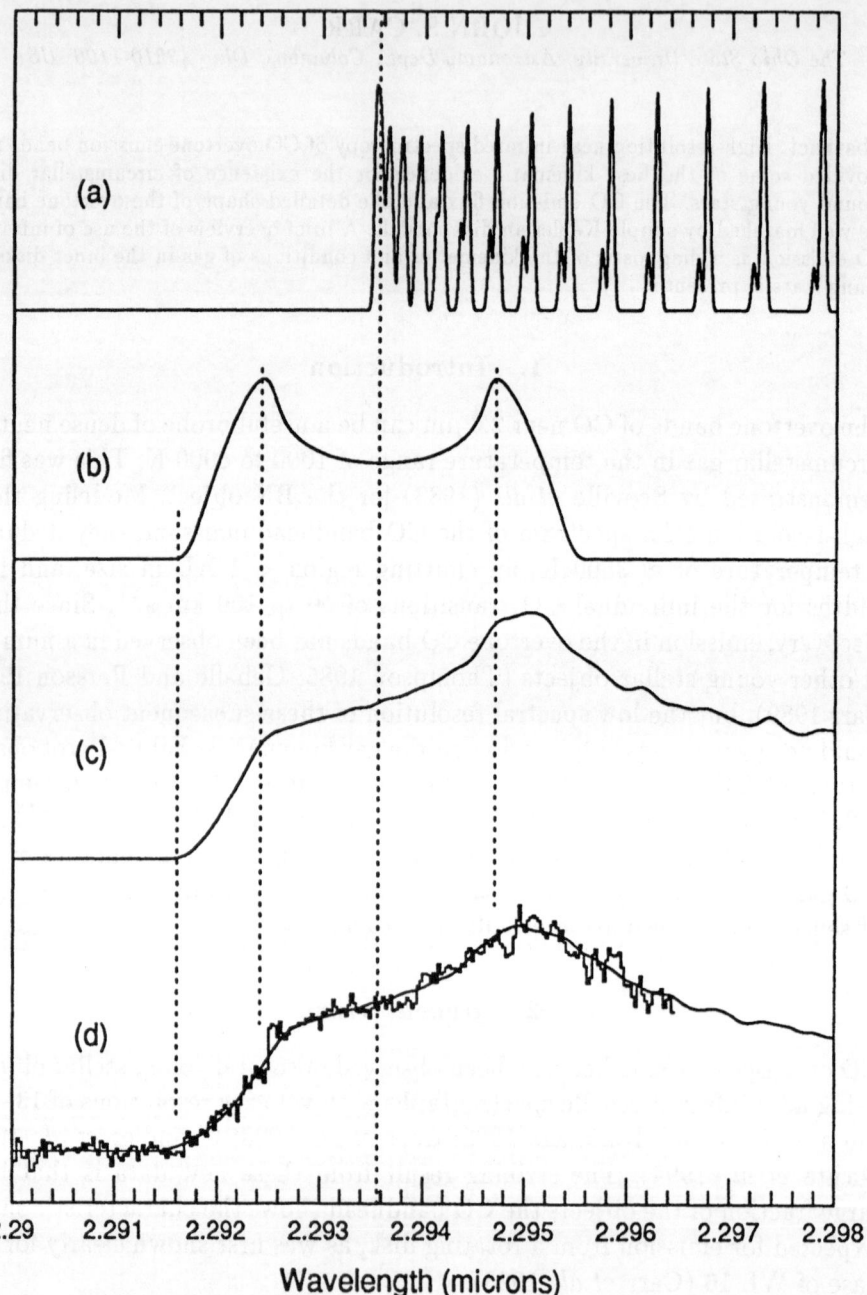

Fig. 1. (a) CO 2-0 bandhead spectrum of hot gas with a velocity dispersion of a few km s^{-1}. (b) Profile for a single emission line from a Keplerian disk with a maximum projected velocity of 250 km s^{-1}. (c) The CO 2-0 bandhead profile for emission from the same disk. (d) The observed CO 2-0 bandhead in the young star WL 16. The smooth line is a Keplerian disk model calculated for WL 16.

emission line from a rotating Keplerian disk will be double-peaked (Fig. 1b), and the line profile will depend upon the radial intensity distribution of that line and the projected velocities at the inner and outer radii of the emission. The emission profile of the 2 − 0 bandhead from a disk can be obtained by convolving the disk profile for a single line with the CO bandhead spectrum. The resulting spectrum (Fig. 1c) has a blue wing, a shoulder to the blue of the bandhead rest wavelength, and a redshifted intensity peak. The maximum velocity in the blue wing corresponds to emission from gas at the minimum radius of the CO emission. The velocity separation between the blue shoulder and the red peak is approximately equal to the separation of the peaks in a single emission line, and the projected velocity at the outermost radius of CO emission is roughly half of this velocity separation. Figure 1d compares the CO 2 − 0 bandhead spectrum from WL 16 to a Keplerian disk model, showing that a disk provides an excellent fit to the shape of the observed profile.

The CO overtone bands are composed of a large number of lines with different excitation potentials and line strengths, and the relative line intensities depend upon the CO excitation temperature and column density. Because these quantities are expected to vary as a function of radius in the disk, the shapes and relative fluxes of the overtone bands will depend upon both the gas kinematics and the range of gas temperatures and column densities that are present. Hence, by modeling the shape and flux of the CO emission bands, it is possible to place constraints on the radial properties of the emitting gas.

Accretion disk models for CO emission have been calculated by Carr (1989) and Calvet *et al.* (1991). In the models of Carr (1989), the emission occurs in inner regions of the disk which are optically thin in the continuum. In Calvet *et al.* (1991), the disk is assumed to be optically thick at all radii, and the CO emission originates in the upper disk atmosphere in which a temperature inversion is produced by stellar irradiation. Both of these models predict CO emission fluxes and velocity widths consistent with the observations, but they also predict CO emission to occur only in stars with relatively low mass accretion rates. In the models in Carr (1989), the inner disk will not be optically thin at high accretion rates; in Calvet *et al.* (1991), accretion heating will dominate stellar radiative heating at high accretion rates and a temperature inversion will not appear. These results contradict the fact that CO emission has been observed only in T Tauri stars with very high accretion rates, such as DG Tau and AS 353A.

Instead of attempting to fit our high-resolution data with physical accretion disk models (such as those above), we have modeled the spectra with a simple parameterized two-layer disk model. This allows us to test whether a rotating disk may plausibly explain the observations and to easily investigate the range of parameters required to fit the data, *i.e.* the gas temperature and surface density and their variation with radius, the physical radii of the emission, the projected Keplerian (or non-Keplerian) velocities, etc. Any physical model would have to reproduce these conditions.

In this model, the CO line emission occurs in a single vertically homogeneous disk layer in which the radial variation of temperature and surface density are characterized by power laws. The line emitting layer lies above a cooler optically thick continuum-producing layer characterized by a different temperature power law. This situation represents an optically thick accretion disk with a temperature inversion in the upper disk atmosphere (as in Calvet *et al.* 1991), but the mechanism for heating the gas is not specified. If the continuum layer is omitted from the model, then we have the case of an optically thin disk (as in Carr 1989). Further details of the modeling are given in Najita *et al.* (1994).

This model has been applied in detail to WL 16, the probable Herbig Ae/Be star 1548C27, and the T Tauri star DG Tau. In the case of late-type stars such as DG Tau, the stellar CO absorption spectrum must be modeled as well. Disk models are able to reproduce the shape of the CO bands in these three objects very well. The exponents in the temperature power law range from 0.4 to 0.8 with maximum CO temperatures between 3000 and 5000 K. The column density in the line emitting layer ranges from about 1 to 1000 g cm^{-2}. The inner radius for the CO emission is just a few times the stellar radius in each case, and the outer emission radius is 4 to 7 times the inner radius.

Acknowledgements

My collaborators in this research are J. Najita, A. Glassgold, A. Tokunaga and F. Shu. I have received support for this work from NSF grant AST 9124940 and from a Columbus Fellowship at The Ohio State University.

References

Calvet, N., Patiño, A., Magris C., G. & D'Alessio, P.: 1991, Astrophys. J. **380**, 617.
Carr, J. S.: 1989, Astrophys. J. **345**, 522.
Carr, J. S. & Tokunaga, A. T.: 1992, Astrophys. J. **393**, L67.
Carr, J. S., Tokunaga, A. T., Najita, J., Shu, F. H. & Glassgold, A. E.: 1993, Astrophys. J. **411**, L37.
Chandler, C. J., Carlstrom, J. E., Scoville, N. Z., Dent, W. R. F. & Geballe, T. R.: 1993, Astrophys. J. **412**, L71.
Dent, W. R. F. & Geballe, T. R.: 1991, Astron. & Astrophys. **252**, 775.
Geballe, T. R. & Persson, S. E.: 1987, Astrophys. J. **312**, 297.
Najita *et al.* : 1994, in preparation.
Scoville, N., Kleinmann, S. G., Hall, D. N. B. & Ridgway, S. T.: 1983, Astrophys. J. **275**, 201.
Thompson, R. I.: 1985, Astrophys. J. **299**, L41.

LOW-MASS PROTOSTARS AND PROTOSTELLAR STAGES

An Observational Perspective

PHILIPPE ANDRÉ
CEA, Service d'Astrophysique,
Centre d'Etudes de Saclay,
F-91191 Gif-sur-Yvette Cedex, France.

Abstract. The use of sensitive receivers on large ground-based radiotelescopes such as the JCMT, the IRAM 30 m MRT, and the VLA has recently yielded significant progress in our observational understanding of low-mass protostars. Submillimeter continuum observations suggest that the youngest stellar objects detected in the near-/mid-IR range – the so-called Class I sources or "infrared protostars" – have only residual amounts of circumstellar material and are thus relatively evolved. At the same time, a smaller number of colder and more obscured YSOs – designated "Class 0" – characterized by virtually no emission below 10 μm but strong submillimeter emission have been identified. These Class 0 or "submillimeter protostars" have not yet assembled the bulk of their final stellar mass, and correspond to the youngest protostar stage known to date (probable age $\lesssim 10^4$ yr). Direct evidence for gravitational infall has been found in some of these sources confirming their protostellar nature. However, most (if not all) Class 0 protostars already drive highly collimated CO outflows.

Key words: Protostars – Evolution – Infall – Jets

1. Background: What is a Protostar ?

The quest for protostars has often been described as the "Holy Grail" of infrared and submillimeter astronomy (Wynn-Williams 1982). One source of confusion is that various definitions of the word "protostar" exist in the literature, the most common of which being: (1) "an object deriving more than half its energy from accretion of infalling material" (*e.g.* Beichman *et al.* 1986), (2) "a stellar object in the process of assembling the bulk of the material it will contain when it ultimately reaches the main sequence" (*e.g.* Lada 1991; André, Ward-Thompson & Barsony 1993, hereafter AWB), and (3) "an isothermal cloud fragment undergoing gravitational collapse" (*e.g.* Chini *et al.* 1993; Güsten 1994). All these definitions are *a priori* acceptable, and probably correspond to different phases of the star formation process. The real challenge to observers is not so much to find a "true protostar", but to constrain possible theories by identifying a range of "protostellar sources" at various evolutionary stages and estimating the relevant associated physical parameters such as timescales.

According to current theoretical ideas on (isolated) star formation, low-mass stars form from the inside-out collapse of dense cloud cores (*e.g.* Shu *et al.* 1993). After a probably very short isothermal phase (*cf.* definition 3 above), during which the liberated gravitational energy is freely radiated

Astrophysics and Space Science **224**: 29–42, 1995.
© 1995 *Kluwer Academic Publishers.*

away (*e.g.* Larson 1969; Henriksen 1994), an opaque stellar object forms at the center and starts heating up, while continuing to build up its mass ($M_\star$) from a surrounding infalling envelope or cocoon (of mass M_{env}). The youngest protostars thus have $M_{env} \gg M_\star$ (which is the basis for definition 2 above). Their luminosity is well approximated by the infall luminosity $L_{inf} \approx GM_\star \dot{M}/R_\star$, where $\dot{M} \approx a_{eff}^3/G$ is the infall rate depending primarily on the effective sound speed a_{eff} of the cloud (*e.g.* Shu *et al.* 1993; Foster & Chevalier 1993). When protostars have accumulated most of their final, main sequence mass, they become pre-main sequence (PMS) stars (*e.g.* Stahler & Walter 1993) which evolve approximately at fixed mass (although accretion of residual amounts of material may still occur through an accretion disk).

Essentially two types of strategies have been used to search for protostars. The first approach (*e.g.* § 2) consists in looking for young stellar objects (YSOs) which are still deeply "self-embedded" in their circumstellar envelopes (as opposed to merely embedded in interstellar cloud material). The second approach (§ 3) is to try and identify cloud cores which are both cold and dense enough to collapse against their own internal pressure (according to the Jeans criterion). The youngest protostars yet observed (§ 4) were actually found by combining these two approaches.

This field has seen rapid observational progress in the last few years thanks to the advent of sensitive bolometers (including arrays in some cases) on large ground-based radiotelescopes such as the JCMT or the IRAM 30 m MRT. Because dust emission remains optically thin at $\lambda \sim 1$ mm up to extremely high column densities ($N_{H_2} \lesssim 10^{26}$ cm^{-2}), submillimeter continuum observations provide a very sensitive way to measure the masses of both circumstellar structures (envelopes and/or disks) around YSOs (*e.g.* § 2) and high density prestellar/protostellar clumps (*e.g.* § 3).

2. "Class I" or "Infrared" Protostars

In the near-/mid-infrared, three broad classes of YSOs can be distinguished based on the slope $\alpha_{IR} = \mathrm{dlog}(\lambda F_\lambda)/\mathrm{dlog}(\lambda)$ of their spectral energy distributions (SEDs) longward of 2.2 μm, and are interpreted in terms of an evolutionary sequence (*e.g.* Lada 1991). Going backward in time, Class III ($\alpha_{IR} < -1.5$) and Class II ($-1.5 < \alpha_{IR} < 0$) sources correspond to PMS stars (*e.g.* T Tauri stars) surrounded by optically thin and optically thick circumstellar disks, respectively (see André & Montmerle 1994; hereafter AM). The youngest YSOs detected at 2 μm are the Class I sources (also called "IRAS dense cores"), which are characterized by $\alpha_{IR} > 0$ and the close association with dense molecular gas (*e.g.* Myers *et al.* 1987). The SEDs of these sources are successfully modeled in the framework of the "standard" protostellar theory of Shu and collobarators (*e.g.* Adams, Lada & Shu 1987; Kenyon *et al.* 1993). This suggests Class I sources derive most of their luminosity from accretion and are protostars in the sense of the first definition of § 1.

Dust continuum maps of ρ Ophiuchi YSOs with the IRAM 30 m telescope show that Class I sources are extended at 1.3 mm, confirming they are protostellar objects "self-embedded" in spheroidal circumstellar envelopes (AM). (In contrast, Class II YSOs are point sources at the resolution of the 30 m telescope, consistent with disk-dominated 1.3 mm emission – AM.) These observations also support the general evolutionary sequence Class I → Class II → Class III by determining that the *total* (disk + envelope) circumstellar mass decreases by a factor $\sim$ 5–10 on average from one SED class to the next.

However, interferometric millimeter observations show that the typical *disk mass* at the Class I protostellar stage is not large, perhaps even less than the typical disk mass during the T Tauri phase (*e.g.* Terebey *et al.* 1993; Ohashi *et al.* 1994). Furthermore, AM have found that the circumstellar envelope mass of ρ Oph Class I YSOs is small, at most $\sim$ 0.1-0.3 $M_\odot$ (assuming a dust opacity per mass column density of gas + dust $\kappa_{1.3} \simeq$ 0.01 cm^2 g^{-1} at 1.3 mm, a value known to within a factor of 2 in dense cloud cores; *e.g.* Ossenkopf & Henning 1994). These circumstellar masses are significantly smaller than both the typical masses of T Tauri stars ($\sim$ 0.3-2 $M_\odot$) and the masses derived for *pre–protostellar* condensations ($\sim$ 0.5-3 $M_\odot$) using a similar dust continuum technique (Ward-Thompson *et al.* 1994). This suggests that most (if not all) Class I sources have only remnant infalling envelopes, and are thus "evolved" protostars which have accumulated a large fraction of their final stellar mass (in that sense they do not match the second definition of § 1). Nevertheless, they clearly make up the bulk of the "self-embedded" protostellar population observed in molecular clouds, with a typical age estimated to be $\sim 10^5$ yr from statistical comparison with T Tauri stars (*e.g.* Wilking, Lada & Young 1989; Kenyon *et al.* 1990).

3. Cold Protostellar vs. Pre-Protostellar Clumps

A posteriori, the failure to find an object with $M_{env} > M_\star$ among near-IR sources is perhaps not too surprising since detailed numerical calculations of protostellar collapse suggest that the spectral appearance of the youngest protostars should closely resemble a blackbody at 10–30 K and peak at 100–300 μm (*cf.* Boss & Yorke 1990; Yorke, Bodenheimer & Laughlin 1994). The search for such cold objects is therefore best conducted at submillimeter or longer radio wavelengths.

In the last few years, dust continuum mapping of molecular cloud cores with large radiotelescopes has indeed revealed a number of strong submillimeter clumps with such cold, blackbody-like SEDs (see reviews by

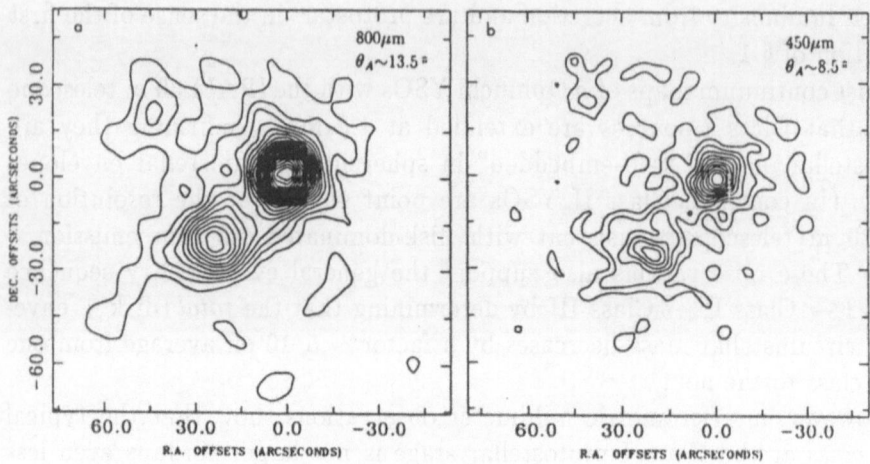

Fig. 1.   JCMT 800 μm (a) and 450 μm (b) maps showing the strong submm protostellar clumps NGC 1333-IRAS4A and NGC 1333-IRAS4B (from Sandell *et al.* 1991).

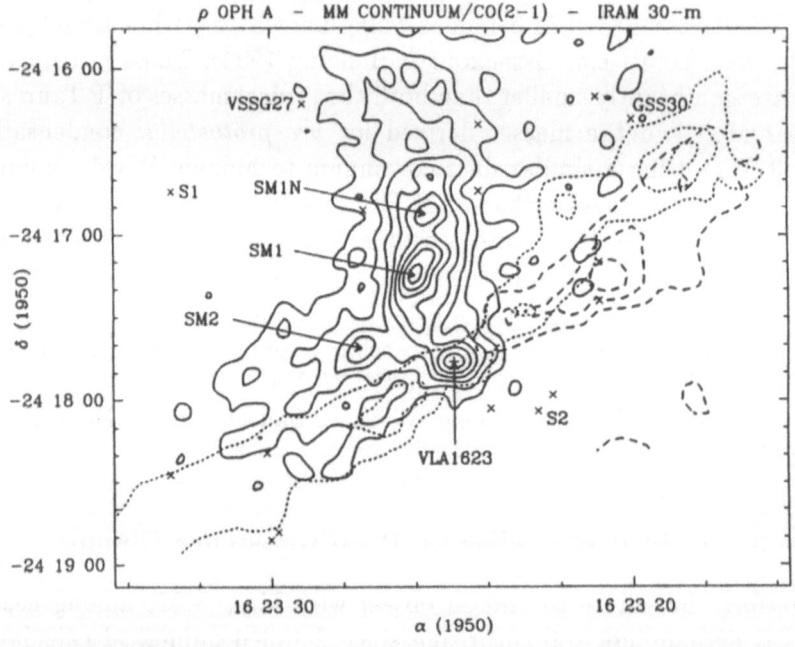

Fig. 2.   Overlay of an IRAM 1.3 mm continuum map of the ρ Oph A cloud core with the jet-like CO(2-1) outflow driven by the prototypical Class 0 protostar VLA 1623 (adapted from AWB and André *et al.* 1990).

Mezger 1994 and Güsten 1994). The most spectacular of these centrally condensed "protostellar" clumps, all invisible at infrared wavelengths (shortward of 20 μm at least), are those of NGC 2024 (*e.g.* Mezger *et al.* 1992a),

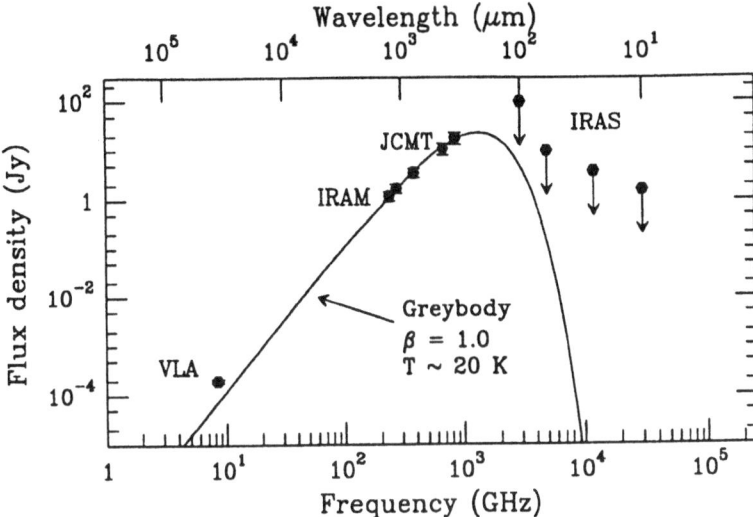

Fig. 3. Spectral energy distribution of HH24MMS along with a graybody fit (adapted from Ward-Thompson *et al.* 1995 – WCK). The excess VLA emission at 8.4 GHz (Bontemps *et al.* 1994) is most likely due to free-free radiation, indicating a Class 0 protostar.

NGC 1333 (Sandell *et al.* 1991; Fig. 1), ρ Oph A (*e.g.* AWB; Fig. 2), Serpens (Casali *et al.* 1993; Hurt *et al.* , this volume), and HH24MMS in Orion B (Chini *et al.* 1993; see Fig. 3). They range in size from $\leq 3.0 \times 10^{16}$ cm to 9.0×10^{16} cm and in mass from ≈ 0.5 M$_\odot$ to 20 M$_\odot$, with average gas densities $\gtrsim 10^7$-10^8 cm^{-3}. Some of these sources have been claimed to be isothermal collapsing protostars on the basis of their compact sizes, high densities, and cold (T $\approx$ 15-30 K) dust temperatures (*e.g.* Mezger *et al.* 1992a; Chini *et al.* 1993). However, they are generally found to be very close to virial equilibrium (*e.g.* AWB), and assessing their nature and dynamical state on the basis of submillimeter dust observations alone is difficult. Besides isothermal protostars, they may also be "pre-protostellar" condensations on the verge of collapse and at the end of their quasi-static ambipolar diffusion phase (*e.g.* Ward-Thompson *et al.* 1994), or dense circumstellar cocoons collapsing from the inside-out onto an already formed, hydrostatic protostellar object (*e.g.* AWB; see § 4).

4. "Class 0" or "Submillimeter" Protostars

It has recently been realized that some of the protostellar condensations of § 3 are associated with formed (*i.e.* hydrostatic) protostars. This is the case for the ρ Ophiuchi source VLA 1623, first detected by the VLA at 6 cm, and subsequently identified as the driving source of a jet-like CO(2-1) outflow (André *et al.* 1990; see Fig. 2). Another example is the source HH24MMS in the LBS 23 cloud core, originally interpreted as an "isothermal protostar"

(Chini *et al.* 1993), and more recently detected by the VLA at 3.6 cm
(Bontemps *et al.* 1994; see Fig. 3). In both VLA 1623 (Fig. 6a of AWB)
and HH24MMS (Fig. 3; Ward-Thompson *et al.* 1995 – WCK), the VLA
detection lies above the cold (T $\sim$ 20 K) dust spectrum responsible for the
submm emission, and most likely traces free-free radiation from ionized gas
at T $\sim 10^4$ K. This clearly suggests that the brief initial isothermal phase
of cloud collapse has been passed, and that a condensed, hydrostatic object
is present in the center. Thus, both VLA 1623 and HH24MMS appear to be
two extremely young YSOs or protostars.

TABLE I

Approximate properties of known candidate Class 0 protostars.

Source	L_{bol} (L$_\odot$)	M_{env} (M$_\odot$)	T_d (K)	$S_{1.3mm}^{160pc}(int)/L_{bol}$ (Jy/L$_\odot$)	Reference
L1448-N	10	2.3	–	0.8	see AWB
L1448-C	9	1.4	–	0.4	Bachiller *et al.* '91
NGC1333-IRAS4A	14	7.	37	1.4	Sandell *et al.* '91
NGC1333-IRAS4B	14	2.7	27	0.6	"
L1527	2	0.4	–	0.5	Ladd *et al.* '91
RNO43-MM	5.	0.6	33	0.5	Zinnecker *et al.* '92
NGC2024-FIR5	$\gtrsim 10$	15.	20	$\lesssim 4.4$	Mezger *et al.* '92a
NGC2024-FIR6	$\gtrsim 15$	6.	20	$\lesssim 1.3$	"
HH24MMS	5.	4.	20	2.0	WCK
VLA 1623	1	0.6	20	1.5	AWB
IRAS 16293	23	2.3	39	0.4	Walker *et al.* '86
Serpens-SMM4	2-30	2.	15-20	0.2-2.3	Casali *et al.* '93
L723-MM	3	0.6	–	0.5	Cabrit & André '91
B335	3	0.8	25	0.7	Chandler *et al.* '90
S106-SMM	$\gtrsim 24$	$\lesssim 10$	$\gtrsim 20$	$\lesssim 1.2$	Richer *et al.* '93

4.1 DEFINITION AND OBSERVATIONAL PROPERTIES OF CLASS 0 SOURCES

AWB have shown that, together with the most extreme YSOs detected by
IRAS, objects such as VLA 1623 and HH24MMS make up an entirely new
class of protostellar sources, designated "Class 0". Observationally, this class
is characterized by the following empirical properties:

• (1) Indirect evidence for a central YSO, as traced for instance by the
detection of compact centimeter continuum VLA emission or the presence
of a collimated CO outflow.

• (2) SED resembling a single temperature blackbody at T $\sim$ 15-30 K,
implying a high L_{submm}/L_{bol} ratio and virtually no emission at wavelengths

shorter than ~ 10 μm.

More specifically, AWB defined Class 0 sources as those self-embedded YSOs which have $L_{submm}/L_{bol} > 5 \times 10^{-3}$, where L_{submm} is the submillimeter luminosity measured longward of 350 μm and L_{bol} the bolometric luminosity (see § 4.2 for a justification of this number). While property (1) above distinguishes Class 0 sources from pre-protostellar clumps/condensations such as SM1, SM1N, SM2 (Fig. 2; AWB) or those studied by Ward-Thompson *et al.* (1994), property (2) distinguishes them from Class I near-IR sources and pre-main sequence objects.

In practice, Class 0 sources are best separated from Class I YSOs in diagrams plotting integrated millimeter flux (or better circumstellar mass) against bolometric luminosity (see Fig. 10 of AM and Fig. 5 below). While $S_{1.3mm}^{int}$ is well correlated with L_{bol} for the majority of (Class I) embedded YSOs (see Reipurth *et al.* 1993), Class 0 sources clearly stand out as objects with excess (sub)millimeter emission. [The "border line" between Class 0s and Class Is is roughly $S_{1.3mm}^{int}(d/160pc)^2/L_{bol} \sim 0.2$ Jy/L$_\odot$; see Fig. 5].

We also argue that the lack of emission from Class 0 sources at $\lambda < 10$ μm does not result from insufficient sensitivity, but reflects a more fundamental, qualitative difference with Class I near-IR sources. Since the cold, dense circumstellar cocoon responsible for the submm emission of Class 0 sources is probably optically thick below ~ 100 μm (*cf.* AWB), some of these objects may be detectable in *absorption* at mid-IR wavelengths against a diffuse background; this hypothesis may be tested in the near future by sensitive observations with ISOCAM (see Puget 1992).

4.2 AGE ORDERING AND INTERPRETATION OF CLASS 0 SOURCES

The locations of Class 0 sources in the various evolutionary diagrams that have been proposed recently for embedded YSOs all point to extremely young objects/protostars.

Adams (1990) suggested to use the internal visual extinction A_V as an indicator of YSO evolution, and plotted evolutionary tracks in the L_{bol}-A_V diagram based on the Terebey, Shu & Cassen (1984) collapse models. In this diagram, younger objects are generally less luminous and have higher internal A_V extinctions than older objects. Class 0 sources, which typically have $A_V \gtrsim 1000$, are underluminous compared to the models, suggesting they are at most $\sim 3 \times 10^4$ yr old and have not yet started to form a disk. A problem of this diagram, however, is that A_V is not a directly observable, model-independent quantity.

Myers & Ladd (1993) introduced the concept of "bolometric temperature" T_{bol}, defined as the temperature of a blackbody having the same mean frequency as the observed SED, and proposed the L_{bol}-T_{bol} diagram as a direct analog to the H–R diagram. In this diagram, general YSO evolution occurs from low to high T_{bol}. Class 0 sources appear as the youngest

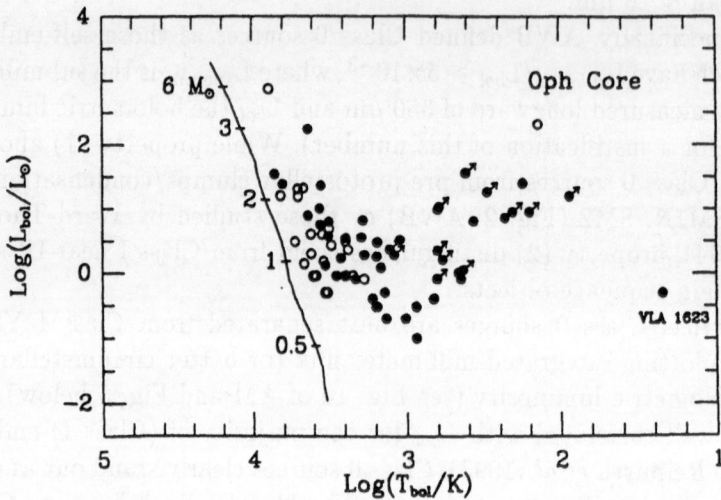

Fig. 4. L_{bol}-T_{bol} diagram for YSOs in the ρ Oph core (adapted from Chen *et al.* 1994).

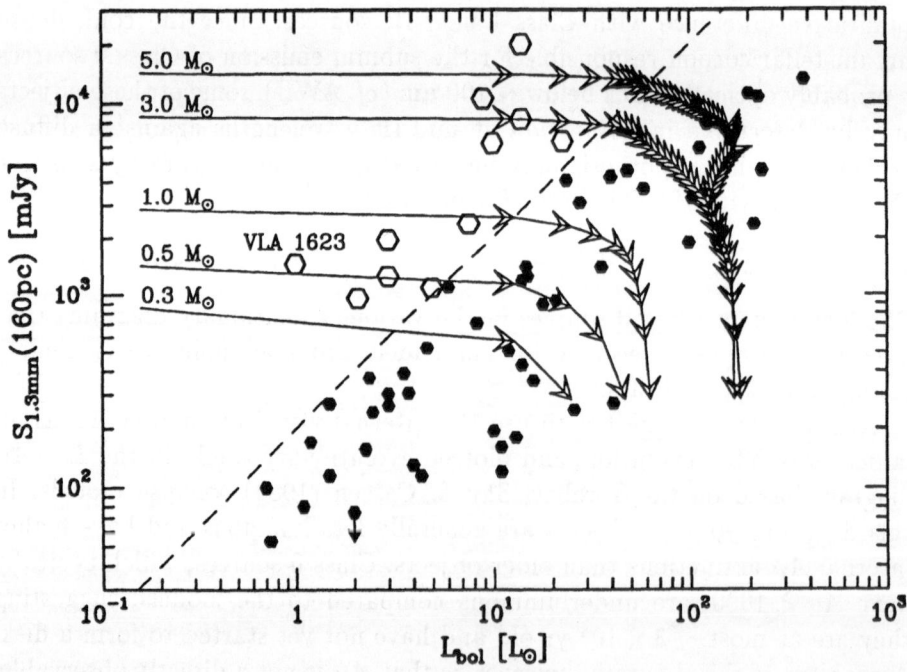

Fig. 5.   Integrated millimeter flux (scaled to a distance of 160 pc) against bolometric luminosity for a sample of Class I (filled circles) and Class 0 sources (open circles). The dashed line is the formal "border line" between Class I and Class 0 sources (AWB). Indicative protostellar evolutionary tracks for a (constant) mass infall rate $\sim 10^{-5}\ M_\odot\ yr^{-1}$ and various initial cloud core masses are shown. The arrows on the tracks indicate constant time steps of 10^4 yr (adapted from Saraceno *et al.* 1995).

objects as they are characterized by the lowest T_{bol} values of all YSOs: they all have $T_{bol} \lesssim 80$ K (see, *e.g.* Fig. 4). A shortcoming of the L_{bol}-T_{bol} diagram is that, because of anisotropic distributions of circumstellar matter, the SEDs of embedded protostellar sources, hence their bolometric temperatures, depend on viewing angles (*e.g.* Yorke *et al.* 1994).

Another approach is to use the circumstellar mass inferred from millimeter continuum maps as an indicator of evolutionary state. In a statistical sense at least, larger amounts of circumstellar material are expected to surround younger stellar objects. Also, because submm dust emission is generally optically thin, it is almost insensitive to viewing angle. In this spirit, AWB proposed an age ordering of embedded YSOs based on the L_{submm}/L_{bol} ratio. While the (integrated) submillimeter luminosity L_{submm} of a protostellar source provides a relative measure of its (total) circumstellar mass M_{env}, the bolometric luminosity L_{bol} may be used to infer the central stellar mass $M_\star$ on the basis of existing mass–luminosity relations for protostars (see AM for details). In the youngest sources, dominated by an envelope rather than a disk, L_{submm}/L_{bol} should thus tend to reproduce the variations of the mass ratio $M_{env}/M_\star$, which is expected to decrease with protostellar age. The formal boundary between Class 0 and Class I sources ($L_{submm}/L_{bol} \sim 5 \times 10^{-3}$) was set so as to correspond to a mass ratio $M_{env}/M_\star = 1$, assuming the most plausible relations between L_{bol} and $M_\star$ on the one hand and between L_{submm} and M_{env} on the other hand (see AWB for details). Thus, not only are Class 0 YSOs the youngest sources according to this diagnostic, they also appear as *conceptually* different: *they are protostars which have yet to accrete the bulk of their final stellar mass.* To some extent, the L_{submm}/L_{bol} and T_{bol} age indicators are related since, for instance, any blackbody colder than T ~ 80 K will have $L_{submm}/L_{bol} < 5 \times 10^{-3}$. The advantage of the L_{submm}/L_{bol} criterion is that it exploits the fact that the submillimeter part of a YSO SED is optically thin and traces circumstellar mass. Based on this diagnostic, Saraceno *et al.* (1995) plot evolutionary tracks in the $S_{1.3mm}$ vs. L_{bol} diagram (see Fig. 5), consistent with ages of $\sim 10^4$ yr for Class 0 sources.

Note also that Class 0 sources are rare compared to both pre-protostellar condensations (Ward-Thompson *et al.* 1994) and Class I sources. For example, in the ρ Oph central region where the IRAM 30 m mapping study of Mezger *et al.* (1992b) provides an essentially complete survey for Class 0 sources, VLA 1623 is the *only* good candidate (AWB) while there are more than a dozen near-IR Class I sources with $L_{bol} \gtrsim 1\ L_\odot$ (*e.g.* Greene *et al.* 1994). The lifetime of Class 0 sources is thus at least an order of magnitude shorter (*i.e.* $\lesssim 10^4$ yr) than the estimated lifetime of Class I sources ($\lesssim 10^5$ yr; *cf.* § 2). This again is in agreement with the interpretation of Class 0 sources as very young protostars (see Barsony 1994 for a further discussion of this point).

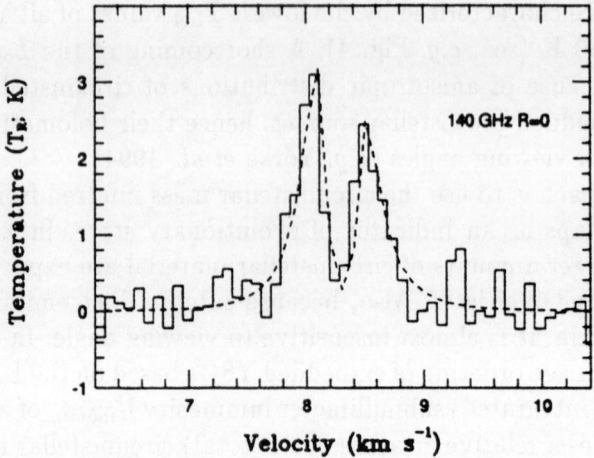

Fig. 6. $H_2CO(2_{12}-1_{11})$ line profile observed toward the center of B335 plotted along with a collapse model fit (adapted from Zhou *et al.* 1993).

4.3 DIRECT EVIDENCE FOR INFALL

Rather convincing spectroscopic signatures of protostellar collapse have been reported recently around some Class 0 protostars. Ongoing collapse can be traced by moderately optically thick lines ($\tau \sim 1$) which should show *local* red-shifted self-absorption with a stronger blue peak (see, *e.g.* Fig. 6). Unfortunately, interpretation is often complicated by the simultaneous presence of rotation and/or outflow (*e.g.* Menten *et al.* 1987; Walker *et al.* 1994). However, if line maps are available in both an optically thick line (*e.g.* $H_2CO(3_{12} - 2_{11})$ or CS(3-2)) and an optically thin line (*e.g.* $C^{18}O(2-1)$), the former showing the characteristic collapse asymmetry and the latter being symmetric and peaking at the absorption dip of the other, then the infall interpretation is reasonably secure (*e.g.* Zhou & Evans 1994). To date, three Class 0 sources at least (B335, IRAS 16293, L1527) meet these requirements (Zhou *et al.* 1993; Walker *et al.* 1994; Zhou *et al.* 1994). Another case in point seems to be the protostellar condensation GF9-2 (Güsten 1994), although very little has been published about this source yet. The present lack of spectroscopic evidence for collapse in the other known Class 0 sources (*e.g.* VLA 1623) can be attributed to heavy depletion of molecules onto grains during the earliest protostellar stages (*e.g.* Mundy, this volume). An interesting conclusion was reached by Zhou *et al.* (1994) in a recent $C^{18}O$ and H_2CO survey of dense cores in Taurus: among the five cores with embedded sources they studied, they found evidence for collapse in only one object, L1527, which was the only Class 0 source in their sample (the others being Class Is). Although infall is probably present at the Class I stage, the global kinematics of Class I circumstellar envelopes may be dominated by (broad) outflow motions rather than infall motions (*cf.* Fuller *et al.* 1995 and Cabrit

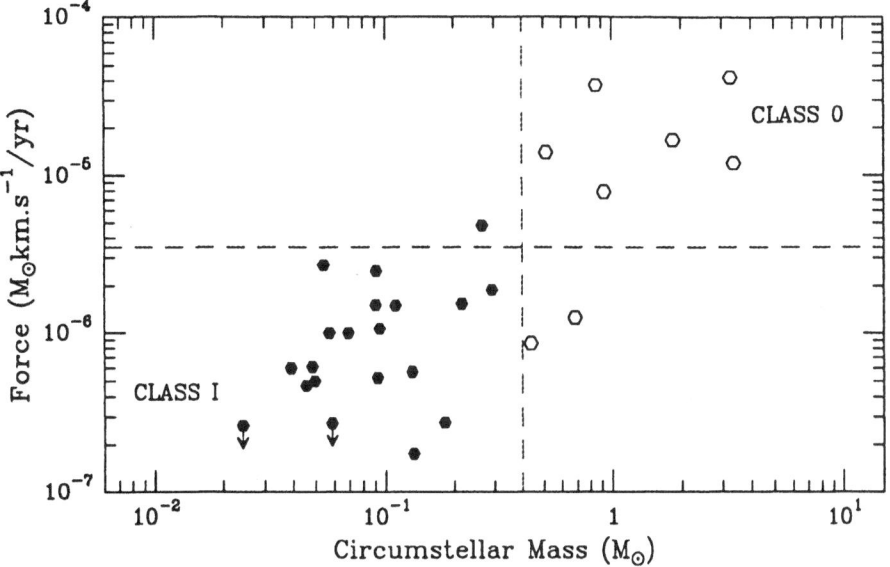

Fig. 7. Outflow momentum flux against circumstellar mass of the driving source for a sample of nearby, low-luminosity Class I and Class 0 YSOs (from Bontemps *et al.* 1995).

et al. 1995 for detailed studies of L1551-IRS5 and HL Tau, respectively).

4.4 JET–LIKE OUTFLOWS

Class 0 protostars tend to drive highly collimated or "jet–like" CO molecular outflows (see, *e.g.* Fig. 2 and Bachiller & Gómez-González 1992). These outflows are relatively fast (with maximum velocities $V_{char} \gtrsim 50$ km s^{-1}), apparently very young (with dynamical timescales $t_{dyn} << 10^4$ yr), and powerful (with mechanical powers approaching ~ 50 % of the bolometric luminosity of the central sources). (The associated mass-loss rates are however quite typical; *e.g.* $\dot{M} \sim 2 \times 10^{-6}$ M$_\odot$ yr^{-1} for VLA 1623.) In contrast, while some outflow activity probably exists throughout the embedded phase (*e.g.* Parker *et al.* 1991), the CO outflows from Class I sources tend to be poorly collimated, slower, and much less energetic than those from Class 0 sources (see Fig. 7). This evolution of outflow characteristics from Class 0 to Class I is consistent with the most recent theoretical views on the structure of molecular outflows, according to which the observed CO outflows represent ambient gas that has been progressively entrained by an underlying jet directly originating in the central star and/or circumstellar structure (*e.g.* Stahler 1993; Masson 1994). In this picture, the youngest CO outflows are indeed expected to be the fastest and most highly collimated flows, *i.e.* to appear " jet–like". Evidence for the driving jet has recently been found in the form of shock-excited molecular hydrogen emission (Bally *et al.* 1993; Davis *et al.* 1994; Dent *et al.* 1994) or Herbig-Haro objects (Eiroa *et al.* 1993).

5. Open Problems

Since Class 0 YSOs appear to characterize the beginning of protostar evolution ($M_{env} \gg M_\star$), they should still be dominated by the effects of their circumstellar cocoon and are likely to retain detailed information about their genesis. Further studies of these candidate young protostars may thus shed light on the star formation process itself, and, in particular, on the initial conditions required for protostellar collapse. A few important unanswered questions are listed below and may be viewed as a plan for future work.

• Do all Class 0 protostars have outflows ?
The fact that most known Class 0 sources (*e.g.* VLA 1623) drive energetic bipolar flows shows that the outflow phase starts extremely early in protostar evolution, perhaps as soon as a hydrostatic core forms at the center of a collapsing dense cloud fragment. However, it is at present unclear whether *all* Class 0 objects have outflows (see, *e.g.* Krügel & Chini 1994 for the case of HH24MMS).

• Is there already a disk at the Class 0 stage ?
The presence of a disk may be expected in at least those Class 0 sources which drive an outflow. However, this is still a matter of debate, and a recent MHD 'cored-apple' model shows that a thin disk may not be necessary to generate an outflow (Henriksen & Valls-Gabaud 1994). While it has been suggested that no significant disk has developed yet in the case of VLA 1623 (AWB), observations with the Owens Valley mm interferometer point to the presence of two massive disks in the (perhaps more evolved) protobinary IRAS 16293 (Mundy *et al.* 1992).

• Magnetic field structure
Submillimetre-wave polarimetry observations on JCMT have just started to probe the magnetic field morphology in cloud cores and Class 0 protostars, with only preliminary conclusions at this stage. While the average magnetic field direction in the dense circumstellar "apple" surrounding VLA 1623 appears to be *perpendicular* to its jet-like CO outflow (Holland *et al.* 1994), the opposite result is found in the case of IRAS 16293 and NGC1333-IRAS4 (*e.g.* Tamura *et al.* 1993 and this volume).

• Density structure of Class 0 envelopes
Modeling of the submm continuum maps and SEDs of some Class 0 sources such as VLA 1623 suggests that, at least in some cases, stars may form from dense clumps with flat density gradients but finite sizes (*e.g.* AWB; Barsony 1994), which are not appropriately described by the scale-free, singular isothermal sphere of the standard protostar theory from Shu *et al.* If

this is the case, the mass-infall rate would be time-dependent (*e.g.* Foster and Chevalier 1993; Henriksen 1994). However, since other sources seem to follow the standard theory (*e.g.* Ladd *et al.* 1991), more studies are required to draw general conclusions.

• Nature of the centimeter radio continuum emission

Although the centimeter radio emission from Class 0 sources is consistent with partially optically thick free-free radiation, the physical mechanism responsible for the ionization of the emitting gas is not entirely clear yet. Some of the emission probably arises in a shock-ionized jet at the base of the outflow when present (*e.g.* Rodríguez *et al.* 1989). However, another plausible scenario invokes photoionization by the soft X-rays which are produced in an accretion shock at the surface of the hydrostatic core (see, *e.g.* Bertout 1983).

References

Adams, F.C.: 1990, Astrophys. J. **363**, 578.

Adams, F.C., Lada, C.J. & Shu, F.H.: 1987, Astrophys. J. **312**, 788.

André, P., Martín-Pintado, J., *et al.* : 1990, Astron. & Astrophys. **236**, 180.

André, P. & Montmerle, T.: 1994, Astrophys. J. **420**, 837 **(AM)**.

André, P., Ward-Thompson, D. & Barsony, M.: 1993, Astrophys. J. **406**, 122 **(AWB)**.

Bachiller, R., André, P. & Cabrit, S.: 1991, Astron. & Astrophys. **241**, L43.

Bachiller, R. & Gómez-González, J.: 1992, Astron. & Astrophys. Rev. **3**, 257.

Bally, J., Lada, E.A. & Lane, A.P.: 1993, Astrophys. J. **418**, 322.

Barsony, M.: 1994, in: *Clouds, Cores, and Low Mass Stars*, eds: D. Clemens & R. Barvainis, A.S.P. Conference Series, **65**.

Beichman, C.A., Myers, P.C., Emerson, J.P. *et al.* : 1986, Astrophys. J. **307**, 337.

Bertout, C.: 1983, Astron. & Astrophys. **126**, L1.

Bontemps, S., André, P. & Ward-Thompson, D.: 1994, Astron. & Astrophys. in press.

Boss, A.P. & Yorke, H.W.: 1990, Astrophys. J. **353**, 236.

Cabrit, S. & André, P.: 1991, Astrophys. J. **379**, L25.

Cabrit, S., Guilloteau, S., André, P., Bertout, C., Montmerle, T. & Schuster, K.: 1995, Astron. & Astrophys. submitted.

Casali, M.M., Eiroa, C. & Duncan, W.D.: 1993, Astron. & Astrophys. **275**, 195.

Chandler, C.J., *et al.* : 1990, Mon. Not. of the RAS **243**, 330.

Chen, H., Myers, P.C., Ladd, E.F. & Wood, D.O.S.: 1994, in: *Clouds, Cores, and Low Mass Stars*, eds: D. Clemens & R. Barvainis, A.S.P. Conference Series, **65**.

Chini, R., Krügel, E., Haslam, C.G.T., Kreysa, E., Lemke, R., Reipurth, B., Sievers, A. & Ward-Thompson, D.: 1993, Astron. & Astrophys. **272**, L5.

Davis, C.J., *et al.* : 1994, Mon. Not. of the RAS **266**, 933.

Dent, W.R.F., Walther, D.M. & Matthews, H.: 1994, The JCMT Newsletter **3**, 45.

Eiroa, C. *et al.* : 1994, Astron. & Astrophys. **283**, 973.

Fuller, G.A. *et al.* : 1995, Astrophys. J. submitted.

Foster, P. & Chevalier, R.A.: 1993, Astrophys. J. **416**, 303.

Greene, T.P., Wilking, *et al.* : 1994, Astrophys. J. **434**, 614.

Güsten, R.: 1994, in: *The Cold Universe*, eds: T. Montmerle, C.J. Lada, I.F. Mirabel & J. Trân Thanh Vân, Editions Frontières, Gif-sur-Yvette, p169.

Henriksen, R.N.: 1994, in: *The Cold Universe*, eds: T. Montmerle, C.J. Lada, I.F. Mirabel & J. Trân Thanh Vân, Editions Frontières, Gif-sur-Yvette, p241.

Henriksen, R.N. & Valls-Gabaud, D.: 1994, Mon. Not. of the RAS **266**, 681.

Holland, W.S., Greaves, J.S. & Ward-Thompson, D.: 1994, The JCMT Newsletter **3**, 51.

Kenyon, S.J., Calvet, N. & Hartmann, L.: 1993, Astrophys. J. **414**, 676.

Kenyon, S.J., Hartmann, L.W., Strom, K.M. & Strom, S.E.: 1990, Astron. J. **99**, 869.

Krügel, E. & Chini, R.: 1994, Astron. & Astrophys. **288**, 947.

Lada, C.J.: 1991, in: *The Physics of Star Formation and Early Stellar Evolution*, eds. C.J. Lada & N.D. Kylafis, Kluwer, Dordrecht, p329.

Ladd, E.F., Adams, F.C., Casey, S., Davidson, J.A., Fuller, G.A., Harper, D.A., Myers, P.C. & Padman, R.: 1991, Astrophys. J. **382**, 555.

Larson, R.B.: 1969, Mon. Not. of the RAS **145**, 271.

Menten, K.M., *et al.* : 1987, Astron. & Astrophys. **177**, L57.

Mezger, P.G.: 1994, Astrophys. J., Suppl. **212**, 197.

Mezger, P.G., Sievers, A.W., Haslam, C.G.T., Kreysa, E., Lemke, R., Mauersberger, R. & Wilson, T.L.: 1992a, Astron. & Astrophys. **256**, 631.

Mezger, P.G., Sievers, A.W., Zylka, R., Haslam, C.G.T., Kreysa, E. & Lemke, R.: 1992b, Astron. & Astrophys. **265**, 743.

Mundy, L.G., *et al.* : 1992, Astrophys. J. **385**, 306.

Myers, P.C., Fuller, G.A., Mathieu, R.D. *et al.* : 1987, Astrophys. J. **319**, 340.

Myers, P.C. & Ladd, E.F.: 1993, Astrophys. J. **413**, L47.

Ohashi, N., *et al.* : 1994, Astrophys. J., Suppl. **212**, 239.

Ossenkopf, V. & Henning, T.: 1994, Astron. & Astrophys. in press.

Parker, N. D., Padman, R. & Scott, P. F.: 1991, Mon. Not. of the RAS **252**, 442.

Puget, J.-L.: 1992, in: *Infrared Astronomy with ISO*, eds: Th. Encrenaz & M.F. Kessler, Nova Science, New York, p331.

Reipurth, B., Chini, R., Krügel, E., Kreysa, E. & Sievers, A.: 1993, Astron. & Astrophys. **273**, 221.

Richer, J.S., Padman, R., Ward-Thompson, D., Hills, R.E. & Harris, A.I.: 1993, Mon. Not. of the RAS **262**, 839.

Rodríguez, L.F., *et al.* : 1989, Astrophys. J. **347**, 461.

Sandell, G., Aspin, C., Duncan, W.D., Russell, A.P.G. & Robson, E.I.: 1991, Astrophys. J. **376**, L17.

Saraceno, P., André, P., Ceccarelli, C., Griffin, M. & Molinari, S.: 1995, Astron. & Astrophys. submitted.

Shu, F., Najita, J., Galli, D., Ostriker, E. & Lizano, S.: 1993, in: *Protostars and Planets III*, eds: E.H. Levy & J.I. Lunine, University of Arizona Press, Tucson, p3.

Stahler, S.W.: 1993, in: *Astrophysical Jets*, eds: M. Livio *et al.* , Cambridge U. Press, New York, p183.

Stahler, S.W. & Walter, F.M.: 1993, in: *Protostars and Planets III*, eds: E.H. Levy & J.I. Lunine, University of Arizona Press, Tucson, p405.

Tamura, M., Hayashi, S.S., Yamashita, T., Duncan, W.D. & Hough, J.H.: 1993, Astrophys. J. **404**, L21.

Terebey, S., Chandler, C.J. & André, P.: 1993, Astrophys. J. **414**, 759.

Terebey, S., Shu, F.H. & Cassen, P.: 1984, Astrophys. J. **286**, 529.

Walker, C.K., Lada, C.J., Young, E.T., Maloney, P.R. & Wilking, B.A.: 1986, Astrophys. J. **309**, L47.

Walker, C.K., Narayanan, G. & Boss, A.P.: 1994, Astrophys. J. **431**, 767.

Ward-Thompson, D., Scott, P.F., Hills, R.E. & André, P.: 1994, Mon. Not. of the RAS **268**, 276.

Ward-Thompson, D., Chini, R., *et al.* : 1995, Mon. Not. of the RAS submitted (**WCK**).

Wilking, B.A., Lada, C.J. & Young, E.T.: 1989, Astrophys. J. **340**, 823.

Wynn-Williams, C.G.: 1982, Ann. Rev. of Astron. & Astrophys. **20**, 587.

Yorke, H.W., Bodenheimer, P. & Laughlin, G.: 1994, Astrophys. J. in press.

Zhou, S. & Evans II, N.J.: 1994, in: *Clouds, Cores, and Low-Mass Stars*, eds: D. Clemens & R. Barvainis, A.S.P. Conference Series, **65**.

Zhou, S., Evans II, N.J., Kömpe, C. & Walmsley, C.M.: 1993, Astrophys. J. **404**, 232.

Zhou, S., Evans II, N.J., Wang, Y., Peng, R. & Lo, K.Y.: 1994, Astrophys. J. **433**, 131.

Zinnecker, H., Bastien, P., *et al.* : 1992, Astron. & Astrophys. **265**, 726.

AMMONIA IMAGES NEAR OBJECTS WITH CLASS 0
SPECTRAL ENERGY DISTRIBUTIONS

ALWYN WOOTTEN

NRAO*, 520 Edgemont Road,
Charlottesville, Virginia 22903-2475 U. S. A.

Abstract. A shroud of cold dust shapes the 'Class 0' spectral distribution of the youngest stars. We have imaged a number of these objects in the inversion lines of ammonia to investigate the molecular circumstellar environment on thousand AU scales. Although the dust spectra of these objects resemble one another, a variety of physical characteristics occur in the sample. In some circumstellar environments, NGC1333IRAS4A and S106FIR, ammonia bearing gas clearly traces outflowing warm gas and a dense warm elongated structure immediately surrounding the young star. More commonly ammonia bearing gas shows only a weak association with any structure in immediate contact with the dusty near-stellar environment (NGC1333IRAS4B, IRAS4C, IRAS16293-2422 A and B) and in some cases no evidence can be found that ammonia occurs in the near-stellar environment in association with the dust at all (VLA1623, SM-2).

The ammonia abundance in all objects must lie two orders of magnitude or so below its value in more extended cloud regions. The high molecular excitation conditions which mark the spectra of other molecules, such as formaldehyde, are invisible in ammonia spectra taken with beamsizes of projected diameters of $\sim$1000 AU.

Key words: Star formation – Protostars – Circumstellar Matter – Class 0 – ammonia

1. Introduction

The coldest cores have been shown to be propitious locales in which to seek the youngest stars. These objects, with spectral energy distributions of 'Class 0', have been targeted in an ammonia survey, which has now obtained VLA maps of seven or so of the about a dozen known objects. In these maps, differences among the objects become quite apparent, though in few of them could one recognize the region of the protostar from any aspect of the ammonia emission. In more luminous objects, such as NGC1333IRAS4A and S106FIR, warm ammonia correlates well with the dust emission, and shows up in outflowing gas as well. In another group of objects, including IRAS16293A, NGC1333IRAS4B and VLA1623-2418, very cold ammonia exists, but shows little correlation with cold dust structures enveloping the star. In objects so young that no bipolar flow has commenced (thus, strictly speaking, even younger than 'Class 0') such as IRAS16293B or NGC1333IRAS4C, ammonia structures are readily discerned though excitation is low. It appears that ammonia becomes severely depleted, perhaps through freezing onto grains, in the final stages before stellar ignition.

* The National Radio Astronomy Observatory is operated by Associated Universities, Inc., under cooperative agreement with the National Science Foundation.

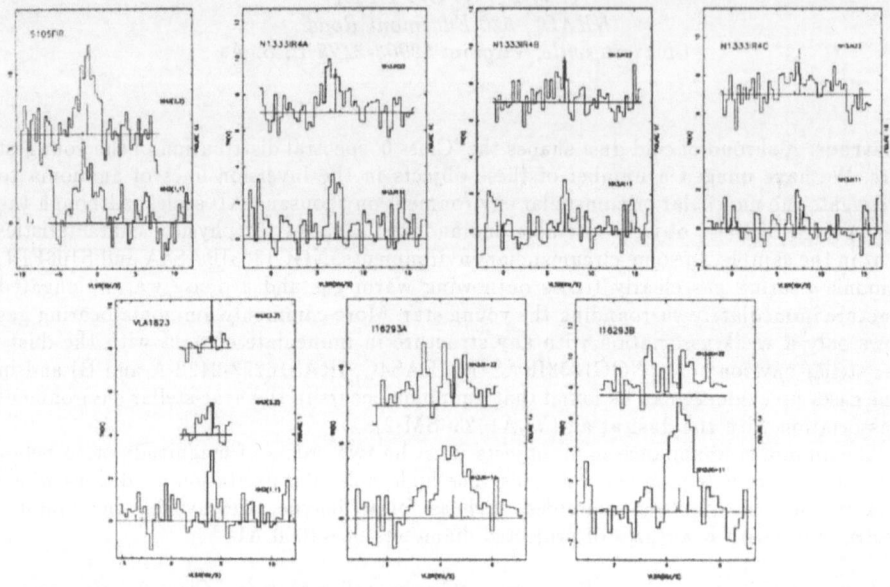

Fig. 1.   VLA (D array) spectra of NH$_3$ (J,K)=(1,1) and (2,2) (lower and upper spectra, respectively in each panel) emission in a single beam toward objects with 'Class 0' spectral energy distribution, as labelled. Locations of hyperfine components are denoted by vertical slashes, whose heights represent their optically thin intensities.

2.   Near-protostellar Ammonia

S106FIR. Determination of the luminosity of the far-infrared source in S106 depends upon establishment of the dust temperature. In dense regions, the gas should be thermally well-coupled to the dust by collisions, and the gas temperature may be used as an estimate of the dust temperature. In Figure 1, profiles are shown of the NH$_3$ (1,1) and (2,2) lines in a 3".4 VLA beam. These profiles occur toward the peak of the 10"× 6" region containing S106-FIR where integrated intensity in the (2,2) line is strong (Wootten & Mangum, in preparation). Using the analysis described in Mangum, Wootten & Mundy (1991) one finds an ammonia excitation temperature of $T_{ex} \sim 70K$ and column density $6 \times 10^{14} \text{cm}^{-2}$. Applying this temperature to the dust measurements of Richer *et al.* (1993) and using a distance of 1200 pc (Rayner 1994) we find a dust mass of $16 M\odot$, and a dust column density of $1 \times 10^{25} \text{cm}^{-2}$. The abundance of ammonia must be X(NH$_3$)=6 ×10^{-11} in the dust-emitting region of the luminous (and therefore not strictly 'Class 0') source.

NGC1333IRAS4. Of the three obscured objects in the 'IRAS4' region of NGC1333, object 'A' contains the most mass, with about 9.2 $M\odot$ (Sandell *et al.* 1991) for a column density of 2×10^{25}cm^{-2}, based on submillimeter measurements of dust emission. In 3".5 resolution VLA maps of integrated intensity in the (2,2) line, a structure of size 8400 × 4600 AU coincides with the submillimeter source. A model of the NH$_3$ spectra toward the protostar (Figure 1, Wootten & Mangum 1994) requires a temperature of T_{ex}=55K for a column density of 1.2×10^{14}cm^{-2}. The ammonia abundance must lie near X(NH$_3$)=6 ×10^{-12} in this archetypical 'Class 0' source. Lying only 33" southeast of 'A', object 'B' has a little less than half its neighbor's dust mass, with a column density of 4×10^{24}cm^{-2}. No clearly defined ammonia structure occurs near this object at all, and a model of the unimpressive emission suggests a temperature less than T_{ex}=45K for an ammonia column density of 7×10^{13}cm^{-2}. The ammonia abundance toward this protostar must lie below X(NH$_3$)=2 ×10^{-11}. A third infrared object , which we call 'C', lies northeast of A and B in the submillimeter maps displayed by Sandell *et al.* ; it shines clearly in the NH$_3$ (1,1) maps also. From the published maps, we estimate a solar mass of dust emitting a few solar luminosities at most can reproduce the emission, suggesting a column density of material of 7×10^{23}cm^{-2}. The ammonia emission requires cold (T_{ex}=25K) gas with a column density of 6×10^{13}cm^{-2}. Our ammonia and dust models then imply that X(NH$_3$)=9 ×10^{-11}.

VLA 1623. The prototypical 'Class 0' object is VLA 1623 in the ρ Ophiuchi molecular core 'A' (André, Ward-Thompson & Barsony (AWB) 1993). Also lying within the core are several dense clumps which have not yet formed stars, SM1, SM1-N and SM2. None of these objects coincides with a peak of ammonia emission in either the (1,1), (2,2) or (3,3) line, though they all locate strong submillimeter emission peaks. In fact, in the ortho-ammonia lines, emission peaks between the submillimeter maxima, which lie on saddle points of the ammonia ridge connecting them. Toward VLA 1623, the submillimeter emission suggests a column density of 1×10^{24}cm^{-2} while the ammonia in that direction measured with a 6" beam suggests a column of 2×10^{14}cm^{-2} of ammonia at a temperature of T_{ex}=17 K. The ammonia peak actually lies 11" northeast of VLA 1623, so the abundance at the protostar must lie below X(NH$_3$)=2 ×10^{-10}. Similar abundances obtain at the other three submillimeter sources.

IRAS16293-2422. Lying within the L1689 portion of the ρ Ophiuchi molecular cloud is the binary protostar IRAS16293-2422, consisting of components A and B. Component A alone powers a fabled bipolar flow (Wootten 1993) and clearly shows all the features expected of a Class 0 protostar. Although the column density of material toward the peak of its dust emission lies near 6×10^{24}cm^{-2}, its ammonia distribution (Mundy *et al.* 1990, 1992) straddles the submillimeter source, which lies in a saddle point of the ammonia distribution. Although the (1,1) line reaches a local maximum in width at 'A', the ammonia emitting gas appears cool and much more quiescent, kinematically, than for example the $C^{18}O$ emitting gas. At a temperature $T_{ex}=25K$, a model for the ammonia emission necessitates a column density of only $3 \times 10^{14} cm^{-2}$. The abundance of ammonia at IRAS16293A must be near $X(NH_3)=5 \times 10^{-11}$. Component B of IRAS16293 shows no currently active flow; this object may give us a glimpse of a protostar at a stage before outflow commences. Emission from B has not been resolved on any scale at any wavelength. From observations at 1.3 mm and 2.7 mm (Mundy, private communication, and Mundy et $al.$ 1992) we estimate the total column density of material to be $8 \times 10^{24} cm^{-2}$. At the position of the object, the (1,1) ammonia line is strong and narrower than at A, at a somewhat more positive velocity. There is little, if any, emission in the (2,2) line. Here, the ammonia-emitting gas appears cold, $T_{ex}=16$ K at a slightly lower column density than near A, $1.8 \times 10^{14} cm^{-2}$. The abundance of ammonia is consequently slightly smaller, $X(NH_3)=2.3 \times 10^{-11}$.

Models of the ammonia emission toward the known submillimeter protostar positions show that the ammonia-emitting gas there is cold and quiescent. The abundance of ammonia gas within the central few hundred AU diameter region of the accretion envelope of these protostars must be several orders of magnitude below its levels in lower density gas.

Acknowledgements

The ammonia images discussed here were obtained in collaboration with J. Mangum, L. Mundy, B. Wilking, P. André, D. Despois and A. Sargent.

References

André, P., Montmerle, T. & Feigelson, E.D.: 1989, Astron. J. **93**, 1182.

Mangum, J., Wootten, A. & Mundy, L.: 1991, Astrophys. J. **378**, 576.

Mundy, L. G., Wootten, A. & Wilking, B. A.: 1990, Astrophys. J. **352**, 159.

Mundy, L. G., Wootten, A., Wilking, B., Blake, G. & Sargent, A.: 1992, Astrophys. J. **385**, 306.

Sandell, G., Aspin, C., Duncan, W., Russell, A. & Robson, I.: 1991, Astrophys. J. **376**, L17.

Rayner, J.: 1994, *The Young Star Clusters in S106*, in *Infrared Astronomy with Arrays: The Next Generation*, ed: I.S. McLean, Kluwer: Dordrecht, p185.

Richer, J., Padman, R., Ward-Thompson, D., Hills, R. & Harris, A.: 1993 Mon. Not. of the RAS **262**, 839.

Wootten, A.: 1993, *Water Maser Monitoring of IRAS16293-2422*, in *Astrophysical Masers*, eds: A. Clegg & G. Nedoluha, Springer: Berlin, p315.

Wootten, A. & Mangum, J.: 1994, BAAS **25**, 1367.

Wootten, A., André, P., Despois, D. & Sargent, A.: 1994, *Ammonia Distribution on 1000 AU Scales near VLA 1623*, in *Clouds,Cores, and Low Mass Stars*, eds: D. Clemens and R. Barvainis, ASP: San Francisco, p294.

SUBMILLIMETRE CONTINUUM OBSERVATIONS OF PRE-PROTOSTELLAR CORES

D. WARD-THOMPSON[1], P. F. SCOTT[2], R. E. HILLS[2] and P. ANDRÉ[3]

[1] *Royal Observatory, Blackford Hill, Edinburgh EH9 3HJ, UK.*

[2] *Mullard Radio Astronomy Observatory, Cavendish Laboratory, Madingley Road, Cambridge CB3 0HE, UK*

and

[3] *Service d'Astrophysique, Centre d'Etudes de Saclay, F-91191 Gif-sur-Yvette, France*

Abstract. Results are outlined of a JCMT submillimetre continuum survey of Myers cores that have no known infrared associations – the so-called 'starless cores'. Detailed parameters are calculated, such as temperature, mass, luminosity and radial density dependence. On the basis of lifetime and luminosity arguments, the cores are found to be pre-protostellar in nature, undergoing the ambipolar diffusion phase prior to protostellar collapse. The cores do not follow the r^{-2} density dependence predicted by the standard model, but are consistent with a recent model of magnetic support of cloud cores.

Key words: stars: formation – ISM: dust

1. Introduction and Results

Submillimetre observations have recently identified a class of objects believed to be true protostars, the 'Class 0' objects (André, Ward-Thompson & Barsony 1993). We have recently carried out a survey of cold cores in dark clouds to search for even younger objects at the pre-protostellar phase, prior to protostellar collapse (Ward-Thompson *et al.* 1994 – WT94). The most extensive survey so far carried out of cold cores is that of Myers and co-workers (Benson & Myers 1989 – BM, and refs therein). Roughly half were found to have associated IRAS sources (Beichman *et al.* 1986), with two-thirds having no optical counterpart and one-third being associated with T Tauri stars (TTS). Therefore the invisible IRAS sources must be either embedded TTS or objects at an earlier stage of evolution. It is hypothesized that the cores with IRAS sources have already formed young stellar objects at their centres, while those without IRAS sources are designated 'starless cores'. In this paper we outline some of the results of a comprehensive submillimetre continuum study of starless cores. We refer the reader to WT94 for a more detailed discussion.

Fig 1 shows an 800-μm continuum isophotal contour map of L1689A. This object is typical of our sources, which all have a geometric mean FWHM greater than twice the beamwidth. This can be compared with the equivalent parameters for cores with IRAS sources (Ladd *et al.* 1991), where in a sample of cores at similar distances to ours, with similar resolution, 24 out

Astrophysics and Space Science **224**: 47–50, 1995.

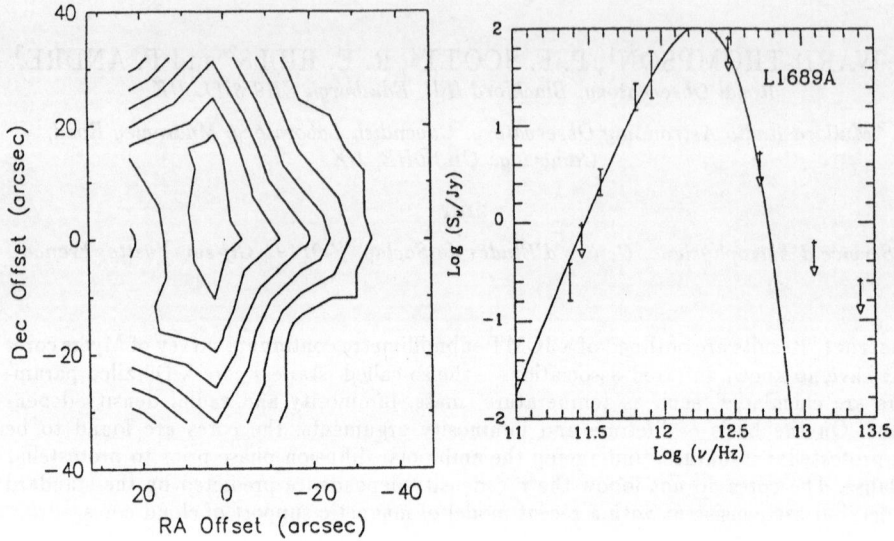

Fig. 1.   800-μm isophotal contour map & SED of the pre-protostellar core L1689A.

of 27 cores were seen to have FWHM less than twice the beamwidth. Thus starless cores tend to be more extended than cores with IRAS sources. This is consistent with the view that starless cores are at a pre-protostellar stage. Fig 1 also shows the spectral energy distribution (SED) of L1689A as a plot of $\log(S_\nu)$ versus $\log(\nu)$. Approximate upper limits corresponding to *IRAS* non-detections are also shown. The solid curve is a greybody consistent with the data, of the form $S_\nu = \Omega_s B_{\nu,T}(1-e^{-\tau(\nu)})$, where symbols take their usual meanings. The dust temperature thus derived is consistent with the gas temperature (BM).

Fig 2(a) shows the azimuthally averaged radial flux density profile versus projected radius, θ, for L1689B on a log-log scale, normalized to the peak flux density. The horizontal 'error bars' indicate the FWHM of the beamsize. The vertical 'error bars' represent the total deviation from spherical symmetry observed. On a log-log plot a power-law radial flux density profile is simply a straight line. The two solid straight lines shown correspond to $S_\nu(\theta)\propto \theta^{-1}$ and $S_\nu(\theta)\propto \theta^{-0.25}$. L1689B is typical in that none of the cores was found to be consistent with a single power-law profile throughout its full extent. In each case the outer regions (at radii greater than 20 arcsec) are consistent with the θ^{-1} profile, but extrapolation of this profile into the centre overestimates the peak flux density by more than a factor of five.

To convert 2-D radial (θ) intensity profiles into 3-D radial (r) density profiles, standard techniques show that, if temperature $T(r) \propto r^{-q}$, density $\rho(r) \propto r^{-p}$ and flux density $S_\nu(\theta) \propto \theta^{-m}$, then the indices are related by the simple expression $m = p + q - 1$. We assume isothermal cores, there-

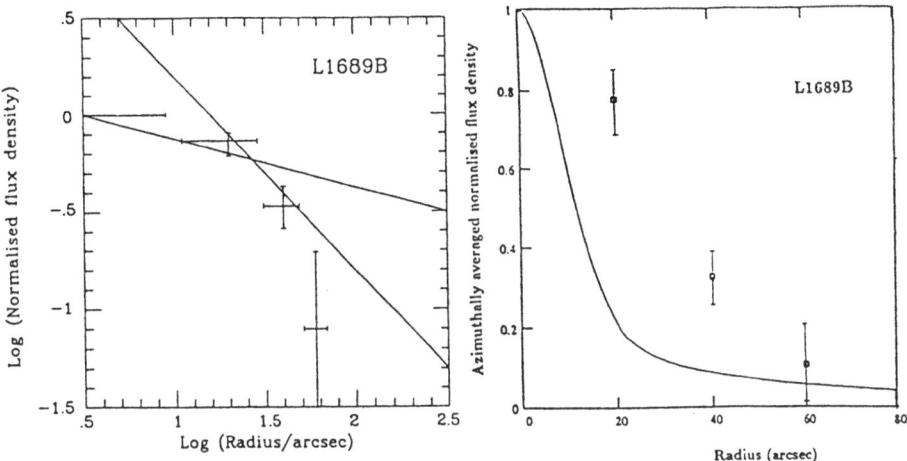

Fig. 2. Azimuthally averaged normalised radial flux density profile of L1689B: (a) on a log-log scale compared to $\rho(r) \propto r^{-2}$ and $\rho(r) \propto r^{-1.25}$ profiles; (b) on a linear scale compared to a $\rho(r) \propto r^{-2}$ convolved with the JCMT beam.

fore $q = 0$, and thus $p = m + 1$. So we see that the θ^{-1} and $\theta^{-0.25}$ flux density profiles translate into r^{-2} and $r^{-1.25}$ space density profiles. The standard model predicts that the precursor cores in which star formation occurs should have a standard isothermal sphere profile of $\rho(r) \propto r^{-2}$ (Shu 1977). All cores appear to have radial density profiles flatter than this in their inner regions, and only their edges are satisfactorily fitted by such a steep radial density profile. Fig 2(b) shows the same data on a linear scale, compared to the $\rho(r) \propto r^{-2}$ profile, convolved with the JCMT beam, to test whether the observed 'flattening' is an instrumental effect. Again it is seen that the inner part of the core is flatter than r^{-2}. A recent model (Ciolek & Mouschovias, 1994) predicts radial profile calculations based on a thermally and magnetically supported cloud undergoing ambipolar diffusion. The initial magnetohydrostatic model has a flat density profile in the core, and a steep profile in the envelope, consistent with our observations.

2. Discussion

The parameters of four of the cores are listed in Table 1. The temperature can be used to calculate the mass of each core, using the formula $M_{\mathrm{gas}} = S_\nu D^2 / \kappa_\nu B_{\nu,T}$ (WT94). In all cases, the masses we derive are lower than the masses derived by BM for the entire (larger) NH_3 cores. However, the mean density for each continuum clump given in Table 1 is an order of magnitude greater than the mean density derived for the whole NH_3 core. This is consistent with the continuum clumps representing true density peaks

TABLE I

Measured and derived parameters of four of the cores.

Source	FWHM (pc)	$S_{800\mu m}^{FWHM}$ (Jy)	T_{gb} (K)	M_{FWHM} ($M_\odot$)	M_{vir} ($M_\odot$)	$n(H_2)$ (cm^{-3})	L_{Tot} ($L_\odot$)
L183	0.044×0.029	1.8	8^{+4}_{-3}	2.2	0.8	2.0×10^6	0.05–0.2
L1689A	0.047×0.039	1.8	20^{+5}_{-8}	0.5	3.1	2.4×10^5	1.7–2.25
L1689B	0.047×0.047	3.0	15^{+3}_{-7}	1.2	1.2	4.7×10^5	0.9–1.8
L63	0.047×0.035	2.4	8^{+2}_{-3}	3.3	0.8	2.0×10^6	0.15–0.35

within the NH_3 cores. The virial masses within the FWHM listed in Table 1 show that all of the clumps, except L1689A (which has a much larger velocity dispersion than the others), have virial masses lower than or equal to their observed masses (within errors), as expected during the ambipolar diffusion stage.

The luminosities of the sources are listed in Table 1. The lower limits quoted are measured under greybody curves, while the upper limits are calculated by treating the IRAS upper limits as if they were detections. All of the objects have a luminosity of ≤ 2 $L_\odot$, which corresponds to a protostar mass of ≤ 0.02 $M_\odot$ if the luminosity is due solely to accretion, using standard assumptions (Shu *et al.* 1987). However, this cannot be the case, since at typical accretion rates this would make the age of each protostar only $\sim 10^3$–10^4 yr. This is clearly too young, given the number of objects seen – WT94 detected 17 out of 21 cores surveyed and derived lifetimes of $\sim 10^6$ yr – and therefore the cores are probably pre-protostellar in nature, and contain no central condensed source on to which matter is accreting.

The ambipolar diffusion time-scale for a typical clump is given by $t_{AD} \sim 8 t_{dyn}$ (Shu *et al.* 1987), where $t_{dyn} \sim R_{core}/v_A$. Taking v_A, the Alfvén velocity, to be $\sim \delta v$, the velocity dispersion within a clump (~ 0.6 km s^{-1}), after Shu *et al.* , we obtain $t_{AD} \sim 1.3 \times 10^6$ yr. This is consistent with the lifetimes derived for the starless cores. Thus all evidence points to the explanation that the cores we have observed in the submm continuum are magnetically supported, undergoing ambipolar diffusion prior to protostellar collapse.

References

André, P., Ward-Thompson, D., Barsony, M.: 1993, Astrophys. J. **406**, 122.
Beichman, C. A., *et al.* : 1986, Astrophys. J. **307**, 337.
Benson, P. J., Myers, P. C.: 1989, Astrophys. J., Suppl. **71**, 89 **(BM)**.
Ciolek, G. E., Mouschovias, T. C.: 1994, Astrophys. J. **425**, 142.
Ladd, E. F., *et al.* : 1991, Astrophys. J. **382**, 555.
Shu, F.: 1977, Astrophys. J. **214**, 488.
Shu, F. H., *et al.* : 1987, Ann. Rev. of Astron. & Astrophys. **25**, 53.
Ward-Thompson, D., *et al.* : 1994, Mon. Not. of the RAS **268**, 276 **(WT94)**.

THE DENSEST PRESTELLAR CONDENSATIONS:
H_2CO STUDIES OF ρ OPH B1

M. BARSONY

University of California, Riverside

D. SASSELOV

Harvard-Smithsonian Center for Astrophysics

S. RUCINSKI

Institute for Space and Terrestrial Science

E. BLOEMHOF

Harvard-Smithsonian Center for Astrophysics

and

L.-A. NYMAN

Swedish-European Southern Observatory Submillimeter Telescope

Abstract. The ρ Oph B1 cloud core region is unique in that it exhibits 2-cm H_2CO line emission over a substantial area (indicating gas densities $\geq 10^6$ cm^{-3}), while lacking any known embedded infrared sources. Subsequent H_2CO studies at the VLA and the JCMT have not provided a consistent model of the gas structure and kinematics. Here, we present the first high-velocity resolution maps in the 218 GHz and 352 GHz transitions of H_2CO. Our multi-transition study, at velocity and spatial resolutions comparable to existing VLA data, has proven crucial to forming a coherent picture of the gas kinematics.

Key words: star formation – pre-stellar cores – formaldehyde

1. Introduction

The ρ Oph clouds are among the closest (d=120–160 pc) and most well-studied star-forming regions in our Galaxy, so it is not surprising that they are the site of the densest known pre-stellar condensation (Sasselov & Rucinski 1990).

The condensation known as ρ Oph B1 was discovered in an early survey of the 2-cm line of H_2CO which usually appears in absorption against the 3K background (Gottlieb *et al.* 1978). It was a great surprise, then, to discover a $1' \times 2'$ *emission* region in this line, since the 2-cm line goes into emission only for $n_{H_2} \geq 10^6$ cm^{-3} (Garrison *et al.* 1975).

2. Observations and Results

We acquired our 218 GHz H_2CO map of ρ Oph B1 in 1993 February at the JCMT, and the 352 GHz H_2CO map in February of 1994 and 1993. A high-velocity resolution 226 GHz H_2CO map (not shown here, but useful in the

Astrophysics and Space Science **224**: 51–54, 1995.
© 1995 *Kluwer Academic Publishers.*

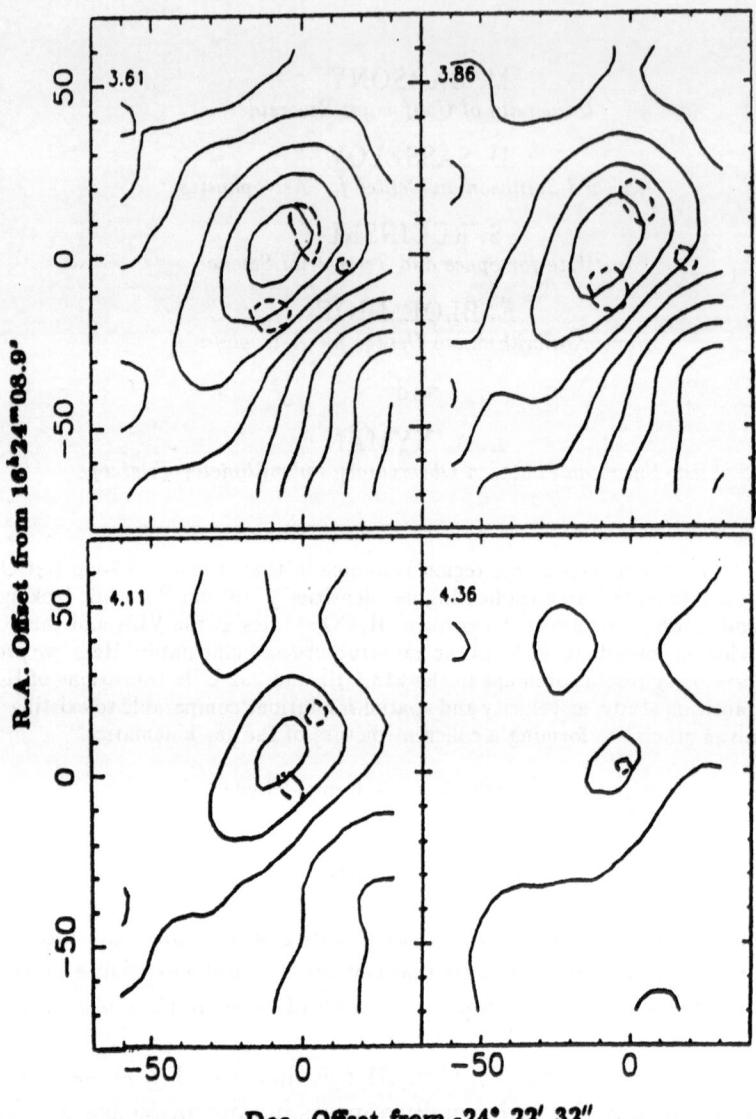

Fig. 1.   218.22 GHz H_2CO channel maps of ρ Oph B1, with the 2-cm VLA clumps indicated by the dashed contours. The V_{LSR} values (in km s^{-1} units) are indicated in the top left of each panel.

analysis) was obtained with SEST in April of 1992 during the commissioning run of the SIS 230 GHz receiver.

The possibility of the 218.22 GHz line being optically thick can be suspected by comparison of the SEST 225.7 GHz (H_2CO (3_{12}–2_{11}) "ortho-" line strengths with the 218.222 GHz (3_{03}–2_{02}) "para-" line strengths. In particular, since both lines lie at nearly the same energy above ground, in the optically thin case, one would expect their line strengths to be in the ratio of their statistical weights, namely a ratio of 3:1. The observed line ratios are 1:1, indicating high optical depths in the 225.7 GHz line, and leading one to suspect high optical depth in the 218.22 GHz line as well.

Figure 1 shows the emission from the 218.22 GHz (H$_2$CO (3_{03}–2_{02}) line, with the emission from the 14.49 GHz (2-cm) (H$_2$CO (2_{12}–2_{11}) superposed (from Wadiak *et al.* 1985). Note the lack of correspondence between the maps. Since we know the 2-cm line to be optically thin, this is clear evidence that the 218.22 GHz line must be optically thick.

The upper level of the 351.8 GHz (H$_2$CO (5_{15}–4_{14}) transition is 63 K above ground, compared with kinetic temperature determinations of $\sim$ 16 K from (larger beam) NH$_3$ observations (Martin-Pintado *et al.* 1983). We also obtained JCMT Canadian service time observations in April and May of 1994 of the 362.74 GHz (H$_2$CO (5_{05}–4_{04}) and 365.4 GHz (H$_2$CO (5_{23}–4_{22}) transitions, whose line ratio gives a good T$_{kin}$ determination (Mangum & Wootten 1993). The observed 363 GHz/365 GHz line ratio gives a preliminary determination of T$_{kin}$ = 50K$^{+10K}_{-30K}$ for the position of the dust continuum peak of ρ Oph B1 (Mezger *et al.* 1992).

Figure 2 shows the emission from the 351.8 GHz (H$_2$CO (5_{15}–4_{14}) line, with emission from the 14.49 GHz (2-cm) (H$_2$CO (2_{12}–2_{11}) line superposed. Comparison of the 352 GHz map with the VLA 14.5 GHz map allows identification of the same spatial/velocity structures in both transitions, leading one to suspect that the 352 GHz H$_2$CO transition may be optically thin. Differences in source structure in the two lines may be attributed to the insensitivity of the interferometer to extended structures and absorption of the 2-cm line, due to the presence of less dense, cool foreground gas. The 352 GHz line, due to its relatively high excitation energy, is not likely to suffer from foreground absorption, as the 14.5 GHz line clearly does. This difference may explain the discrepancy in the sense of the velocity gradient of the southeastern condensation detected in both line maps–foreground absorption could result in the apparent velocity peak of the 2-cm line differing from its true velocity peak. For this reason, we lend more credence to the sense of the velocity gradient in this structure as detected by the JCMT at 352 GHz. In magnitude, this velocity gradient corresponds to roughly a 0.75 km s^{-1} velocity shift over 12″, corresponding to a velocity gradient of about 90 km s^{-1} pc^{-1}. This is comparable to the 40 km s^{-1} pc^{-1} found by Wadiak *et al.* Assuming a clump radius of 5″ (or 0.0034 pc at a distance of 140 pc), this amounts to a rotational velocity of 0.3 km s^{-1}, or a rotation rate of $\Omega \sim 2.9 \times 10^{-12}$ rad s^{-1}. The inferred included mass for the southeastern pre-stellar clump is then $\sim$ .07 M$_\odot$. This leads to a mean mass density of 3×10^{-17} gm cm^{-3}, or a hydrogen molecule number density of $\sim 1.2 \times 10^7$, in good agreement with previous constraints from H$_2$CO line measurements (Sasselov & Rucinski 1990).

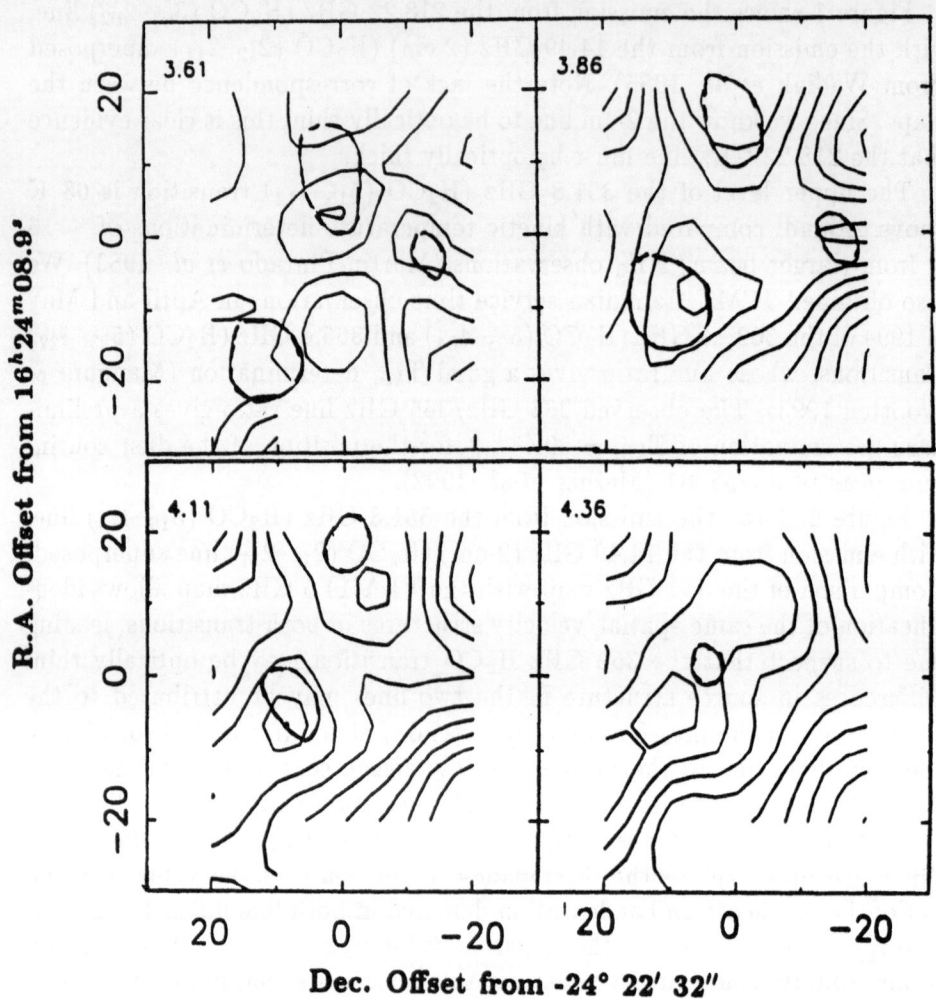

Fig. 2.   351.78 GHz H_2CO channel maps of ρ Oph B1, with the 2-cm VLA clumps indicated by the dashed contours. The V_{LSR} values (in km s^{-1} units) are indicated in the top left of each panel.

References

Garrison, *et al.* : 1975, Astrophys. J. **200** L175.
Gottlieb, *et al.* : 1978, Astrophys. J. **219** 77.
Mangum & Wootten, A.: 1993, Astrophys. J., Suppl. **89** 123.
Martin-Pintado, *et al.* : 1983, Astron. & Astrophys. **117** 145.
Mezger, *et al.* : 1992, Astron. & Astrophys. **265** 743.
Sasselov, D. & Rucinski, S.: 1990, Astrophys. J. **351** 578.
Wadiak, *et al.* : 1985, Astrophys. J. **295** 43.

MASERS IN STAR-FORMING REGIONS

R. J. COHEN

University of Manchester,
Nuffield Radio Astronomy Laboratories, Jodrell Bank,
Macclesfield, Cheshire SK11 9DL, UK

Abstract. Maser action in the interstellar medium produces the brightest and most spectacular molecular lines that radio astronomers can study. Strong maser action was first detected in OH (at 1.6 GHz) and water (at 22 GHz) in star-forming regions, but with improvements in mm and submm-wave technology, and improved laboratory data, many new maser transitions are being identified. For methanol alone over 20 maser transitions have been identified in star-forming regions. This review summarizes recent observational developments. Masers provide the most readily detectable indicators of the formation of massive young stars, and offer our best prospect for making a complete census of star-forming regions in the Galaxy. Using radio interferometers the structure of the regions can be probed on angular scales of 1 milliarcsecond. The use of masers as probes of the physical conditions in these regions is discussed.

Key words: masers – molecular lines – star-formation

1. Introduction

It is almost thirty years since the first interstellar masers were discovered (Weaver *et al.* 1965). The existence of natural masers was so unexpected that for a while it was thought that a new substance, mysterium, had been discovered. In retrospect the discovery is not so surprising however. Maser action is produced by stimulated emission, which increases in importance relative to spontaneous emission as we go to longer wavelengths. Furthermore the energy levels involved in microwave transitions are so close together that only a small transfer of population, for example one percent, is necessary to set up a significant population inversion. This can happen naturally in space. The molecular transition is then strongly amplified to produce an intense beam of radiation. Gain factors of masers in star-forming regions can exceed one billion, whichever definition you choose for the billion.

Maser sources have several characteristics which distinguish them from normal (thermal) molecular line sources. The most fundamental properties are the intensity of the maser lines, and the strongly non-equilibrium line ratios. If we see these we can be sure we are looking at a maser. In addition the maser amplification usually leads to line-narrowing and to strong beaming of the radiation. It may also lead to polarization in the presence of a magnetic field, and to variability, as the maser output is often a strongly nonlinear function of the physical parameters involved. All this makes it inherently difficult to recover the physical conditions in the maser regions. However the beauty of masers is their enormous intensity, which enables us to study small regions of excited molecular gas close to young stars in

Astrophysics and Space Science **224**: 55–62, 1995.

regions wh' h could not otherwise be studied, and to study these regions at milliarcsecond (mas) resolution.

The main molecular species which produce masers in star-forming regions are hydroxyl (OH), water (H_2O), silicon monoxide (SiO), methanol (CH_3OH), ammonia (NH_3) and methaladyne (CH). In addition to these molecular masers, hydrogen recombination line masers have been detected in the source MWC349 (Martin-Pintado et al. 1989, 1994). In this review I will discuss the use of masers as signposts of star-formation, and their use as diagnostics of the source structure, kinematics, and physical conditions. The review concentrates on recent developments, mainly within the last three years. Earlier work is reviewed elsewhere (e.g. Cohen 1989, 1991, and references therein). These references also discuss the use of masers to measure magnetic fields in star-forming regions, an important topic which could not be included in the present review for reasons of time and space.

2.   New Masers

One of the exciting developments in recent years has been the great increase in the number of maser transitions identified. These have been mainly new maser transitions of previously known maser species. The processes which lead to population inversion in a particular transition of a molecule usually cause inversions in several other transitions as well. For example the inversion of the 22 GHz transition of water involves excitation by any means, followed by radiative decay, which occurs preferentially down the ladders of constant J. This leads to overpopulation of the so-called backbone levels at the bottom of each J–ladder (de Jong 1973). These levels contain the bulk of the population and become over-populated with respect to other levels of similar energy, provided that the collision rate is low compared to the radiative decay rate. This process explains the well-known 22 GHz maser, which is the 6_{16}–5_{23} transition from the backbone level of the J=6 ladder of ortho-water, but moreover the same physics predicts a whole family of inversions. Each maser transition occurs between states of similar energy which both lie well above the ground state, and the energy difference can give a transition wavelength anywhere in the microwave to far-infrared bands. Many new submillimetre water maser transitions have been identified recently, despite the obvious difficulty of absorption by the atmospheric water at the same frequencies (Menten et al. 1990a, 1990b, Melnick et al. 1993). Many of the new transitions are both strong and widespread. The fact that these masers were searched for on the basis of theoretical predictions (e.g. Neufeld & Melnick 1991) indicates that we have come a long way since the days of mysterium.

A more surprising development has been the detection of interstellar ammonia masers. Although ammonia was the first laboratory maser it had been thought until recently to be a reliable tracer of physical conditions in interstellar clouds. However many maser transitions have been discovered in recent years, including one with a brightness temperature of 10^{13}K (Pratap et al. 1991). The number of known ammonia maser sources is also growing apace. These developments have been reviewed by Wilson & Schilke (1993).

Even more spectacular has been the progress on the study of methanol masers, which is treated separately below.

3. Masers as Signposts

Masers act as beacons which signpost the birthplaces of stars within our Galaxy and nearby galaxies. The high photon rates make masers among the most readily detectable indicators of star-formation. If you could place a star-forming region at progressively larger distances then the last detectable signal would be the water masers, in most cases, followed by the methanol masers, and then perhaps the far-infrared emission and the OH masers (e.g. Cohen 1989). The masers would remain visible long after other (thermal) molecular lines tracing outflows and discs had disappeared below the instrumental noise.

The intense maser spots have characteristic sizes of order 1 milliarcsec, but the spots are frequently broadened by interstellar scattering. They cluster on scales of typically 1 milliparsec (mpc). Maser clusters are found preferentially near the densest and most massive young stars. There is a rough correlation, seen in many masers, between the maser luminosity and the stellar luminosity as traced by far-infrared emission. For example Wilking et al. (1994) recently confirmed such a correlation for the 22 GHz water masers associated with low-mass stars. Their correlation diagram is remarkable in that their data follow closely the same empirical correlation found for high-mass stars by Felli et al. (1992), for which the luminosities involved were four orders of magnitude higher. This is a real correlation which does not go away if we plot the observed flux densities rather than the derived luminosities.

It is clear from their kinematics that masers are associated with the outflow from young stars, and in this sense they are evolutionary markers as well as beacons. Radio interferometry has provided direct evidence of proper motions of H_2O masers and OH masers. The expansion of H_2O masers has been known for some time. By comparing the expansion proper motions with the radial velocities and modelling the source the distance can be obtained (e.g. Reid et al. 1988; Gwinn et al. 1992). This technique, which can be extended to external galaxies, is now set to make a significant contribution to measuring the galactic distance scale. The detection of OH proper motions is more recent (Bloemhof et al. 1992; Migenes et al. 1992), and settles a long-standing controversy over the nature of these sources. In both sources which have been studied the OH masers are expanding.

4. Methanol

The most striking advances in the last three years have been in the study
of methanol masers. Methanol has more detected radio lines than any other known interstellar molecule. During the 1980's it became apparent that
many of these could be excited as masers, including some which are strong
and widespread (Cohen 1989 and references therein). A turning point was
the discovery of the strong 2_0-3_{-1} transition of E-type methanol (Batrla *et
al.* 1987). With this discovery it also became clear that the methanol sources
can be divided into two classes, now called Class I and Class II, according
to which transitions are excited. The rapid growth in the subject since then
is exemplified by the case of the even stronger 5_1-6_0 transition of A-type
methanol. This was discovered only a few years ago by Menten (1991), and
has already risen to become the second strongest interstellar maser, with
more sources known than OH which has been studied for almost thirty
years.

There has been good progress in understanding the pumping of methanol
masers, despite the complexity of the molecular energy level diagrams. Cragg
et al. (1992) discuss pumping by collisional excitation and radiative decay,
and find a simple explanation for the two classes. Class I masers occur when
the temperature of the continuum radiation is lower than the kinetic temperature of the gas. The strongest maser is then predicted to be the 7_0-6_1
transition of A-methanol at 44 GHz, followed by the 4_{-1}-3_0 transition of
E-methanol at 36 GHz, as is observed. Class II masers on the other hand
occur when the temperature of the continuum radiation exceeds the kinetic temperature of the gas. The model then predicts the strong 12.2 GHz
and 6.7 GHz masers that are observed, although it does not get their relative intensities right, predicting that 12.2 GHz should be the stronger. The
model may thus need refinement. However the results to date are certainly
encouraging. The inversion is robust, and appears to depend entirely on the
radiative selection rules. The pumping models also predict anti-inversion of
some of the Class II transitions in Class I regions (*i.e.* anomalous absorption), as indeed has been observed (Walmsley *et al.* 1988).

The representative transition for Class I sources is the 44.07 GHz line.
A large scale survey of this transition has been completed recently by Slysh
et al. (1994) using the Parkes radio telescope. They searched 250 sources
and found 55 new methanol maser sources, bringing the total known in this
transition to over 100. The intensities are typically 10-100 Jy. Slysh *et al.*
find that 82 percent of the detected sources are associated with 6.7 GHz
methanol masers, to within 15 arcsec, and suggest that interferometry is
needed to investigate this association further. Some sources are pure Class I,

e.g. NGC2264, some are pure Class II, *e.g.* Cepheus A, but most sources are mixtures. An inverse correlation is found between 44 GHz and 6.7 GHz flux densities. It is already known from interferometry of DR21 that Class I and Class II masers can be excited in different regions on the scale of 1 pc (Plambeck & Menten 1990), but the results of Slysh *et al.* (1994) suggest that there is also competition between Class I and Class II pumping within different regions of the same source on a much smaller scale.

The other major methanol survey recently completed is the survey by Caswell *et al.* (1994) of the 6.7 GHz line, which is the representative transition for Class II sources. Like the survey by Slysh *et al.* (1994) this was a targetted search of previously known maser sources. Together with the results of Menten (1991) and the surveys at Hartebeesthoek (MacLeod *et al.* 1992; Schutte *et al.* 1993) this brings the total number of 6.7 GHz sources in the Galaxy to over 250. Caswell *et al.* estimate that at least the same number again can be found in an unbiassed survey of the galactic plane to 1 Jy. This is a project which could be done using a medium-sized radio telescope. Methanol has many advantages over OH and H_2O for such an unbiassed survey for star-forming regions in the Galaxy. The 6.7 GHz line is strong, exceeding thousands of Janskies in some cases, and widespread, it does not suffer confusing absorption as OH does, and most important of all it is excited only in star-forming regions. Both OH and H_2O suffer from the problem that they are also easily excited in circumstellar envelopes. An unbiassed survey of the galactic plane in OH or H_2O would thus yield many circumstellar sources, and there would be no way to distinguish them from interstellar sources in most cases, without time-consuming follow-up work. Methanol masers on the other hand can only be interstellar.

Synthesis imaging of Class II methanol masers has begun in earnest. Norris *et al.* (1993) have mapped 14 sources at 6.7 GHz, and find that in many cases the maser spots lie along lines or arcs. These are interpreted as evidence for protoplanetary discs. In some sources the data were compared with previous maps of 12.2 GHz masers, and close agreement was found between the maser spot positions in some cases. Class II methanol masers are known to be strongly associated with OH masers, and indeed there are only a handful of Class II sources without any known OH. Very-long-baseline-interferometry of the prototypical Class II source W3(OH) has been carried out at 12.2 GHz by Menten *et al.* (1988) and at 6.7 GHz by Menten *et al.* (1992). They find that the methanol masers are spatially coincident with OH maser spots on a scale of 0.1 mpc. Hartquist *et al.* (1994) explain this spatial coincidence, and the enhanced densities of OH and methanol necessary to account for the observed maser intensities, in terms of grain mantle evaporation. In their model the grains are heated by the passage of shocks, and this releases methanol and water into the gas. Photodissociation then allows substantial amounts of OH and CH_3OH to coexist.

Methanol masers have other possible uses as astrophysical probes which have yet to be exploited. For example they offer excellent prospects for proper motion studies. They have a sufficiently high frequency to give good results on a timescale of a year or two, they are stronger than OH masers in general, and their velocities are not distorted by the Zeeman effect, so kinematics can be reliably determined. Furthermore they do not present the degree of variability which can bedevil attempts to measure proper motions of H_2O masers. Another area where they have something new to offer is in studies of interstellar scattering, where the known coincidences between maser spots at 12.2 GHz and 6.7 GHz could be harnessed to study variations of spot size with frequency (in combination with a suitable maser model).

5.   Physical Conditions

In principle masers offer a unique diagnostic of the physical conditions in compact regions close to young stars. Recent work by Magnum & Wootten (1994) for example shows how an observed inversion of the $NH_3(3,3)$ maser but not the (1,1), (2,2) and (4,4) transitions constrains the physical conditions about as well as they are usually constrained by observations of thermal molecular line emission. In the case of OH, masers models by Cesaroni & Walmsley (1991) and Gray *et al.* (1992) are able to account for much of the general OH maser phenomenology, and in some cases it has even been possible to model individual OH maser spots (*e.g.* Masheder *et al.* 1994). However there are many practical difficulties, and I would like to finish by summarizing some of them.

What we observe is a set of maser spots whose apparent sizes may be affected by interstellar scattering and by beaming within the source, as well as by our instrumental angular resolution. These effects are all frequency-dependent, so we must take greater care than usual in interpreting maser line ratios, intensities, linewidths and other observables. In particular if we wish to then derive information about physical conditions we must have a good model of the maser source and the maser excitation. Maser modelling has its own difficulties. It needs accurate collision cross-sections, which are not always available. For example most published models of H_2O masers are based on collisions with He rather than with H_2. The H_2O inversion is robust, but in other cases small differences like this can have major consequences. Flower *et al.* (1990) have shown for example that collisions of NH_3 with ortho- and para-H_2 are not equivalent, and that the difference is crucial for the NH_3 (3,3) maser. A maser model needs to treat radiative transfer within the source in a self-consistent way. In this respect it is encouraging to see moves away from the large-velocity-gradient approximation, for example the work presented by Yates *et al.* in these proceedings. Further effects which come into play are beaming and saturation, as beams propagating in different directions in the source compete for the available inverted molecules (*e.g.* Elitzur 1994). There are also subtle effects such as line-overlap which must be considered. Furthermore the variability which is such a striking feature of maser observations is seldom included in maser modelling. Finally I refer you to the recent paper by Watson (1994) to discover for yourselves that the topic of maser polarization is still controversial.

References

Batrla W., Matthews H.E., Menten K.M. & Walmsley C.M.: 1987, Nature **326**, 49.

Bloemhof E.E., Reid M.J. & Moran J.M.: 1992 Astrophys. J. **397**, 500.

Caswell J.L., Vaile R.A., Ellingsen S.P., Whiteoak J.B. & Norris R.P.: 1994, Mon. Not. of the RAS in press.

Cesaroni R. & Walmsley C.M.: 1991, Astron. & Astrophys. **241**, 537.

Cohen R.J.: 1989 Rep. Prog. Phys. **52**, 881.

Cohen R.J.: 1991 in: *Molecular Clouds*, eds: R.A. James & T.J. Millar, Cambridge University Press, p 189.

Cragg D.M., Johns K.P., Godfrey P.D. & Brown R.D.: 1992, Mon. Not. of the RAS **259**, 203.

de Jong T.: 1973, Astron. & Astrophys. **26**, 297.

Elitzur M.: 1994, Astrophys. J. **422**, 751.

Felli M., Palagi F. & Tofani G.: 1992, Astron. & Astrophys. **255**, 293.

Flower D.R., Offer A. & Schilke P.: 1990, Mon. Not. of the RAS **244**, 4P.

Gray M.D., Field D. & Doel R.C.: 1992, Astron. & Astrophys. **262**, 555.

Gwinn C.R., Moran J.M. & Reid M.J.: 1992, Astrophys. J. **393**, 149.

Hartquist T.W., Menten K.M., Lepp S. & Dalgarno A.: 1994, Mon. Not. of the RAS in press.

MacLeod G.C., Gaylard M.J. & Nicolson G.D.: 1992, Mon. Not. of the RAS **254**, 1P.

Mangum J.G. & Wootten A.: 1994, Astrophys. J. **428**, L33.

Martin-Pinatado J., Bachiller R., Thum C. & Walmsley C.M.: 1989 Astron. & Astrophys. **215**, L13.

Martin-Pintado J., Neri R., Thum C., Planesas P. & Bachiller R.: 1994, Astron. & Astrophys. **286**, 890.

Masheder M.R.W., Field D., Gray M.D., Migenes V., Cohen R.J. & Booth R.S.: 1994, Astron. & Astrophys. **281**, 871.

Melnick G.J., Menten K.M., Phillips T.G. & Hunter T.: 1993, Astrophys. J. **416**, L37.

Menten K.M.: 1991, Astrophys. J. **380**, L75.

Menten K.M., Melnick G.J. & Phillips T.G.: 1990a, Astrophys. J. **350**, L41.

Menten K.M., Melnick G.J., Phillips T.G. & Neufeld D.A.: 1990b, Astrophys. J. **363**, L27.

Menten K.M., Reid M.J., Moran J.M, Wilson T.L., Johnston K.J. & Batrla W.: 1988, Astrophys. J. **333**, L83.

Menten K.M., Reid M.J., Pratap P., Moran J.M. & Wilson T.L.: 1992, Astrophys. J. **401**, L39.

Migenes V., Cohen R. J. & Brebner G.C.: 1992, Mon. Not. of the RAS **254**, 501.

Neufeld D.A. & Melnick, G.J.: 1991, Astrophys. J. **368**, 215.

Norris R.P., Whiteoak J.B., Caswell J.L., Wieringa M.H. & Gough R.G.: 1993, Astrophys. J. **412**, 222.

Plambeck R.L. & Menten K.M.: 1990 Astrophys. J. **364**, 555.

Pratap P., Menten K.M., Reid M.J., Moran J.M. & Walmsley C.M.: 1991, Astrophys. J. **373**, L13.

Reid M.J., Schnepps M.H., Moran J.M., Gwinn C.R., Genzel G., Downes D. & Ronnang B.: 1988, Astrophys. J. **330**, 809.

Schutte A.J., van der Walt D.J., Gaylard M.J. & Macleod G.C.: 1993, Mon. Not. of the RAS **261**, 783.

Slysh V.I., Kalenskii S.V., Val'tts I.E. & Otrupek R.: 1994, Mon. Not. of the RAS **268**, 464.

Walmsley C.M., Batrla W., Matthews H.E. & Menten K.M.: 1988, Astron. & Astrophys. **197**, 271.

Watson W.D.: 1994, Astrophys. J. **424**, L37.

Weaver H., Williams D.R.W., Dieter N.H. & Lum W.T.: 1965, Nature, **208**, 29.

Wilking B.A., Claussen M.J., Benson P.J., Myers P.C., Tereby S. & Wootten A.: 1994, Astrophys. J. **431**, L119.

Wilson T.L. & Schilke P.: 1993 in: *Astrophysical Masers*, eds: A.W. Clegg & G.E. Nedoluha, Springer-Verlag, Heidelberg, p 123.

THEORETICAL MODELLING OF SIO MASER LINESHAPES

M.D. GRAY, E.M.L. HUMPHREYS and D. FIELD

School of Chemistry, University of Bristol, Cantock's Close, Bristol. BS8 1TS, UK

Abstract. Various factors affecting the lineshapes of SiO masers have been considered: saturation, competitive gain, velocity redistribution and properties of the circumstellar envelope supporting the masers. The lineshape of a single maser spot is likely to be dominated by conditions of complete velocity redistribution (CVR) for low to moderate amplification, changing to NVR (no velocity redistribution) only for strongly saturating masers. Both CVR and NVR regimes can produce peculiar lineshapes. When many maser 'cells' were considered, distributed at random in the stellar envelope, the individual lineshapes were combined to produce synthetic spectra. Maser action is predicted in higher rotational transitions than those observed to date, for example the J=7-6 transition of v=2.

1. Factors Affecting Maser Lineshapes

Unlike their OH and H_2O counterparts, SiO masers are associated with both the initiation of mass loss, at the inner edge of the maser zone and the formation of dust at the outer edge of the maser zone (Cohen 1989, Elitzur 1992). SiO maser lineshapes are affected by processes associated with emission by a single maser 'cell' and by the effects of combining the emission from many 'cells' from different places in the stellar envelope. Single cell lineshapes are affected by narrowing due to amplification: rebroadening occurs only if the effects of velocity redistribution among molecular velocity subgroups can be ignored for strong saturation (NVR case). Competitive gain between masers of different frequency can alter lineshapes considerably. Velocity gradients in the maser column can lead to the appearance of new maser features, separated in velocity from the original. Lineshapes from an assembly of cells depend, in addition to the above factors, on the central velocities and gain lengths of the individual cells.

2. Velocity Redistribution under NVR

We have performed calculations as present only under the two extreme cases of NVR and CVR. The intermediate case is extremely difficult to solve rigorously. The propagation code uses a semi-classical theory for the coupling of maser radiation to the SiO kinetic scheme, based on an earlier model for OH (Field & Gray, 1988). Propagation under NVR yields lineshapes shown in Figures 1a,b. There is initial narrowing on amplification and a re-broadening, dramatically complicated by competition between the two maser transitions displayed. An important feature of the lineshape evolution is the decay of the original line centres through absorption. The final result is lineshapes with a symmetrical structure of a broad base supporting two narrow peaks.

Astrophysics and Space Science **224**: 63–68, 1995.

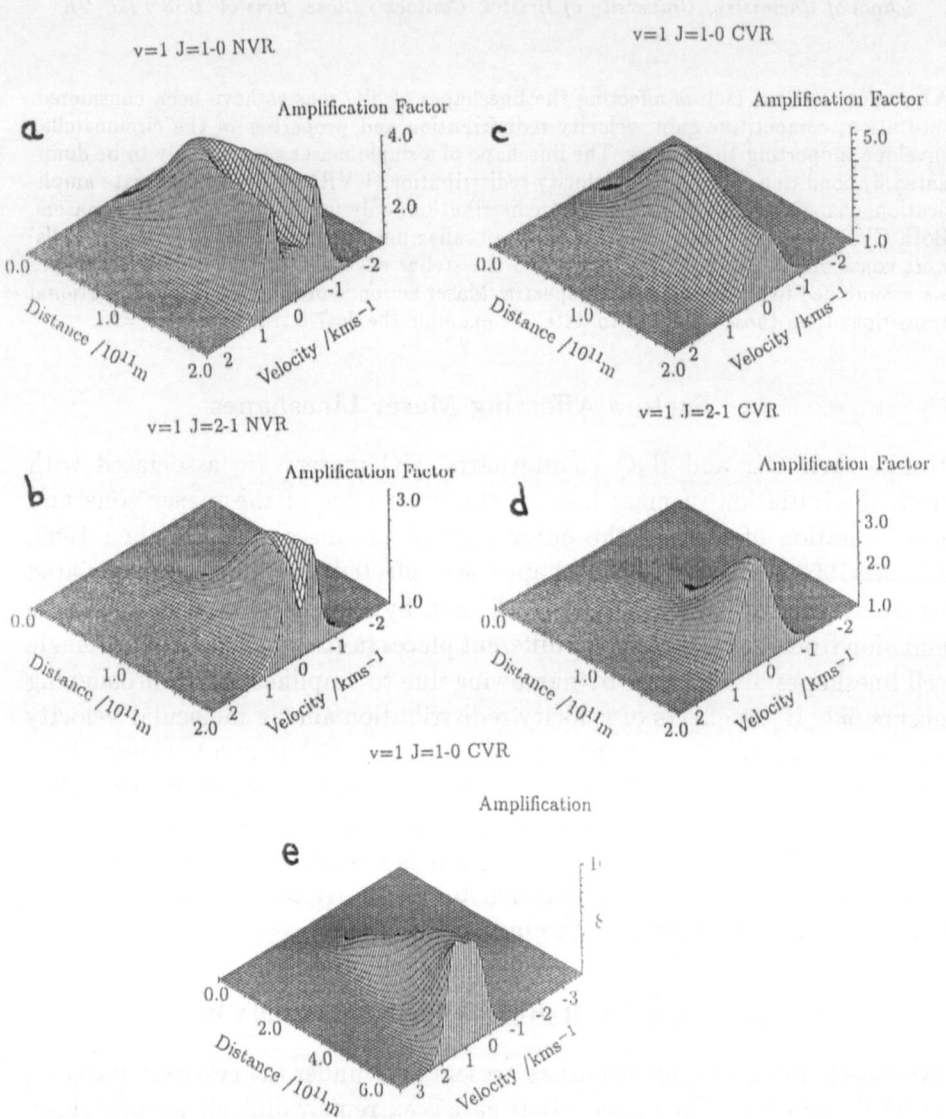

Fig. 1.   Competitive propagation of SiO masers under CVR and NVR

The physical conditions for the model used to produce Figure 1a,b are: dust temp.=1500K (dilution=0.01); stellar temp.=2500K (dilution=0.1); kinetic temp.=1500K; H_2 no. density=6×10^9 cm^{-3}; [SiO]/[H_2]=10^{-4}; Sobolev velocity gradient = 3×10^5 kms^{-1}(pc)$^{-1}$.

2.1 CVR

Figures 1c,d show the same pair of lines under the alternative extreme of CVR. A small velocity gradient in the direction of propagation is present here, but does not affect the lineshapes significantly. Note the increased peak amplification values over the NVR case. In the CVR limit, lineshapes act as a whole and the double peaked structure is absent. At the end of the model, the J=1-0 line is in absorption everywhere. In Figure 1e, we see the effect of a large velocity gradient on the lineshape. Initial saturation at $1.5-2.0 \times 10^{11}$m is followed by an interval of secondary gain as population is replenished by fresh molecules from outside the original velocity range. A catastrophic release of saturation as the molecular and radiation profiles separate leads to the end of growth in the old line and the appearance of a new emission wing. This effect is less distinct in SiO than in OH or H_2O (Nedoluha & Watson, 1988, Field *et al.* 1994). Conditions for Figure 1e are as for Figures 1a-d except that: H_2 no. density=5×10^9cm^{-3}; velocity gradient parallel to maser propagation= 5×10^4kms^{-1}(pc)$^{-1}$.

2.2 DEVELOPMENT OF REDISTRIBUTION

It is crucial to know how important velocity redistribution is in SiO masers. Data in Table 1 show how redistribution develops over a typical maser propagation calculation. Column 1 is the line-centre amplification factor of the maser, column 2 is the ratio of the unsaturated to the saturated population inversion and column 5 the propagation distance required to achieve these conditions. The more important columns 3 and 4 show the redistribution and maser rates respectively. It can be seen that the maser progresses from CVR conditions in the unsaturated state to NVR conditions after high amplification and strong saturation. Physical conditions for the model which produced the data in Table 1 are: dust temp.=200K (dilution factor=0.9); stellar temp.=2500K (dilution=0.1); kinetic temp.=1500K; H_2 no. density=2×10^8cm^{-3}; $[SiO]/[H_2]=10^{-4}$; Sobolev velocity gradient = 3×10^3 kms^{-1}(pc)$^{-1}$.

Amplification	ζ	P /s^{-1}	R /s^{-1}	x /m
1.000	1.0	0.841	0.0	0.0
12.35	1.13	0.841	0.055	7.5×10^{11}
32.34	2.91	0.841	0.807	1.0×10^{12}
284.9	18.71	0.841	7.485	1.5×10^{12}

3. Composite Lineshapes

Figures 2 & 3 show composite synthetic spectra made up from 47 cells, distributed at random in the inner regions of a model circumstellar envelope. Transitions are shown for the v=1,2 and 3 vibrational states and for upper

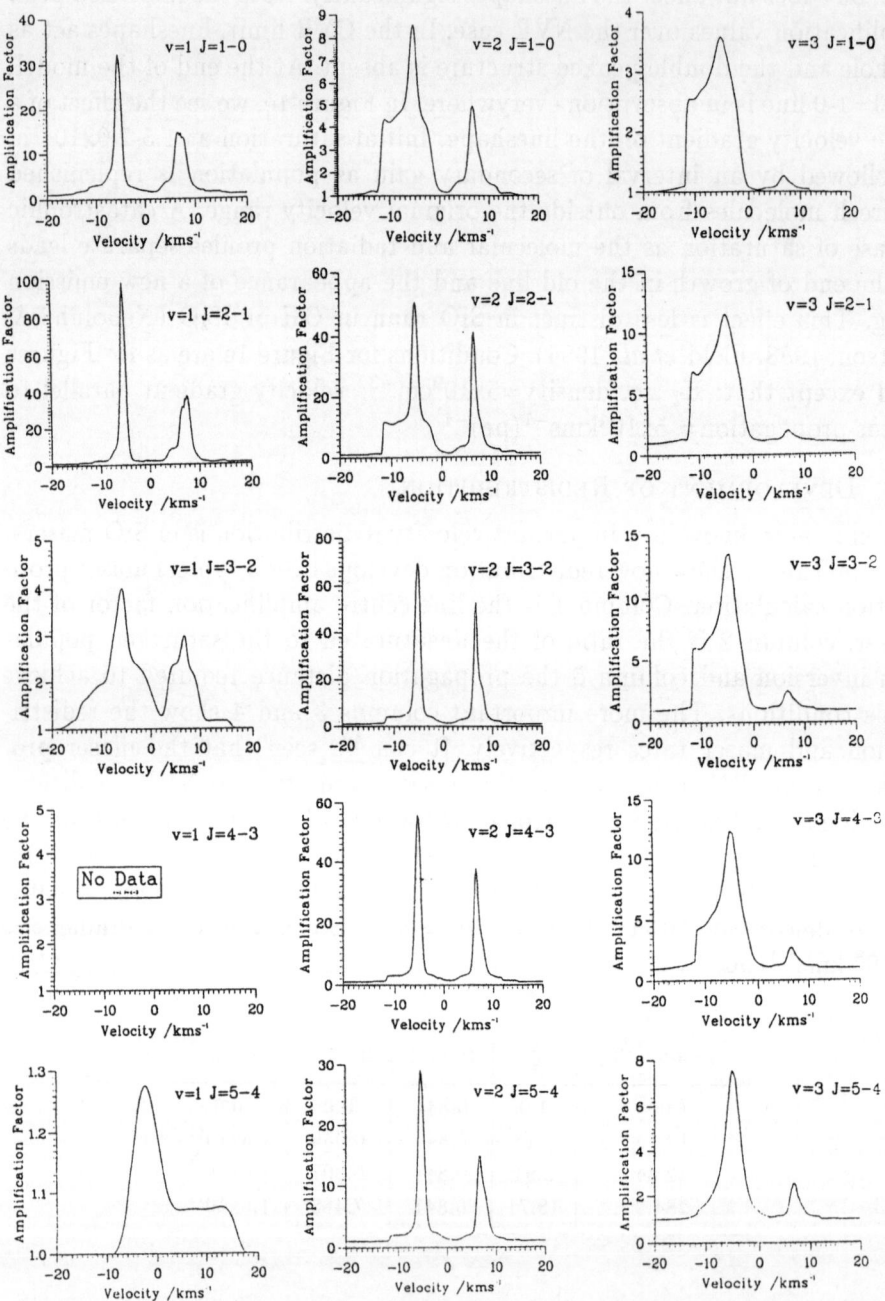

Fig. 2. Synthetic spectra produced by a composition of SiO propagation models.

J states up to J=5. Masers in higher rotational transitions may be present but are not shown. Physical conditions for each cell were drawn from a Mira atmosphere model by Willson 1987. All cells used CVR and a propa-

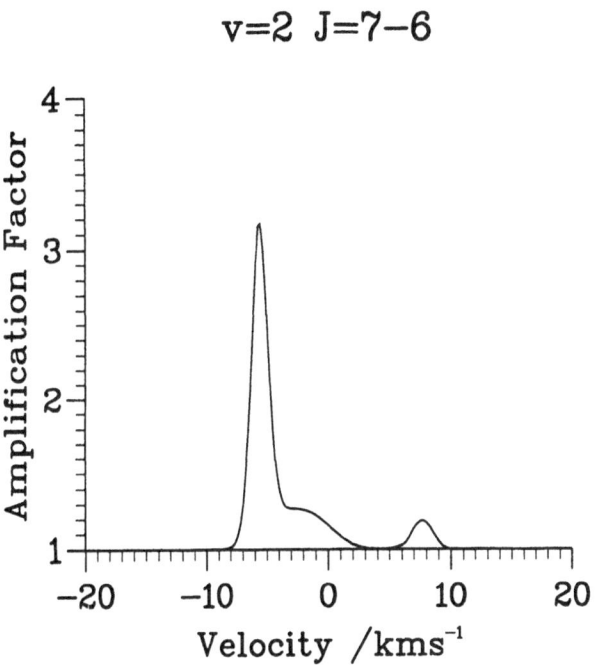

Fig. 3.   Synthetic spectra produced by a composition of SiO propagation models.

gation length of 2.5×10^{11}m. Several lines are predicted, which have not been observed to date, particularly the J=7-6 line in v=2 (Figure 3). Almost all emission comes from cells within 2 stellar radii of the photosphere. There appear to be two types of emission: strong spikes of typical amplification factor >10 in the lower J-state lines and weaker emission forming a broad plateau covering 15-20 kms^{-1}.

4.   Discussion

The synthetic spectra compare well with many observed SiO maser spectra in terms of lineshapes and the set of transitions found to be masing. The strong maser spikes appear to come from 'cells' which have low to moderate velocity gradient and can reach saturating intensity. The broad emission plateaux arise from cells with high velocity gradients in the line of sight, spreading maser emission across a wide velocity range. There remain a number of problems with the modelling: the model star is smaller than the size indicated by observation, leading to a higher velocity spread in the spectra than is usual in observations. Also the v=2, J=2-1 line appears as a bright maser in the models, whilst observationally it is anomalously weak. Future models should also incorporate a general treatment of redistribution for all intermediate conditions between CVR and NVR.

Acknowledgements

The authors would like to thank the Royal Society for the award of a 1983 University Research Fellowship to MDG and for funding EMLH as a vacation student.

References

Cohen, R.J.: 1989, Rep. Prog. Phys. **52**, 88.
Elitzur, M.: 1992, *Astrophysical Masers*, Kluwer.
Field D. & Gray M.D.: 1988, Mon. Not. of the RAS **234**, 353.
Field D., Gray M.D. & de St. Paer P.: 1994, Astron. & Astrophys. **282**, 213.
Nedoluha G.E. & Watson W.D.: 1988, Astrophys. J. **335**, L19.
Willson L.A.: 1987, in: *Late Stages of Stellar Evolution*, eds: S. Kwok & S.R. Pottasch, Reidel, p253.

UNRAVELLING CEPHEUS A

V. A. HUGHES

Astronomy Group, Department of Physics, Queen's University, Kingston,
Ontario K7L 3N6, Canada

and

R. J. COHEN and S. GARRINGTON

University of Manchester, Nuffield Radio Astronomy Laboratories, Jodrell Bank,
Macclesfield, Cheshire SK11 9DL, U.K.

Abstract. The monitoring of Cep A over a 13 year period has shown it to contain interesting aspects of star forming regions. In particular, there are two highly time-dependent radio sources which may be evidence for proto-stars losing their magnetic fields, and a 'bullet' which appears to be the head of a high velocity jet which causes the entrainment of molecular outflows.

Key words: Star formation – radio sources –molecular outflows – jets

1. Introduction

Information on Cepheus A has been accumulating over the past fifteen years. It is a highly obscured star-forming region situated in a dense CO condensation (Sargent 1979) which contains numerous molecules (*e.g.* Moriarty-Schieven *et al.* 1991). Also, there are OH and H_2O masers (Lada *et al.* 1981; Cohen *et al.* 1984), molecular outflows (Rodríguez *et al.* 1980), a strong IR source (Evans *et al.* 1981), and Herbig-Haro objects (Lenzen 1988). Radio observations indicate two lines of sources, roughly in the shape of a Y on edge, each source being consistent with it being a Strömgren sphere produced by a B3 star (Hughes & Wouterloot 1984). Most of the sources have spectra with $\alpha \sim$ -0.1, consistent with optically thin thermal regions, although a few have values of α between 0.4 and 0.6 which could be due to stars undergoing mass loss. There were doubts regarding the large number of stars since not only was it strange that stars should form in lines, but most HII regions had dimensions of about 1,000 AU and were not expected to last for long, let alone to all form in a very short time interval.

These concerns led to the monitoring program and revealed the presence of two highly time-dependent sources. Their spectra indicated gyrosynchrotron radiation. Also detected was a 'bullet' which had a cross speed of 280 km s^{-1} (Hughes 1993), and appears to be the working surface of a high speed jet which is causing the entrainment of a molecular outflow. A further possible 'bullet' has also been found. The individual new phenomena to date will be described; the re-analysis is continuing.

Astrophysics and Space Science **224**: 69–72, 1995.
© 1995 *Kluwer Academic Publishers.*

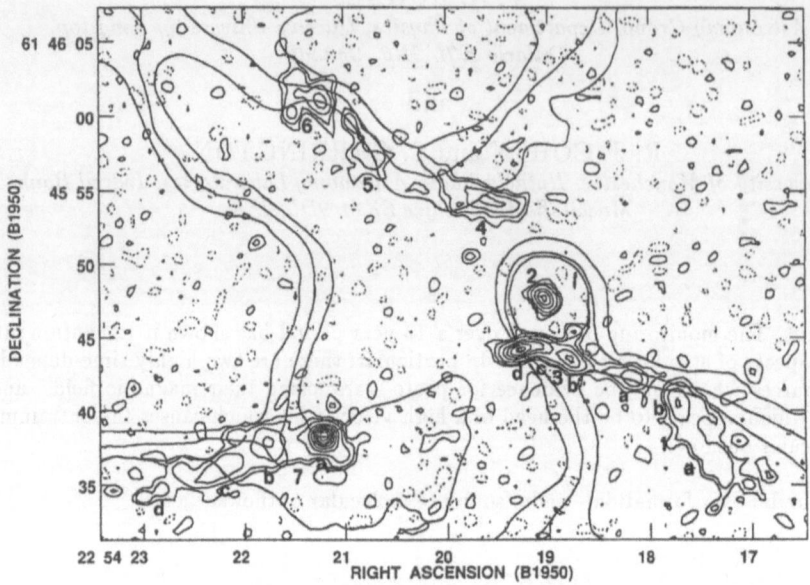

Fig. 1.   VLA image of Cepheus A East at 20-cm, showing the structure of the source, together with the first three contour levels of NH$_3$ by Torrelles *et al.* (1993) which show the two "bays" associated with molecular outflows. (Image from Hughes *et al.* (1994).)

2.   The Highly Time-dependent Sources

The monitoring program was carried out using the VLA of the National Radio Astronomy Observatory*, from 1988 Nov 2 to 1990 May 8. Observations were made at the beginning and end of two "A" configuration modes (Hughes 1991), and showed the new variable Sources 8 and 9. The spectrum of Source 9, determined at 20-, 6- and 2-cm, peaked at about 6-cm, and an increase in flux density was accompanied by a shift of the peak to lower wavelengths. If we ignore plasma oscillations, and assume the radiation to be due to either thermal, synchrotron or gyro-synchrotron mechanisms, the rising part of the spectrum corresponds to the optically thick region, and from this we obtain the relationship that $T_bD^2 \approx 2.5 \times 10^8$ K AU^2, where T_b(K) is the brightness temperature, and D(AU) is the diameter of the assumed circular source. We eliminate thermal radiation since the size is clearly < 100 AU which gives too high a value for T_b; synchrotron radiation requires electron energies ≥ 1 MeV and thus dimensions ~ 0.1 AU. In both cases there is also the difficulty of fitting a spectrum. The most compelling mechanism is gyro-synchrotron radiation from a region where 10^7 K < T_b < 10^9 K, D ~ 1 AU, and B ~ 100 G. In addition, flaring could be associated

* The NRAO is operated by Associated Universities Inc., under cooperative agreement with the National Science Foundation

with an increase in B.

To test these derived parameters, observations were made using MERLIN*
(Hughes, Cohen & Garrington 1994) where a resolution of $0\rlap{.}''03$ was achieved
at 6-cm, corresponding to a diameter < 20 AU. This leads to $T_b \geq 6 \times 10^5$
K and lends support to the gyro-synchrotron mechanism.

It was suggested that the radiation was due to a proto-star ridding itself
of its magnetic field by winding it up, thereby increasing the flux and leading
to the acceleration of electrons. However, other mechanisms are possible and
perhaps associated with variable outflows from adjacent stars.

3. The Number of Stars

The fact that there were less than 14 stars in the region was suggested by
the fact that 2-cm images showed only Sources 2, 3(c) and 3(d), together
with 8 and 9, as point sources. A plot of log(Luminosity) as determined from
the IRAS spectra, vs log(Ionizing Flux) from radio data, was interpreted as
implying the presence of 4 B1.5 stars (Hughes & MacLeod 1993). However,
using the MERLIN data on sources 2, 3(c) and 3(d) and assuming mass-
loss, there could be present about 3 B0.5 stars, presumably blanketed by
outflowing gas to simulate B1.5 stars. It is of interest that Rodríguez et al.
(1994) interpreted their data to indicate that Source 2 was a B0.5 star.

4. The Outflows and 'Bullet'

There appear to be two well defined outflows in CO, one to the NE and one
to the SE, which are shown as "bays" in the NH_3 images by Torrelles et al.
(1993). It was discovered that Source 7(c)(ii) had a proper motion of $0\rlap{.}''7$ in
about 10 years, in the direction of the SE outflow, giving a cross-speed of 280
km s^{-1} (Hughes 1993; Hughes, Cohen & Garrington 1994). By degrading
the angular resolution at 3.6- and 6.1-cm to $0\rlap{.}''42$, a spectral index of $\alpha =$
-0.37 ± 0.25 was measured, and a size $\sim 0\rlap{.}''36 \times 0\rlap{.}''18$ at PA 108°, indicating
that the 'bullet' contained a non-thermal component. It appears to be the
"working surface" of a high speed jet which is causing the entrainment of the
molecular outflow. Henriksen, Ptuskin & Mirabel (1991) have shown that
such an object is expected to have a mostly thermal spectrum if the jet is
propagating into a dense region, and a mostly non-thermal spectrum if into
a less dense region. The presence of CS (J=3-2) and CS (J=7-6) ahead of
the SE jet (Moriarty-Schieven et al. 1991) indicate densities of $\geq 10^5$ cm^{-3},
and is consistent with the former. The origin of the jet is not well defined,
but does appear to be in the region of Sources 2 or 9. If so, it is collimated
to within $0\rlap{.}''9$ and has had a lifetime of ~ 290 years.

---

* MERLIN is a UK national facility operated by the University of Manchester on behalf
of PPARC.

The possibility that Source 2 is the origin of the NE outflow led to the reduction of 20- and 6-cm data with the resolution degraded to $2''.35$. One additional source was apparent at RA 22^h 54^m 28^s, Dec $61°$ $46'$ $19''$, coincident in position with an IR source by Lenzen (1988). The spectral index of the integrated flux density showed a strong non-thermal spectrum with $\alpha \sim -0.9$. This is as expected if the jet is propagating into to a lower density region. If this is a working surface, the angular size of $\sim 2''$ leads to a collimation angle for the jet $< 1°.5$, assuming an origin near to Source 2. If the cross-speed is 280 km s^{-1}, its lifetime is ~ 870 years.

5.   Summary

Cepheus A has shown a number of aspects associated with star formation, in particular two highly time-dependent radio objects, and at least one 'bullet' which is evidence for a high speed jet which causes the entrainment of a molecular outflow. This is the first case where strong evidence exists for the latter association which is predicted in the model by Masson & Chernin (1993). The two time-dependent radio sources could be examples of protostars reducing their magnetic fields.

No other objects similar to Cepheus A have yet been found, but this is likely due to the lack of any long term monitoring of possible HII regions. The paper has concentrated chiefly on Cepheus A East. Future work is being directed to the origins of the outflows, and to a more detailed study of Cepheus A West.

References

Cohen, R. J., Rowland, P. R. & Blair, M. M.: 1984, Mon. Not. of the RAS **210**, 425.
Evans, N. J. *et al.* : 1981, Astrophys. J. **244**, 115.
Henriksen, R. N., Ptuskin, V. S. & Mirabel, L. F.: 1991, Astron. & Astrophys. **248**, 221.
Lada, C. J., Blitz, L., Reid, M. A. & Moran, J. M.: 1981, Astrophys. J. **243**, 769.
Lenzen, R.: 1988, Astron. & Astrophys. **190**, 269.
Masson, C. R. & Chernin, L. M.: 1993, Astrophys. J. **414**, 230.
Moriarty-Schieven, G. H., Snell, R. L. & Hughes, V. A.: 1991, Astrophys. J. **374**, 169.
Hughes, V. A.: 1991, Astrophys. J. **383**, 280.
Hughes, V. A.: 1993, Astron. J. **105**, 331.
Hughes, V. A., Cohen, R. J. & Garrington, S.: 1994, Mon. Not. of the RAS in press.
Hughes, V. A. & MacLeod, J. M.: 1993, Astron. J. **105**, 1495.
Hughes, V. A. & Wouterloot, J. G. A.: 1984, Astrophys. J. **276**, 204.
Rodríguez, L. F., Garay, G., Curiel, S., Ramïrez, S., Torrelles, J. M., Gómez, Y. & Velázquez, A.: 1994, Astrophys. J. **430**, L65.
Rodríguez, L. F., Ho, P. T. P. & Moran, J. M.: 1980, Astrophys. J. **240**, L149.
Sargent, A. I.: 1979, Astrophys. J. **218**, 736.
Torrelles, J. M., Verdes-Montenegro, L., Ho, P. T.P., Rodríguez, L. F. & Cantó, J.: 1993, Astrophys. J. **410**, 202.

STAR FORMATION IN L 1199

M. KUN

Konkoly Observatory of Hungarian Academy of Sciences,
H-1525 Budapest, P. O. Box 67, Hungary

Abstract. Lynds 1199 is an extended dark cloud of moderate opacity in the upper Cepheus region. The B3V type star HD206135 illuminating the reflection nebula DG 175 is associated with this cloud. In this paper the nature of the cloud and its associated young stellar objects is studied on the basis of ^{13}CO data obtained with the 4 m millimeter wave telescope of Nagoya University, photographic observations taken with the 60/90 cm Schmidt telescope of Konkoly Observatory, as well as IRAS data.

1. Location and Physical Properties of L 1199

The Cepheus region above $b=+10°$ contains a large number of dark clouds (Lynds, 1962). Lebrun's (1986) CO observations revealed a giant molecular cloud complex between $l=100°$ and $116°$ and above the latitude of $10°$. The Cepheus Flare molecular cloud complex at an average latitude of $+15°$ is a most prominent feature of the local interstellar medium. At a probable average distance of 400 pc the mass of the cloud complex is about $2×10^5 M⊙$. The molecular gas in the upper Cepheus region consists of two components of different velocity (Grenier et al., 1989). These are supposed to represent clouds of different distance. The velocity of the nearer component is centered around $0 \, \mathrm{kms}^{-1}$ whereas the other component around $-10 \, \mathrm{kms}^{-1}$ is located at a distance of 800 pc, where, at lower latitudes, the OB associations Cepheus OB2 and Cepheus OB3 are found.

L 1199 covers an area of 0.235 square degrees in the region of Cepheus Flare. Its opacity class is 2 (Lynds, 1962). Its galactic coordinates are: $l=106.6°$, $b=+12.5°$. At the southern border of the cloud the reflection nebula DG 175 is illuminated by the B3V type star HD 206135, indicating that this star was born in the cloud.

L 1199 is of special interest for two reasons.

1. Low resolution CO observations of Grenier et al. (1989) have shown that L 1199 belongs to the more negative velocity component of the molecular gas. In this case the cloud is supposed to have the same distance as the neighbouring NGC 7129 and together with this compact star forming region it probably forms the outermost part of the association Cepheus OB 2. The distance modulus of 9.5 mag derived by Racine (1968) for HD206135 appears to support this view. In this case the distance of the cloud from the galactic plane is nearly 200 pc. Formation mechanism of massive star forming clouds near the edge of the galactic gaseous disc deserves a closer study.

2. Though star formation in the Cepheus Flare molecular cloud complex is poorly studied, it is obvious that no high mass star formation can be

Astrophysics and Space Science **224**: 73–76, 1995.

observed in this region of the sky. Probably the most luminous young star in the region is the Herbig Be star HD200775 (Thé et al., 1994). Further B and A type stars born in the Cepheus Flare molecular clouds are the members of the Cepheus R2 association (Racine, 1968). HD206135 is the most luminous member of Cep R2. After correcting its distance modulus given by Racine (1968) for the interstellar absorption on the basis of the UBV data of the star a distance value of 450 pc is obtained. In this case the B3V spectral type of HD206135 indicates that this star is one of the most massive objects formed in this molecular complex. Study of its environment therefore may reveal the differences between the formation sites of stars of different masses.

In order to study the properties of the cloud L 1199 was observed in the ^{13}CO line with the 4 m radio telescope of Nagoya University in April 1994. The observations show that the velocity of the molecular gas is about -10 kms^{-1}. After a preliminary reduction of data it can be seen that the cloud consists of three clumps. Figure 1a shows the distribution of the ^{13}CO.

Cloud properties can be revealed by studying the dust component using IRAS data. L 1199 is one of the most prominent features of the 100 μm IRAS image of the Cepheus Flare region. In order to get an insight into the dust distribution a 100 μm optical depth image was constructed using the ISSA images. This is shown in Figure 1b. The peak optical depths coincide with the ^{13}CO peaks. Further peaks can be seen which were not covered by the millimeter observations.

TABLE I

Physical properties of clumps of L 1199 derived from IRAS data

No.	RA(1950)	D(1950)	$\tau_{max}(10^{-4})$	r(pc)	A_V	$M(M_\odot)$	n(cm^{-3})
1	21 21 42.0	69 15 44	2.83	1.05	5.4	780	3.0×10^3
2	21 27 12.6	67 54 39	1.71	0.56	3.5	135	3.3×10^3
3	21 27 50.5	68 33 07	1.64	1.23	3.4	245	2.3×10^3
4	21 29 24.0	68 45 28	1.62	1.18	3.4	158	2.8×10^3
5	21 33 15.8	68 11 09	2.6	1 32	5.0	1130	2.2×10^3

The τ_{100} image was used for deriving the physical properties of the cloud following the methods applied by Wood et al. (1994). The results are listed in Table I. Sizes and masses of the clumps were computed adopting a distance of 450 pc. The physical properties derived from IRAS data favour this distance value. At a distance of 800 pc the clump masses were unreasonably high for that height above the galactic plane. The density values obtained are smaller than those typical of star forming cores, and suggest that no further star

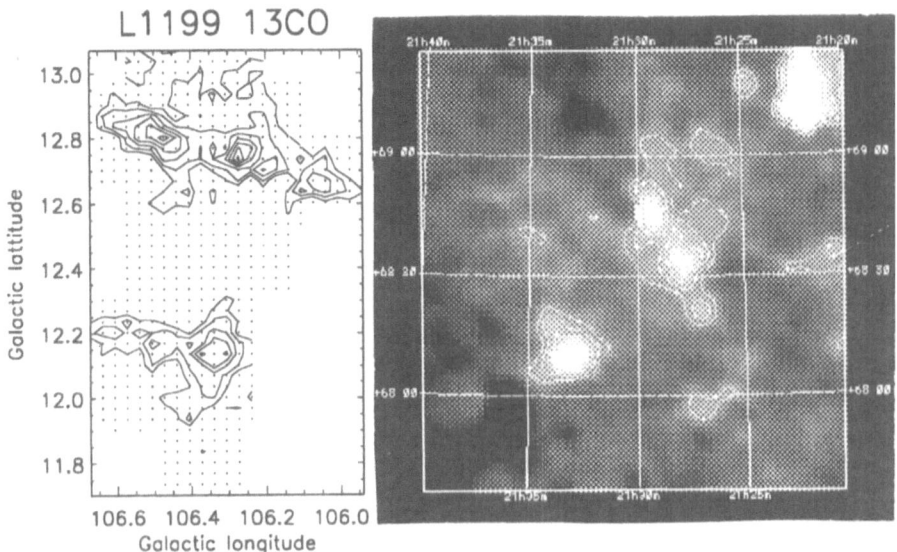

Fig. 1.   a: Distribution of the ^{13}CO emission in L 1199 (in galactic coordinates); b: 100 μm optical depth image of the same cloud (in equatorial coordinates)

formation is expected in these clumps.

2.   Young stellar objects in L 1199

Star formation usually occurs in groups. We may expect to find low luminosity young stellar objects born in L 1199 among IRAS point sources. Optically visible pre-main sequence star candidates can be found by objective prism Schmidt surveys.

A few years ago I started an objective prism search for optically visible Hα emission stars in the Cepheus Flare region using the 60/90 cm Schmidt telescope of Konkoly observatory. The aim of this work is to identify T Tauri star candidates associated with the clouds. The survey includes L 1199. In order to identify T Tauri star candidates born in L 1199 three objective prism plates were obtained. I found seven Hα emission stars near HD 206135. Their real nature has to be clarified via spectroscopic observations.

There is a group of IRAS point sources in the cloud region which probably represent young stellar objects. The most luminuous is associated with HD 206135. The IRAS Point Source Catalog contains eight more objects of nonstellar flux density distribution in the cloud area. None of them has optical counterparts on the POSS plates. The IRAS Faint Source Catalog contains five objects in the vicinity of HD 206135. All of them have 12 and 25 μm flux values with F(25)>F(12). Except for the faintest one, they coincide in position with visible stars. They are good low-mass pre-main sequence star candidates. Properties of IRAS sources are summarized in Table II.

TABLE II

YSO candidates in L 1199. The final column indicates: i = invisible; s = star; n = star+nebula

Name	RA(1950)	D(1950)	F_{12}	F_{25}	F_{60}	F_{100}	$L(L_\odot)$	
P21225+6908	21 22 30.1	69 08 52	<0.25	<0.25	0.834	<10.20	>0.24	i
P21321+6836	21 32 08.2	68 36 47	<0.25	<0.25	0.429	5.14	>0.68	i
F21331+6749	21 33 11.5	67 49 02	0.144	0.163	1.304	12.52	2.02	s
F21332+6745	21 33 12.5	67 45 40	0.130	0.133	<2.36	<22.32	>0.23	s
F21334+6802	21 33 27.9	68 02 25	0.101	0.106	0.803	6.863	1.17	s
F21335+6753	21 33 30.6	67 53 08	0.107	0.157	<1.81	<30.16	>0.21	s
P21341+6816	21 34 10.7	68 16 32	<0.25	<0.25	0.743	10.50	>1.82	i
P21351+6803	21 35 06.8	68 03 22	<0.25	<0.25	1.020	6.720	>1.48	i
F21351+6804	21 35 11.6	68 04 10	0.075	0.115	<2.128	<18.94	>0.15	i
P21360+6756	21 36 00.1	67 56 50	0.268	1.823	30.24	77.77	18.51	n
P21368+6802	21 36 48.6	68 02 04	<0.25	<0.25	1.180	9.89	>1.88	i
P21371+6820	21 37 10.3	68 20 30	<0.25	<0.25	0.798	<10.200	>0.70	i
P21375+6831	21 37 31.3	68 31 15	<0.25	<0.25	0.538	<6.040	>0.15	i

These results indicate that in addition to HD206135 a small group of lower luminosity stars was born in L 1199. There is a large amount of molecular material associated with these objects, which probably have too low density to form further stars.

Acknowledgements

This research was supported by the Hungarian grants OTKA No. T4731 and OTKA No. T7438. I am indebted to the staff of Radio Astronomy Laboratory of Nagoya University for the help in radio observations and data reduction. My visit to Nagoya was supported by JSPS. My thanks are due to Péter Ábrahám for his help in radio data reduction.

References

Grenier, I. A., Lebrun, F., Arnaud, M., Dame, T. M. & Thaddeus, P.: 1989, Astrophys. J. **347**, 231.
Lebrun, F.: 1986, Astrophys. J. **306**, 16.
Lynds, B. T.: 1962, Astrophys. J., Suppl. **7**, 1.
Racine, R.: 1968, Astron. J. **73**, 233.
Thé, P.S., De Winter, D. & Pérez, M. R.: 1994, Astron. & Astrophys. Suppl. **104**, 315.
Wood, D. O. S., Myers, P. C. & Daugherty, D. M.: 1994, Astrophys. J., Suppl. in press.

MODELING FAR-INFRARED OBSERVATIONS OF YOUNG STELLAR OBJECTS

H. M. BUTNER

Department of Terrestrial Magnetism, Carnegie Institution of Washington, 5241 Broad Branch Road N.W., Washington, D.C. 20015, USA

G. H. MORIARTY-SCHIEVEN

Dominion Radio Astrophysical Observatory, P. O. Box 248, Penticton, BC V2A 6K3, Canada

and

M. E. RESSLER and M. W. WERNER

Jet Propulsion Laboratory, California Institute of Technology, 4800 Oak Grove Drive, Pasadena, CA 91109, USA

Abstract. We discuss high-resolution, far-infrared spatial observations of low mass embedded young stellar objects in the Taurus molecular complex. By comparing the observed spatial extent with the results of radiative transport models, we can estimate the density distribution of the envelopes. We compare the initial results for 4 objects (L1551-IRS 5, L1489-IR, L1551-NE, and L1527-IR) with star formation theories.

Key words: star formation:theory – far infrared:observations – radiative transport:models – Young stellar objects:Taurus

1. Spectral Energy Distributions

In the 1980s, star formation theories underwent a tremendous advance. First, there was the realization (Lada 1987) that young stellar objects (YSOs) might be classified by their spectral energy distributions (SED) into an evolutionary sequence between deeply embedded objects (Class I) through visible objects with little remnant material (Class III). Terebey, Shu & Cassen (1984, henceforth TSC) developed the idea that a slowly rotating isothermal sphere would undergo inside-out collapse. The collapsing sphere's density gradient would go from $n(r) \propto r^{-2}$ (the isothermal case) to $n(r) \propto r^{(-1.5)}$ (the infall case). The collapse begins at the center, and as the collapse proceeds, material falling inwards encounters a centrifugal barrier due to the slow rotation of the original cloud. This mechanism leads naturally to a disklike geometry on solar system size scales for protostars of one solar mass. The TSC model makes specific predictions about the expected density gradient at different stages of the collapse, making it testable by observations.

Adams, Lada & Shu (1987) combined the inside-out collapse model with a radiative transport model. Dust in the envelope reprocesses most of the radiation from the central source, shifting the flux to longer wavelengths. The result is a SED that is similar to the Class I sources. Adams, Lada &

Astrophysics and Space Science **224**: 77–80, 1995.
© 1995 *Kluwer Academic Publishers.*

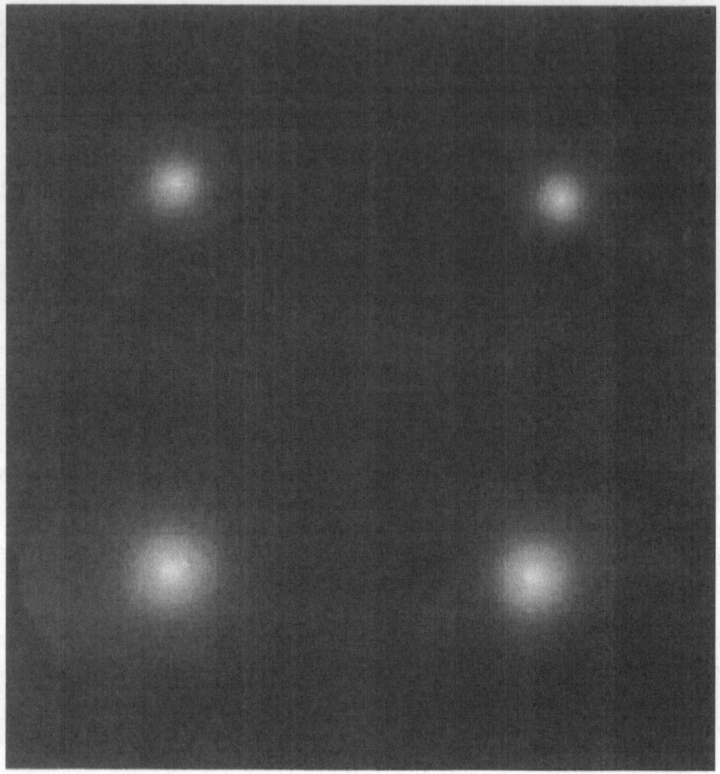

Fig. 1. L1551-IRS 5 100 μm and 160 μm images from the Yerkes Camera. The upper plot is 100 μm, the lower 160 μm. The source is on right, the point source profile is on left. Size scale is 5 arcminutes per side, and no correction has been applied for field rotation.

Shu modeled the SEDs of 6 Class I YSOs in detail, including L1551-IRS 5 in Taurus. They found excellent matches between the predictions of theory and infrared observations. More recent work, however, has revealed that the SED is not necessarily unique to a particular source model. Changing the dust opacity law (Butner et al. 1991), or the source geometry (Kenyon et al. 1993) could produce SEDs that would match as well. In some very optically thick cases like B335, even very flat density gradient models could produce the same SED as a TSC-type model (Zhou et al. 1990). Herbig Ae/Be stars seem to have flatter density gradients around them than the TSC model might predict (Natta et al. 1993, Di Francisco et al. 1994). Thus, the question remains: how general is the TSC model?

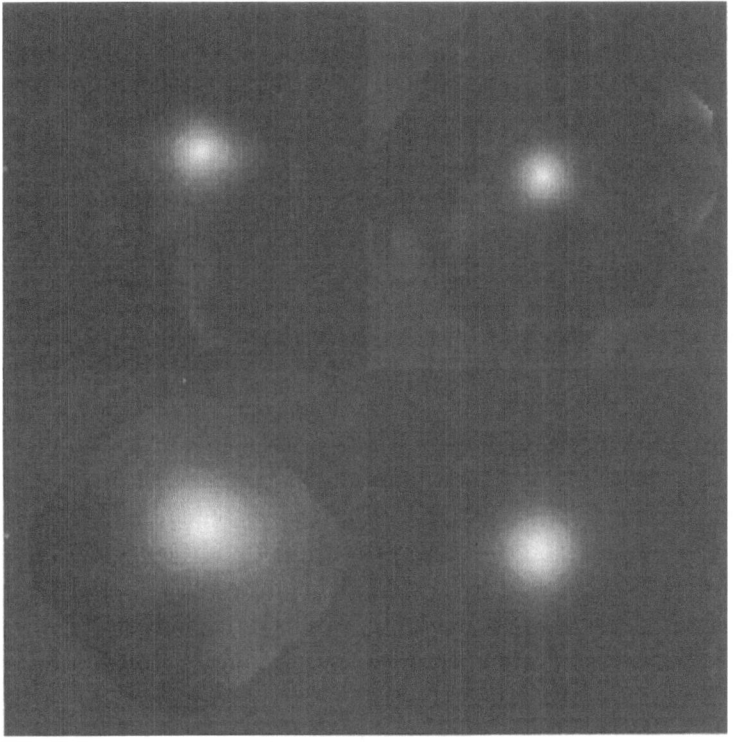

Fig. 2.   L1527-IR 100 μm and 160 μm images. Same as Figure 1.

2.   Spatial Observations

One way to distinguish between different star formation models, even those which produce similar SEDs, is to use high resolution, far-infrared spatial observations. If an envelope has a steep density gradient, it will have a smaller spatial extent than a flat density gradient model with the same envelope optical depth and the same total source luminosity. The technique has been used successfully for a number of sources such as L1551-IRS 5 (Butner *et al.* 1991), NGC 2071 (Butner *et al.* 1990) and Herbig Ae/Be stars (Natta *et al.* 1993, Di Francisco *et al.* 1994).

We have undertaken a study of a group of low mass YSOs in Taurus identified by Tamura *et al.* (1991). These sources have luminosities less than 30 $L_\odot$, implying $M_{YSO} \leq 1$ $M_\odot$. Because of their low luminosities, they are difficult to resolve at 100 μm from the Kuiper Airborne Observatory (KAO). However, models indicate that by observing at longer wavelengths (160 μm or 200 μm), it is possible to distinguish between models with different density gradients. Therefore, we used the Yerkes Observatory 60-channel

Far-Infrared Camera on board the KAO, which images at 60, 100, 160 and 200 μm near the diffraction limit of the KAO (52 arcseconds at 200 μm). In September 1993, we observed 4 sources (L1551-IRS 5, L1489-IRS, L1527-IR, and L1551-NE). We present images from two of the sources (L1551-IRS 5 and L1527). We have not removed the background low-level extended emission from the two images.

3.   Results: Two Geometries?

We find that our sources split into two groups: L1551-IRS 5 and L1489 fall in one group, with TSC-like density gradients. The full-width-half-maximum sizes are consistent with either $n(r) \propto r^{-2}$ or $n(r) \propto r^{-1.5}$ and optical depths of $A_V \geq 40$. L1527-IR and L1551-NE are in the other group, with a shallower density gradient, but similar optical depths. In addition, L1527-IR and L1551-NE are flattened, suggesting that their geometries are not spherical, as assumed in the TSC model. It is interesting to note that it is the "Class 0" sources, L1527-IR and L1551-NE, that are flattened. Such behavior may represent a different starting condition for the collapse or a different collapse pathway than envisioned by the TSC picture such as the slab model of Hartmann *et al.* (1994). Within the Taurus star formation region, we find a range of density gradients and geometries among the low mass YSOs.

Acknowledgements

This work was supported in part by grants from the NASA Origins of the Solar System Program (NAGW-4097) and from the Kuiper Airborne Astronomy Program.

References

Adams, F. C., Lada, C. J., & Shu, F. H.: 1987, Astrophys. J. **312**, 788.

Butner, H. M., Evans, N. J., II, Harvey, P. M., Mundy, L. G., Natta, A., & Randich, M. S.: 1990, Astrophys. J. **364**, 164.

Butner, H. M., Evans, N. J. II., Lester, D. F., Levreault, R. M., & Strom, S. E.: 1991, Astrophys. J. **396**, 696.

Di Francesco, J., Evans, N. J. II., Harvey, P. M., Mundy, L. G., & Butner, H. M.: 1994, Astrophys. J. **432**, 710.

Hartmann, L., Boss, A., Calvet, N., & Whitney, B.: 1994, Astrophys. J. **430**, L49.

Lada, C.J.: 1987 in: *Star Forming Regions: Proc. IAU Symp. 115*, eds: M. Peimburt & J. Jugaku, Reidel: Dordrecht, p1.

Kenyon, S. J., Calvet, N. & Hartmann, L.: 1993, Astrophys. J. **414**, 676.

Natta, A., Palla, F., Butner, H. M., Evans, N. J. II, & Harvey, P. M.: 1993, Astrophys. J. **406**, 374.

Tamura, M., Gatley, I., Waller, W., & Werner, M. W.: 1991, Astrophys. J. **374**, L25.

Terebey, S., Shu, F.H. & Cassen, P.: 1984, Astrophys. J. **286**, 529.

Zhou, S. Evans, N. J. II, Butner, H. M., Kutner, M. L., Leung, C. M. & Mundy, L. G.: 1990, Astrophys. J. **363**, 168.

OBSERVATIONS OF CHEMICAL PROCESSING IN THE CIRCUMSTELLAR ENVIRONMENT

L.G. MUNDY and J.P. MCMULLIN

Astronomy Department, Univ. of Maryland

and

G.A. BLAKE

Division of Geological and Planetary Sciences, Caltech

Abstract. High resolution interferometer and single-dish observations of young, deeply embedded stellar systems reveal a complex chemistry in the circumstellar environments of low to intermediate mass stars. Depletions of gas-phase molecules, grain mantle evaporation, and shock interactions actively drive chemical processes in different regions around young stars. We present results for two systems, IRAS 05338-0624 and NGC 1333 IRAS 4, to illustrate the behavior found and to examine the physical processes at work.

Key words: Chemistry – Depletions – Shocks – IRAS 05338-0624 – NGC 1333 IRAS 4

1. Introduction

The richness and diversity of molecular emission observable from molecular clouds makes it possible, in principle, both to derive the current physical state of the gas and dust in the young stellar environment and to place constraints on the energetic processes and timescales associated with stellar birth. In practice, the complex interactions between the physical processes active in the stellar environment and the chemistry of gas-phase and grain surface reactants make extensive physical interpretation of observed molecular emission difficult. The simultaneous presence of infalling molecular cloud material, high-velocity outflowing winds, and photon heating in the natal environment insure that the chemistry is both time and position dependent within the nebula. Photo-reactions, shock induced high temperature reactions, the introduction of gas-phase atomic species through stellar winds and grain destruction, grain surface reactions, and the adsorption and desorption of molecules from grain surfaces can all influence the local molecular abundances and hence the observed morphology of the nebulae.

To fight this complexity, one must approach the study of the chemistry of young stellar objects armed with a willingness to simplify. One way to simplify such studies is to identify systems, or discrete portions of systems, where single physical processes dominate the current chemistry and molecular morphology. In this paper we present results for two young stellar environments: one where shock processes can clearly be identified and another where the depletion of gas-phase molecules dominates all other chemical

Astrophysics and Space Science **224**: 81–84, 1995.

considerations. In the last section, we will provide an overall outline for the chemical evolution of low-mass young stellar environments.

2. The Chemistry of Shocks: IRAS 05338-0624

The region around the deeply embedded young stellar object (YSO) IRAS 05338-0624 provides an excellent example of a stellar outflow interacting with its ambient core. IRAS 05338 is a typical YSO with a surrounding core of dense gas (Takaba *et al.* 1986; Harju, Walmsley & Wouterloot 1991), a strong large-scale bi-polar outflow (Wilking *et al.* 1990), and compact millimeter wavelength dust and centimeter wavelength ionized gas emission. Millimeter interferometer maps of CS J=2-1, as well as single-dish maps of SO 5_6-4_5 and CS J=5-4 emission, show red and blue lobes of the molecular outflow on either side of the YSO position (McMullin *et al.* 1994). A strong localized area of interaction between this outflow and the ambient core is clearly evident in both the spectra and maps. In interferometric maps with the BIMA array, the brightest peaks in CS J=2-1 and SiO J=2-1 emission are located 14″ west and 8″ south of the YSO position. Maps of the SO and CH_3OH emission from the CSO 10.4 m telescope also peak to the southwest of the YSO position. Kinematically, spectra of CO, HCN, CS, CH_3OH, SO, and SiO show strong red wings toward the southwest position. The SiO emission exhibits the most extreme behavior with little to no emission at the stellar position and broad emission extending out nearly 10 km s^{-1} from the cloud line center at the southwest position.

The combination of kinematic and spatial separation between the shock interaction region and the ambient core makes it possible to derive cleanly relative abundances in the shock material. Figure 1 summarizes the derived abundance ratios relative to CO based on emission in the V_{LSR} range of 10 - 15 km s^{-1}. For comparison, relative abundances for the gas in the YSO cocoon (velocity interval 5 - 10 km s^{-1}) and the TMC-1 condensation are also shown. The relative abundances of SO, CH_3OH, and SiO are elevated by factors of 4 - 1000 relative to both the surrounding condensation and TMC-1. Such molecular enhancements enable shocks to "light-up" selected regions of interaction and significantly alter the appearance of circumstellar nebulae even though such regions contain small amounts of gas. In the future, these molecular probes can serve as dyes of the shock activity and help reveal many of the details of outflow-cloud interactions.

3. Depletion of Gas-Phase Molecules: NGC 1333 IRAS4

There is growing evidence for gas phase depletions of molecules in the dense environments of YSO's (Mausersberger *et al.* 1992; Blake *et al.* 1995 and references therein). The YSO NGC 1333 IRAS4 presents one of the most

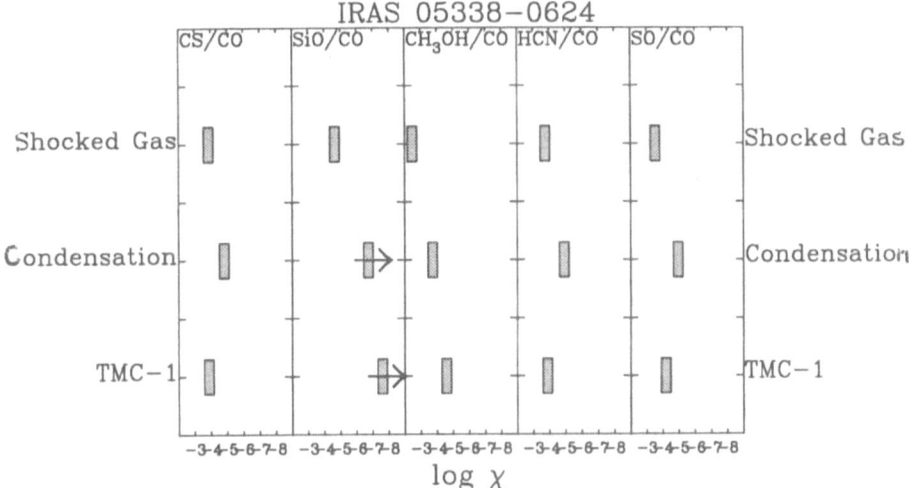

Fig. 1. Log-abundances of selected molecular species relative to CO in the shocked gas in IRAS 05338-0624, the condensation surrounding IRAS 05338-0624 (McMullin *et al.* 1994, and the TMC-1 condensation (van Dishoeck *et al.* 1992). The width of each bar represents the uncertainty in the ratio; the arrows on the SiO/CO ratios indicate an upper limits.

extreme examples of molecular depletions discovered to date. NGC 1333 IRAS4 itself is strong sub-millimeter wavelength double source (4A and 4B) with a separation of roughly 30″, located in the NGC 1333 molecular cloud (Sandell *et al.* 1991). Both sources exhibit molecular outflows and are associated with extremely large masses of circumstellar material; Sandell *et al.* estimate 9 M$_\odot$ around 4A and 4 M$_\odot$ around 4B. The combined luminosity of the two sources is estimated to be at most 30 L$_\odot$.

A recent molecular line survey of the IRAS 4A and 4B sources found them to be surprisingly weak molecular line sources (Blake *et al.* 1995). On size scales of 5,000 to 10,000 AU (20″ - 30″), depletions factors of 10 - 20 compared to normal molecular cloud abundances are derived for all molecules, including CO. On smaller scales, new interferometer observations covering wavelengths from 1 cm to 1 mm suggest that ∼3 M$_\odot$ of material with a scale size of 3000 - 4000 AU surrounds the central YSO IRAS 4A (Mundy *et al.* 1995). Observations of the C^{18}O J=1-0 line on similar scales find weak emission. Analysis of these data suggest that CO is depleted by at least a factor of 20 in this inner region (McMullin, 1994). In fact, all of the depletion factors may seriously underestimates the true depletions in the bulk of the gas as a small amount of "normal" abundance gas (from shock interactions with the outflow or an outer skin of un-depleted gas) can easily mask more severe depletions.

Evolutionary Phase	Molecular Abundances	Example Region
Early Prestellar Condensation	"Normal" abundances	TMC-1
Contracting Condensation	Growing depletions	
Formation of Early Protostar	Increasing depletions	NGC 1333 IRAS4
Onset of Strong Wind Activity	Depletion throughout core	IRAS 05338-0624
	Rising abundances at center	
Developing Wind Activity	Full liberation of grain	Orion IRc2
	near star, shocked regions	
Late Embedded Period	"Normal" abundances	

4. A Chemical Evolutionary Sequence for Young Stellar Environments

As our knowledge increases, it will hopefully become possible to construct a chemical evolutionary sequence for young stellar systems which is descriptive of both the appearance of the nebulae and the physical processes shaping that appearance. From the studies that we have done to date, it is clear that shock interactions and molecular depletions play very major roles in modifying the chemistry of the circumstellar environment. Table 1 is a crude outline of the evolution driven by these two processes. One key point of this sequence is that the early evolution is characterized by strong molecular depletions on circumstellar scales which will make high resolution observations of these objects very challenging.

References

Blake, G.A., *et al.* : 1995, Astrophys. J. March 10.
Harju, J., Walmsley, C.M. & Wouterlout, J.G.A.: 1991, Astron. & Astrophys. **245**, 643.
Mausersberger, R., *et al.* : 1992, Astron. & Astrophys. **256**, 640.
McMullin, J.P.: 1994, Ph.D. Thesis, Univ. of Maryland.
McMullin, J.P., Mundy, L.G. & Blake, G.A.: 1994, Astrophys. J. December 10.
Mundy L.G., *et al.* : 1995, in preparation.
Sandell, G., *et al.* : 1991, Astrophys. J. **376**, L17.
Takaba, H., *et al.* : 1986, Astron. & Astrophys. **166**, 276.
van Dishoeck, E.F. *et al.* : 1992 in: *Protostars and Planets III*, eds: E.H.Levy, J.I. Lunine & M.S. Matthews, University of Arizona: Tucson, p163.
Wilking, B.A., *et al.* : 1989, Astrophys. J. **345**, 257.

SUB-MM AND FAR-INFRARED CONTINUUM OBSERVATIONS OF HERBIG-HARO EXCITING STARS AND OTHER YSO'S

W.R.F. DENT and H.E. MATTHEWS
Joint Astronomy Centre, Hilo, Hawaii

and

D. WARD-THOMPSON
Royal Observatory Edinburgh

Abstract. We show initial results from a survey of the mm to far-IR continuum spectra of ∼ 30 YSO's known to be exciting Herbig-Haro objects. The data are also compared with line intensities of $C^{18}O$ and H_2CO. We include in this analysis results from other sub-mm continuum surveys of compact HII regions, T-Tauri stars and class 0 YSO's. The results provide a statistical sample of the long-wavelength dust spectra of ∼ 60 Young Stellar Objects. Data are displayed on mm to FIR colour-colour diagrams, with the aim of trying to discriminate between different stages of star formation through general spectral characteristics, rather than detailed model fits to individual sources.

Key words: Young Stellar Objects - Dust

1. Observational Method

The 1100, 800 and 450 μm observations were made using the UKT14 bolometer on the James Clerk Maxwell Telescope in Hawaii. In all cases where accurate coordinates were not known (such as those from IRAS), we peaked up on the source before carrying out the photometry. The detailed results will be given in a future paper. Far infrared data is taken mostly from the IRAS PSC. Other published mm and sub-mm data has been obtained mostly from Sandell (1994), Weintraub *et al.* (1989), Barsony & Kenyon (1992), Dent *et al.* (1989), André *et al.* (1993), Barsony & Chandler (1993), Reipurth *et al.* (1993) and Chandler *et al.* (1990). We have also used fluxes of main-sequence Vega-excess stars from Sylvester *et al.* (1994).

2. Colour-colour diagrams

The FIR to mm-wave fluxes are displayed on colour-colour diagrams (Fig. 1), which have the advantage of being concise, as well as distance-invariant. We also display vectors of temperature (the crosses correspond to 10K intervals increasing upwards, with 30K being the central fiducial point), spectral index β (the crosses show intervals of 0.5, from 0.0 to 2.0 increasing to the right), and FIR optical depth (optically thin at the fiducial point, and optically thick to the left of the arc).

Astrophysics and Space Science **224**: 85–88, 1995.
© 1995 *Kluwer Academic Publishers.*

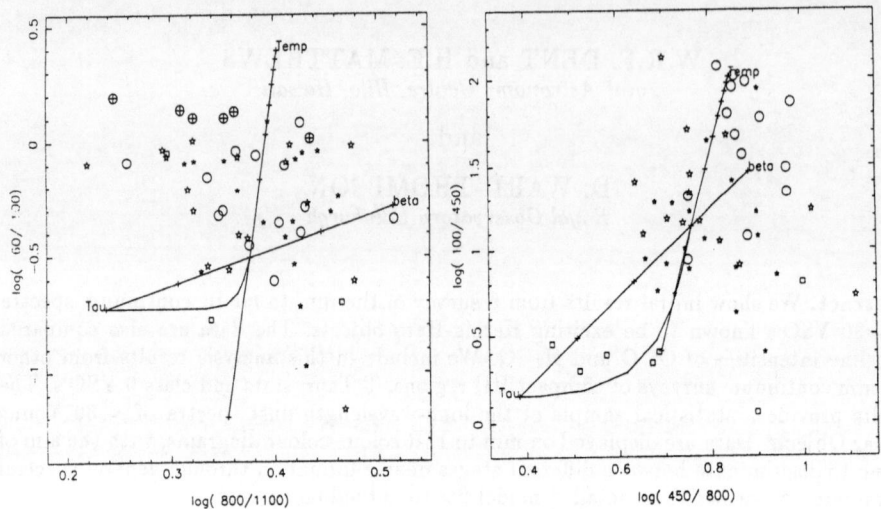

Fig. 1.   Colour-colour diagrams at various wavelengths. Filled and open stars are
HH-exciting and T-Tauri stars, open circles are CHII regions, open squares are class 0
YSO's and circles with crosses are Vega-excess stars. (a) Shows Vega-excess stars have
warm dust with low β (fiducial point has β=1). (b) By using only sub-mm wavelengths,
a clear separation of the objects is apparent (fiducial β=1.5). See text for details.

When fluxes at $60\mu m$ or shorter are included, there is no clear correlation
nor separation of the different types of YSO's (Fig. 1a), suggesting that the
dominant FIR source may *not* be the same as the source of the sub-mm flux.
However, several of the Vega-excess stars do appear on this diagram and
are concentrated in the high-temperature and low spectral index ($\beta \approx 0.5$)
region. As $1100\mu m$ is included in this diagram, most of the CHII regions are
displaced to the left because of excess emission from free-free radiation at
this longer wavelength.

Fig. 1b shows the $100 - 800\mu m$ colour-colour diagram. This has a clear
separation of different YSO's, and can be divided up into three main regions.
The upper parts of the temperature and β vectors contain the CHII regions;
these are luminous, so emission is dominated by warm ($\approx 60K$) dust. A
spectral index of ≈ 2.0 fits most objects. The middle part contains the lower-
luminosity HH-exciting and T-Tauri stars. Emission is from dust at 25-40K
with $\beta \approx 1.5$ - somewhat different from the CHII regions. Class 0 YSO's
appear only in the lower regions of Fig. 1b, implying dust temperatures of
$\approx 20K$. They are also concentrated to the left of the temperature vector,
indicating either β is decreased to ~ 1.0 or the optical depths are high.
Better continuum spectra are required, although the diagrams tend to favor
a lower β for these sources. Note that a few objects may be misclassified,

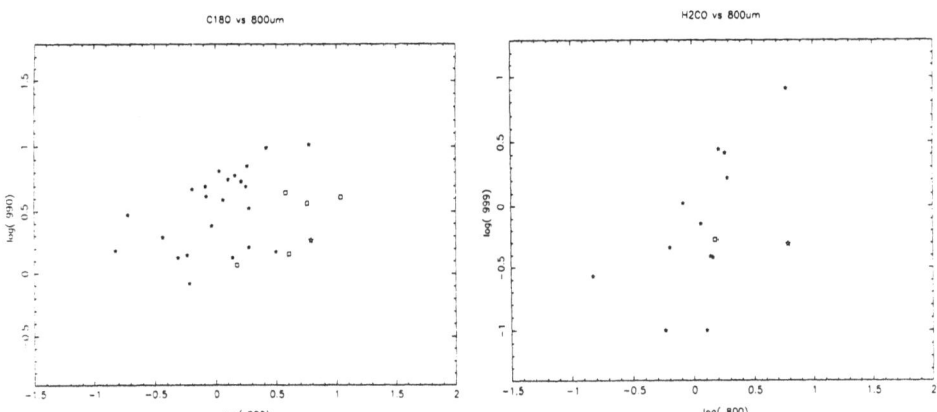

Fig. 2.   Plot of $800\mu m$ continuum against integrated line intensity. (a) is $C^{18}O$ and (b) is H_2CO.

resulting in increased scatter in the diagram, but it is clearly possible to identify YSO's based solely on the sub-mm colours.

3.   Comparison of dust and line characteristics

We have obtained central spectra of 25 of the sources in the J=2-1 $C^{18}O$ transition at 219.5GHz, and a smaller number in the 3(0,3)-2(0,2) line of H_2CO at 218.2GHz. Fig. 2a shows the variation of $C^{18}O$ integrated intensity with $800\mu m$ continuum flux. A weak correlation is seen, with a slope of ≈ 0.4; this is also found at $100\mu m$, however, $25\mu m$ shows very little correlation. This may be explained if the region contributing most to the $C^{18}O$ column density lies in an envelope *outside* the continuum-emitting centrally-heated core; furthermore the envelope mass is *not* highly dependent on the core mass. $C^{18}O$ may therefore not be a good tracer of dense cores.

Fig. 2b shows the variation of H_2CO integrated intensity with the continuum flux at $800\mu m$. Although there are fewer points, the slope appears to be ≈ 1.0, suggesting that the emitting regions are co-located. We intend to extend this H_2CO study in the near future to obtain temperatures using multiple transitions and compare this with the dust temperatures.

4.   Discussion

The above results suggest that placing YSO's on the sub-mm colour-colour diagram can be used as a method of classification. We derive the following general dust characteristics:

1. Compact HII regions: $T_d \approx 60K$, $\beta \approx 2.0$

2. T-Tauri and HH-exciting stars: $T_d \approx 30K$, $\beta \approx 1.5$
3. Class 0 YSO's: $T_d \approx 20K$, possibly $\beta \approx 1.0$
4. Vega-excess stars: $T_d \approx 60K$, $\beta \approx 0.5$

These spectral differences imply that dust characteristics *are* somewhat dependent on the environment. Larger grains are predicted to have lower values of β (*e.g.* Krügel & Siebenmorgen, 1994). This would imply that grains around main-sequence stars and possibly also class 0 YSO's are large (μm to mm sizes). T-Tauri stars have intermediate and compact HII regions the smallest (sub-μm) grains, similar to the MRN model of ISM dust (Mathis, Rumpl & Nordseick, 1977).

Comparison with line observations suggests that $C^{18}O$ emission arises more from an envelope than from the warm continuum-emitting core, and that envelope mass is not very dependent on the core mass. However, H_2CO may be co-existent with the warm dust, and thus could be a better core tracer.

References

André, P., Ward-Thomson, D. & Barsony, M.: 1993, Astrophys. J. **406**, 122.
Barsony, M. & Chandler, C.J.: 1993, Astrophys. J. **406**, L71.
Barsony, M. & Kenyon, S.J.: 1992, Astrophys. J. **384**, L53.
Chandler, C.J., *et al.* : 1990, Mon. Not. of the RAS **243**, 330.
Dent, W.R.F., *et al.* : 1989, Mon. Not. of the RAS **238**, 1497.
Krügel, E. & Siebenmorgen, R.: 1994, Astron. & Astrophys. **288**, 929.
Mathis, J.S., Rumpl, W., & Nordseick, K.H.: 1977, Astrophys. J. **217**, 425.
Reipurth, B., *et al.* : 1993, Astron. & Astrophys. **273**, 221.
Sandell., G.: 1994, Mon. Not. of the RAS in press.
Sylvester, R.J., Barlow, M.J. & Skinner, C.J.: 1994, Astrophys. & Space Sci. **212**, 261.
Weintraub, D.A., Sandell, G. & Duncan, W.D.: 1989, Astrophys. J. **340**, L69.

EXTENDED STRUCTURES AROUND YSO
IN MID-IR BROAD EMISSION FEATURES

LYNNE K. DEUTSCH

FCAD/University of Massachusetts, Amherst, MA 01003

JOSEPH L. HORA

Institute for Astronomy, 2680 Woodlawn Drive, Honolulu, HI 96822

HAROLD M. BUTNER

DTM - Carnegie Institution of Washington, Washington, D.C. 20015

WILLIAM F. HOFFMANN

Steward Observatory, University of Arizona, Tucson, AZ 85721

and

GIOVANNI G. FAZIO

Smithsonian Astrophysical Observatory, 60 Garden St., Cambridge, MA 02138

Abstract. We present high-spatial-resolution, mid-IR images of young stellar objects (YSO). Images were obtained with MIRAC (Mid-InfraRed Array Camera) at wavelengths of grain continuum and feature emission near 10 and 20 μm. Three of the objects observed (WL 16, TY CrA and MWC 1080) show structure and extended emission (on the order of 1200 AU or greater) at wavelengths of the UIR (Unidentified InfraRed) emission features. Mid-IR imaging of other sources in these clouds which do not exhibit UIR features (Elias 29 and R CrA) shows these sources to be point-like. We have produced morphological models based on the mid-IR images for the observed extended structures.

Key words: young stellar objects: individual (WL 16) – Herbig Ae/Be stars: individual (TY CrA, MWC 1080) – young stellar objects: General – young stellar objects: Infrared

1.   Introduction

We have undertaken a mid-IR survey of young stellar objects (YSO) to identify structural and spectral characteristics common to the circumstellar environments of young stars and to understand observed differences in terms of their formation and subsequent evolution. We have focused on the Unidentified InfraRed (UIR) emission features seen in the near- and mid-IR, which are thought to be produced by IR fluorescence from large hydrocarbon molecules. The YSO with feature emission are unusual in that few in a given molecular cloud exhibit these spectral features, making the features a unique diagnostic of the circumstellar environments of nascent stars.

2.   WL 16

WL 16 is a low-mass protostar in the ρ Oph cloud. No CO outflow, H_2, radio or mm continuum has been detected from this object, but near-IR CO

Astrophysics and Space Science **224**: 89–92, 1995.

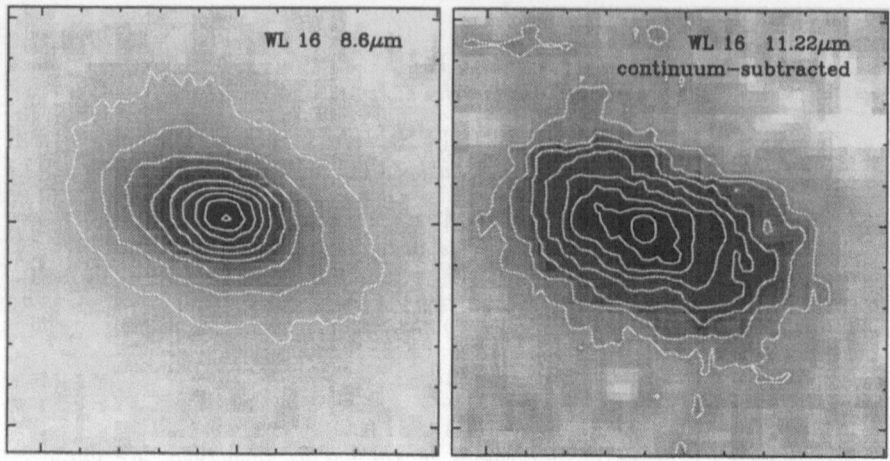

Fig. 1.　MIRAC grayscale images of WL 16 with contours overlaid. The figure on the left is the 8.6 μm image including both UIR and continuum emission, with evenly-spaced contours from 0.13 to 1.18 Jy arcsec^{-2}. The figure on the right is a continuum-subtracted image of the 11.22 μm feature emission, with evenly-spaced contours from 0.16 to 0.53 Jy arcsec^{-2}. The tick mark separation is 1$''$. North is up and east is to the left.

bandhead and strong UIR feature emission (Hanner et $al.$ 1992) are observed. Mid-IR images of WL 16 were obtained with the UA/SAO/NRL MIRAC system (Hoffmann et $al.$ 1993) at the NASA IRTF. In August 1993, we used a 1.8% CVF to acquire images at UIR feature wavelengths of 8.6, 11.22 and 12.7 μm and at a continuum wavelength of 10.3 μm. The 8.6 μm image in Figure 1 shows elliptical, extended emission of $\sim$8$''$ ($\sim$1300 AU) along its major axis. The morphology is suggestive of a torus of material around a central source. We have produced a geometric model which is able to reproduce the observed 2-D morphology and profile. The 10.3 μm continuum image (not shown) is significantly less extended and fainter, and its major axis is rotated with respect to that of the three UIR feature images. The continuum-subtracted 11.22 μm image in Figure 1 shows that the extended emission is primarily UIR feature emission, suggesting that UIR feature carriers are present and excited well beyond the region of classical grain emission near the central heating source. We also imaged Elias 29, a nearby YSO in ρ Oph which has no detected UIR emission or CO outflow and found it to be unresolved at all mid-IR wavelengths.

3. TY CrA

TY CrA is a Herbig Ae/Be star in the R CrA molecular cloud. This object is a known eclipsing binary with no reported outflows and very strong UIR feature emission (Roche et $al.$ 1991). We obtained images of this object with

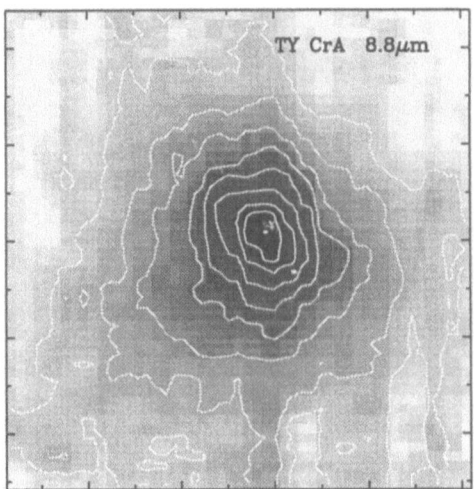

Fig. 2. MIRAC grayscale image with overlaid contours of TY CrA at 8.8 μm. Contour levels are evenly-spaced from 0.03 to 0.24 Jy arcsec^{-2}. North is up and east is to the left. Tick mark separation is 1$''$.

MIRAC at the IRTF in August 1993 using discrete narrowband (10%) filters at wavelengths covering the UIR features. The 8.8 μm image presented in Figure 2 shows an asymmetric morphology with extension on the order of 8$''$ ($\sim$1000 pc) in both dimensions, though there appears to be elongation in declination toward the center of the object. Our mid-IR images of the neighbor source R CrA, which has no UIR feature emission, show a point source at all mid-IR wavelengths.

4. MWC 1080

MWC 1080 is a Herbig Ae/Be source in Cepheus with observed CO outflow, extended 100 μm emission (Di Francisco *et al.* 1994) and UIR emission. We observed this source at the IRTF in August 1994 with the upgraded MIRAC2 system (with a new 128 x 128 pixel Rockwell BIB array). The 8.8 μm image in Figure 3 shows extended emission of $\sim$ 4$''$, or about 4000 AU in diameter. The elliptical morphology is somewhat like that seen in WL 16. The eastern extended emission appears to be much fainter than the western and it is possible that the eastern side is actually behind the brighter central and western emission with the object tipped toward the observer.

5. Conclusions

We have observed the YSO WL 16, TY CrA and MWC 1080 at mid-IR wavelengths. All three of the observed YSO with UIR emission were significantly extended (1000-4000 AU) at feature wavelengths, while neighboring

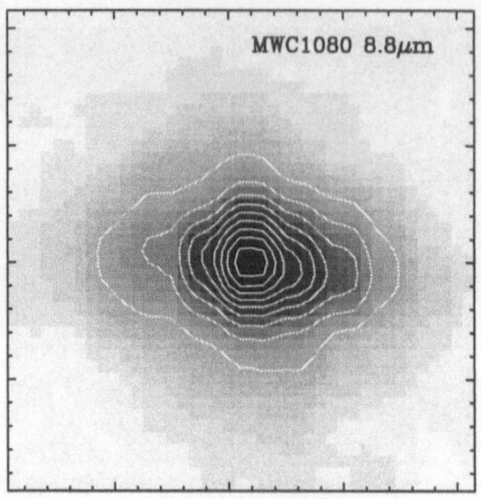

Fig. 3.   MIRAC2 grayscale image with overlaid contours of MWC 1080 at 8.8 μm. The image is uncalibrated and smoothed; contour levels are evenly spaced in digital processing units. North is up and east is to the left. Major tick mark separation is $1''$.

objects in these clouds which do not exhibit UIR emission were unresolved at all mid-IR wavelengths. The three UIR-emitting sources range from 50 to 40,000 $L_\odot$. Continuum emission in WL 16 was also found to be fainter, smaller in extent and to have a different symmetry axis than the UIR emission. Both WL 16 and MWC 1080 showed elliptical morphologies similar to those seen in outflow objects, though only MWC 1080 has a detected outflow. Surprisingly, this morphology is similar to that seen in MIRAC images of proto-planetary nebulae with bipolar outflows. These young and evolved objects also share many spectral features in the near- and mid-IR, including the broad emission features and CO bandhead emission. That such physical similarities can be produced in objects with such different ages and environments suggests possible similarities in the underlying processes responsible for the observed properties.

References

Di Francisco, J., Evans. N.J., Harvey, P.M., Mundy, L.G. & Butner, H.M.: 1994, Astrophys. J. **432**, 710.

Hanner, M.S., Tokunaga, A.T. & Geballe, T.R.: 1992, Astrophys. J. **395**, L111.

Hoffmann, W. F., Fazio, G. G., Shivanandan, K., Hora, J. L., & Deutsch, L. K.: 1993, in *Infrared Detectors & Instrumentation*, ed: A.M. Fowler, Proc. SPIE 1946, p449.

Roche, P.F., Aitken, D.K. & Smith, C.H.: 1991, Mon. Not. of the RAS **252**, 282.

ABSORPTION OF X-RAY EMISSION OF T TAURI STARS BY CIRCUMSTELLAR MATERIAL

RALPH NEUHÄUSER, MICHAEL F. STERZIK and
JÜRGEN H.M.M. SCHMITT
Max-Planck-Institut für Extraterrestrische Physik, D-85740 Garching, Germany

Abstract. The study of star forming regions (SFR) allows us to observe many young stellar objects with both the same metallicities and distances but with different masses. Because of its close distance ($\sim$ 140 pc) Taurus–Auriga is one of the best studied SFR with more than 100 well-studied, low-mass, pre-main sequence stars, T Tauri stars (TTS). A motivation for studying X-ray emission of T associations is to understand the origin of X-rays and coronal activity. The large sample observed with the ROSAT All-Sky Survey (RASS) also enables us to compare different types of young stars. Other primary goals include star formation efficiency and the interaction of young stars with their intermediate environment (probed by absorption of X-rays). RASS detection rates are comparable with *Einstein Observatory* results: 43 out of 65 (66%) weak-lined TTS (WTTS) and 9 out of 79 (11%) classical TTS (CTTS) exhibit X-ray emission above RASS detection limit. A strong correlation between X-ray surface flux and stellar rotation indicates that WTTS are intrinsically more X-ray active than CTTS, because WTTS rotate faster. However, rotation is not the only parameter that determines X-ray activity. Also, we compare Taurus–Auriga TTS with TTS of southern SFR like ScoCen, Lupus, Chamaeleon, and CrA. A new result is that CTTS and WTTS can be discriminated reliably by their X-ray spectral hardness ratios. X-ray emission of CTTS appears to be harder, partly because of circumstellar absorption. Spectral fits give results consistent with Raymond-Smith spectra and emission temperatures of $\sim$ 1.0 keV for both WTTS and CTTS. However, we find that CTTS and WTTS have significantly different X-ray luminosity functions. Medians of absorption corrected X-ray luminosities ($\log L_X$ in cgs units) are 29.701 ± 0.045 for WTTS and 29.091 ± 0.032 for CTTS. WTTS are intrinsically more luminous than CTTS, most likely because WTTS rotate on average faster than CTTS and are less absorbed. This paper concentrates on differences between CTTS and WTTS and indirect clues to be drawn from X-ray absorption and hardness ratios about circumstellar material around TTS.

Key words: T Tauri stars – X-ray emission

1. Introduction

T Tauri stars (TTS) are young, low-mass, late-type, pre-main sequence stars (PMS) with $H\alpha$ in emission and Lithium in absorption. TTS with strong $H\alpha$ emission are called classical TTS (CTTS) and often show strong infrared and (sub-)mm excess emission contributed to dusty disks and/or envelopes. The *Einstein Observatory* (EO) discovered many new TTS with strong X-ray emission (*e.g.* Walter *et al.* 1988). These TTS have $H\alpha$ equivalent width smaller than $\sim$ 10 Å (or $H\alpha$ in absorption) and are called weak-emission line TTS (WTTS). Most WTTS do not show infrared excess and may be clear of dusty disks. According to the so-called disk paradigm (*e.g.* Beckwith *et al.* 1990), CTTS with disks evolve into diskless WTTS. However, it

Astrophysics and Space Science **224**: 93–96, 1995.
© 1995 *Kluwer Academic Publishers.*

is not clear, whether all WTTS have been CTTS before. While most WTTS show strong X-ray emission, only a few CTTS have been detected with EO. We have evaluated ROSAT All-Sky Survey (RASS) observations of TTS in several different star forming regions (SFR) with particular emphasis on Taurus–Auriga, a near-by ($\sim 140~pc$) SFR with ~ 150 TTS known prior to the ROSAT mission. There are several hundred more TTS in and around Taurus–Auriga as our follow-up observations of unidentified RASS sources show (Krautter *et al.* 1994; Neuhäuser *et al.* 1994c). Our RASS flux limit is $\sim 10^{13}$ erg/sec/cm^2. RASS observations have the advantage of being spatially unbiased, so that TTS X-ray properties and correlations with other stellar parameters can be studied statistically on a large sample.

2. ROSAT observations

TTS rotate fast with periods down to a few days. Rotation periods can be measured photometrically from light curves due to spots on stellar surfaces (*e.g.* Bouvier *et al.* 1993). Fast rotation supports and perturbs the internal magnetic field and, if the dynamo effect works in the convective zone of TTS, magnetic field lines can pop out of the stellar surface, thereby producing spots. Hot gas confined in tubes along the magnetic field lines emit X-rays.

We study the area roughly confined by $\alpha = 3^h$ to 4^h and $\delta = 0°$ to $40°$. RASS detection rates are 43 out of 65 WTTS (66%) and 9 out of 79 CTTS (11%). CTTS and WTTS can be discriminated from each other and from other X-ray emitting objects by X-ray spectral properties (Neuhäuser *et al.* 1994a,b,c). Hardness ratios (HR) are X-ray colors defined as follows:

$$HR1 = \frac{H - S}{H + S} \quad \text{and} \quad HR2 = \frac{H2 - H1}{H2 + H1}$$

with count rates S in the soft band ($0.1 \ldots 0.4$ keV), H in hard band ($0.5 \ldots 2.0$ keV), $H1$ in hard band 1 ($0.5 \ldots 0.9$ keV), and $H2$ in hard band 2 ($0.9 \ldots 2.0$ keV). As shown in Fig. 1a & b, all CTTS have $HR1 = 1$, while most WTTS have $HR1 < 1$. Utilizing the full Raymond-Smith spectral code, we can fit to the observed pair of ($HR1, HR2$) the spectral properties (X-ray emission energy T_X and absorbing column density N_H) for any individual source detected with RASS and identified with a TTS counterpart. Fit results T_X and N_H yield together with ROSAT response matrix the individual energy conversion factor, which together with the count rate gives flux (as well as luminosity and surface flux) individualy corrected for spectral parameters. X-ray luminosity functions of WTTS and CTTS are significantly different. A correlation between X-ray surface flux and rotation period as found by Bouvier (1990) can be confirmed (96% confidence level) with RASS data. WTTS emit intrinsically more X-rays than CTTS, because

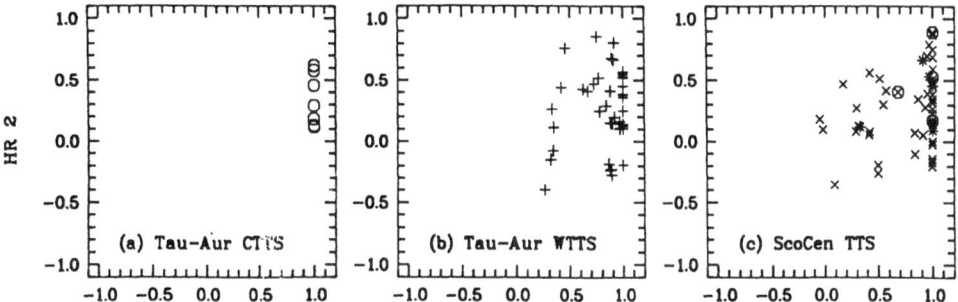

Fig. 1. Plotted are hardness ratios $HR2$ versus $HR1$ for (a) CTTS and (b) WTTS, all in Taurus–Auriga and known prior to ROSAT, as well as for pre-ROSAT WTTS (+), CTTS (o), and newly discovered WTTS (x) in ScoCen (c).

WTTS rotate faster than CTTS (Neuhäuser *et al.* 1994b). X-ray luminosity is correlated with stellar luminosity, mass, age, and effective temperature.

3. X-ray hardness ratios

There are several different reasons that may be responsible for different HR for CTTS and WTTS: Different X-ray emission temperatures of WTTS and CTTS and/or different absorption either by ISM in front of the SFR clouds and/or by intercloud gas and/or by circumstellar material.

Having fitted individual X-ray spectra, we test whether CTTS and WTTS are drawn from the same sample and find that CTTS and WTTS are not different as far as both $HR\,2$ and T_X are concerned. Hence, different HR are not due to different emission temperatures. CTTS and WTTS are significantly different as far as $HR\,1$ is concerned ($> 99\%$ confidence).

Different absorption by foreground ISM can easily be excluded as main reasons for different HR, because Taurus–Auriga is near-by and WTTS populate by parts the same regions as CTTS do. Extinction can very largely even among stars with low angular separation (Cohen & Kuhi, 1979).

We have studied HR of RASS-detected CTTS (4) and WTTS (57) in Sco-Cen (Fig. 1c), too, and find similar results as in Taurus–Auriga. Many new WTTS have already been identified in ScoCen with follow-up observations of unidentified RASS sources. Since there seems to be at least almost no CO gas left in ScoCen, we conclude that intercloud gas is not the main reason for different HR of CTTS and WTTS. We find similar results in Perseus, Lupus, Orion, Chamaeleon, and R CrA, too.

If WTTS and CTTS would have intrinsically the same X-ray emission, *i.e.* if different HR are mainly due to absorption, HR should be correlated with visual extinction. We do find such a trend and conclude that different X-ray spectra are observed, because soft X-rays are absorbed by circumstellar material, so that CTTS appear to harder than WTTS.

Most WTTS have low absorption and low HR and since those two properties are correlated. Hence, CTTS with low absorption should also have low HR. However, while most WTTS have low absorption, there are only a few such CTTS. WTTS emit intrinsically more X-rays than CTTS. Hence, most WTTS are detected with RASS, so that there appear to be many WTTS with low $HR1$, while most CTTS (both with low and high absorption) fall just below the RASS detection limit. The reason for the fact that we observe a difference in $HR1$ between CTTS and WTTS is a combination of absorption of soft X-rays in circumstellar environment of CTTS and the low X-ray emission level, hence, low RASS detection rate of CTTS.

Since WTTS can be discrimated from most other X-ray emitting objects, we can pre-select the sample of unidentified RASS sources via HR in order to limit follow-up observations. Follow-up observations performed so far enable us to estimate the selection reliability and to extrapolate the total number of WTTS in the vicinity of Taurus–Auriga. By comparing the ratio of WTTS to CTTS ($\sim 10:1$) with the typical PMS life-time, one can obtain an estimate of the typical disk life-time.

Acknowledgements

We would like to thanks Gregor Morfill, Juan Alcalá, Michael Kunkel, Rainer Wichmann, and Fred Walter for stimulating discussions and providing data prior to publication. R.N. wishes to thank the Hanns-Seidel-Stiftung and the federal ministry for education and science for financial support.

References

Beckwith S.V.W., Sargent A.I., Chini R.S. & Güsten R.: 1990, Astron. J. **99**, 924.
Bouvier J.: 1990, Astron. J. **99**, 946.
Bouvier J., Cabrit S., Fernandez M., Martin E.L. & Matthews J.M.: 1993, Astron. & Astrophys. **272**, 176.
Cohen M. & Kuhi L.V.: 1979, Astrophys. J., Suppl. **41**, 743.
Krautter J., Alcalá J.M., Wichmann R., Neuhäuser R. & Schmitt J.H.M.M.: 1994, in *Symposium on Stars, Gas and Dust in the Galaxy* to honor Eugenio E. Mendoza, eds: P. Pismis & S. Toores-Peimbert, Rev. Mex. Astron. Astrofis. **29**, 41.
Neuhäuser R., Sterzik M.F. & Schmitt J.H.M.M.: 1994a, in *8th Cambridge Workshop on Cool Stars, Stellar System, and the Sun*, ed: J.P. Caillault, San Francisco, p.113.
Neuhäuser R., Sterzik M.F., Schmitt J.H.M.M., Wichmann R. & Krautter J.: 1994b, Astron. & Astrophys. submitted.
Neuhäuser R., Sterzik M.F., Schmitt J.H.M.M., Wichmann R. & Krautter J.: 1994c, Astron. & Astrophys. submitted.
Walter F.M., Brown A., Mathieu R.D., Myers P.C., Vrba F.J.: 1988, Astron. J. **96**, 297.

SESSION B:

PRE-MAIN SEQUENCE OUTFLOWS

PRE-MAIN-SEQUENCE OUTFLOWS

COLIN R. MASSON

Center for Astrophysics, 60 Garden Street, Cambridge, MA 02138, USA

Abstract. In the 3 decades since winds from young stars were discovered, there have been many observations of bipolar molecular flows and ionized jets, and it has been recognized that outflows are intimately linked to star formation. Despite many observational clues and theoretical ideas, we still do not have a fully coherent picture of the outflow process.

Key words: Stars: Formation – Stars: Pre-Main-Sequence – ISM: Jets and Outflows

1. Introduction

Winds from Pre-Main-Sequence (PMS) stars were first noted by Herbig (1962), and molecular outflows from PMS stars were recognized in 1975 when Kwan & Scoville (1976) and Zuckerman, Kuiper & Kuiper (1976) found CO emission in OMC1 at velocities of 75 km s^{-1}, too high to be due to any gravitational motion. At about the same time, Herbig-Haro objects were shown by Schwartz (1975) to be due to shock excitation by supersonic outflows from PMS stars, and Gautier *et al.* (1976) found H$_2$ emission, indicating the presence of high-velocity molecular shocks.

More outflows were rapidly found, including a strikingly bipolar outflow from L1551–IRS5 (Snell, Loren & Plambeck 1980). By mid-1984, Lada (1985) catalogued 67 molecular outflows, many of which were shown to be bipolar. Optical images showed highly collimated jets (Graham & Elias 1983; Mundt & Fried 1983), and radio emission was detected from ionized gas near the PMS stars (Cohen, Bieging & Schwartz 1982).

In the ensuing decade, the number of known outflows has increased, and images at all wavelengths have improved dramatically in quantity and quality. Theoretical understanding has also improved, although current models still cannot reproduce realistic outflows.

2. Observed Properties

Figure 1 shows the well-known object HH46/47, which illustrates many of the properties of outflows. The greyscale optical image shows a collimated and wiggly blue-shifted jet emerging from a dark cloud. The redshifted jet is hidden by the cloud, except where its tip emerges at the far side. The star itself is also obscured. The contours show the associated molecular flow, which is aligned with the optical, but less well collimated. There is little blueshifted molecular material, since the jet rapidly breaks out of the cloud, but more redshifted material where the redshifted jet bores through the

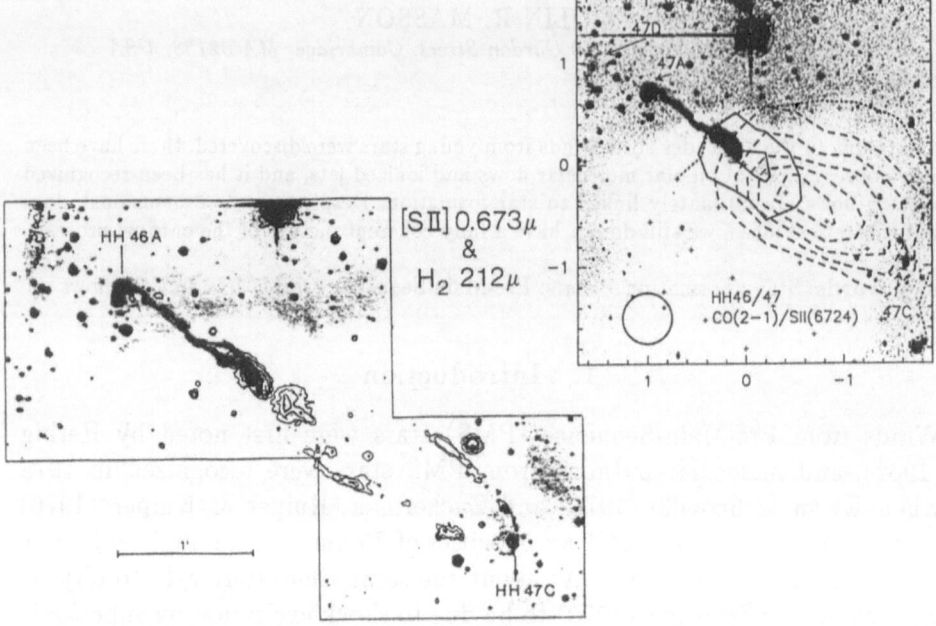

Fig. 1. a) Molecular outflow in HH46/47 overlaid on optical image. Solid contours show redshifted gas and dashed contours show blueshifted. b) Shocked molecular hydrogen (contours) overlaid on optical image (Eislöffel *et al.* 1993).

body of the cloud (Chernin & Masson 1991). Figure 1b is an infrared picture which shows scattered light on either side of the central source, as well as a redshifted jet, and long arcs which indicate the position of shocks impacting the ambient molecular cloud (Eislöffel *et al.* 1993).

2.1 MOLECULAR OUTFLOW PROPERTIES

Reviews of molecular outflows are given by Lada (1985), Bachiller & Gomez-Gonzalez (1992), Fukui *et al.* (1993), and references therein. Surveys give a high detection rate of flows amongst PMS stars (*e.g.* Parker *et al.* 1991; Fukui *et al.* 1993), showing that molecular outflows are very common and that perhaps all PMS stars produce outflows. The same surveys also imply that outflows have lifetimes comparable with that of the embedded phase $(1 \sim 2 \times 10^5$ years).

The outflows are massive $(0.1 \sim 100 \text{ M}_\odot)$, indicating that the molecular material is swept up from the ambient cloud rather than emanating directly from the star. Most outflows are bipolar, with some of the exceptions being due to the environment as in the case of HH46/47. In at least some flows, the outflow lobes appear to be partly hollow, *e.g.* L1551 (Moriarty-Schieven & Snell 1988), and in all cases they are clumpy or irregular. The outflows can be highly collimated, up to 10:1 (*e.g.* VLA 1623, Fig. 2), although they are always wider than the associated optical jets, if any.

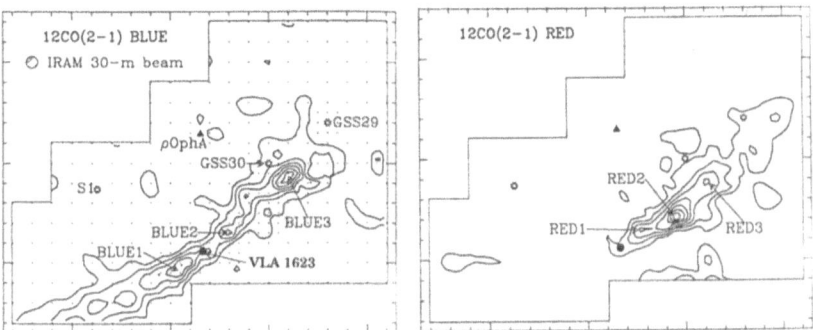

Fig. 2.   Highly collimated molecular outflow in VLA 1623 (André *et al.* 1990).

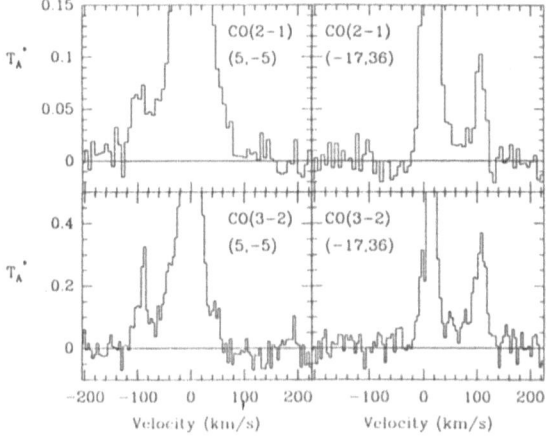

Fig. 3.   CO emission from EHV features in HH7-11

The molecular velocities are highly supersonic (and super-Alfvénic), and can extend to more than 100 km s^{-1}, significantly greater than the shock velocity required to dissociate molecules. Most of the mass is found at low velocities, with a falloff roughly as $m(v) \propto v^{-2}$. There is therefore no characteristic velocity, although in some flows extremely high velocity (EHV) features are found, having narrow velocity extents ($\Delta v \sim 10$ km s^{-1}), and high velocities ($50 \sim 100$ km s^{-1}) in compact regions (*e.g.* Bachiller *et al.* 1990; Masson, Mundy & Keene 1990, Fig. 3). Since the separation between red and blueshifted lobes in many flows requires a high collimation in velocity as well as space, the velocity vectors must be directed outward along the flow axes, with transverse components typically less than half of the longitudinal components (Cabrit & Bertout 1990, Meyers-Rice & Lada 1991).

Herbig-Haro objects and jets are often associated with molecular outflows, although they are sometimes obscured and visible only in the infrared.

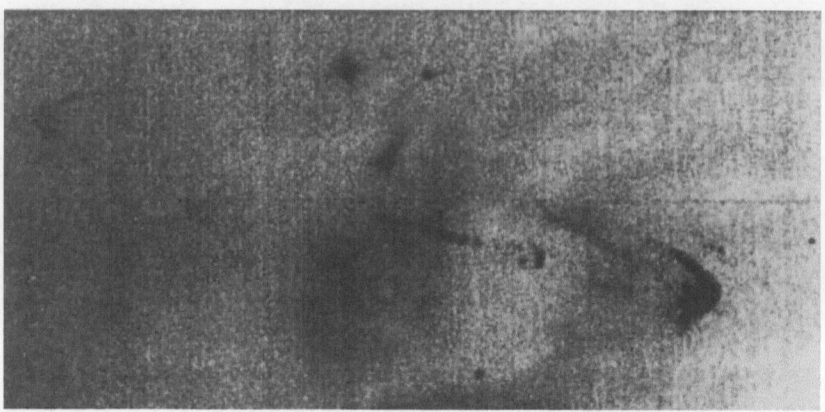

Fig. 4.   HH34 jet and bowshock, rotated so that South is to the right. (Reipurth &
Heathcote 1992)

2.2 OPTICAL JET PROPERTIES

Reviews of optical jets can be found in Mundt, Brugel & Buhrke (1987),
Reipurth (1991), Edwards, Ray & Mundt (1993), and references therein.
Optical jets are highly collimated, typically 10 ~ 100 : 1. They are probably all bipolar, although in many cases the redshifted lobe is obscured (*cf.*
HH46/47). The visible material is ionized, with the jets appearing bright at
shocks and faint elsewhere. Sometimes the jet cools until it is invisible, as in
the case of HH34, Fig. 4, where the jet disappears 30″ south of the star, but
the distant bow shocks indicate that it extends farther. Even in the shocked
gas, the ionization is low, ~ 5%, (Raga, Binette & Canto 1990; Hartigan,
Morse & Raymond 1994). Analysis of the shocked gas shows that jets are
dense ($100 \sim 1000$ cm^{-3}), comparable with molecular clouds. However, the
optical jets are usually detected outside clouds, where the ambient medium
is much less dense. Near the star, many jets are dense enough to be visible as radio continuum sources. A high resolution VLA map of L1551–IRS5
shows that the jet is collimated on a scale less than 100 AU (Fig. 5).

Jet velocities are typically ~ 300 km s^{-1}, comparable with the escape
velocities of PMS stars, suggesting that the jets are accelerated very close to
the stellar surface. Proper motions have been measured for knots in many
jets and HH objects, showing that these are moving at the jet velocity (*e.g.*
Cudworth & Herbig 1979; Eislöffel & Mundt 1992). The transit time for
optical jets is ~ 1000 yrs, but the statistics indicate that they must live much
longer (Mundt *et al.* 1987). This is consistent with the discovery that many
jets are running into material which is already moving rapidly, probably as
a result of earlier interaction (Hartigan *et al.* 1987; Morse *et al.* 1992).

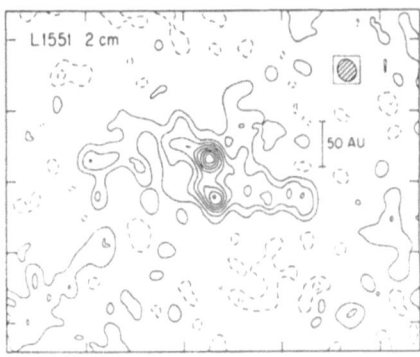

Fig. 5. VLA image of L1551–IRS5 (Rodriguez *et al.* 1986).

3. What Drives Molecular Outflows?

Since a molecular flow consists of swept-up ambient cloud material, it must be driven by some kind of underlying wind or jet. When this interacts with the ambient material, the shock could be *adiabatic*, creating a pressure-driven bubble, where all the kinetic energy of the wind is extracted and converted into kinetic energy, or *radiative*, where only momentum is conserved. Shocks in dense molecular clouds are likely to be radiative, producing momentum-driven flows (Levreault 1988), and infrared radiation shows evidence of the energy losses in L1551–IRS5 (Clark & Laureijis 1986; Edwards *et al.* 1986). Further confirmation is given by the demonstration by Meyers-Rice & Lada (1991) that the observed axial direction of velocity vectors is inconsistent with the pattern produced by a pressure-driven bubble (see also Masson & Chernin 1993).

Momentum, as the only conserved quantity, is the most important parameter in molecular flows. Early estimates appeared to show that the ionized material in jets and cm radio sources had inadequate momentum to drive molecular outflows (Bally & Lada 1983; Mundt *et al.* 1987), leading to postulation of a slower, denser wind surrounding the observed jets, which would provide sufficient momentum. More recently, the discrepancy has disappeared with the realization that jets are more dense than originally thought, and outflows longer lived (Raga *et al.* 1990; Raga 1991; Hartigan *et al.* 1994; Padman & Richer 1994). It appears that there is no need for any driving wind beyond the observed jets, although there are still very few cases where both optical jet and molecular flow have been analyzed in detail. In a few sources, HI has been detected (*e.g.* Bally & Stark 1983; Bally 1986; Lizano *et al.* 1988), but it is unclear whether this material is part of the original wind from the star, or dissociated ambient gas.

It is possible to discriminate between highly collimated jets and poorly

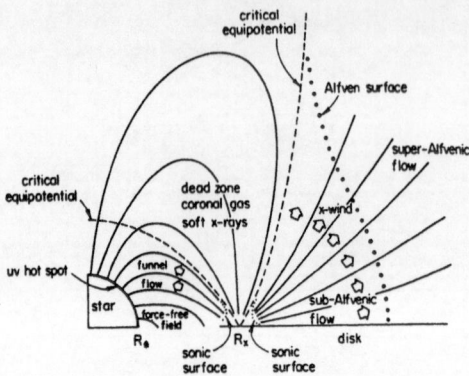

Fig. 6. Model for wind driven by interaction between star and disk (Shu *et al.* 1994)

collimated wide-angle winds. Shu *et al.* (1991) presented an analysis of the
pattern of outflow momentum-driven by a wide-angle wind, which Masson &
Chernin (1992) used to show that such an outflow would contain too much
material at extreme velocities, because most material would be swept up
along the flow axis, where the highest velocities are found. They concluded
that molecular outflows must be driven by jets which can sweep aside the
ambient material in bow shocks as they drive it. Thus, even if there is a
second (neutral) component to the wind, it must also be collimated.

4.   Where Do Jets Come From?

Two deductions about the origin of jets, or indeed any driving winds, can be
made from observations of the molecular flows. First, it was shown by Bally
& Lada (1983) that radiation pressure could not provide enough momentum
to drive molecular outflows, leading to the conclusion that the flows were
ultimately driven by the gravitational binding energy of the stars.

Second, it is possible to estimate the total mass of the jet from the
observed molecular momentum (Masson & Chernin 1994). For a wide range
of outflows, roughly 5-10% of the mass accreted on to the star must have
been ejected at 300 km s^{-1} to account for the observed momentum, in general agreement with FU Ori data (Hartmann, Kenyon & Hartigan 1993).
Since 300 km s^{-1} is comparable with the escape velocity, a similar percentage of the binding energy must be extracted to drive the jet. This implies
that the ejection process must be efficient, and must take place within 10-20
stellar radii. It also casts doubt on any model attempting to drive outflows
with slower winds, since more mass would be required. At 10 km s^{-1}, for
example, the ejected mass would typically have to exceed the stellar mass.

Most models for the generation of protostellar jets involve accretion disks

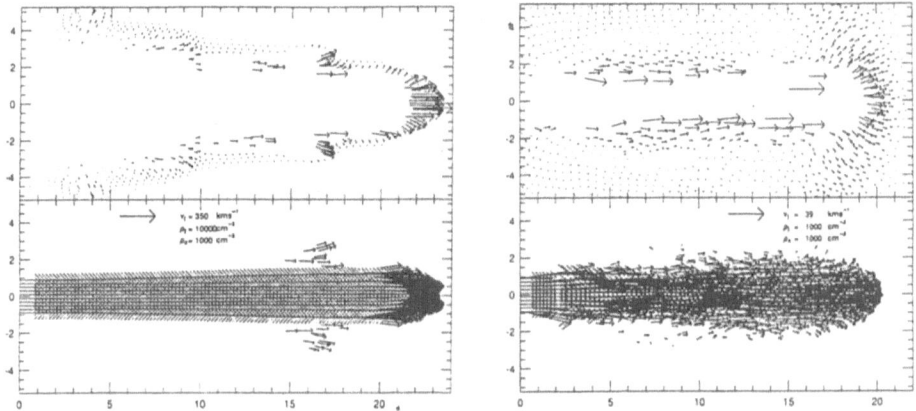

Fig. 7.   SPH simulations (Chernin *et al.* 1994); Upper panels: ambient, Lower panels: jet;
a) High M jet, showing prompt interaction; b) Low M jet, showing steady-state interaction.

and magnetic fields. If gas is ejected along lines of a magnetic field anchored
in a rotating disk, the combination of disk rotation and the expansion of the
gas gives rise to a toroidal magnetic field which collimates the jet. Models
invoking a flow directly from the surface of an accretion disk are reviewed by
Pudritz *et al.* (1991) and Königl & Ruden (1993), while Shu *et al.* (1994)
have recently presented a model with an interaction between the disk and the
magnetic field of the star (Fig. 6). As yet, there is no decisive observational
test of these models, but spectroscopic observations are beginning to probe
the critical region (*e.g.* Calvet *et al.* 1993).

5.   How Do Jets Interact with Molecular Clouds?

As described by DeYoung (1986), there are two forms of interaction between
jets and ambient material. In the *prompt* interaction (Fig. 7a), a strong bow
shock at the head of the jet pushes the ambient material strongly away from
the jet axis, creating a dense swept-up shroud, with a low density cocoon
surrounding the jet, inhibiting interaction along the sides of the jet. This
is the pattern expected for high Mach number jets (M>>1). The *steady-
state* interaction takes place where there is no bow shock, or only a weak
one which is unable to clear away the ambient material to form a protective
cocoon (Fig. 7b). Instabilities rapidly develop along the sides of the jet at
the interface with the ambient material, entraining ambient material and
weakening the jet. After some distance, the jet becomes subsonic, and the
flow is completely turbulent (Stahler 1993). This is the interaction expected
for slow jets (M~1). These expectations are borne out by simulations of

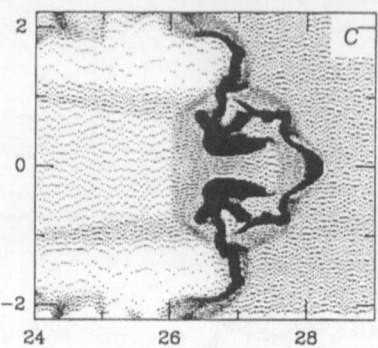

Fig. 8.   Bowshock simulation (Blondin *et al.* 1990).

both adiabatic and radiative jets (DeYoung 1986; Chernin *et al.* 1994).

Models using appropriate parameters for protostellar jets show that the prompt interaction should dominate over the steady state. The cooling material in the bowshock may reform molecules to create the EHV features (Chernin *et al.* 1994), and it may fragment as shown in Fig. 8, perhaps producing H_2O masers. However, models with straight jets produce flows which are too narrow and times which are too short. The solution may be the same as that suggested for extragalactic radio sources, wandering of the jet, carving out a larger volume of the cloud. It is possible that we are misled by the straight jets which are often observed, if the less variable jets are more likely to be observed because they break out of their clouds, while the more variable ones churn up larger molecular flows and remain obscured.

Alternatively, Raga *et al.* (1993) have suggested that a jet with variable velocity may generate internal shocks which eject material from the jet and couple it to the external medium, creating a steady-state type of flow. These internal shocks may produce the molecular EHV features in the same manner as suggested by Chernin *et al.* for bow shocks (Raga & Cabrit 1993). Such shocks may be observed in L1448 (Bachiller *et al.* 1990), where the high symmetry of the EHV features indicates that they must be nearly unaffected by the ambient medium, unless it is symmetrical about the star.

Thus, although it appears likely that molecular outflows are driven by jets, all of the current models have some problems. It is still conceivable that a second type of wind really does drive the outflows, but the constraints described above severely limit the nature of such a wind, and there is no unambiguous observational evidence for a wind which is dynamically distinct from the observed jets.

6. Outflows and Star Formation

The ubiquity of outflows, and the large energy required for their production indicate that they play a major role in star formation. One obvious effect is that an outflow could carry off the angular momentum of the accretion disk, allowing the remaining mass to fall into the star (Hartmann & MacGregor 1982; Pudritz & Norman 1983). If this is the case, then outflows and inflows are intimately linked, and one could not occur without the other. It may be no coincidence that the youngest PMS stars already possess outflows.

A second role ascribed to outflows is the stirring up of turbulence in the ambient cloud, possibly regulating star formation in general, or limiting the mass of each star by pushing back the infalling material. While these ideas appear arguable on the basis of the available energy (Margulis & Lada 1986; Levreault 1988), high collimation limits the effect of an outflow on the local infalling material. However, the rapid ejection of a significant mass in an outflow may decrease the gravitational potential and have some effect on limiting the mass accretion (Masson & Chernin 1994).

Although the jets and outflows do not tell us directly about the accretion of mass on to protostars, they act as signposts to the very youngest stars, and provide a long-term record which, properly understood, may tell us much about the development of PMS stars.

Acknowledgements

I am grateful for many useful discussions with Lawrence Chernin.

References

André, P., Martin-Pintado, J., Despois, D. & Montmerle, T.: 1990, Astron. & Astrophys. **236**, 180.

Bachiller, R., Cernicharo, J., Martin-Pintado, J., Tafalla, M. & Lazareff, B.: 1990, Astron. & Astrophys. **231**, 174.

Bachiller, R. & Gomez-Gonzales, J.: 1992, Astron. & Astrophys. Rev. **3**, 257.

Bally, J.: 1986, in *Masers, Molecules and Mass Outflows in Star Forming Regions*, ed: A.P. Haschick, Haystack Observatory, p179.

Bally, J. & Stark, A. A.: 1983, Astrophys. J. **266**, 61.

Bally, J. & Lada, C. J.: 1983, Astrophys. J. **265**, 824.

Blondin, J.M., Fryxell, B. A. & Königl, A.: 1990, Astrophys. J. **360**, 370.

Cabrit, S. & Bertout, C.: 1990, Astrophys. J. **348**, 530.

Calvet, N., Hartmann, L. & Kenyon, S. J.: 1993, Astrophys. J. **402**, 623.

Chernin, L. M. & Masson, C. R.: 1991, Astrophys. J. **382**, L93.

Chernin, L. M. & Masson, C. R.: 1992, Astrophys. J. **396**, L35.

Chernin, L. M., Masson, C.R., Gouveia Dal Pino, E. M. & Benz, W.: 1994, Astrophys. J. **426**, 204.

Clark, F. O. & Laureijis, R. J.: 1986, Astron. & Astrophys. **154**, L26.

Cohen, M., Bieging, J. H. & Schwartz, P. R.: 1982, Astrophys. J. **253**, 707.

Cudworth, K. M. & Herbig, G. H.: 1979, Astron. J. **84**, 548.

DeYoung, D. S.: 1986, Astrophys. J. **307**, 62.

Edwards S., Strom, S., Snell, R., Jarrett, T., & Beichman, C.: 1986, Astrophys. J. **307**, L65.

Edwards, S., Ray, T. & Mundt, R.: 1993, in *Protostars and Planets III*, eds: E. H. Levy & J. I. Lunine, Univ. Arizona : Tucson, p567.

Eislöffel, J. & Mundt, R.: 1992, Astron. & Astrophys. **263**, 292.

Eislöffel, J., Davis, C. J., Ray, T. P. & Mundt, R.: 1993, Astrophys. J. **422**, L91.

Fukui, Y. Iwata, T., Mizuno, A., Bally, J. & Lane, A.: 1993, in *Protostars and Planets III*, eds: E.H. Levy & J.I. Lunine, Univ. Arizona: Tucson, p603.

Gautier, T. N., Fink, U., Treffers, R. R. & Larson, H. P.: 1976, Astrophys. J. **207**, L129.

Graham, J. A. & Elias, J. H.: 1983, Astrophys. J. **272**, 615.

Hartigan, P., Morse, J. & Raymond, J.: 1994, Astrophys. J. in press.

Hartigan, P., Raymond, J. & Hartmann, L.: 1987, Astrophys. J. **316**, 323.

Hartmann, L. & MacGregor, K. B.: 1982, Astrophys. J. **259**, 180.

Hartmann, L., Kenyon, S. & Hartigan, P.: 1993, in *Protostars and Planets III*, eds: E.H. Levy & J.I. Lunine, Univ. Arizona: Tucson, p497.

Herbig, G.: 1962, Adv. Astron. & Astrophys. **1**, 47.

Konigl, A. & Ruden, S. P.: 1993, in *Protostars and Planets III*, eds: E.H. Levy & J.I. Lunine, Univ. Arizona : Tucson, p641.

Kwan. J. & Scoville, N. Z.: 1976, Astrophys. J. **210**, L39.

Lada, C. J.: 1985, Ann. Rev. of Astron. & Astrophys. **23**, 267.

Levreault, R. M.: 1988, Astrophys. J. **330**, 897.

Lizano *et al.* : 1988, Astrophys. J. **328**, 763.

Margulis, M. & Lada, C. J.: 1986, Astrophys. J. **309**, 87.

Masson, C. R. & Chernin, L. M.: 1992, Astrophys. J. **387**, L47.

Masson, C. R. & Chernin, L. M.: 1993, Astrophys. J. **414**, 230.

Masson, C. & Chernin, L. M.: 1994, in *Clouds, Cores and Low-Mass Stars*, eds: D. Clemens & R. Barvainis, A.S.P. Conference Series **65**, p350.

Masson, C. R., Mundy, L., G. & Keene, J.: 1990, Astrophys. J. **357**, L25.

Meyers-Rice, B. & Lada, C. J.: 1991, Astrophys. J. **368**, 445.

Moriarty-Schieven, G. H. & Snell, R. L.: 1988, Astrophys. J. **338**, 952.

Morse, J., Hartigan, P., Cecil, G., Raymond, J. & Heathcote, S.: 1992, Astrophys. J. **399**, 231.

Mundt, R., Brugel, E. W. & Buhrke, T.: 1987, Astrophys. J. **319**, 275.

Mundt, R. & Fried, J. W.: 1983, Astrophys. J. **274**, L83.

Padman, R. & Richer, J. S.: 1994, preprint.

Parker, N. D., Padman, R. & Scott, P. F.: 1991, Mon. Not. of the RAS **252**, 442.

Pudritz, R., Pelletier, G. & Gomez de Castro, A.: 1991, in *The Physics of Star Formation and Early Stellar Evolution*, eds: C. Lada & N. Kylafis, Kluwer:Dordrecht, p539.

Pudritz, R. & Norman, C.: 1983, Astrophys. J. **274**, 677.

Raga, A. C., Binette, L. & Canto, J.: 1990, Astrophys. J. **360**, 612.

Raga, A. C.: 1991, Astron. J. **101**, 1472.

Raga, A. C. & Cabrit, S.: 1993, Astron. & Astrophys. **278**, 267.

Raga, A., Canto, J., Calvet, N., Rodriguez, L., & Torrelles, J.: 1993, Astron. & Astrophys. **276**, 539.

Reipurth, B.: 1991, in *The Physics of Star Formation and Early Stellar Evolution*, eds: C.J. Lada & N.D. Kylafis, Kluwer:Dordrecht, p497.

Reipurth, B. & Heathcote, S.: 1992, Astron. & Astrophys. **257**, 693.

Rodriguez, L. F., Canto, J., Torrelles, J. M. & Ho, P. T. P.: 1986, Astrophys. J. **301**, L25.

Schwartz, R. D.: 1975, Astrophys. J. **195**, 631.

Shu, F. H., Ruden, S. P., Lada, C. J. & Lizano, S.: 1991, Astrophys. J. **370**, L31.

Shu, F., Najita, J., Ostriker, E., Wilkin, F., Ruden, S., & Lizano, S.: 1994, Astrophys. J. **429**, 781.

Snell, R. L., Loren, R. B. & Plambeck, R. L.: 1980, Astrophys. J. **239**, L17.

Stahler, S. W.: 1993, Astrophys. J. **422**, 616.

Zuckerman, B., Kuiper, T. B. H. & Kuiper, E. N. R.: 1976, Astrophys. J. **209**, L137.

THE SMALL-SCALE OUTFLOW STRUCTURE OF
EMBEDDED SOURCES IN TAURUS

C. J. CHANDLER

National Radio Astronomy Observatory*
Socorro, New Mexico, USA

S. TEREBEY

Infrared Processing and Analysis Center, Jet Propulsion Laboratory and California
Institute of Technology, Pasadena, California, USA

M. BARSONY

Physics Department, University of California
Riverside, California, USA

and

T. J. T. MOORE

Physics Department, University College, ADFA
Canberra, ACT, Australia

Abstract. High resolution interferometric CO $J=1-0$ observations of the outflows from two young embedded sources, TMC1 and TMC1A, show the high-velocity gas to have a conical structure, with a constant opening angle of $\sim 45°$ extending to within 1000 AU of the central stars. The correspondence of near-infrared reflection nebulosity at K band with blueshifted CO emission in both objects suggests the lobes are partially evacuated, as do position-velocity diagrams from single-dish CO $J=2-1$ data. We suggest that the outflows are driven by jets which impart momentum to the ambient medium through shocks, rather than through the entrainment of molecular material along the edges of the jet.

Key words: jets and outflows – reflection nebulae – TMC1 – TMC1A

1. Introduction

All young stars are thought to go through a phase of energetic mass-loss, most commonly observed as high-velocity molecular outflows. Current models (see Masson, this volume, for a review) suggest that the molecular outflows can be driven by the jets often seen at optical and radio wavelengths, but high-resolution observations of the molecular gas have only recently been available to compare with predictions.

To try to distinguish between models for the driving mechanism and the acceleration of the ambient molecular gas, we have carried out a high-resolution study of two young embedded sources in the nearby Taurus star-forming region: TMC1 and TMC1A. Both objects were originally reported not to have outflows in large-beam, single-dish surveys, but were shown to have high-velocity, compact emission in the interferometer snapshot survey

* The NRAO is operated by Associated Universities, Inc., under cooperative agreement with the National Science Foundation

Astrophysics and Space Science **224**: 109–112, 1995.
© 1995 *Kluwer Academic Publishers.*

of Terebey *et al.* (1989). They are therefore good candidates for having young, compact outflows, and offer the opportunity to investigate the earliest stages of the interaction between young stars and their environment.

2.   Observations and results

We have imaged the central $1' \times 1'$ around TMC1 and TMC1A in CO(1–0) using the Owens Valley Millimeter Array with a resolution of $6.9''$, and we have mapped a slightly larger area in CO(2–1) with the James Clerk Maxwell Telescope (JCMT) to investigate the outflow kinematics. Both sources were also imaged in near-infrared continuum at H and K band using a 128×128 HgCdTe array on the Hale 5m telescope.

Figure 1 displays the integrated blue and redshifted CO(1–0) emission from the Owens Valley array in both sources, overlaid on greyscale images of the K band continuum. The high-resolution interferometer data show conical outflow lobes in both sources, with the apex located at the position of the millimeter continuum source. The emission in the JCMT data shows the outflow lobes to be more extended over $\sim 1'$, but the interferometer is insensitive to structures larger than $\sim 20''$.

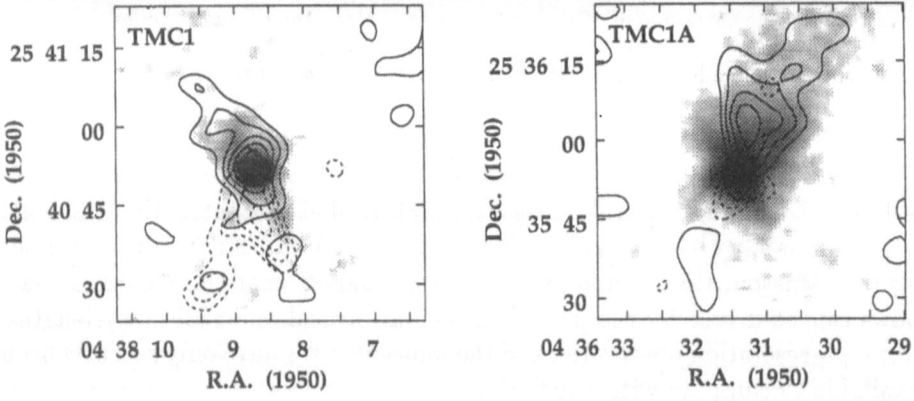

Fig. 1.   Integrated blueshifted (solid contours) and redshifted (dashed contours) CO(1–0) emission from TMC1 and TMC1A, overlaid on greyscale images of the K band nebulosity.

Perhaps the most spectacular result is the correspondence between the infrared nebulosity and the the blueshifted outflow lobe in the interferometer maps. In particular, at some distance from the central source, there is nebulosity only where there is high-velocity gas. Since these are low-luminosity objects ($L_{bol} = 0.7\ L_\odot$ for TMC1, and 2.4 $L_\odot$ for TMC1A), it is highly unlikely that they are able to heat dust to a few 100 K at distances of 5000 AU. The nebulosity is more likely to be due to reflection, tracing the loca-

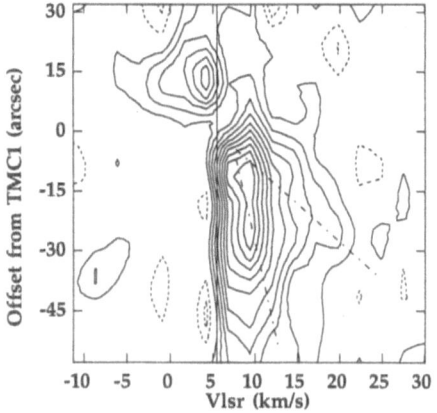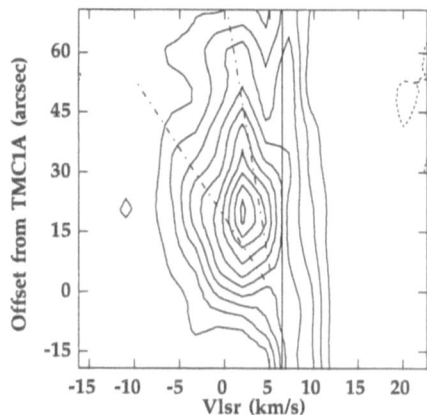

Fig. 2. Position-velocity diagrams along the outflow axes of TMC1 and TMC1A from CO(2–1) observations using the JCMT. The solid line marks the systemic velocity for each source. The dot-dash lines indicate the features in the P-V diagram associated with the front and back sides of the lobe in each case.

tion of ambient material with a direct line of sight to the central star. This strongly suggests that the lobes are evacuated to some degree by whatever is driving the molecular outflow.

There are probably two effects which prevent similar nebulosity from being associated with the red lobe. First, there will be higher extinction to the red lobe: less than 1 $M_\odot$ of material spread over $\sim 20''$ in front of the red lobe is needed to hide the nebulosity at K band. Secondly, the dust grains may preferentially scatter radiation from the star in a forward direction, which in the case of the red lobe is away from our line of sight.

The JCMT CO(2–1) maps cover the main extended lobe in each source, i.e., the blue lobe in TMC1A and the red lobe in TMC1. Figure 2 shows a position-velocity (P-V) diagram along the outflow axis for each object. They show components associated with the front and back sides of each lobe, also suggesting an evacuated cavity rather than the filled lobes modelled by Cabrit & Bertout (1986, 1990).

3. Comparison with outflow models

We consider three outflow models to try to explain our data. The first is the jet-entrainment model most recently discussed by Stahler (1994), in which the molecular outflow is driven by a jet, and the acceleration of molecular gas is due to the turbulent entrainment of ambient material along the edges of the jet. This model predicts filled lobes, and conical outflow structure.

The second model is the swept-up outflow proposed by Shu *et al.* (1991) and analysed by Masson & Chernin (1992). The outflow is driven by a radial stellar wind into the surrounding medium, and the shape of the outflow is defined by the driving pressure and column density of swept-up material. Shu *et al.* showed that this model could result in the elongated outflows commonly observed, but Masson & Chernin's analysis suggests that for reasonable distributions of the wind and cloud core the swept-up outflow model has problems reproducing the observed line profiles.

The third model we consider is the jet-driven outflow proposed by Masson & Chernin (1993), where a highly-collimated jet drives the outflow and transfers momentum to the ambient medium through shocks, leaving behind a cocoon of shocked gas which sweeps up ambient material as it expands away from the jet axis. The wide opening angles observed in some sources are caused by the wandering or precession of the jet.

The conical structure of the outflow lobes in Figure 1 are consistent with jet entrainment, but both the infrared reflection nebulosity, and the P-V diagrams, indicate that the lobes are not filled with molecular gas. We therefore think it unlikely that the ambient gas is accelerated by entrainment as described by Stahler (1994). Both TMC1 and TMC1A show projected opening angles of $\sim 45°$, as does B335 (Chandler & Sargent 1993). With a monotonically decreasing density gradient in the ambient cloud, it is difficult to explain a constant opening angle, or conical structures with distinct, sharp edges, by a wide-angle stellar wind sweeping up ambient material. An analysis of the swept-up outflow model similar to that carried out by Masson & Chernin (1992) shows that to obtain a 45° opening angle and to account for the line profile requires the cloud to be flattened with an axial ratio of over 10^5:1. In the context of the models we consider, it is therefore most likely that a collimated jet component drives the flow, and momentum is transferred to the molecular gas through shocks, as in the model described by Masson & Chernin (1993).

The next step is the determine the excitation conditions in the highest velocity gas, to see whether the temperatures have been elevated by the passage of a shock through the molecular material.

References

Cabrit, S. & Bertout, C.: 1986, Astrophys. J. **307**, 313.
Cabrit, S. & Bertout, C.: 1990, Astrophys. J. **348**, 530.
Chandler, C. J. & Sargent, A. I.: 1993, Astrophys. J. **414**, L29.
Masson, C. R. & Chernin, L. M.: 1992, Astrophys. J. **387**, L47.
Masson, C. R. & Chernin, L. M.: 1993, Astrophys. J. **414**, 230.
Shu, F. H., Ruden, S. P., Lada, C. J. & Lizano, S.: 1991, Astrophys. J. **370**, L31.
Stahler, S. W.: 1994, Astrophys. J. **422**, 616.
Terebey, S., Vogel, S. N. & Myers, P. C.: 1989, Astrophys. J. **340**, 472.

PHYSICAL PROPERTIES OF THE OUTFLOW SOURCES IN TAURUS

N. HIRANO[1], T. HASEGAWA[2], M. HAYASHI[3], M. TAMURA[3] and
N. OHASHI[4]*

[1] *Laboratory of Astronomy and Geophysics, Hitotsubashi University*

[2] *Institute of Astronomy, University of Tokyo*

[3] *National Astronomical Observatory*

and

[4] *Nobeyama Radio Observatory*

Abstract. We have searched for CO outflows in eight embedded IRAS sources located in the Taurus molecular cloud using the 45m telescope of Nobeyama Radio Observatory. We have detected CO wing emission in four of these sources. CO outflow associated with TMC1A (04365+2535) is strong and spatially compact (radius $\leq$ 0.04 pc). The dynamical timescale of $\sim$ 2.5 x 10^3 yr suggests this outflow is the youngest one in Taurus.

We have combined our data with previously published survey data and have analyzed the physical properties of the outflow sources. We found that 12 out of 16 embedded sources ($\sim$ 75 %) have CO outflows associated with them; this indicates that almost all stars experience a phase of molecular outflow in their embedded stage. The IRAS color of the outflow sources suggests that the outflows appear in considerably early phase of the evolution of YSOs, that is, as early as YSOs became observable with IRAS and that visible outflow sources are in a transient phase of evolution between embedded sources and visible T Tauri stars without outflow. Visible outflow sources are systematically more luminous than visible no-outflow sources, while embedded outflow sources have comparable luminosities with visible no-outflow sources. Such luminosity function suggests that the YSOs with outflow undergo mass accretion and increase their stellar mass as they progress from embedded sources to visible outflow sources. Typical mass accretion rate derived from the bolometric luminosity is $\sim$ 2 x$10^{-6} M_\odot yr^{-1}$. The timescale for mass accretion to acquire typical stellar mass, 0.5 - 0.8 $M_\odot$, is $\sim$ 2.5 - 4 x 10^5 yr.

1. Introduction

CO outflows are commonly found toward the young stellar objects (YSOs) in their deeply embedded phase. In one current model of star formation, the YSOs with CO outflows are the protostars in which mass inflow and outflow are taking place simultaneously (*e.g.* Adams, Lada & Shu, 1987). However, it is not clear whether all stars experience a phase of outflow, at what stage in a protostellar evolution the outflow begins and what the role of the outflow is in a star-formation process. In this paper we will discuss the physical properties of the outflow driving YSOs in the Taurus molecular cloud on the basis of our CO outflow search with high-angular resolution.

We have selected 59 IRAS sources as a sample of YSOs. The selected sources satisfy the color criteria $\log[F_\nu(12\mu m)/F_\nu(25\mu m)] < 0.0$ and

Present address: Harvard-Smithonian Center for Astrophysics

Astrophysics and Space Science **224**: 113–116, 1995.
© 1995 *Kluwer Academic Publishers.*

$\log[F_\nu(25\mu m)/F_\nu(60\mu m)] < 0.3$. Our sample consists of 43 optically visible sources (visible sources) and 16 optically invisible sources (embedded sources). Among 16 embedded sources, only five were known to possess CO outflows and have been mapped (Heyer *et al.* 1987, Myers *et al.* 1988, Terebey *et al.* 1990, Ohashi 1991). Thus, we have selected eight embedded sources which were not known as outflow sources (see Table 1) and have searched for evidence of CO outflow using the 45 m telescope of the Nobeyama Radio Observatory. Our beamsize of 16″ corresponds to 0.01 pc at the distance of Taurus molecular cloud (140 pc).

2. Results and Discussion

2.1 CO WING EMISSION

The CO line parameters are summarized in Table 1. Four of the observed sources, 04169+2702, 04239+2436, TMC1A and L1527, have CO line wings whose total width above the noise level exceeds 10 km s^{-1}. Especially, the blueshifted emission in TMC1A is remarkably strong ($T_R^*(wing) \sim 2$ K) and broad ($\Delta V_{blue} \sim 17$ km s^{-1}). The spatial extent of this blueshifted emission is less than 1′ (0.04 pc). The dynamical timescale of the outflow is estimated to be ~ 2.5 x 10^3 yr; this means that the CO outflow in TMC1A is the youngest one in Taurus molecular cloud.

TABLE I

CO line parameters.

IRAS Name	Other ID	L_{bol} ($L_\odot$)	V_{LSR} (km s^{-1})	ΔV_{total} (km s^{-1})	Outflow
04108+2803	L1495N	0.74	6.24	9.1	No
04154+2823		0.33	7.25	8.6	No
04169+2702		1.3	7.75	11.3	Yes
04239+2436		1.4	6.75	11.2	Yes
04248+2612	B217	0.53	7.25	7.0	No
04295+2251	L1536	0.64	5.74	4.9	No
04365+2535	TMC1A	2.4	6.40	22.6	Yes
04368+2557	L1527	2.1	6.24	10.8	Yes

2.2 NUMBER OF THE OUTFLOW SOURCES

We have combined our CO data with previously published outflow surveys (Heyer *et al.* 1987, Myers *et al.* 1988, Terebey *et al.* 1989, 1990, Ohashi 1991, Moriarty-Schieven *et al.* 1992) to characterize the physical properties of the outflow sources. The combined data show that 12 out of 16 embedded sources are outflow or promising outflow sources; five are previously known outflow sources, four detected by this work and the other three found by recent CO (J=3-2) observations (Moriarty-Schieven *et al.* 1992). This means that 75% of the embedded sources are associated with CO outflows and that almost all stars experience a phase of molecular outflow in their embedded stage. On the other hand, total number of the visible outflow sources is eight, which is comparable to that of the embedded outflow sources. This implies that the duration timescale of visible outflow phase is comparable to that of embedded outflow phase.

2.3 INFRARED PROPERTIES OF THE OUTFLOW SOURCES

In Figure 1-a we present a two-color diagram of our sample sources. This diagram shows that four embedded sources without outflows have rather high dust-color temperature. This suggests that the embedded sources without outflows are not the youngest ones of pre-outflow phases. Since these four sources have quite low luminosities, less than 1 $L_\odot$ (see Table 1 or Fig. 1-b), we interpret that these sources do not have enough power to drive detectable molecular outflow because of their low luminosities. In other words, outflows are associated with the coldest, youngest embedded sources. This implies that the outflows appear in considerably early phase of the evolution of YSOs, that is, as early as they became observable with IRAS. In this two-color diagram visible outflow sources are located at the boundary between embedded and visible sources. This indicates that visible outflow sources are in a transient phase of protostellar evolution between embedded sources and T Tauri stars without molecular outflows.

2.4 LUMINOSITY FUNCTION OF THE OUTFLOW SOURCES

In Fig. 1-b we plot the bolometric luminosities of our sample sources as a function of the infrared color. It is prominent that the visible outflow sources have large luminosities, 6.9 ± 1.7 $L_\odot$ (mean ± standard error of mean), which is ∼ 2.5 times larger than 2.9 ± 0.9 $L_\odot$ for visible no-outflow sources. According to the theoretical models of protostars (*e.g.* Stahler, Shu, & Taam 1980) the greater luminosity of visible outflow sources is interpreted that these sources are still accreting mass. However, the luminosity of embedded outflow sources, 4.1 ± 2.3 $L_\odot$ is not large compared to that for visible no-outflow sources, although the IRAS color as well as optical appearance indicates that these sources are younger than visible outflow sources.

One plausible explanation for this discrepancy is the luminosity evolution due to the growth of the central star. The total luminosity released by mass accretion is related to the mass of the protostar itself M_*, mass accretion rate $\dot{M}$ and a radius of the protostar R_* (Stahler, Shu & Taam 1980),

$$L_{acc} = GM_*\dot{M}/R_* \tag{1}$$

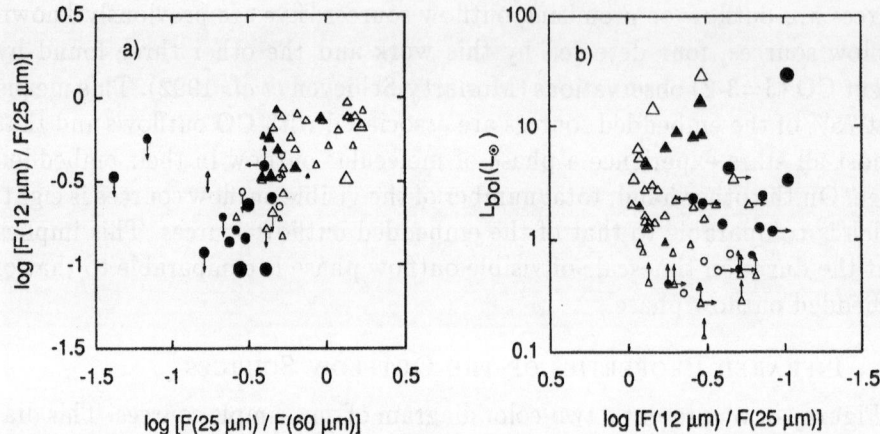

Fig. 1. a) A two-color diagram for 59 IRAS sources selected in Taurus. b) Bolometric luminosity plotted against the 12/25 μm color diagram for 59 IRAS sources. *Filled circles:* embedded outflow sources. *Open circles:* embedded no-outflow sources. *Filled triangles:* visible outflow sources. *Open triangles:* visible no-outflow sources.

where G is the gravitational constant. If $\dot{M}$ and R_* do not change largely during the stage of mass accretion, L_{acc} is simply proportional to the mass of the protostar, M_*. As a result, a protostar will increase its luminosity as it increases its mass through accretion. The large luminosities found in visible outflow sources suggest that these sources have already acquired considerable mass and still continue mass accretion. Unlike visible outflow sources, embedded sources in earlier evolutionary stage have not yet acquired enough mass to create large accretion luminosities. If visible outflow sources have already acquired the typical mass of T Tauri stars in Taurus, ~ 0.5 - 0.8 $M_\odot$, then the mass accretion rate estimated from equation (1) is $\dot{M} \sim 1$ - $2 \times 10^{-6} M_\odot yr^{-1}$ for the average luminosity of ~ 7 $L_\odot$ and $R_* \sim 4R_\odot$. In order to acquire the typical stellar mass of ~ 0.5 - 0.8 $M_\odot$ with this constant mass accretion rate, the mass accretion should last for ~ 2.5 - 4×10^5 yr.

References

Adams, F.C., Lada, C.J. & Shu, F.H.: 1987, Astrophys. J. **312**, 788.
Heyer, M.H., Snell, R.L., Goldsmith, P.F., & Myers, P.C.: 1987, Astrophys. J. **321**, 370.
Moriarty-Schieven, G.H., Wannier, P.G., Tamura, M., & Keene, J.: 1992, Astrophys. J. **400**, 260.
Myers, P.C., Heyer, M., Snell, R.L. & Goldsmith, P.F.: 1988, Astrophys. J. **324**, 907.
Ohashi, N.: 1991, Ph.D. thesis, Nagoya University.
Stahler, S.W., Shu, F.H., & Taam, R.E.: 1980, Astrophys. J. **241**, 637.
Terebey, S., Vogel, S.N., & Myers, P.C.: 1989, Astrophys. J. **340**, 472.
Terebey, S., Beichman, C.A., Gautier, T.N., & Hester, J.J.: 1990, Astrophys. J. **362**, L63.

SUBMILLIMETER CO SPECTROSCOPY OF
LOW MASS YOUNG STELLAR OBJECTS

K.-F. SCHUSTER, A.P.G. RUSSELL** and A.I. HARRIS***

Max-Planck-Institut für extraterrestrische Physik
Postfach 1603, 85740 Garching, Germany

Abstract. We report submillimeter CO(6-5) observations around 15 nearby young stellar objects of low mass. The correlation between linewidth and peak temperature indicates shock heating of dense gas, presumably at the origin of molecular outflows.

Key words: Submm. CO spectroscopy, YSO, shocks

1. Introduction

Submillimeter CO(6-5) observations trace interstellar and circumstellar material at higher temperature and density than lower level CO lines. Typical temperatures of the detected gas range from 40 to a few 100 Kelvin with densities from 10^4 to $10^7 cm^{-3}$. Recent development of heterodyne receivers using SIS-technology and the good surface accuracy of the JCMT allow us to observe faint sources in low mass star forming regions at the CO(6-5) frequency of 691 GHz. We observed a number of nearby young stellar objects including typical embedded sources as well as classical T-Tauri stars. The observations were done during a series of runs at the JCMT with the MPE Schottky- and lately with the MPE SIS-heterodyne receiver (Harris *et al.* (1987), Schuster *et al.* (1993b) and Harris *et al.* (1994)). Even with the increased sensitivity of the new system observing of faint sources at this frequency remains restricted to very good weather conditions. We summarize here the observational results of this survey and give some arguments for the mechanism which leads to the observed emission.

2. Results

The sample consists of 15 well known low mass young stellar objects in nearby cloud complexes (mainly Taurus and ρ-Oph.). We detected CO(6-5) emission in 10 sources. Table I gives an overview of the measured line parameters. These parameters were obtained by single Gaussian fits and in case of self-absorption a single absorption component was included.

In some sources the emission is surprisingly strong and the linewidths are much broader than anything expected from quiescent clouds or gravitation-

---

* Now IRAM Grenoble, France
** Now ROE Edinburgh, Scotland
*** Now FCRAO UMass. Amherst, USA

Astrophysics and Space Science **224**: 117–120, 1995.
© 1995 *Kluwer Academic Publishers.*

ally bound material. In particular, the emission is strong in all cases where a large circumstellar column density coexists with mass-loss activities (see also Schuster *et al.* (1993a)). Radiative heating by the central stars cannot explain the measured line temperatures. It is more likely that the CO(6-5) emission arises in dense shock heated gas generated by mass-loss activities (Fig. 1).

TABLE I

CO(6-5) Emission of 15 low mass Young Stellar Objects (T_R^*).

Source	Region	Peak Int. (K)	FWHM (km/s)	VLSR (km/s)
IRAS-16293	ρ-Oph	17.1	12.9	5.0
VLA-1623	"	22.7	6.8	3.2
WL16	"	<3	-	-
Elias29	"	4.5	4	1.0
WL21	"	<3	-	-
L43	Sco OB	4.8	2.4	1.0
L1551	Taurus	6.1	2.9	6.0
T-Tau	"	24	5.1	7.6
HL-Tau	"	8	5.3	7.5
DG-Tau	"	10	4.0	6.7
GG-Tau	"	<5	-	-
RY-Tau	"	<5	-	-
N1333	Perseus	<3	-	-
B335	-	5.4	5.2	8.7
L1262	-	7.7	6.5	3.8

3.   The case of IRAS-16293

A typical example of an active embedded young stellar object with low mass is IRAS-16293 which is also known to be the origin of a powerful outflow. Fig. 2 shows the CO(6-5) spectrum measured at the center position. The line is very broad and indicates a dense warm outflowing component near the source center. The line is heavily self absorbed at the line center. This absorption is due to an extended foreground layer of quiescent cold material (Menten *et al.* 1987). Another component which is probably due to gravitationally bound material around the object can be traced with isotopic CO lines (see Fig. 2). The origin of the CO(6-5) emission becomes more obvious when the emission is mapped. Fig. 3 shows the CO(6-5) emission which is blueshifted and redshifted in respect to the self-absorption feature.

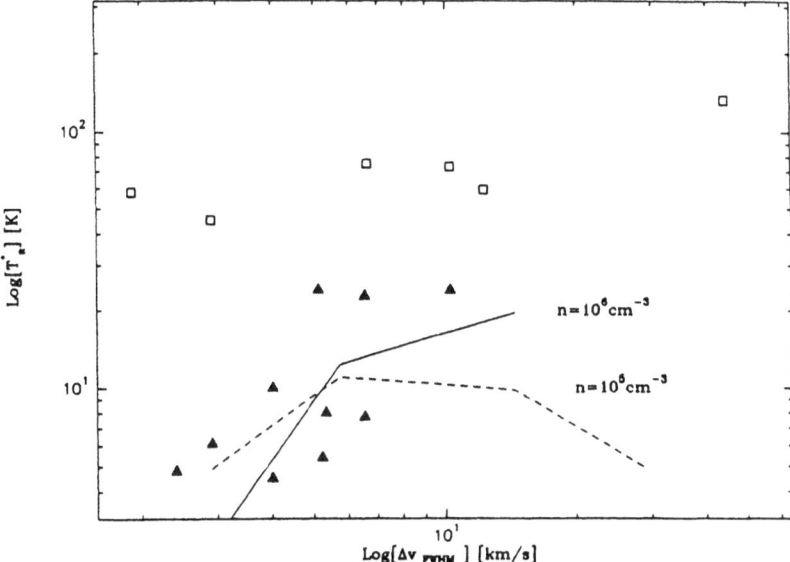

Fig. 1. Measured CO(6-5) peak temperature against linewidth for the low mass sources from Table I (filled triangles). A clear correlation is visible which indicates the dynamical origin of the gas heating. The dashed and solid lines are derived from the shock models of Draine & Roberge (1984) for different preshock densities. The open squares are CO(6-5) line temperatures from typical high mass star forming regions (Graf 1991) with little variation of temperature against linewidth (mainly PDR's).

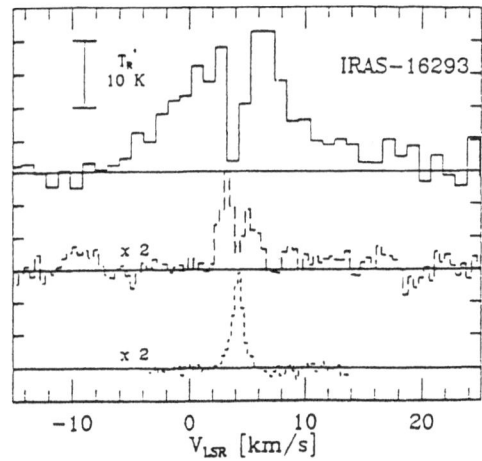

Fig. 2.   CO(6-5) spectrum of IRAS-16293 (top). ^{13}CO(3-2) spectrum (middle) and C^{18}O(1-0) (Menten *et al.* 1987) (bottom).

The emission is distributed around the central object with a spatial extension of about 10 to 20 arcsec.

There is a clear bipolarity which coincides in direction with the structure of a large scale molecular outflow consisting of colder material (Mizuno *et*

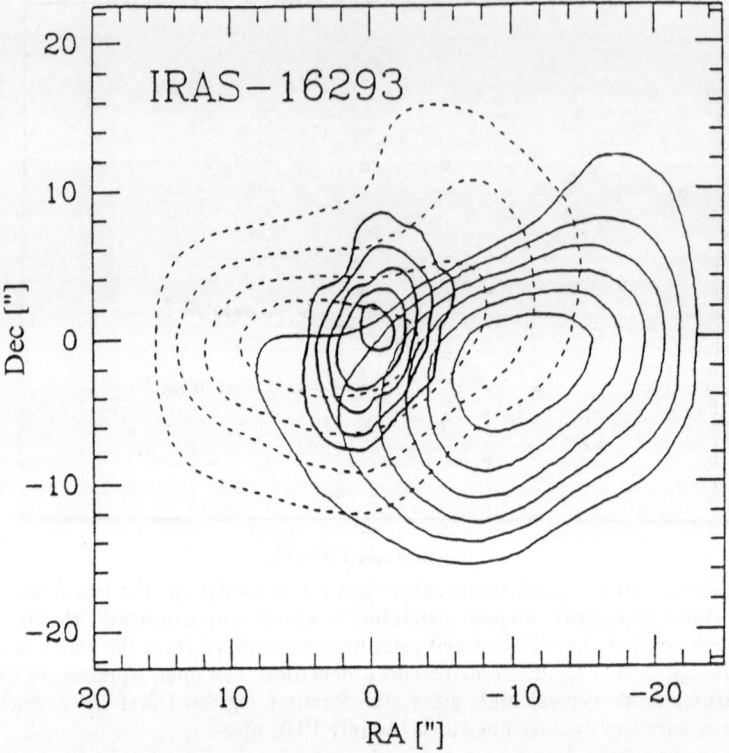

Fig. 3. CO(6-5) map around IRAS-16293. Dashed contours blueshifted emission (-11 to 3 km/s), fine solid contours redshifted emission (5 to 24 km/s). The contours are spaced by 20 K km/s starting from 20 K km/s for the blueshifted and from 80 K km/s for the redshifted emission.The strong solid contours indicate the $C^{18}O$ disk-like structure found by Mundy *et al.* (1990).

al. 1990). It is therefore evident that the gas traced by the CO(6-5) emission reflects the active central part of the large scale molecular outflow.

References

Draine, B.T. & Roberge, W.G.: 1984, Astrophys. J. **282**, 491.

Graf, U.U.: 1991, PhD Thesis, Ludwigs-Max. Universität München.

Harris, A.I., Schuster, K.-F., Genzel, R., Plathner, B. & Gundlach, K.-H.: 1994, Int. J. Infrared & MM Waves **15**, No. 9.

Harris, A.I., Jaffe, D.T., Stutzki, J. & Genzel, R.: 1987, Int. J. Infrared & MM Waves **8**, No. 8.

Menten, K.M., Serabyn, E., Güsten, R. & Wilson, T.L.: 1987, Astron. & Astrophys. **177**, L57.

Mizuno, A., Kufui, Y., Iwata, T., Nozawa, S. & Tokono, T.: 1990, Astrophys. J. **356**, 184.

Mundy, L.G., Wootten, H.A. & Wilking, B.A.: 1990, Astrophys. J. **352**, 159.

Schuster, K.-F., Harris, A.I., Anderson, N. & Russell, A.P.G.: 1993a, Astrophys. J. **412**, L67.

Schuster, K.-F., Harris, A.I. & Gundlach, K.-H.: 1993b, Int. J. Infrared & MM Waves **15**, No. 14.

NEAR-INFRARED IMAGING OF H₂
IN MOLECULAR OUTFLOWS FROM YOUNG STARS

C.J. DAVIS and R. MUNDT

Max-Planck-Instit für Astronomie, Königstuhl,
D-69117 Heidelberg, Germany.

and

J. EISLÖFFEL

Laboratoire d'Astrophysique, Observatoire de Grenoble,
Université Joseph Fourier, B.P. 53X, F-38041 Grenoble Cedex, France.

Abstract. In an attempt to identify the molecular shocks associated with the entrainment of ambient gas by collimated stellar winds from young stars, we have imaged a number of known molecular outflows in H₂ v=1-0 S(1) and wide-band K. In each flow, the observed H₂ features are closely associated with peaks in the CO outflow maps. We therefore suggest that the H₂ results from shocks associated with the acceleration or entrainment of ambient, molecular gas. This molecular material may be accelerated either in a bow shock at the head of the flow, or along the length of the flow through a turbulent mixing layer.

Key words: Shock waves – ISM: jets and outflows – ISM: molecules – Infrared

1. Introduction

The 2.121 μm v=1-0 S(1) ro-vibrational line of H₂ is a much-used tracer of low-velocity shocks in outflows from young stars. H₂ emission often delineates the oblique shocks in bow shock wings. H₂ shocks may also be associated with turbulent mixing layers. Near-IR H₂ emission may thus be an ideal tracer of the shocks that accelerate ambient gas and so produce molecular (CO) outflows. Below we show images in H₂ v=1-0 S(1) of four outflow systems where evidence for a relationship between H₂ emission features and peaks in the molecular (CO) flows is seen.

2. VLA 1623

The VLA 1623 outflow was mapped in CO J=2-1 by André *et al.* (1990). The flow is orientated NW–SE, it lies close to the plane of the sky and is extremely well collimated. André *et al.* infer a dynamical age of only 3000 years for the flow. In Fig. 1 we present an H₂ image of the blue-shifted molecular flow. The VLA 1623 exciting source is marked. Two very bright, compact knots of H₂ emission are observed towards the end of the blue CO lobe, while faint H₂ features are seen along the flow axis. Knot A in particular appears bow-shaped in morphology, and notably lies only 20″ ahead of (downstream of) the strongest peak in the CO flow (labelled BLUE 3 by

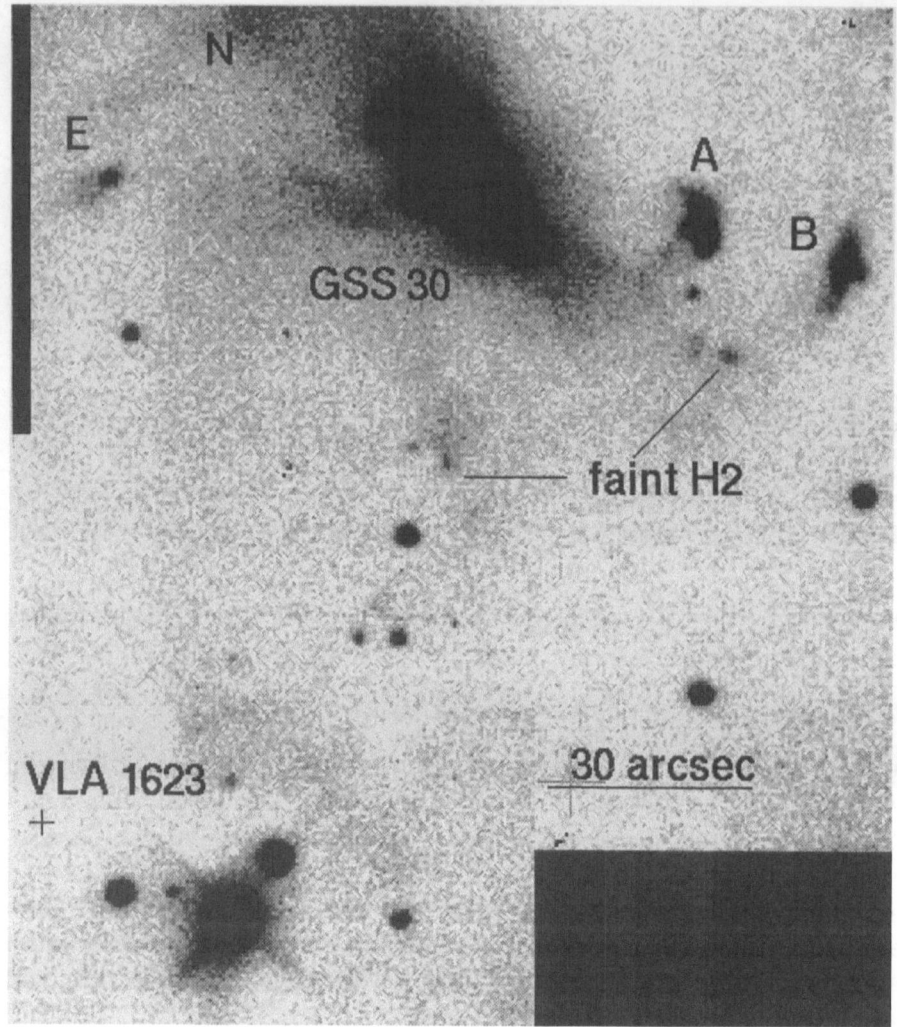

Fig. 1. Narrow-band, H_2 + continuum image of the north-western, blue CO outflow lobe. The flow is powered by the embedded, low-mass radio source VLA 1623.

André *et al.* 1990). It seems that the CO peak, BLUE 3, and knot A are associated. Indeed, BLUE 3 is situated in the wake of the H_2 "bow shock".

2.1 NGC 6334I

The molecular outflow associated with the infrared peak I in the NGC 6334 complex of molecular clouds has been mapped by Bachiller & Cernicharo (1990). NGC 6334I is resolved into three IR sources: IRS 1, labelled in Fig. 2, contains OB stars which power a cometary-shaped HII region known as NGC 6334F. IRS 1 also lies at the centre of the NE–SW outflow. Like VLA 1623, the NGC 6334I flow is probably only $\sim 10^3$ years old.

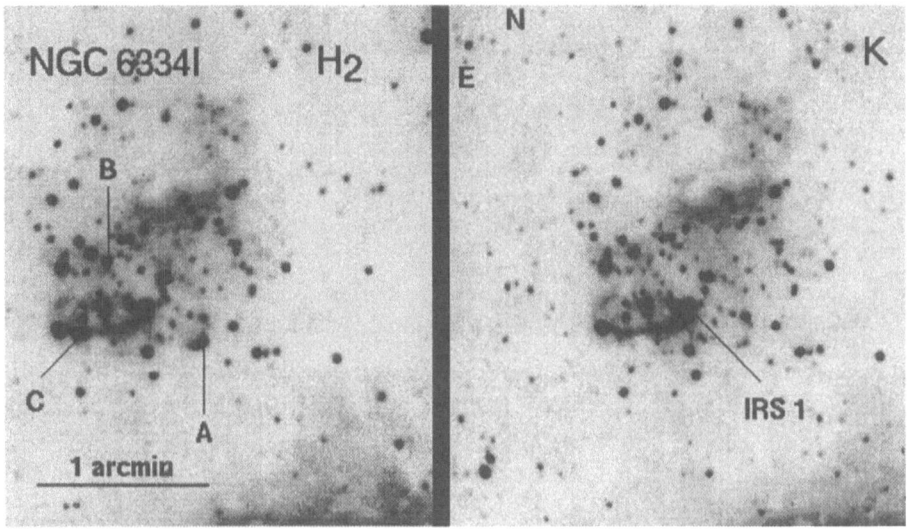

Fig. 2.   Narrow-band H₂ + continuum and K-band images of the CO outflow in
NGC 6334I. IRS1, a cometary HII region, lies at the center of the outflow.

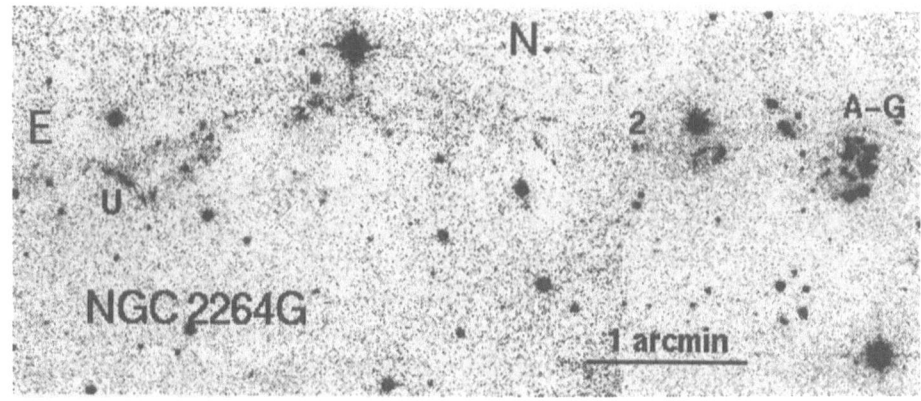

Fig. 3.   Narrow-band H₂ + continuum image of the NGC 2264G molecular outflow in
Monoceros OB1. VLA2 (marked) powers the east-west outflow.

Our IR images of the NGC 6334I region (Fig. 2) reveal a complex cluster
of IR sources and filamentary structures. Many of these structures are continuum features, though three distinct peaks appear in the H₂ image that
are not seen in the scaled K image (only continuum features appear equally
as bright in both). Two of these bow-shaped H₂ line-emission features, A
and B, lie along the outflow axis; the leading heads of both A and B face
away from IRS 1, as one would expect if they were bow shocks associated

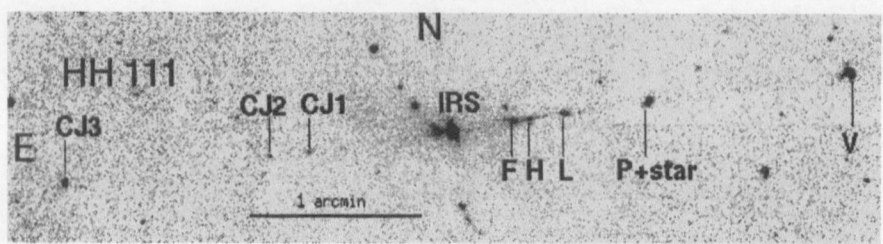

Fig. 4.   Narrow-band H_2 + continuum image of the Herbig-Haro jet HH 111. Note the three newly discovered knots in the counter-jet east of the source, CJ1, CJ2 and CJ3.

with the outflow. Moreover, knot B coincides exactly with the peak in the blue CO lobe, while knot A lies only 10″ SW of the red CO peak.

2.2   NGC 2264G

The high-velocity molecular outflow NGC 2264G has been mapped in CO J=2-1 by Margulis *et al.* (1990). The outflow extends E–W, is powered by the radio source VLA 2, and lies almost in the plane of the sky. In our H_2 image (Fig. 3) a compact cluster of line-emission knots is observed ~80″ west of VLA 2. Again, these knots, labelled A–G, coincide precisely with the peak in the blue CO lobe. East of the source, a faint, filamentary structure is also seen in H_2 aligned with the northern edge of the red CO flow lobe.

2.3   HH 111

Fig. 4 shows a H_2 image of the entire extent of the HH 111 bipolar jet (Davis *et al.* 1994). The VLA source of the flow is situated a few arcseconds east of the IRS peak. The brightest knots in the optical HH 111 jet, knots F, H, L, P and V, are all observed in H_2. Three new knots are also discovered in the counter-jet direction; CJ1 – CJ3. The H_2 knots along the jet may result from internal working surface shocks produced by a varying jet velocity. Entrainment in the wings of these bow shocks could also be the cause of the observed CO flow associated with the jet (Reipurth & Olberg 1991).

References

André, Ph., Martín-Pintado, J., Despois, D. & Montmerle, T.: 1990, Astron. & Astrophys. **236**, 180.
Bachiller, R. & Cernicharo, J.: 1990, Astron. & Astrophys. **231**, 174.
Davis, C.J., Mundt, R. & Eislöffel, J.: 1994, Astrophys. J. in press.
Margulis, M. *et al.* : 1990, Astrophys. J. **352**, 615.
Reipurth, B. & Olberg, M.: 1991, Astron. & Astrophys. **246**, 535.

MOLECULAR SHOCKS IN STAR FORMING REGIONS

P. W. J. L. BRAND
Institute for Astronomy, University of Edinburgh, Royal Observatory Edinburgh EH9 3HJ

Abstract. The role of excited molecular hydrogen as a powerful observational tool is examined in the context of shock phenomena in molecular clouds, particularly in star forming regions. Conclusions that may be drawn from line intensities and line profiles, and the properties of J and C shocks in a bow shock structure are discussed.

Key words: Molecular hydrogen – shocks – outflows

1. Introduction

The realization that forming stars produce powerful supersonic molecular wind, often collimated into bipolar jets, came from mapping CO line profiles (*cf.* Lada 1985).

Soon afterwards the discovery of shocked molecular hydrogen in Orion (Gautier *et al.* 1976) led a series of papers describing shock models. The observation (Nadeau & Geballe 1979) of wide line profiles confirmed the dynamic origin of the emission, but raised problems for the models which have not yet been satisfactorily overcome. This investigation of shocks using molecular hydrogen lines which is still in progress, is the subject of this article. The remainder of this paper will develop current understanding of shocks by considering evidence from Orion (section 2), Herbig-Haro 7 (section 3) and other sources (section 4). Section 5 lists briefly the current situation and expected progress.

The theory of interstellar shocks has been recently reviewed by Draine & McKee (1993) and that review contains important insights into the physics of molecular shocks, and references thereto.

2. Orion

2.1 PREAMBLE

The shocked outflow region in the Orion molecular cloud just behind the Trapezium-lit HII region was mapped by Beckwith *et al.* in 1978.

The first line ratio measurements (Beckwith *et al.* 1979) showed that the line emission was thermalized, and hydrodynamic shock models (Hollenbach & Shull 1977, Kwan 1977, London *et al.* 1977) were able to explain the H_2 emission. It was quickly realized (Kwan *ibid.*) that since H_2 was completely thermally dissociated by a non-magnetic shock travelling faster than 22 km s^{-1} through the gas, the wide velocity profiles observed by Nadeau & Geballe (1979) were a problem for the models. The introduction by Draine (1980, also Draine, Roberge & Dalgarno 1983) of 'C shocks', which are able to travel at >40 km s^{-1} through H$_2$ without completely dissociating it, led to several models that were consistent with the contemporary measurements of H$_2$ line intensities, and also gave a better prediction of the amount of hot CO (Draine & Roberge 1982).

By 1988 (Brand *et al.*) the H$_2$ line ratio measurements for Orion had been extended sufficiently to rule out the plane parallel C shock models. Relatively too much emission was observed from the lines of highest excitation to match the models. A brief discussion of model properties, needed to demonstrate the differences, follows.

2.2   J AND C MODELS

Draine (1980) used the term J shock (jump shock) to denote the hydrodynamic jump followed by a cooling zone which constitutes a shock in a non-magnetic medium. In contrast, the C shock (continuous shock) arises in a magnetised medium consisting predominantly of neutral material mixed with a small population of ions. These ions are taken with the magnetic field to be a magnetohydrodynamic fluid which supports Alfvén and magnetosonic waves. The neutral fluid interacts with the magnetised fluid through ion-neutral collisions, which produce dynamic friction and heating of the neutral fluid. A plane pressure wave travelling through the magnetised fluid perpendicular to the field lines drives a submagnetosonic compression ahead of it into the undisturbed magnetised fluid, and the resulting flow of ions through the neutrals heats them and accelerates them in the direction of propagation. The disturbance is submagnetosonic with respect to the ion fluid because of the low ion density, but is supersonic as far as the neutral fluid is concerned. If the coupling between ions and neutrals is sufficient, the neutral fluid will have been heated and accelerated sufficiently by ion/neutral drag to achieve its 'shocked' condition (required by conservation of momentum and mass) without passing through a hydrodynamic 'jump'. This structure is called a C shock, extensively investigated by Draine *et al.* (1983).

The properties of these two shock types are contrasted in Fig. 1. If the post-shock density is high enough, then the gas will be locally in thermodynamic equilibrium. This appears to be the case in most of the shocks which have been observed.

The temperature profile of the neutral gas is of particular interest. In the J shock it rises virtually immediately to its maximum value, and then drops as the gas cools, under nearly isobaric conditions. The rate of cooling is determined by the cooling function of the neutral gas.

Then the column density of each level of excited H$_2$ in the cooling zone will depend on the run of temperature with column density in the shock model and the ratio of these column densities will depend only on the cooling

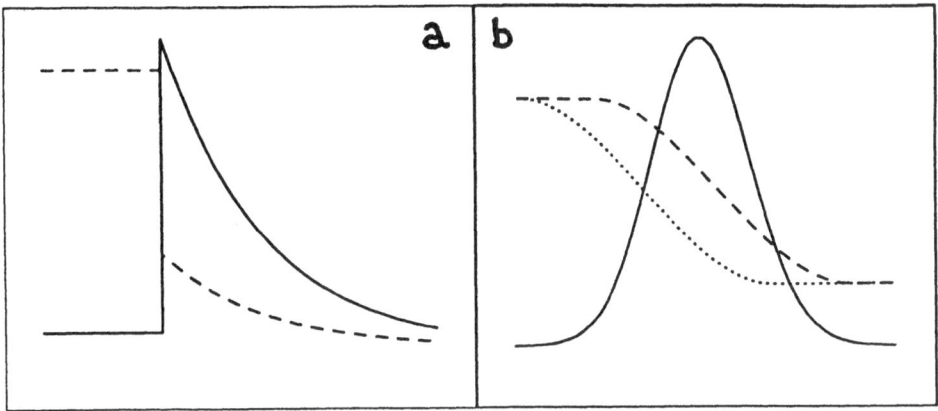

Fig. 1. T profile of *(a)* a J shock and *(b)* a C shock. The dashed line denotes velocity of the neutral gas, and the dotted line denotes ion velocity.

function. Cooling at the highest temperatures in these shocks in molecular clouds is due to H_2 dissociation, and causes complete dissociation in shocks travelling faster than 22 km s^{-1} through dense gas (Kwan 1977, Brand 1992). At lower temperatures ($< 3500K$) H_2 line cooling and then cooling by molecules dominates.

In contrast, the run of temperature of the neutrals in a C shock is a smooth function, rising to a maximum close to the region where drag is greatest, and then subsiding in a cooling zone. The temperature profile can be roughly approximated by a 'top-hat' distribution, whose peak value is determined by the frictional heating rate (which depends, *inter alia*, on the velocity). Thus the H_2 emission in this case will look more like that from a slab of gas at a uniform temperature set by the shock parameters, but not depending on the cooling function.

Fig. 2 shows the run of column density in H_2 versus energy of the level predicted by a naive J shock. The concave curve is characteristic, mapping the cooling rate as a function of T in such a shock. A constant T distribution would be represented by a straight line in such a plot (for example the 2000K line indicated in the figure by a dashed line).

Relative to such a straight line, the J shock model predicts excess high and low excitation emission. This is emphasised in Fig. 3 which displays the results from Peak 1 in the Orion outflow. Here the plot has been skewed by dividing column density values by those from a Boltzmann distribution at 2000K to accentuate those departures from a Boltzmann distribution. The data are well-fit by a naive J shock model. The dashed lines represent the predictions from plane C shock models (Draine & Roberge 1982, Chernoff *et al.* 1982), and these, having the expected flat distribution, are ruled out

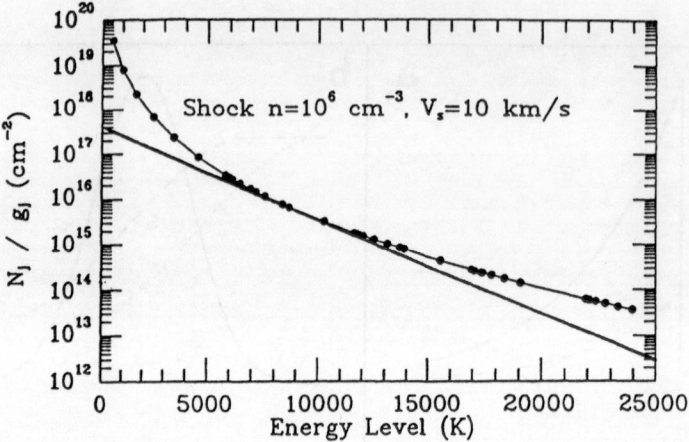

Fig. 2. Post-shock cooling: logarithm of column density in each energy level of H_2 versus energy (after Burton (1992).

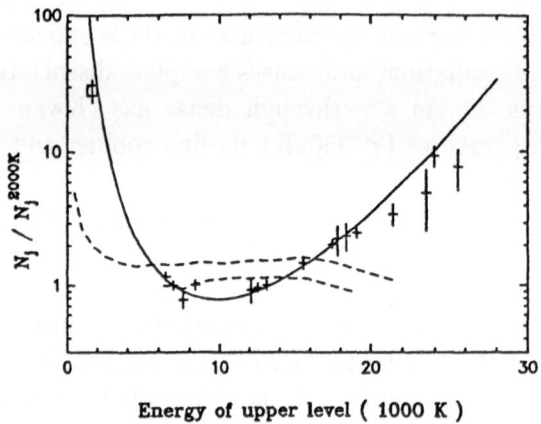

Fig. 3.   Logarithm of column density in each energy level versus energy, inferred from observations of Peak 1 in the Orion molecular outflow.

by the data. Alas, that is not the end of the story.

2.3 VELOCITY PROFILES AND BOW SHOCK STRUCTURES

The velocity profile of a bright H_2 line at the same position in Orion was measured to have a line width of 140 km s^{-1} FWZI (Brand *et al.* 1989a) which cannot be produced in a single plane J shock, with its constraints on velocity. In the first place, the shock has a maximum velocity of 22 km s^{-1} before complete dissociation. In the second place the velocities predicted by J shocks for H_2 lines are all close to the final gas velocity, producing a very small range of velocities. In the case of a C shock the limits are relaxed,

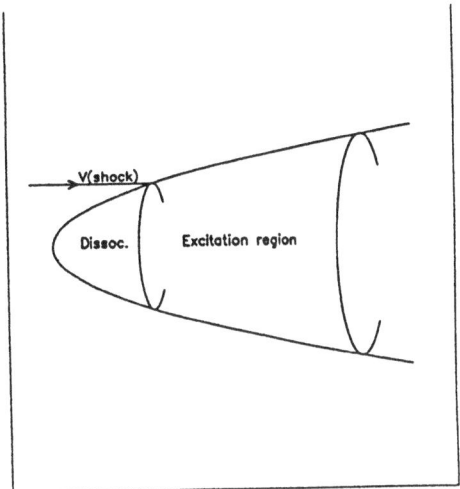

Fig. 4.   Bow shock structure used to explain the H_2 spectra.

but neither the increased maximum (to 40 km s^{-1}) nor the possible range of velocities will be sufficient to explain a profile such as that at Peak 1 in Orion.

The observation requires for its explanation a large range of shock speeds in the same small region of sky which is observed. The profile was obtained in a 5 arcsecond aperture (with the UK Infrared Telescope and its CGS II spectrometer) — a projected size of 3×10^{16} cm at Orion. The easiest way to meet this tight constraint is by modelling a bow shock in the molecular material (Fig. 4).

In this structure, gas flows through the curved surface, with a variety of normal velocities and therefore of shock strengths. Provided the structure is small enough, the observed emission will be the sum (or rather the integral) of all of these differently directed shocks. How does this change the plane shock predictions? This has been well-treated for atomic emission by Hartigan *et al.* (1987). They demonstrate that a range of velocities equal to the velocity of the structure through the gas is seen in line profiles taken from the whole shocked region and demonstrate how to infer this velocity and the angle with respect to the line of sight of the supersonic flow.

Fig. 4 shows the model used to investigate the line profile. The shape of the bow shock is prescribed (in fact it is determined by the behaviour of the shocked gas and its interaction with whatever is driving the shock, but this is considered still to be ill-determined). The shock structure may be considered to be driven by a jet or a bullet through stationary gas. This is not critical for the model — a simple velocity shift can be applied to allow for a 'wind hitting an obstruction' scenario. The shocked gas is given an impulse perpendicular to the shock surface, and radiates line emission into the corresponding part of the line of sight velocity profile. The velocity profile of a line is then the integral of each of the elements of shocked gas over the shock surface, each element with its own line of sight velocity.

Molecular gas near the head of the structure will meet the shock more or less head on, and (assuming a high velocity for the gas through the shock structure) the molecules will be dissociated by the shock. Only further down the flank, with a more oblique shock, will molecular gas survive and radiate.

The importance of this model is obvious. It provides the possibility of explaining the wide velocity profiles. It also re-opens the issue of line intensity ratios from J and C shocks. It will be recalled that plane J shocks naturally reproduce the curvature seen in the plot (Fig. 2 or Fig. 3) of log (column density) versus energy level for H_2 line emission, whereas plane C shocks tend to produce straight (Boltzmann-like) plots. However, a moment's consideration will show that the sum of two or more differently excited plane C shocks will produce a concave plot! But by hypothesis we *are* looking simultaneously at shocks of different strengths when we observe a bow structure. Thus, we cannot now choose between J and C shocks on the basis of this property.

2.4   EXCITATION CONDITIONS

There is one further result from the Orion shocked region which bears on the problem (Brand *et al.* 1989b). This is shown in Fig. 5, and is a plot of the ratio of two lines (0-0 S(13) and 1-0 O(7)) which have nearly the same wavelength (and so are not subject to differential extinction), but originate from different energy levels. Thus, their ratio determines the particular curve of column density versus energy level at the point of observation. The plot is of this ratio versus line intensity, from points all over the region originally mapped by Beckwith *et al.* (1978).

It is clear from Fig. 5 that the ratio does not vary significantly, and that therefore it is independent of position throughout the map. Apparently all of the shocks in this region have the same excitation state. What does this signify?

In the case of plane J shocks, it implies that the cooling curve is the same everywhere, but is insensitive to other shock properties. Only if all cooling processes have the same density dependence will this condition be met. This would be so if the important coolants have the same density dependence. But for instance H_2 dissociational cooling per unit volume depends on density squared, while H_2 line radiation cooling depends (at these densities) on density linearly. Thus, if as expected these are major contributors to the cooling function, different strength shocks will have different cooling curves, and it becomes hard to claim that plane J shocks can produce the result in Fig. 5. (However, a recent result by Smith (1994) which is discussed in section 4.4 may modify this conclusion).

A fortiori, C shock excitation is strongly dependent on velocity and so

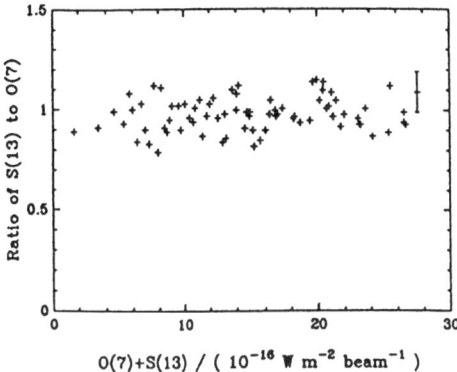

Fig. 5. The ratio of the H_2 lines 0-0 S(13) to 1-0 O(7) as a function of line intensity in the region of the Orion outflow mapped by Beckwith *et al.* (1978).

the constancy in Fig. 5 implies all shocks have the same speed at least — an even tighter constraint.

Now, as pointed out by Smith *et al.* (1991a) the bow shock structure provides a potential cure. The leading part of the bow structure consists of a completely dissociating shock, and only further down the structure does H_2 survive to radiate. The range of excitation observed is that arising from between this survival zone and the back of the structure where the shock is too weak to excite the observed H_2 lines. This region, which I will call the excitation region, exists where perpendicular shock velocities range from v_{max} to 10 km s^{-1} (the velocity necessary to strongly excite the 1-0 S(1) line). The velocity v_{max} for C shocks may be 50 km s^{-1} (or greater: *cf.* Smith *et al.* 1991b). For any given incident velocity of gas on shock structure — bow shock speed — typically up to several hundred km s^{-1}, the excitation region will encompass a range of incident perpendicular velocities. This range will remain the same for different bow shock speeds although the excitation region will move further aft as the bow shock speed is increased. The contribution to intensity and velocity profile will be determined by the bow shock shape in the excitation region.

Then different velocity structures will give the same excitation, provided that the shape of the bow structures are all the same and approximately paraboloidal, in a nearly scale-free manner. This seems an easier constraint but (Smith *et al.* 1991a) it requires that we observe the entire bow structure, and that certain other conditions are met.

2.5 High Magnetic Field Bow Shocks, Bullets, and Other Matters

The ability of bow shock structures to explain not just the *excitation* but also the *velocity profiles* of H_2 which are observed depends on the ability of such a structure to turn the emitting gas in a variety of directions.

It was shown (Brand *et al.* 1989a) that non-magnetic J shocks are unable to do so effectively enough. Even without the tight constraint on maximum velocity before complete dissociation (22 km s^{-1}), the profiles carry too little power in the wings to replicate the observations. The problem lies in the ratio of observed velocity width (and hence total structure velocity) to the velocity corresponding to maximum emission of H_2. If this is high, then that fraction of the gas in the flow that contributes significantly is turned through a small angle, and will therefore not contribute much to the profile outside velocities of one or two tens of km s^{-1}.

C shocks can turn the gas without destroying the H_2 because the Alfvénic Mach number is relatively small if the magnetic field is high. Smith *et al.* (1991b) showed that both the line ratios and velocity profiles of H_2 can be replicated in Orion (Peak 1) by a bow-shock structure in which the shock transition is by C shock, and the magnetic field is particularly — but perhaps not unrealistically — high.

There are several lingering doubts. In the first case, the C shock is oblique, and as shown by Wardle & Draine (1987) and Wardle (1991) such shocks can be unstable.

In the second case the constraints mentioned earlier still apply, namely the entire flow is required to contribute to the line intensities and profiles, and the shape and cooling function requirements are rather tight. Thus the situation is still far from clear. Indeed Smith (1994) has shown that a highly magnetized J shock is still a possible contender.

A remaining possibility is that proposed by Graham *et al.* (1991,1992) for H_2 emission seen in the Cygnus Loop, and by Hartigan *et al.* (1989) and by Carr (1993) for emission from Herbig-Haro 7. These authors suggest that the emission is due to a *magnetic precursor* (in effect, the leading edge of a C shock) followed by a standard J shock which completely dissociates the molecular component. This possibility has not been fully explored, and has several attractive features, but doubt as to its stability against rupture to a J shock and its ability to produce sufficient column density have been voiced (*e.g.* Draine & McKee 1993).

A recent and exciting development has been the detailed confirmation by Allen & Burton (1993) that the faint H_2 'fingers' of emission seen to the north-west of the Orion outflow (Taylor *et al.* 1984) are bow shocks driven probably by bullets originating at or near the source IRc2. The iron emission seen at the tips of these bow shock structures is consistent with a model in which the H_2 wakes are the tails of the shock structures.

Work is in progress (Chrysostomou *et al.*, Tedds *et al.*) to decide if a model based on an ensemble of such objects will explain the entire flow, or if these 'bullets' are the exception in a wind or blast that inflates a 'bubble' of gas inside the molecular cloud.

3. Herbig-Haro 7

We turn now to Herbig-Haro 7. This object is clearly the bow shock at the end of a jet apparently originating in the young stellar object SSV 13. Maps by Hartigan *et al.* (1989) show the H_2 emission surrounds the ionic emission associated with the jet, and the investigation by Carr (1993) shows the velocity structure in the H_2 profile map with the characteristics of a bow shock.

The bow structure is resolved (5 arc seconds) and can be used to test predictions of the bow shock models proposed to explain the Orion results.

If any dust in the structure is assumed to be optically thin at the H_2 lines (this is consistent with Carr's measurement *ibid.*), then an arbitrary orientation of the bow shock will imply that each line of sight intersects the structure in two points, with differing shock strengths. Thus as noted at the end of subsection 2.3, a curved distribution of log (column density) versus energy level will be seen.

The outcome of the measurements (Fernandes *et al.* 1995) exceeds expectations. The resulting column density plot (Fig. 6) shows the data for column densities, and a good fit of a model consisting of a bow C shock plus fluorescence, the latter assumed to be due to the high excitation of the bow at its head. If this is fully borne out, then the simultaneous conditions this imposes on the model will lead to a rather exact description of such bow shocks, and the implied UV field - presumably from the hot head of HH7 - will set a lower limit on the degree of ionization in the vicinity which will have implications for magnetic precursor models (and indeed for C shocks themselves).

There remains the possibility that what is being observed is not fluorescence but rather the re-formation of H_2 after dissociation on passage through the shock. A decision on this awaits a fuller data set, and more precise modelling. In any event, identification of the source of the non-thermalised lines remains a top priority.

4. Elsewhere

The data on column densities and line profiles in Orion and Herbig Haro 7 have reached an exciting juncture. But what of other evidence bearing on the J shock/C shock debate?

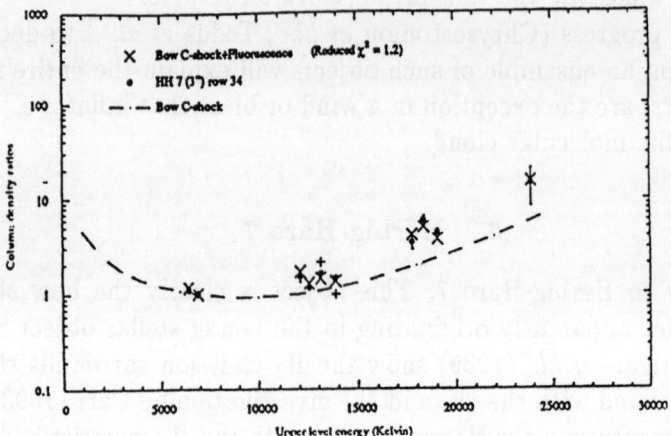

Fig. 6.   Normalized line H_2 column density ratio versus energy level in HH7 (from Fernandes 1995)

I discuss briefly the evidence from a massive outflow DR21, two supernova remnants IC443 and the Cygnus Loop, and two jet-like outflows from young stellar objects HH90A and L1551.

4.1   DR21

This is possibly the most massive outflow source. The hydrogen emission delineates a collimated bipolar flow from a heavily obscured HII region, the emission being highly clumped (Garden *et al.* 1990) in both lobes, the clumps having a range of velocities and velocity widths. In the east lobe the lines are relatively narrow (<20 km s^{-1}), while in the west lobe velocity widths of up to 100 km s^{-1} are seen (Garden *et al.* 1991). Measurements of line ratios in this region (Fernandes *et al.* 1995) have been made and show, over the limited area that has been investigated, that the emission is explained by C shock excitation, but with a small component of fluorescent emission, as in HH7.

4.2   IC443

This extended supernova remnant is associated with a molecular cloud in which the expanding blast wave can be seen (Burton *et al.* 1988) as a ring of shocked H_2. This shock is also seen in HI (Braun & Strom 1986) and CO (DeNoyer 1979, White *et al.* 1987).

Measurement of H_2 column densities (Moorhouse *et al.* 1991) show the characteristic shock signature, and when originally interpreted as a J shock, indicated that multiple shocks had to be involved. Burton *et al.* (1990) find that the [OI] 63μm emission from this region is consistent with C shocks provided special constraints apply to their velocities, or with J shocks provided that H_2O and OH are not present. Neither explanation is satisfying. Interferometric CO observations by Wang & Scoville (1992) of the region lead them to model the CO dynamics by a doubly shocked cloud as did Moorhouse *et al.* (*ibid.*) but with a one shock of C type and the other, the incident supernova blast wave, a J shock. The situation for the IC443 molecular shock is still unclear.

4.3 CYGNUS LOOP

A very surprising result by Graham *et al.* (1991) demonstrates that in the north eastern filamentary optical nebulosity there is also shocked H_2 emission. This appears on the outside of the $H\alpha$ filaments, and so cannot be the strongly cooled post-shock gas. The explanation given is that a magnetic precursor is being observed, and indeed it is hard to envisage any other simple explanation. But as already noted at the end of Section 2 there are concerns about the stability of such structures, especially in a region where ionization is not necessarily small (*cf.* also Elmegreen & Combes 1992).

4.4 HH91A

Gredel *et al.* (1992) measured many H_2 lines in this complex part of an outflow which excites HH90 and HH91. HH90A is a bright knot very close to an infrared source whence the flow possibly originates. The remarkable fact about the H_2 line intensities is that the derived plot of log (column density) versus energy level is straight, corresponding to thermal excitation at T=2750K. From what has been said, it is clear that this cannot arise from J shock excitation, and if it arises from a C shock, it must be a single plane shock. There is a rapid change of intensity with position in HH91A and the dominant component is very small. It is hard to envisage a geometry (possibly spherically symmetric) which would produce such a remarkable column density distribution.

It may be that since the highest energy level observed was at only 14500K, curvature is present and indeed the lowest levels deviate very slightly above the line. A C shocked bow viewed with special geometry might then be a possible explanation. However, new J shock models by Smith (1994) can fit these data. If this is confirmed, this most intense source will be a severe test for theory.

4.5 L1551

As a last example of data which may enable a clearer view of molecular cloud shocks, the CO interferometer map of L1551 (Barsony *et al.* 1994) shows several filaments which are asserted to be shocks, and which coincide with shock-excited $H\alpha$ and [NII] emission (Graham & Heyer 1990). In the brightest of these, a smooth gradient of velocity of 8 km s^{-1} over 10^{16} cm is observed and might be the signature of a C shock.

The range of velocities (8 km s^{-1}) is too small — unless projection effects are significant — to give strong H_2 emission, but this source may be an important testbed given its apparent planarity over 2 arcminutes.

5.   The ups and downs of shocked H_2 measurements

A review of where we are and where we are going in the attempt to understand shocks in star-forming regions turns out to be a mixed bag, and I list here progress points: past, present and future.

a) The jury is still out on J versus C shocks as the exciting mechanism for most H_2 emission. The weight of evidence is swinging towards C shocks, but (*e.g.* Smith 1994) J shocks will not go away. The best data sets (Orion, HH7 and HH91A) each appear to tell different stories.

b) It seems clear that bow shock structures are necessary to produce the nearly omni-present broad velocity profiles. But this explanation requires very high magnetic fields (Smith *et al.* 1991b), and observing apertures big enough to include the entire structure. For the nearest sources, such as Orion, this latter requirement seems to be a tall order, and the constancy of the ratio of column densities in Orion (Brand *et al.* 1989b) tightens further this constraint.

c) Perhaps the most critical question about the structure of these shocks is that of cooling functions. In particular, is water the dominant coolant below 1000K? This question has been hard to answer observationally to date. It is one of the more important goals of the ISO mission to find the answer.

d) The first steps are being taken to produce *detailed* models of *detailed* observations of shock structures. For the nearest sources maps with a resolution of 1 arc second or better, and velocity resolution of the order of 10 km s^{-1} should constrain the models so that the J or C pips squeak. The fortunate fact is that this capability is now becoming available. It is one of the pleasures of this research that as new questions arise, so do the means to seek the answers.

e) In similar vein, the cooling lengths in the shocks being discussed range from 10^{13} cm for J shocks up to 10^{17} cm for some C shocks. This translates to a few milli-arc seconds to several arc seconds in the nearest sources. Examination of the caustic edges of shock structure images with the enhanced resolution becoming available with adaptive optics should give a clear distinction between J and C shocks.

Thus, on the one hand it is frustrating how slow progress has been in the development of our understanding of shock physics in star forming regions. We still do not have a clear idea of how these things work.

On the other hand, current and future observational capabilities are exactly what are required to address the now much clearer questions about the shock mechanisms. I am moderately confident that in three or four years time this subject will have given shock theory the diagnostic power needed to measure the conditions in regions where stars are disrupting their birthplaces.

Acknowledgements

Thanks are due to colleagues who helped to form these ideas, particularly Michael Burton, Tom Geballe, Michael D. Smith, and also to the staff of the United Kingdom Infrared Telescope, a facility of the Particle Physics and Astronomy Research Council.

References

Allen, D. A. & Burton, M. G.: 1993, Nature 363,54.
Barsony, M., Scoville,N.Z. & Chandler, C.J.: 1994 Astrophys. J. 409, 275.
Beckwith S., Persson S.E., Neugebauer, G. & Becklin, E.E.: 1978, Astrophys. J. 223, 464
Beckwith S., Persson, S.E. & Neugebauer, G.: 1979, Astrophys. J. 227, 436.
Braun, R. & Strom, R.G.: 1986, Astron. & Astrophys. Suppl. 63, 345.
Brand, P.W.J.L., Toner, M.P., Geballe, T.R., Webster, A.S., Williams, P.M. & Burton, M.G.: 1989a, Mon. Not. of the RAS 237, 929.
Brand, P.W.J.L., Toner, M.P., Geballe, T.R. & Webster, A.S.: 1989b, Mon. Not. of the RAS 237, 1009.
Brand, P.W.J.L.: 1993, in: Dust and Chemistry in Astronomy, eds: T.J. Millar & D.A. Williams, Institute of Physics, Bristol.
Burton, M.G.: 1992, Aust. J. Phys. 45, 463.
Burton, M.G., Hollenbach, D.J., Haas, M.R. & Erikson, E.F.: 1990, Astrophys. J. 355, 197.
Burton, M.G., Hollenbach, D.J. & Tielens, A.G.G.M.: 1992, Astrophys. J. 399, 563.
Burton, M.G., Geballe, T.R., Brand, P.W.J.L. & Webster, A.S.: 1988, Mon. Not. of the RAS 231, 617.
Carr, J.S.: 1993, Astrophys. J. 406, 553.
Chernoff, D.F., Hollenbach, D.J. & McKee, C.F.: 1982, Astrophys. J. 259, L97.
DeNoyer, L.K.: 1979, Astrophys. J. 232, L165.
Draine, B.T.: 1980, Astrophys. J. 241, 1081.
Draine, B.T. & Roberge, W.G.: 1982, Astrophys. J. 259, L91.
Draine, B.T., Roberge, W.G. & Dalgarno, A.: 1983, Astrophys. J. 264, 485.
Draine, B.T. & McKee, C.F.: 1993, Ann. Rev. of Astron. & Astrophys. 31, 373.
Elmegreen, B.G & Combes, F.: 1992, Astron. & Astrophys. 259, 232.
Fernandes, A.J.L. & Brand, P.W.J.L.: 1995, Mon. Not. of the RAS in press.
Garden, R.R. & Geballe, T.R.: 1990, Astrophys. J. 354, 232.
Garden, R.R. & Geballe, T.R.: 1991, Astrophys. J. 366, 474.
Gautier, T.N., Fink, U., Treffers, R.R. & Larson, H.P.: 1976, Astrophys. J. 207, L129
Graham, J.R. & Heyer, M.H.: 1990, Publ. of the ASP 102, 972.
Graham, J.R., Wright, G.S., Hester, J.J. & Longmore, A.J.: 1991a, Astron. J. 101, 175.
Graham, J.R., Wright, G.S. & Geballe, T.R.: 1991b, Astrophys. J. 372, L21.
Gredel, R., Reipurth, B. & Heathcote, S.: 1992, Astron. & Astrophys. 266, 439.
Hartigan, P., Raymond, J. & Hartmann, L.: 1987, Astrophys. J. 316, 323.
Hartigan, P., Curiel, S. & Raymond, J.: 1989, Astrophys. J. 347, L31.
Hollenbach, D.J., & Shull, M.: 1977, Astrophys. J. 216, 419.
Kwan, J.: 1977, Astrophys. J. 216, 713.
Lada, C.: 1985, Ann. Rev. of Astron. & Astrophys. 23, 267.

London, R., McCray, R. & Chu S-I.: 1977, Astrophys. J. **217**, 442.

Moorhouse, A., Brand, P.W.J.L.,Geballe, T.R. & Burton, M.G.: 1992, Mon. Not. of the RAS **253**, 232.

Nadeau, D. & Geballe, T.R.: 1979, Astrophys. J. **230**, L169.

Simon, M., Righini-Cohen, G., Joyce, R.R. & Simon, T.: 1979, Astrophys. J. **230**, L175.

Smith, M.D., Brand, P.W.J.L. & Moorhouse, A.: 1991a, Mon. Not. of the RAS **248**, 451.

Smith, M.D., Brand, P.W.J.L. & Moorhouse, A.: 1991b, Mon. Not. of the RAS **248**, 730.

Smith, M.D.: 1994, Astron. & Astrophys. **289**, 256.

Taylor, K.N.R., Storey, J.W.V., Sandell, G., Williams, P.M. & Zealey, W.J.: 1984, Nature **311**, 236.

Wang, & Scoville, N.: 1992, Astrophys. J. **386**, 158.

Wardle, M.: 1991, Mon. Not. of the RAS **250**, 523.

Wardle, M. & Draine, B.T.: 1987, Astrophys. J. **321**, 321.

White, G.J., Rainey, R., Hayashi, S.S. & Kaifu, N.: 1987, Astron. & Astrophys. **173**, 337.

OBSERVATIONS OF SHOCKED [FEII] AND H_2 LINE PROFILES IN ORION BULLET WAKES

JONATHAN A. TEDDS & PETER W.J.L. BRAND
Institute for Astronomy, University of Edinburgh, EH9 3HJ, U.K.

MICHAEL G. BURTON
School of Physics, University of New South Wales, Australia.

ANTONIO CHRYSOSTOMOU
Division of Physical Sciences, University of Hertfordshire, U.K.

and

AMADEU J.L. FERNANDES
Centro de Astrofísica, Universidade do Porto, Portugal.

Abstract. Recent near-IR imaging of the Orion molecular cloud has revealed a complex of dense bullets, visible as [FeII] emitting HH-objects at the tips of H_2 wakes, ejected explosively from the cloud core. Having resolved individual bow-shock structures for the first time in this bright source, we have observed [FeII] 1.644μm velocity profiles of selected bullets and H_2 1-0 S(1), 2.122μm velocity profiles for a series of positions along and across the corresponding bow-shock wakes. We present observed profiles for the bullet M42 HH1 and its associated wake and compare with theoretical bow-shock models.

1. Introduction

The nature of molecular shocks, which play an important role in the processes of momentum and energy transfer within starforming molecular clouds (McKee 1989), is still uncertain (Draine & McKee 1993). The H_2 bullet wakes in the Orion outflow emphasise the importance of bow shock morphology. It is now established that (a) planar C(magnetised)-shock models cannot explain the H_2 line intensities in outflows, while planar J-shock models can (Brand *et al.* 1988); (b) neither planar nor bow J-shock models can reproduce the line profiles while highly magnetised C-shock models can (Brand *et al.* 1989); (c) the H_2 excitation conditions throughout the bright part of the Orion outflow appear remarkably constant, not easily explained by bow shock models (Brand *et al.* 1989, McKee 1989).

Recent near-IR imaging of Orion with 0.5″ spatial resolution in the emission lines of H_2 (2.122 μm) and [FeII] (1.644 μm) has clarified our view of the shocked molecular outflow (Allen & Burton 1993). Many new Herbig-Haro objects were revealed, visible as [FeII] 'bullets', at the heads of wakes of H_2 emitting gas. The bullets (originating within 5″ of IRc2) have been ejected over a wide opening angle and we have measured [FeII] line profile widths of up to 400 km s^{-1} (Tedds *et al.* in preparation), in agreement with [OI] 6300Å linewidths (Axon & Taylor 1984) indicating an explosive origin in

Astrophysics and Space Science **224**: 139–142, 1995.
© 1995 *Kluwer Academic Publishers.*

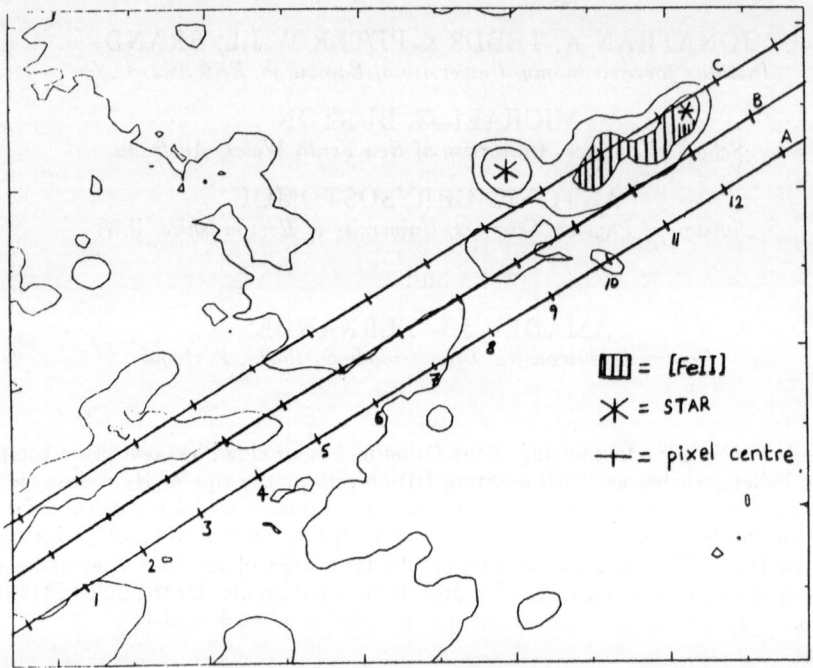

Fig. 1.   H$_2$ 1-0 S(1), 2.122 μm contour plot of M42 HH1 (Allen & Burton 1993). The shaded area indicates the position of the [FeII], 1.644 μm bullet(s). The grid of numbered boxes indicate pixels in 3 line profile slit positions.

the core of Orion within the last 10^3 years.

2.   Observations and Results

We have used the UKIRT near-IR spectrometer CGS4 in its echelle mode to obtain [FeII] line profiles of some of the bullets together with H$_2$ line profiles in the most clearly defined bow shock wakes (Tedds *et al.* in preparation). Figure 1 shows the bullet M42 HH1 imaged in the [FeII] 1.644 μm line together with the associated H$_2$ emission in its wake. Superimposed on this is a grid indicating 3 slit positions observed at 1.5″ steps across the wake region. Figure 2 shows the integrated velocity profile of [FeII] 1.644 μm emission at the central slit position.

The two most important parameters in modelling bow-shock line profiles are the incident shock velocity, v_s and the orientation, ϕ. Observationally, we can directly measure v_s since it is equal to the full width near zero intensity of the integrated emission over an entire bow shock front independent of the shape of the bow-shock, orientation angle, preshock density, elemental abundances and reddening (Hartigan *et al.* 1989). Furthermore, by measuring the maximum and minimum radial velocities for the integrated line

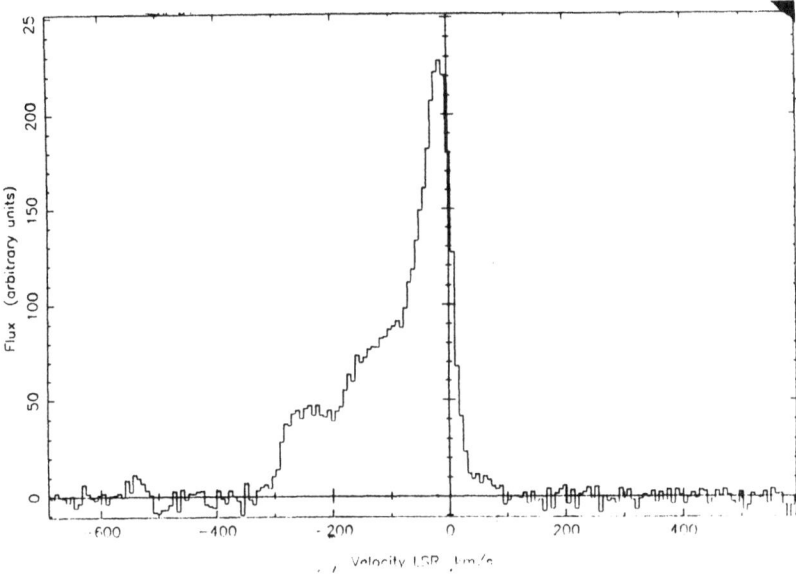

Fig. 2.   [FeII], 1.644 μm integrated velocity profile for M42 HH1.

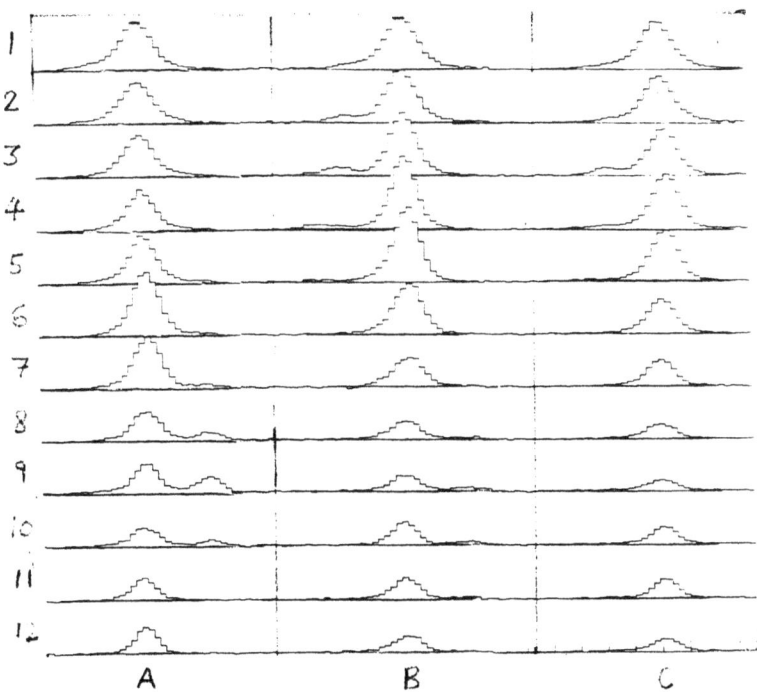

Fig. 3.   H₂ 1-0 S(1), 2.122 μm line profiles at pixels (Fig.1) in wake of M42 HH1.

profile we obtain both v_s and ϕ. In the case of the bullet M42 HH1, the integrated [FeII] profile corresponds to a shock velocity of $v_s = 380 \pm 10$ km s^{-1} and an orientation angle of $\phi = 55 \pm 5°$.

Comparison of the observed [FeII] profile shape (Fig. 2) with radiative bow-shock model predictions (Hartigan $et\ al.$ 1989) shows closest agreement for a $v_s = 400$ km s^{-1}, $\phi = 60°$ bow-shock. We note that the stronger of the two peaks lies near zero radial velocity as predicted in bullet models and also that both the [FeII] and H$_2$ profiles indicate enhancement of this peak.

Shocked H$_2$ 1-0 S(1) 2.122 μm line profiles at each pixel for 3 adjacent slit positions are shown in Figure 3. Clear changes from double-peaked to single-peaked H$_2$ line profiles are observed in stepping across the wake(s) region. This is as expected since at central wake positions two distinct shocked regions on either side of the bow-shock wake will contribute to the emission. If the orientation of the bow to the line-of-sight, $\phi < 90°$, the two components sampled will be at different positions in the wake and hence have different normal velocities - giving rise to a doubly-peaked line profile. However, in moving across a bow the two components will approach the same normal velocity, hence the change to a singly-peaked profile at the edge. The 2 regions of doubly-peaked line profiles, having red and blueshifted secondary peaks respectively, may be explained as separate bow-shocks associated with 2 coincident [FeII] bullets (Tedds $et\ al.$ in preparation).

3. Discussion

We have demonstrated that [FeII] and H$_2$ emission line profiles in Orion are consistent with theoretical bow-shock predictions. By mapping line profiles in this way for a number of different Orion bullet wakes we can test bow-shock models over a range of values of v_s and ϕ and also determine if all of the observed shocked molecular hydrogen emission in this region is dominated by this single event rather than by a steady bubble-type outflow as was previously indicated and commonly observed in starforming regions.

References

Allen, D. A. & Burton, M.G.: 1993, Nature **363**, 54.

Axon, D. A. & Taylor, K.: 1984, Mon. Not. of the RAS **207**, 241.

Brand, P. W. J. L., Moorhouse, A., Burton, M. G., Geballe, T. R., Bird, M. & Wade, R.: 1988, Astrophys. J. **334**, L103.

Brand, P. W. J. L., Toner, M. P., Geballe, T. R. & Webster, A. S.: 1989, Mon. Not. of the RAS **237**, 1009.

Draine, B. T. & McKee, C. F.: 1993, Ann. Rev. of Astron. & Astrophys. **31**, 373.

Hartigan, P., Curiel, S. & Raymond, J.: 1989, Astrophys. J. **347**, L31.

McKee, C. F.: 1989, Astrophys. J. **345**, 782.

THE ORION-IRC2 OUTFLOW OBSERVED IN CO $J=4-3$

A. SCHULZ and C. HENKEL

Max-Planck-Institut für Radioastronomie, D-53121 Bonn

and

L.Å. NYMAN

European Southern Observatory La Silla, Chile

Abstract. A CO $J=4-3$ map of the Orion-IRc2 region (size 80", beam 15") is presented. The outflow has a bipolar structure observed nearly pole-on. We interpret our observations in terms of an hierarchy of "tubes" where the fastest gas proceeds through the narrowest, innermost cone. The gas has an inhomogeneous temperature distribution and must be highly clumped. Most of the properties derived comply with the picture developed for the few known extreme high velocity outflows; however, this outflow contains exceptionally dense gas ($> 10^6$ cm^{-3}).

1. Introduction

CO is the best suited species to study molecular outflows because of its abundance, the optical thickness of its rotational lines and its stability against photodissociation. Complete high angular resolution CO maps of Orion-KL exist only for the two lowest transitions (Wilson *et al.* 1986; Masson *et al.* 1987). Furthermore, no extreme high velocity outflow has yet been mapped in a higher ($J_u > 2$) rotational CO transition. Our high angular resolution CO (4–3) line map of Orion-KL thus provides a detailed view of this prominent outflow in a higher excited CO transition, complementing (2-1) and (1-0) line data and bridging the gap to higher excited CO transitions.

2. Observations and Results

The 460 GHz observations reported here (obtained in Aug. 1989) are the first submm observations performed at the Swedish ESO Submillimetre Telescope (SEST) of the European Southern Observatory, La Silla, Chile. We used the 460 GHz receiver built at the Max-Planck-Institut für Radioastronomie which is a Schottky mixer (noise temperature 1200 K DSB) in conjunction with a 1000-channel acousto-optical spectrometer, (500 MHz total bandwidth). Calibration measurements are reported by Schulz *et al.* (1994).

Our 50-point map (15" beam size, 10" grid steps) is centered on $\alpha(1950) = 5^h32^m46.5^s, \delta(1950) = -5°24'14''$. The CO (4–3) intensities are given in units of effective beam brightness temperature, T_{eb} which for sources not much larger than the full main beam is comparable to the usual main beam brightness temperature, T_{mb}, despite the large error pattern (for details, see Schulz *et al.* 1994).

Astrophysics and Space Science **224**: 143–149, 1995.
© 1995 *Kluwer Academic Publishers.*

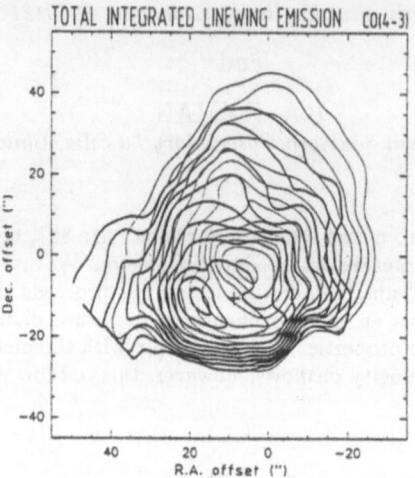

Fig. 1. **Maps of integrated CO (4–3) line wing intensity** (entire velocity range, blue: −60 to +2 km s^{-1} (thin), red: 14 to 70 km s^{-1} (thick)); lowest level: 1200 K km s^{-1}; increment 300 K km s^{-1}; the cross marks IRc2.

The plateau feature of the spectra extends from −70 to +90 km s^{-1}, at least. The extrapolated peak of the plateau emission at $v_{LSR} \sim 5$ km s^{-1} is $T_{eb} = 220 \pm 40$ K. Maps of the emission integrated over the total blue (−60 to +2 km s^{-1}) and total red (+14 to +70 km s^{-1}) line wings (see Fig. ??) show that most of the outflow lobe areas are overlapping, but the lobe maxima show a clear separation of about 14″. Fig. 2a depicts the positions of the maxima of the outflow lobes integrated over 10km s^{-1} wide intervals. For the lower-velocity range there is no significant displacement of the lobes. With increasing flow velocity Δv, we observe a systematic positional shift: (1) an increasing lobe separation up to $\Delta v = 50$ km s^{-1}, and (2) a 'turn over' at very high velocities for the blue lobes towards positions again closer to IRc2. One finds that the offset of IRc2 to the flow axis decreases with increasing velocity; at the high velocities, IRc2 is located very near or on the flow axis. Plotting the mean FWHM sizes (Fig. 2b), we see their sizes decrease with increasing velocity. The collimation of the flow projected onto the sky appears to be poor at all velocities.

3.   Discussion

3.1   The Small Scale Structure

Our (4–3) line is so far the highest-excited CO line in which a complete map spatially resolving the outflow has been obtained. Its general morphology appears to be similar to maps of lower-J CO lines and spectra are similar

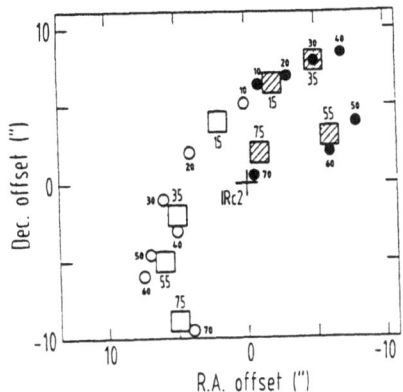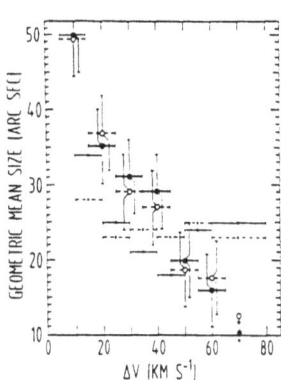

Fig. 2.   a): **Positional distribution of outflow lobe maxima**; lobe intensities integrated over ranges of 10 (circles) or 20 km s⁻¹ (squares); numbers indicate the radial flow
velocity Δv (shaded or filled symbols: blue wings, open symbols: red wings). b): **Mean
lobe sizes of CO emission** integrated over 10 km s⁻¹ wide intervals (large symbols, open
circles: blue CO (4–3) wing lobes, filled circles: red CO (4–3) wing lobes; small symbols
ares the corresponding CO (2–1) values by Wilson *et al.* (1986), their Fig.4).

within the noise for all CO lines up to $J_u = 7$. We thus seem to observe the
same gas in all these CO lines.

For CO (4–3), the bipolar flow angle (*i.e.* the angle between the lines
connecting corresponding red and blue lobe centres with IRc2) opens up
systematically from 0° at low towards ~180° at high flow velocities. At the
highest velocities, IRc2 lies on a straight line between the lobe centres. This
geometry makes it likely that the flow has its origin at IRc2 or at an unknown
object (for which no evidence has been found) very close.

In a scenario of a collimated bipolar flow, the observed high total velocity
range and the poor collimation favour a flow geometry with the observer's
view almost pole-on. Checking the dynamical time scale t_{dyn} we come to a
similar conclusion as Wilson *et al.* (1986): for $t_{dyn} > 10^3$ yrs, the inclination
of the true axis of expansion relative to the line of sight must be $\leq 20°$.
Unless the inclination is very close to zero, the outflow must be extremely
young.

It is difficult to determine the degree of collimation in a 3-dimensional
picture for our object. If the assumption of a well collimated flow holds,
the large lobes in the lower velocity ranges (see Fig. 2b) delineate the outer
walls of the flow cone, being matter swept up from the ambient cloud; Fig. 2b
furtherly implies that the collimated bipolar *higher-velocity gas* proceeds in
tubes or cones of continously decreasing diameter or opening angle with
higher velocity which we observe as smaller lobe sizes. Stahler (1994) suggested a kinematic model which complies with just this interpretation: the

outflow lobes are filled with turbulent (and clumpy, see below) gas entrained by matter propagating with the highest velocity along the central flow axis out to large distances. Then, the material of the inner tubes or cones causes a turbulent drag exerted on to the slower material further apart from the flow axis. Such an acceleration process is consistent with the spatial structure and velocity field of the Orion-KL outflow. Note, however, that the detailed physics of this acceleration process, in particular the momentum transfer from the central jet onto the highest-velocity molecular gas as well as the origin of the jet itself, is not yet fully understood (see Masson & Chernin 1993).

3.2 Gas Densities

3.2.1 a) Low Velocity Gas

Since we can assume from CO isotopic line data that the plateau emission is optically thick at *low flow velocities* (see Wilson *et al.* 1986 and references therein), we can deduce excitation temperatures T_{ex} for the line centres of the various CO plateau features from their brightness temperatures T_b extrapolated in the line centre near $5 \mathrm{km\,s^{-1}}$(Table 1). We recognize that T_{ex} is *not* constant but increases with J-number. This is not consistent with a model involving a smooth gas distribution heated from an interior source. Additionally, the warmer low-velocity material seen in CO (4–3) is more extended than the corresponding material seen in CO (2–1) (see Fig. 2b). Therefore, a highly inhomogeneous gas distribution with steep temperature gradients is the most plausible explanation.

3.2.2 b) High Velocity Gas

With its density and temperature gradients, the entire tube-like flow system must have an extremely 'broken' structure, rendering the derivation of temperatures and densities difficult. Further obstacles to achieve a detailed analysis come from the unfavourable viewing angle. Despite these difficulties, we can roughly confine the average volume density $n\,(\mathrm{H_2})$ of the gas giving rise to the *high velocity* CO emission. For this, we use a standard large velocity gradient (LVG) radiative transfer model, adopting optically thin emission from a homogeneous spherical cloud of constant density and temperature; if existing at all, optically thin emission is found at *high* outflow velocities. According to Watson *et al.* (1985) and Howe *et al.* (1993), $T_{kin} \sim$ 700 K is roughly the upper temperature limit for the bulk of the outflowing material seen in the low/mid-J CO lines. For $T_{kin} = 200-700$ K and realizing that line wing intensities at $\Delta v = 30$ and $50 \mathrm{km\,s^{-1}}$ are rising up to at least $J=7$-6 (see Table 1), we obtain $n\,(\mathrm{H_2}) \geq 10^6\,\mathrm{cm^{-3}}$(radiative excitation of the lower-J CO lines can be neglected; see Schulz *et al.* 1994). Conditions seem to hold for a larger range of velocities since the line ratios are similar for $\Delta v = \pm\,30$ and $\pm 50\,\mathrm{km\,s^{-1}}$.

TABLE I

CO intensities T_{mb} and brightness temperatures T_b

Transition	Calc. T_b [b] [K]	Line Intensity T_{mb} [K] [a]					Ref
		$\Delta v=0$	$\Delta v=+50$	$\Delta v=-50$	$\Delta v=+30$	$\Delta v=-30$	
7–6	380	340	30	40	110	130	1
6–5	320	260	30	30	80	95	2
4–3	260	220	18	13	50	65	-
3–2	150	130	7	9	18	21	3
2–1	130	120	5	6	20	20	4
1–0	90	80	2	2	8	7	4

a) calculated from observations for a 15″ beam assuming a Gaussian source of 35″;
b) assuming a Gaussian source size of 35″.
Ref: 1 – Howe et al. (1993); 2 – Koepf et al. (1982); 3 – Erickson et al. (1982); 4 – Wilson et al. (1986).

3.2.3 Highly Excited Molecular Gas

The outer maxima of shock-excited $2\mu m$ H_2 emission (e.g. Beckwith et al. 1978) are likely to mark the outer shock front of the fast material of the flow (Stacey et al. 1993); the small extent of the outer H_2 maxima supports the idea of 'tube hierarchy', involving a high degree of collimation for the highest velocities.

Highly excited CO is observed in FIR lines up to $J=34$–33. In this case, only integrals over the entire flow area were obtained (Watson et al. 1980; Storey et al. 1981; Stacey et al. 1982; Watson et al. 1985). We interprete these lines indicating a range of temperatures of 700 to 2000 K to originate not only from the outer shock fronts where the material hits ambient cloud gas but also at the clump surfaces within the flow region. The low column densities determined for this highly excited gas (and the very high temperatures of $>700K$) fit well into our scenario involving thin hot cloud layers, the thinner the higher the measured temperatures are (see Table 2).

3.2.4 A Comparison with other Extreme-High-Velocity Outflows

While the viewing angle may be different, the extreme high-velocity (EHV) outflow associated with IRc2 appears to be similar to the few other known EHV outflows (e.g. Bachiller & Gomez-Gonzales 1992 and references therein). Most of the known EHV flows have highly collimated jets with dynamical time scales of order 10^3 yrs, and their degree of collimation is systematically higher at higher outflow velocities. In IRAS 3282 (see Bachiller et al. 1991), the velocity of the 'standard high-velocity' (SHV) outflow increases with increasing distance from the source, while the EHV outflow decelerates, which can be explained as a transfer of momentum and energy from the narrow EHV jet to the ambient gas which generates the SHV outflow

TABLE II

IRc2 outflow gas components

Temp. [K]	Observed Tracer	Derived Column Density [cm^{-2}]
130	low-J CO	1 - 6 10^{22}
700	mid/high-J CO	1 - 5 10^{21}
1000	high-J CO, [OI] (63μm)	2 - 7 10^{20}
2000	H$_2$ (2μm)	1 - 5 10^{19}

References: Wilson *et al.* (1986), Watson *et al.* (1980),
Werner *et al.* (1984), Beckwith *et al.* (1978).

(see Stahler 1994). This is consistent with the small scale structure of the Orion-KL outflow we describe above : its EHV gas seems to be composed of many CO cloudlets inferring strong density and temperature variations along the flow.

The Orion-IRc2 flow viewing angle is optimal to test the 'hollow-shell' model (*e.g.* Bachiller & Gomez-Gonzales 1992). Obviously, this model does not describe the EHV gas in Orion-KL nor in other sources. The cones are filled with clumpy, partly very dense gas. Entrainment of ambient gas and its acceleration to moderate velocities as seen in the Orion-KL outflow seems to be common. Whereas the kinematic properties of Ori-IRc2 are similar to other EHV flows, the very high gradients of densities and temperatures in the IRc2-flow seem to be different. Whether this is exceptional or normal (but not yet found) should be checked by more studies of higher excited CO transitions in EHV flows to search for hot and dense gas components and to relate their properties to those of the central source.

References

Bachiller, R. & Gomez-Gonzales, J.: 1992, Astron. & Astrophys. Rev. **3**, 257.

Bachiller, R., Martin-Pintado, J. & Planesas, P.: 1991, Astron. & Astrophys. **251**, 639.

Beckwith, S., Persson, S.E., Neugebauer, G. & Becklin, E.E.: 1978, Astrophys. J. **223**, 464.

Erickson, N.R., Goldsmith, P.F., Snell, R.L., Berson, R.L., Huguenin, G.R., Ulich, B.L. & Lada, C.: 1982, Astrophys. J. **261**, L103.

Howe, J.E., Jaffe, D.T., Grossman, E.N., Wall, W.F., Mangum, J.G. & Stacey, G.J.: 1993, Astrophys. J. **410**, 179.

Koepf, G.A., Buhl, D., Chin, G., Peck, D.D., Fetterman, H.R., Clifton, B.J. & Tannenwald, P.E.: 1982, Astrophys. J. **260**, 584.

Masson, C.R. & Chernin, L.M.: 1993, Astrophys. J. **414**, 230.

Masson, Lo, K.Y., Phillips, T.G., Sargent, A.I., Scoville, N.Z. & Woody, D.P.: 1987, Astrophys. J. **319**, 446.

Schulz, A., *et al.* : 1994, Astron. & Astrophys. in press.

Schulz, A., Krügel, E. & Beckmann, U.: 1992, Astron. & Astrophys. **264**, 629.

Stacey, G.J., Jaffe, D.T., Geis, N., Genzel, R., Harris, A.I., Poglitsch, A., Stutzki, J. & Townes, C.H.: 1993, Astrophys. J. **404**, 219.

Stacey, G.J., Kurtz, N.T., Smyers, S.D., Harwit, M., Russell, R.W. & Melnick, G.J.: 1982, Astrophys. J. **257**, L37.

Stahler, S.W.: 1994, Astrophys. J. **422**, 616..

Storey, J.W.V., Watson, D.M., Townes, C.H., Haller, E.E. & Hansen, W.L.: 1981, Astrophys. J. **247**, 136.

Watson, D.M., Genzel, R., Townes, C.H. & Storey, J.W.V.: 1985, Astrophys. J. **298**, 316.

Watson, D.M., Storey, J.W.V., Townes, C.H., Haller, E.E. & Hansen, W.L.: 1980, Astrophys. J. **239**, L129.

Werner, W.M., Crawford, M.K., Genzel, R., Hollenbach, D.J., Townes, C.H. & Watson, D.M.: 1984, Astrophys. J. **282**, L81.

Wilson, T.L., Serabyn, E. & Henkel, C.: 1986, Astron. & Astrophys. **167**, L17.

PLATE I: The present-day Royal Observatory from the air, looking South. In the foreground is the '1894 Building' with telescope towers to the East and West with the Library in the middle. The newer building to the West houses the Plate Library and student laboratories. The house in the centre of the grounds was the residence of the Astronomer Royal for Scotland until 1975.

GIANT BOW SHOCK PAIRS ASSOCIATED WITH
HERBIG-HARO JETS

KATSUO OGURA

Kokugakin University, Higashi, Shibuya-ku, Tokyo 150, Japan

Abstract. Two pairs of giant (linear size ∼1 pc) bow shock structures have been discovered, each located symmetrically about HH 1/2 and HH 124. Their Herbig-Haro (HH) natures have been confirmed by narrow band CCD imaging on and off [SII] 6717/6731 and/or slit spectroscopy. Multiple bow shocks are known associated with a few HH objects such as HH 34, and are interpreted as evidence for recurrent outflow activity of the exciting sources. The giant bow shocks associated with HH 1/2 or HH 124 provide further, beautiful examples of this phenomenon and, with dynamical ages of nearly 20000 yr in both pairs, extend its timescale by more than an order of magnitude.

Key words: HH Objects – Bow Shocks – Episodic Outflows – YSOs

1. Introduction

Evidence for recurrent outflow activity of the exciting sources of HH objects has been accumulating in recent years. The simplest is the fact that some HH jets are associated with multiple bow shocks on one side of their outflows; HH 34 (Reipurth & Heathcote 1992) and (probably) HH 136 (Ogura & Walsh 1992) have triple bow shocks, and HH 46/47 (Reipurth & Heathcote 1991) and HH 111 (Reipurth, Raga & Heathcote (1992) have double bow shocks. Very recently Bally & Devine (1994) have suggested the presence of a "superjet" in the HH 34 outflow consisting of chains of six and three bow shocks in each of the lobes.

The second evidence is the well-known discrepancy between the space motion and the shock velocity in HH objects, the latter being estimated from emission line intensity ratios. Raga *et al.* (1988) were the first to discuss this problem in HH 1, where the tangential velocity (which is almost equal to the space motion in this case) is 380 km/s (Eisloeffel *et al.* 1994) and the shock velocity is 150 - 200 km/s (Noriega-Crespo, Boehm & Raga 1989), adopting the results of recent measurements. This discrepancy can be explained by postulating that HH 1 is propagating into a medium which is already flowing at a speed of ∼200 km/s, and this in turn suggests an earlier outflow activity of the exciting source. A similar phenomenon is found in HH 34 ($v_{\rm space}$ ∼310 km/s, $v_{\rm shock}$ ∼150 km/s and $v_{\rm med}$ ∼160 km/s; Heathcote & Reipurth 1992) and HH 111V ($v_{\rm space}$ ∼400 km/s, $v_{\rm shock}$ ∼95 km/s and $v_{\rm med}$ ∼300 km/s; Morse *et al.* 1993).

The third evidence for episodic outflow activities of HH exciting sources is related to the lowest velocity observed among the condensations in a bow shock. Time-dependent bow shock model calculations of Raga *et al.*

Astrophysics and Space Science **224**: 151–154, 1995.

(1988) predicted that condensations far from the stagnation point should be almost stationary, which is not the case in HH 1 with ∼100 km/s in the south-easternmost condensation (Eisloeffel *et al.* 1994). A similar trend is observed in HH 34 (Eisloeffel & Mundt 1992).

The last evidence is the discrepancy between the calculated and observed position-velocity (p-v) diagrams or line profile distributions in bow shocks. Hartigan, Raymond & Meaburn (1990) found that HH 47A has no zero velocity components in its p-v diagram, whilst the model calculations predict that, seen from any angle, bow shocks should always include zero velocity components. This discrepancy can be avoided again by the idea of an already flowing medium. Also, according to the Fabry-Perot observations of Morse *et al.* (1992, 1993) the line profile distributions of HH 34 and HH 111V show offsets of -75 km/s and -30 km/s, respectively, from their detailed bow shock models, which can be reconciled by introducing v_{med} ∼150 km/s and ∼300 km/s, respectively, taking into account the angles of their bow axes from the line of sight.

We have seen above that the medium in front of HH 1 is already flowing with a considerable velocity, a fact which suggests an earlier outflow activity of the HH 1/2 exciting source. This paper reports the discovery of the bow shock structures which are presumably caused by this outflow. Also we have located another bow shock pair of very similar nature, which is associated with HH 124 (Walsh, Ogura & Reipurth 1992) but is caused by an outflow activity prior to it. They are designated as HH 1/2-NW, HH 1/2-SE, HH 124-W and HH 124-E.

2.   Discovery and Follow-up Observations

The discovery has been made on deep UK Schmidt plates in the red. Figures 1a and 1b are unsharp-masked and contrast-enhanced reproductions from them. However all the objects excepting HH 124-E are already visible on the Palomar Observatory Sky Survey red prints. Their HH nature has been confirmed by narrow-band CCD imaging on and off [SII] 6717/6731, using the 105-cm Schmidt telescope at Kiso Observatory. In addition low-dispersion slit spectroscopy has been obtained for the tip of HH 1/2-NW with the RGO spectrograph on the AAT. The spectrum is typical of HH objects of moderate excitation except that [OII] 3727/3729 is quite strong, indicating a very low reddening.

3.   Association with HH 1/2 or HH 124

In testing this, we can rely only on the morphological information owing to the lack of kinematic data on these objects at the moment. First, their bow shapes point toward the immediate vicinities of HH 1/2 or HH 124 as the origines of the outflows. Second, the lines connecting the tips of the pair components nearly or exactly pass the positions of the respective exciting sources, VLA 1 (Pravdo *et al.* 1985) or IRAS 06382+1017 (Walsh, Ogura & Reipurth 1992). Third, the directions of the lines are almost the same as those of the flows. In HH 1/2-NW and HH 1/2-SE the former is in P.A.= 319°, and the proper motion measurements by Eisfoeffel *et al.* (1994) indicate that the latter is in P.A.=317°, excluding the southeastern condensations of HH 1 which belong to a different outflow system (Eisloeffel 1994, private communication). The flow direction in HH 2 is not so well-defined as in HH 1, but the simple mean of the directions of the proper motions of the knots is P.A.=156° (see Eisloeffel *et al.* 1994). No proper motion data are available for HH 124, but the line connecting HH 124-W and HH 124-E almost exactly coincides with the axis of the HH 124 bipolar jet. Finally, the pair components are located at similar distances from their exciting sources (see Table 1).

TABLE I

Properties of Bow Shocks

Object	Linear size (pc)	Distance from source* (pc)	Dynamical age (yr)
HH 1/2-NW	0.8	3.2	20000
HH 1/2-SE	>0.3**	2.7	17000
HH 124-W	0.9	2.8	18000
HH 124-E	>0.3**	2.6	16000

*projected distance,   ** partly obscured by interstellar absorption

4. Properties of the Bow Shocks

Table 1 summarizes the basic properties of the bow shocks. The dynamical ages have been calculated by assuming the propagation velocity of 150 km/s. This value does not contradict the fact that our spectrum of HH 1/2-NW does not show [OIII] 4959/5007 which requires bow shock velocities of 200 km/s or higher to be excited (Morse *et al.* 1992). We also note that probably the angle of the flow direction of HH 124 from the plane of the sky is not large, judging from the fact that the redshifted counter-jet is optically visible (Walsh, Ogura & Reipurth 1992). Their huge sizes of ~1 pc are noteworthy. This fact may be indicative of the poor collimation of these outflows far downstream, contrary to the case of the putative "superjet" in the HH 34 outflow (Bally & Devine 1994) which is very narrow. Also remarkable are

their large (projected) distances from their exciting sources. This in turn
gives the very large dynamical ages, which is in fact 1 - 2 orders of magnitude
larger than those of other HH objects. Thus HH activities can be traces up
to ~20000 yr.

References

Bally, J. & Devine, D.: 1994, Astrophys. J. **428**, L65.
Eisloeffel, J. & Mundt, R.: 1992, Astron. & Astrophys. **263**, 292.
Eisloeffel, J., Mundt, R. & Boehm, K.H.: 1994, Astron. J. in press.
Hartigan, P., Raymond, J. & Meaburn, J.: 1990, Astrophys. J. **362**, 624.
Heathcote, S. & Reipurth, B.: 1992, Astron. J. **104**, 2193.
Morse, J.A., Hartigan, P., Cecil, G., Raymond, J.C. & Heathcote, S.: 1992, Astrophys. J.
    399, 231.
Morse, J.A., Heathcote, S., Cecil, G., Hartigan, P. & Raymond, J.C.: 1993, Astrophys. J.
    410, 764.
Noriega-Crespo, A., Boehm, K.H. & Raga, A.C.: 1989, Astron. J. **98**, 1388.
Ogura, K. & Walsh, J.R.: 1992, Astrophys. J. **400**, 248.
Pravdo, S.H., Rodriguez, L.F., Curiel, S., Canto, J., Torrelles, J.M., Becker,R.H. & Sell-
    gren, K.: 1985, Astrophys. J. **293**, L35.
Raga, A.C., Mateo, M., Boehm, K.-H. & Solf, J.: 1988, Astron. J. **95**, 1783.
Reipurth, B. & Heathcote, S.: 1991, Astron. & Astrophys. **246**, 511.
Reipurth, B. & Heathcote, S.: 1992, Astron. & Astrophys. **257**, 693.
Reipurth, B., Raga, A.C. & Heathcote, S.: 1992, Astrophys. J. **392**, 145.
Walsh, J.R., Ogura, K. & Reipurth, B.: 1992, Mon. Not. of the RAS **257**, 110.

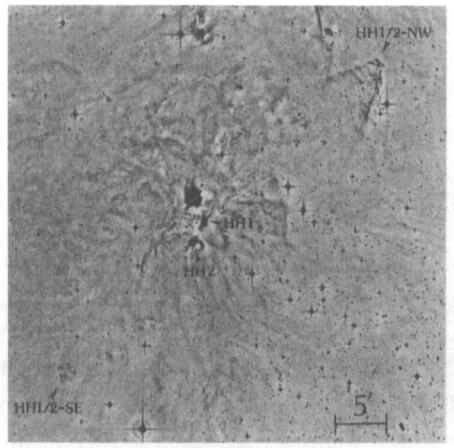
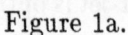
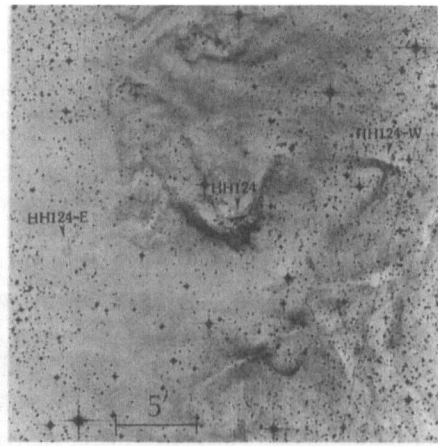

Figure 1a.   Figure 1b.

SESSION C:

HIGH MASS STARS
&
ULTRACOMPACT
HII REGIONS

NEWLY FORMED MASSIVE STARS: ARE THEY GOOD
FOR THE NEIGHBORHOOD?

ED CHURCHWELL

University of Wisconsin

1. Introduction

Newly formed massive stars (NFMS) are still deeply embedded in the molecular cloud cores where they were born. By "massive", I refer to any star hot enough to produce a detectable HII region (*i.e.* B3 or hotter). Because of their fast winds and high luminosities, NFMS have a profound effect on the properties of the matter in their neighborhood, however, due to high obscuration by dust, they are only observable at IR and radio wavelengths. O and B stars are an extreme population among stars and we will find that the circumstellar medium within a radius ≤ 0.1 pc of a NFMS also has an extreme range of physical properties. For example within this radius, gas kinetic temperatures may range from $\sim 10^8$ to $\sim 10^1$ K, densities from $\sim 10^7$ to $\sim 10^0$ cm^{-3}, and gas velocities relative to the ionizing star from ~ 2500 to 2-3 km s^{-1}. Such large changes in physical properties over such small distances are also accompanied by strong shocks, sharp changes in ionization and equations of state. The literature on individual UC HII regions and massive star formation is extensive and I cannot review even a small fraction of it here. Reviews and articles that summarize many of the main themes of research in this area are: Adams (1993 and references therein), Churchwell (1990; 1991; 1993), Hollenbach *et al.* (1993;1994), Mac Low *et al.* (1991), Tenorio-Tagle (1979), Van Buren *et al.* (1990), Welch (1993), Yorke, Tenorio-Tagle & Bodenheimer (1983), and Yorke (1984; 1993). The remainder of this review will concentrate on recent results or hypotheses that were not covered earlier.

2. Morphologies

In this section, I will summarize the models proposed for the morphologies and longevities of UC HII regions. In addition, I will discuss an interesting pattern that seems to be emerging regarding the relative positions of cometary HII regions; hot, dense molecular clumps; and H$_2$O masers.

2.1 HII REGION MORPHOLOGIES

Several models have been proposed to account for the large fraction of cometary and core/halo morphologies among UC HII regions and for their

Astrophysics and Space Science **224**: 157–171, 1995.

© 1995 *Kluwer Academic Publishers.*

longevity which must be $\sim$ 100 times greater than their expansion ages. One of the earliest of these is the "champagne flow" or "blister" model proposed in a series of papers by Tenorio-Tagle and coworkers (see Yorke, Tenorio-Tagle & Bodenheimer 1983 and references therein). The basic idea is that O stars form near the boundary of large molecular clouds and when the ionization front breaks out of the molecular cloud a flow of ionized gas is produced due to lower pressure on the boundary while the flow is confined by the molecular cloud on the opposite side. This produces a bright ionized arc on one side and a "tail-like" feature on the opposite side similar to a cometary structure if viewed from the right perspective. However, when examined in detail, neither the predicted brightness distributions nor the kinematics seem to match the observed morphologies of most cometary UC HII regions (see Churchwell 1991). Garay, Lizano & Gomez (1994) have recently incorporated the effects of a stellar wind to this class of models. There is no doubt that some nebulae are indeed "champagne flows", which must occur whenever an O star exits a molecular cloud. However, this cannot be the explanation for the majority of cometary and core/halo UC HII regions for the simple reason that they evolve on a dynamical time scale which is too short to be consistent with the large number of cometary and core/halo UC HII regions. A variation on this general theme has been proposed by Forster *et al.* (1990), Fey *et al.* (1995) and others who postulate star formation in a medium with a sharp density gradient; however, this also suffers from the short dynamical time scale problem. It is also not clear that the predicted morphologies reproduce the observed ones.

Reid *et al.* (1981) and Ho and coworkers (Zheng *et al.* 1985; Ho & Haschick 1986; Keto, Ho & Reid 1987; Keto, Ho & Haschick 1987,1988) have postulated that UC HII regions could be maintained in a compact state for periods well in excess of their apparent expansion ages due to massive in-fall of molecular gas. Unfortunately, it is not clear how this would account for the observed morphologies and it also requires large in-fall rates for long periods of time (10^{-3} to 10^{-4} solar masses per year for $\sim 10^5$ yr). In the early accretion phase of star formation this idea is surely applicable, but it does not seem to apply to the later UC HII region phase when the star has already initiated nuclear burning.

An interesting recent proposal for UC HII regions is the "photo-evaporating disk" model of Hollenbach *et al.* (1994). Here, the central star slowly photoionizes its own accretion disk and produces a UC HII region by the flow of ionized gas away from the disk. The advantage of this model is that it can naturally account for the large number of UC HII regions and the associated lifetime problem (if the accretion disk is massive enough, $\sim$ 0.3 M_*), it makes no demands on the structure or properties of the natal molecular cloud, and it depends on only 4 parameters (the existence of a massive disk, the stellar UV photon flux, the stellar mass, and the stellar wind mass loss rate). The difficulty with this model is that we do not know if it can produce the observed morphologies. The morphologies resulting from this model were not calculated, but it would appear to be difficult to achieve a cometary morphology if the star and its accretion disk are at rest relative to the ambient medium.

A bow shock hypothesis was initially suggested by Reid & Ho (1985) for G34.26 and by WCa for the class of cometary nebulae. The theoretical details of stellar wind supported bow shocks for NFMS have been presented by Van Buren et al. (1990), Mac Low et al. (1991), and Van Buren & Mac Low (1992). The basic idea is that the molecular fragments out of which massive stars are formed have some motion relative to the molecular cloud core in which they reside. This motion is later apparent as the observed dispersion velocity of OB associations (typically 2-5 km s^{-1}). A NFMS with a motion of a few km s^{-1} relative to the ambient molecular cloud will produce a bow shock with a high Mach number (10-20). The outer bow shock is supported from the inside by the stellar wind which has velocities of 2000-3000 km s^{-1}. Thus one sees a cometary morphology when viewed from the side and a core/halo morphology when viewed head-on or tail-on (Mac Low et al. 1991). A compact cometary or core/halo morphology will persist for as long as the star remains in the cloud core. The observed kinematics of the ionized gas in G29.96 by Wood & Churchwell (1991) are quite consistent with the bow shock models of Van Buren & Mac Low (1992). Shell structured UC HII regions are presumably nebulae that have essentially no motion relative to the ambient cloud core and therefore can expand essentially uniformly in all directions. These objects comprise only 5% of the population, whereas over 60% of all UC HII regions are cometary or core/halo objects which are explicable as bow shocks.

The bow shock hypothesis has been criticized because the stellar velocities implied by the models appear to be too high for the stars to remain long enough in the natal cloud core to be consistent with the required ages based on the number of observed UC HII regions. Typically, dense cloud cores, where massive stars are formed, have sizes of $\sim$ 0.5 pc. At velocities $\leq$ 5 km s^{-1} the crossing time would be $\geq 10^5$ yr which is about the time required by the observed number of UC HII regions. The model calculated for G29.96 by Van Buren & Mac Low (1992) used a stellar velocity of 20 km s^{-1} in order to fit the observations. Although, there may be a few stars with such high relative velocities, they must be the extreme because this is well outside the normal dispersion velocity of OB associations (typically 2-5 km s^{-1}). Perhaps more importantly, it must be recognized that the stellar velocity implied by the model is also a function of the ambient density and the stellar mass loss rate. In principle, the stellar velocity could be reduced by increasing the assumed ambient density, or changing the assumed wind speed and/or mass loss rate, or some combination of all these. Without further data, all these parameters have a fairly large range of possible values. To emphasize this last point, Cesaroni *et al.* (1994) have recently found a very dense molecular clump with a density $\sim 10^7$ cm^{-3} just ahead of the ionized arc toward G29.96. Van Buren & Mac Low (1992) assumed an ambient density of 5.5 x 10^4 cm^{-3}. In conclusion, the main criticism against the bow shock hypothesis does not appear to be fatal.

Perhaps both a photo-evaporating disk and a bow shock may occur simultaneously in a UC HII region. This possibility would be well worth further investigation. The models discussed above are schematically illustrated in Figure 1.

2.2 MASERS, MOLECULAR CLUMPS, AND COMETARY NEBULAE

An interesting pattern is begining to emerge for cometary UC HII regions regarding the relative positions of H$_2$O masers, dense, warm molecular clump, and the ionized arc. In Figure 2, a montage of 6 cometary UC HII regions are shown with the positions of the nearest H$_2$O masers from Hofner & Churchwell (1995) indicated by "+" symbols. Often one or two even more compact HII regions are found in front of the cometary arc (see G19.61 and G34.26 in Fig. 2). One can also see that toward this selection of sources, the H$_2$O masers generally lie a few arc secs in front of the ionized arc in a fairly tight cluster for the most part. In fact, of 9 UC HII regions mapped with the VLA by Hofner & Churchwell (1995) in the 1.3 cm H$_2$O maser line, 7 have H$_2$O masers within a few arcsecs in front of the HII arcs and 2 have H$_2$O masers 10" to 20" to the side of the HII arcs. Although this is small number statistics, the frequency with which this configuration occurs strongly suggests a mechanism for successive massive star formation. Perhaps of more significance is the fact that for those sources where we have high resolution images in the NH$_3$(4,4 and/or 5,5) lines, the H$_2$O masers are coincident with the hot, dense ammonia clump and not with the HII region. This is illustrated in Figure 3 where the HII region, dense ammonia clumps, and water masers are all shown on the same image at the same scale for G9.62. The ammonia clumps are very dense and it is tempting to postulate that they are the site of the next generation of star formation with the water masers perhaps indicating the positions of currently forming stars embedded in the dense ammonia clumps. I will return to this in section IV.

3.   Hot, Wind Shocked Central Cavities

Here, we are interested in the possibility of confirming the presence of a hot, low density, wind-shocked cavity inside cometary and shell-like UC HII regions. A plasma with $T_e \geq 10^7$ K and $n_e \leq 10^2$ cm^{-3} would provide the internal pressure to prevent the cavity from "filling in" on sound crossing time scales (a few thousand years) due to flows from the dense HII shell.

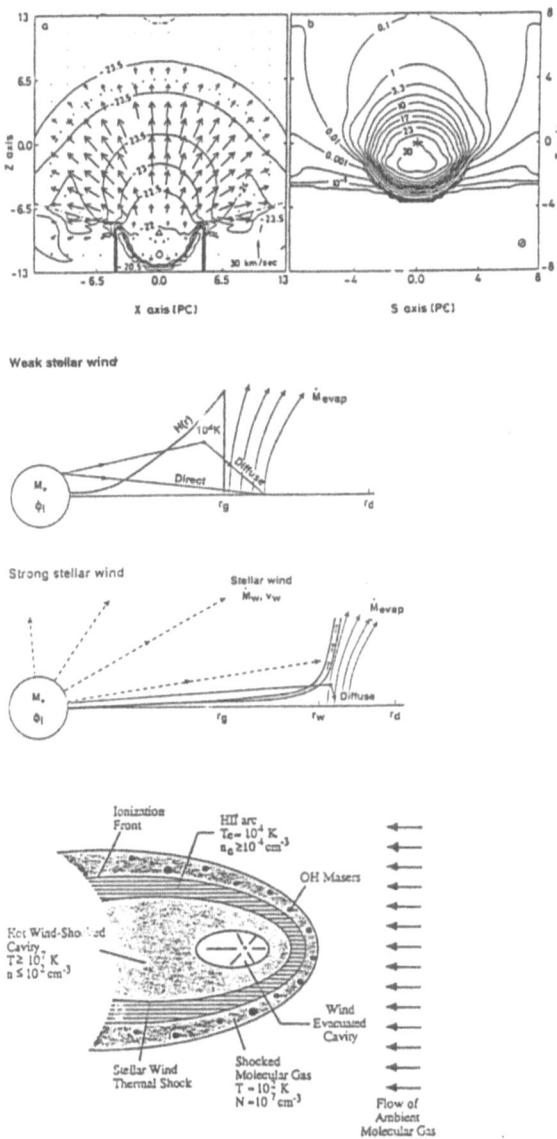

Fig. 1.   Schematics of three suggested models for UC HII regions. Top Panels: A champagne flow model for an O star located close to the boundary of a molecular cloud from York, Tenorio-Tagle & Bodenheimer (1983). The left panel shows the velocity structure of the gas. The triangle is the position of the star and the arrows represent the velocity structure of the gas. The right panel shows the calculated brightness distribution of the champagne flow at 11 cm. Middle Panels: Schematics of a weak wind (top panel) and a strong wind (lower panel) photo-evaporating disk from Hollenbach *et al.* (1994). Bottom Panel: Schematic of a wind supported bow shock from Churchwell (1991).

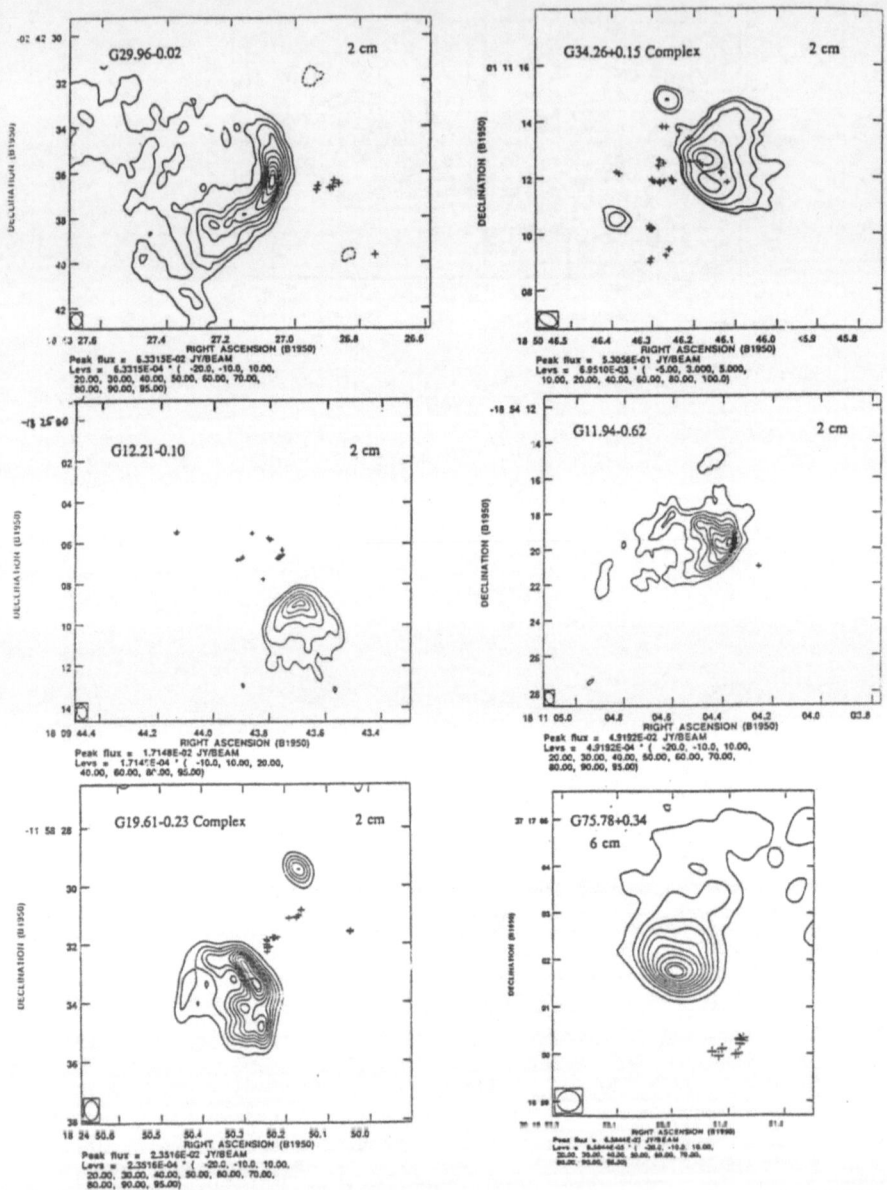

Fig. 2. A montage of cometary UC HII regions showing the 2 or 6 cm continuum structure of the ionized gas solid contours and the positions of water masers by "+" symbols. For the most part, the H₂O masers lie just ahead of the brightest part of the ionized arcs and in some cases an even more compact continuum source is located ahead of the arc. The source name, continuum wavelength, and HPBW are indicated in each panel.

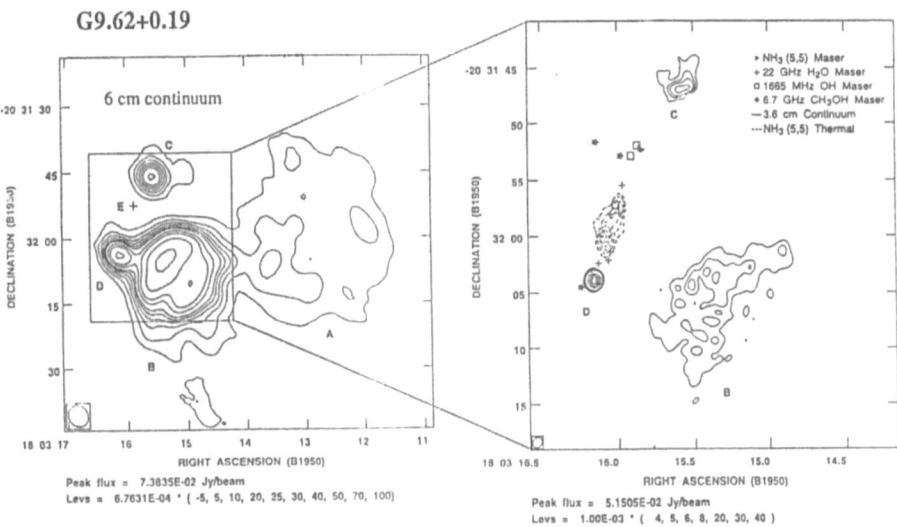

Fig. 3.   Left panel: A 6 cm continuum image of the G9.62+0.19 complex of HII regions. Each HII regions is designated by a letter and the HPBW is shown in the lower left corner. The rectangle enclosing components B, C, D, and E is shown in the right panel with much higher resolution. Right panel: The 3.6 cm continuum is shown for components B, C, D, and E. Component B is not well imaged at this resolution. The "ridge" is composed of very compact HII regions D and C, a very dense and warm NH_3 cloudlet (dashed contours) and a large number of masers which are labeled in the upper right corner of the panel. The ridge appear to be the site of the youngest stars in the complex, including some that probably are still in the process of forming.

Such conditions could be provided by the fast stellar winds from O and B stars.

Yamauchi & Koyama (1993; hereafter YK) have reported extended Fe XXV 6.7 keV line emission from the galactic ridge and the galactic bulge. The integrated luminosity in the 6.7 keV line from the galactic ridge was measured to be $6.5 \pm 1.3 \times 10^{36}$ ergs s^{-1} with a z-scale height of only 100 ± 20 pc. The luminosity in this emission line component accounts for $\sim 7\%$ of that in the 2-10 keV band and it is consistent with gas temperatures in the range $3\text{-}10 \times 10^7$ K. The confinement of gas this hot to such a thin layer in the galactic disk is difficult to understand. YK suggested that this emission might be produced by young supernovae.

It is interesting that the scale height of the Fe line emission is about the same as that of O stars in the galactic plane (see WCa; White, Becker & Helfand 1991; Helfand et $al.$ 1992). Is it possible that a major fraction of the Fe line emission from the galactic ridge may be produced in the hot, wind-shocked central cavities of UC HII regions? Because opacity depends on frequency as ν^{-3} at X-ray frequencies, 6.7 keV photons are not significantly attenuated by the surrounding molecular cloud or hydrogen along the line of sight. In the following, I will estimate the luminosity in the 6.7 keV line from a single UC HII region in two ways and show that the total 6.7 keV luminosity of all UC HII regions combined could produce a major fraction, if not all, of the galactic ridge emission. From Mewe *et al.* (1985), we find that the 6.7 keV line is a blend of 3 Fe transitions, which together have a cooling rate of 5.8 x 10^{-25} erg cm^3 s^{-1} at a temperature of 6.3 x 10^7 K; this represents about 3% of the total cooling rate. Thus, the luminosity in the line is $L_{6.7}$(erg s^{-1}) = 5.8 x 10^{-25} EM$_V$(cm^{-3}) where EM$_V$ is the volume emission measure. From high resolution radio images, we estimate an average hot core radius of $\sim$ 4 x 10^{17} cm and an average density of $\sim$ 100 cm^{-3}, resulting in an EM$_V$ $\sim$ 2.7 x 10^{57} cm^{-3} and $L_{6.7}$ $\sim$ 2 x 10^{33} erg s^{-1} per UC HII region or a total luminosity of the disk from $\sim$ 2000 nebulae of $\sim$ 3 x 10^{36} erg s^{-1}. Within the accuracy of this estimate, this is equivalent to the observed disk luminosity in the 6.7 keV line. A second, independent, method of estimating the 6.7 keV luminosity is to examine the properties of O star winds. Using data tabulated by Lamers & Leitherer (1993) for main sequence O stars, we can estimate wind luminosities L_w = (1/2)$\dot{M}_w$ ν_w^2 where $\dot{M}_w$ is the stellar mass loss rate and ν_w is the wind terminal velocity. For ZAMS stars hotter than O8, typical values are $\nu_w \approx$ 2500 km s^{-1} and $\dot{M}_w \approx 10^{-6}$ solar masses per year, resulting in $L_w \sim$ 2 x 10^{36} erg s^{-1}. We empirically estimate that $\sim$ 5% of L_w goes into heating the cavity to temperatures of several keV from measured X-ray fluxes of the UC HII regions W3B and G25.40 with the IPC aboard the Einstein satellite. About 3% of this is radiated in the 6.7 keV line. Thus we find that $L_{6.7} \sim$ 3 x 10^{33} erg s^{-1} per UC HII region, which is essentially the same as we found from the emission measure of the hot cavity.

A theoretical spectrum at X-ray energies is shown in Figure 4 for the UC HII region W3B. This spectrum was calculated using software provided by the ASCA satellite facility assuming a 40,000 s integration time, a source temperature equivalent to 5 keV, a 1-10 keV X-ray flux of 8.5 x 10^{-11} erg cm^{-2} s^{-1} (based on the spectral type of the ionizing star and its expected average wind strength) attenuated by a H column density of 6 x 10^{23} cm^{-2}. The dominant features in this spectrum is the 6.7 keV line, an underlying 3-10 keV continuum, and several weaker Fe lines between 7-10 keV. If the model assumptions are correct, the 6.7 keV line should be relatively easy to detect from single UC HII regions by the ASCA satellite with its much smaller beam and higher sensitivity than previous X-ray telescopes.

Figure 5 shows the distribution of 6.7 keV ridge emission (b $\leq 2°$) measured by YK with the Ginga satellite. The solid curve is their fit to the Ginga data. We estimate the Fe line flux from UC HII regions (indicated by a diamond in Fig. 5) at l = 27° in a 1° x 2° field of view by assuming

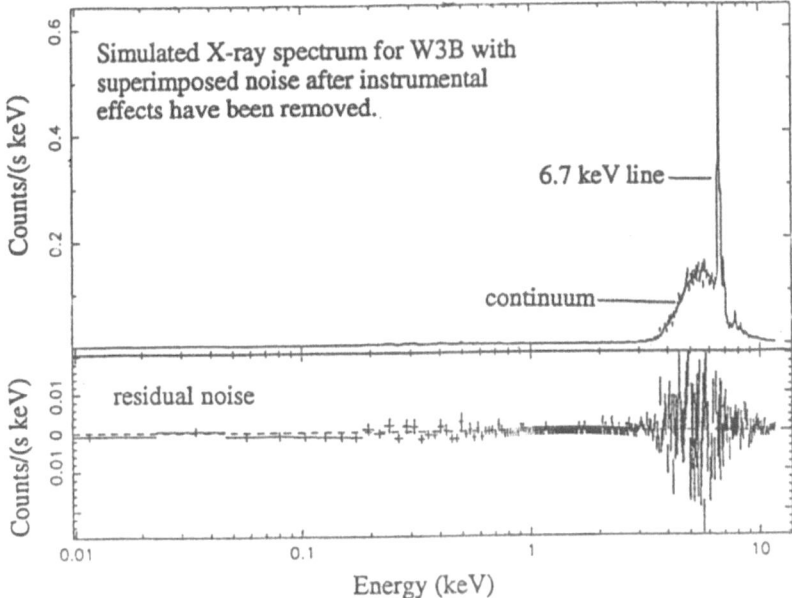

Fig. 4. A simulated X-ray spectrum of W3B for a 40,000 s observation with the ASCA satellite. The source is attenuated by an HI column of 6×10^{23} cm^{-2} and a temperature of 5 keV has been assumed. The statistical noise (source + background) is shown in the lower panel.

that about 2/3 of all UC HII regions (*i.e.* O8 and hotter ionizing stars) have winds strong enough to contribute to the 6.7 keV line; each has a luminosity of 1.6×10^{33} erg s^{-1} in the 6.7 keV line; and, they are uniformly distributed in a ring between 4 to 7 kpc from the galactic center. We find an Fe line flux of ~ 1.8 photons s^{-1} beam^{-1} (see the diamond at $l = 27°$ in Fig. 5), compared to a measured value of ~ 2 photons s^{-1} beam^{-1}. This supports the hypothesis that UC HII regions may produce a large fraction (if not all) of the Fe line flux observed by Ginga in the galactic disk. It also provides an important new tool to study hot central cavities of UC HII regions which has so far only been inferred from indirect evidence but never observed directly. We will be testing this hypothesis using the ASCA satellite during the coming year.

4. G9.62: A Case Study

G9.62 is a complex of compact and UC HII regions accompanied by hot, dense, molecular clumps and OH, H_2O, CH_3OH, and NH_3 masers (see Fig. 3). The only known $NH_3(5,5)$ maser in the entire galaxy resides in the G9.62 complex (Cesaroni, Walmsley & Churchwell 1992; Hofner *et al.* 1994). The following is a summary of the molecular line observations available for this

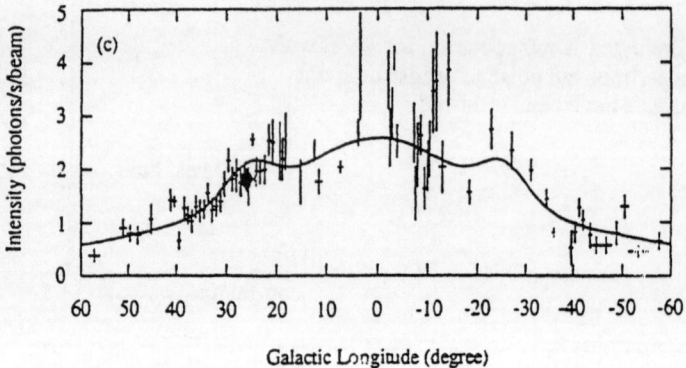

Fig. 5. The 6.7 keV FeXXV line flux as a function of galactic longitude from Yamauchi & Koyama (1993). The solid line is their fit to the observations (points with error bars). The diamond at longitude 27° is our estimate of the Fe line emission fom all UC HII regions ionized by O8 stars or hotter along this line of sight.

source. Observed probes of hot, dense, thermal molecular gas are: CS(7-6) emission by Plume, Jaffe & Evans (1992); CH_3CN (6-5, 8-7, and 12-11) by Olmi, Cesaroni & Walmsley (1993) and Hofner et al. (1995); $NH_3(4,4$ and/or 5,5) by Cesaroni, Walmsley & Churchwell (1992), Cesaroni, et al. (1994), and Hofner et al. (1994); and $C^{18}O$ (1-0) by Hofner et al. (1995). The following masers have been observed toward G9.62: OH and/or H_2O by Kemball, Diamond & Mantovani (1988), Forster & Caswell (1989), and Churchwell, Walmsley & Cesaroni (1990); CH_3OH by Menten (1991), and Norris et al. (1993); and $NH_3(5,5)$ by Cesaroni, Walmsley & Churchwell (1992) and Hofner et al. (1995). HI absorption measurements (Hofner et al. 1994) suggest that G9.62 lies in the "3 kpc arm" at a distance of 5.7 kpc (1" ≈0.028 pc).

Examination of Figure 3 suggests that there are at least 3 successive generations of massive star formation in the G9.62 complex: the oldest being component A (which is not well imaged in Fig. 3 because it is partially resolved out by the VLA), the next oldest being component B just to the west of A. The youngest (and perhaps still forming) generation of stars lies along a narrow, almost linear, ridge comprising the very compact continuum components C, D, and E as well as a dense molecular clump(s) and many masers. The structure and nature of this ridge of diverse objects is the subject of the rest of this section.

Figure 6 shows the 3 mm continuum image of the ridge in G9.62 obtained with the Owens Valley Radio Observatory's millimeter array by Hofner et al. (1995). The synthesized HPBW was 3.1" x 2.7" (0.086 pc x 0.075 pc). Comparing this with the 3.6 cm continuum image in the right panel of Fig. 3 shows that: components C and D are detected at both cm and mm wavelengths, although C is barely detected at 3 mm; continuum emission is

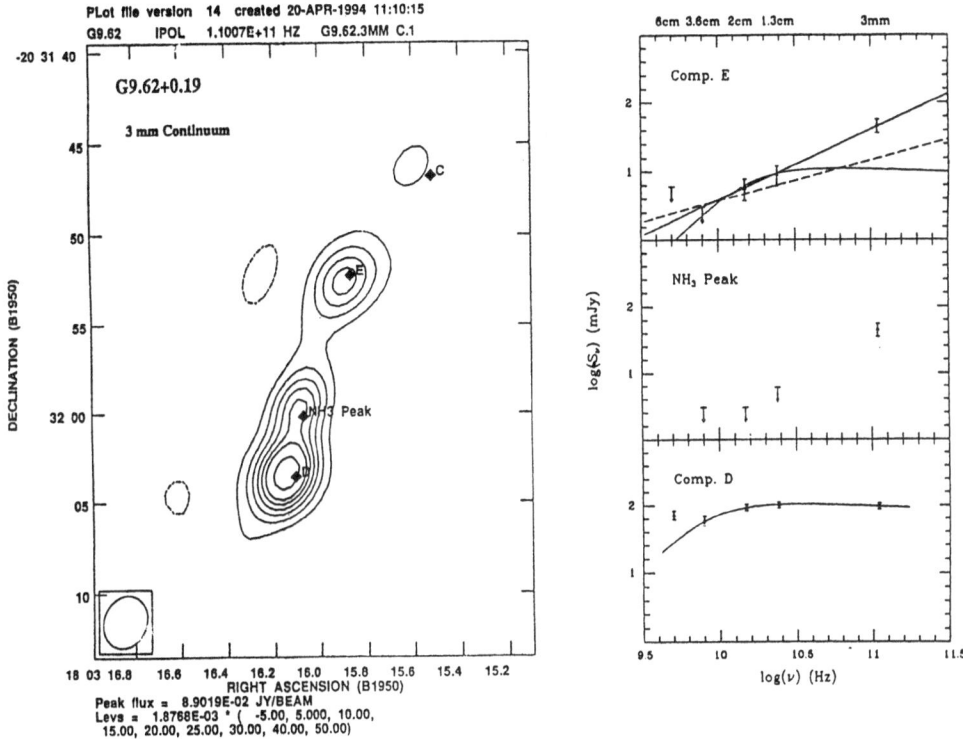

Fig. 6.   Left panel: A 3 mm continuum image of the "ridge" components in the G9.62 complex. Right panel: The measured flux densities of the components D, E, and the NH_3 peak with possible model fits to the observations (Hofner et al. 1995).

detected at 3 mm from the NH_3 clump but not at $\lambda \geq 1.3$ cm; continuum emission at 3 mm is much brighter from component E than at cm wavelengths (actually not detected at $\lambda > 2$ cm). The flux density distributions for components D, E, and the NH_3 peak are shown in the left panel of Fig. 6. Component D is consistent with free-free emission that becomes optically thick at $\lambda \geq 2$ cm. A fit to the continuum spectrum of component D (assuming a homogenious, spherical nebula) implies a diameter ≤ 0.014 pc, $T_e \approx 8000$ K, $n_e(\text{rms}) \approx 1.8 \times 10^5$ cm^{-3}, emission measure of $\approx 4.5 \times 10^8$ pc cm^{-6}, and ionization by a B0.5 ZAMS star.

Since continuum emission has so far only been detected at 3 mm from the NH_3 peak, its spectral index is unknown. If it were a power law from 2 cm to 3 mm, then the spectral index, a, would be ≥ 1. It is interesting that the two compact continuum sources in front of the arc toward G34.26 also have spectral indices of ~ 1 (Gaume, Fey & Claussen 1994). With only the continuum data now available, we cannot be sure whether the 3 mm continuum from this source is due to warm thermal emission from dust or

to free-free emission from ionized gas. Either way, energetic arguments favor
a central source of heat or ionization rather than an external one.

Component E appears to have a power spectrum from ~ 2 cm to 3 mm.
The spectrum could possibly be compatable with a UC HII region having
excess thermal dust emission at 3 mm, or with two UC HII regions with one
having a turn-over wavelength at ~ 1.3 cm and the other between 1.3 cm
and 3 mm. If the spectrum is a power law from 3 cm to 3 mm, it could be
compatable with a massive stellar wind having a partially ionized envelope
(Simon $et\ al.$ 1983) with a density structure

$$n_e(r) \approx 1.6 \times 10^{-2} \left[\frac{r}{pc}\right]^{-2.7} cm^{-3} \qquad (1)$$

where r is the distance from the ionizing star (Hofner $et\ al.$ 1995). A partially
ionized wind envelope can produce a steeper density profile than expected
for a fully ionized constant velocity wind. A spectral index of 0.6 expected
for a fully ionized constant velocity stellar wind is excluded by the data.
Energetic considerations strongly suggest that the continuum components
D, E, and probably the NH_3 peak are ionized from within by newly emerging
massive stars.

In Figure 7 we show images from Hofner $et\ al.$ (1995) of the ridge in
3 mm continuum, CH_3CN (6-5), and $C^{18}O$ (1-0) line emission. Unlike the
hot, thermal $NH_3(5,5)$ emission (see Fig. 3) which does not coincide with
continuum componets D or E, CH_3CN (6-5) encompasses the D, E, and the
NH_3 peak components and peaks at the latter position. The CH_3CN (6-5)
emission is clearly confined to the ridge and it is unresolved perpendicular
to the ridge. It is interesting that CH_3CN (6-5) maximum is precisely on
the $NH_3(5,5)$ peak where a very faint continuum source seems to be just
emerging. The coincidence of CH_3CN maximum emission with the $NH_3(5,5)$
peak and the supposedly embedded faint continuum source suggests that
CH_3CN may be a better probe of newly emerging massive stars than excited
NH_3. The standard LTE, optically thin (column density vs excitation plots)
analysis for CH_3CN is not valid toward any of the clumps in the ridge
because the lower K-transitions are optically thick. Consequently, Hofner
$et\ al.$ (1995) derived properties of the molecular gas using a large velocity
gradient, statistical equilibrium analysis of the CH_3CN emission. The results
are given in Table 1. The most striking results of this analysis are the large
CH_3CN column densities inferred for each clump.

The $C^{18}O$ emission shown in Fig. 7 is also closely confined to the ridge,
but it is more broadly distributed perpendicular to the ridge than CH_3CN
emission or 3 mm continuum emission. Hofner $et\ al.$ (1995) find that the
$C^{18}O$ emission measured with the OVRO array represents about 20% of the
total CO emission in this direction. They infer a total H_2 mass associated
with the ridge of ~ 1700 solar masses from the $C^{18}O$ emission assuming

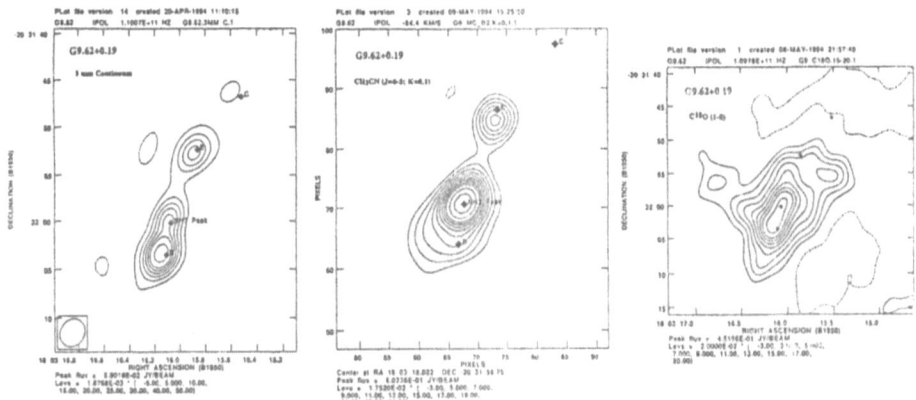

Fig. 7. Images of the "ridge" in 3 mm continuum emission, CH_3CN (J=6-5; K=0,1) line emission, and $C^{18}O$ (J=1-0) line emission from Hofner *et al.* (1995) are shown. The locations of the continuum sources C, D, E, and the NH_3 peak are designated by diamonds.

thermalization at the kinetic temperature of 50 K and $[C^{18}O]/[H_2] \sim 1.7$ x 10^{-7} (Frerking *et al.* 1982). They find a virial mass of about 1300 solar masses.

Perhaps the most interesting aspect of the G9.62 complex is that at least 3 generations of star formation can be identified. From Fig. 3, one sees that component A is larger and more diffuse than all the other sources, component B is brighter and more compact than component A, and the sources in the ridge are the most compact and are the only ones associated with maser emission. The continuum source embedded in the NH_3 peak appears to be a massive star still in the process of forming! The length, thinness, and straightness of the ridge is puzzling. A pre-existing ridge of neutral molecular gas which has been impacted by a shock from component B could have possibly triggered star formation in the ridge, but this seems unlikely since the ridge and I-front of component B are separated by almost 0.2 pc. What has caused the successive generations of star formation to propagate from west to east in this complex? Although we now have a much better picture of this complex than before, it is clear that many questions remain to be answered.

5. Acknowledgements

Much of the work reported in this review are results obtained by Peter Hofner, Deborah Shepherd, Riccardo Cesaroni, and Malcolm Walmsley who have generously shared data prior to publication. I especially thank Peter Hofner and Deborah Shepherd for critically reading this review and giving suggestions that improved the content and presentation.

References

Adams, F. C.: 1993, in *Massive Stars: Their Lives in the Interstellar Medium*, eds: J. P. Cassinelli & E. B. Churchwell, A. S. P. Conf. Series **35**, p56.

Cesaroni, R., Churchwell, E., Hofner, P., Walmsley, C. M., Kurtz, S.: 1994, Astron. & Astrophys. in press.

Cesaroni, R., Walmsley, C. M., Churchwell, E.: 1992, Astron. & Astrophys. **256**, 618.

Churchwell, E.: 1990, Astron. & Astrophys. Rev. **2**, 79.

Churchwell, E.: 1991, in *The Physics of Star Formation & Early Stellar Evolution*, eds: C. J. Lada & N. D. Kylafis, Kluwer Academic Publ., p221.

Churchwell, E.: 1993, in *Massive Stars: Their Lives in the Interstellar Medium*, eds: J. P. Cassinelli & E. B. Churchwell, A. S. P. Conf. Series **35**, p35.

Churchwell, E., Walmsley, C. M., Cesaroni, R.: 1990, Astron. & Astrophys. Suppl. **83**, 119 (**CWC**).

Fey, A. L., Gaume, R., Nedoluha, G. E., Claussen, M. J.: 1995, Astrophys. J. Nov. 1 issue.

Forster, J. R., Caswell, J. L.: 1989, Astron. & Astrophys. **213**, 339.

Forster, J. R., Caswell, J. L., Okumura, S. K., Hasegawa, T., Ishiguro, M.: 1990, Astron. & Astrophys. **231**, 473.

Frerking, M. A., Langer, W. D., Wilson, R. W.: 1982, Astrophys. J. **262**, 590.

Garay, G., Lizano, S., Gomez, Y.: 1994, Astrophys. J. **429**, 268.

Gaume, R. A., Fey, A. L., Claussen, M. J.: 1994, Astrophys. J. Sept.10 issue.

Helfand, D. J., Zoonematkermani, S., Becker, R. H., White, R. L.: 1992, Astrophys. J., Suppl. **80**, 211.

Ho, P. T. P., Haschick, A. D.: 1986, Astrophys. J. **304**, 501.

Hofner, P., Churchwell, E.: 1995, Astrophys. J., Suppl. in preparation.

Hofner, P., Kurtz, S., Churchwell, E., Walmsley, C. M., Cesaroni, R.: 1994, Astrophys. J. submitted.

Hofner, P., Kurtz, S., Churchwell, E., Cesaroni, R., Walmsley, C. M.: 1995, Astrophys. J. in preparation.

Hollenbach, D., Johnstone, D., Shu, F.: 1993, in *Massive Stars: Their Lives in the Interstellar Medium*, eds: J. P. Cassinelli & E. B. Churchwell, A. S. P. Conf. Series **35**, p26.

Hollenbach, D., Johnstone, D., Lizano, S., Shu, F.: 1994, Astrophys. J. **428**, 654.

Kemball, A. J., Diamond, P. J., Mantovani, F.: 1988, Mon. Not. of the RAS **234**, 713.

Keto, E. R., Ho, P. T. P., Haschick, A. D.: 1987, Astrophys. J. **318**, 712.

Keto, E. R., Ho, P. T. P., Haschick, A. D.: 1988, Astrophys. J. **324**, 920.

Keto, E. R., Ho, P. T. P., Reid, M. J.: 1987, Astrophys. J. **323**, L117.

Lamers, H. G. L. M., Leitherer, C.: 1993, Astrophys. J. **412**, 771.

Mac Low, M.-M., Van Buren, D., Wood, D. O. S., Churchwell, E.: 1991, Astrophys. J. **369**, 395.

Menten, K. M.: 1991, Astrophys. J. **380**, L75.

Mewe, R., Gronenschild, E. H. B. M., van den Oord, G. H. J.: 1985, Astron. & Astrophys. Suppl. **62**, 197.

Norris, R. P., Whiteoak, J. B., Caswell, J. L., Wieringa, M. H., Gough, R. G.: 1993, Astrophys. J. **412**, 222.

Olmi, L., Cesaroni, R., Walmsley, C. M.: 1993, Astron. & Astrophys. **276**, 489.

Plume, R., Jaffe, D. T., Evans, N. J.: 1992, Astrophys. J., Suppl. **78**, 505.

Reid, M. J., Haschick, A. D., Burke, B. F., Moran, J. M., Johnston, K. J., Swenson, G. W. Jr.: 1981, Astrophys. J. **239**, 89.

Reid, M. J., Ho, P. T. P.: 1985, Astrophys. J. **288**, L17.

Simon, M., Felli, M., Cassar, L., Fischer, J., Massi, M.: 1983, Astrophys. J. **266**, 623.

Tenorio-Tagle, G: 1979, Astron. & Astrophys. **71**, 59.

Van Buren, D., Mac Low, M.-M.: 1992, Astrophys. J. **394**, 534.

Van Buren, D., Mac Low, M.-M., Wood, D. O. S., Churchwell, E.: 1990, Astrophys. J. **353**, 570.

Welch, W. J.: 1993, in *Massive Stars: Their Lives in the Interstellar Medium*, eds: J. P. Cassinelli & E. B. Churchwell, A. S. P. Conf. Series **35**, p15.

White, R. L., Becker, R. H., Helfand, D. J.: 1991, Astrophys. J. **371**, 148.

Wolfire, M. G., Churchwell, E.: 1994, Astrophys. J. **427**, 889.

Wood, D. O. S., Churchwell, E.: 1989a, Astrophys. J., Suppl. **69**, 831 (**WCa**).

Wood, D. O. S., Churchwell, E.: 1991, Astrophys. J. **372**, 199.

Yamauchi, S. Koyama, K.: 1993, Astrophys. J. **404**, 620 (**YK**).

Yorke, H. W., Tenorio-Tagle, G., Bodenheimer, P.: 1983, Astron. & Astrophys. **127**, 313.

Yorke, H. W.: 1984, in *Proceedings of the Workshop on Star Formation*, ed: R. D. Wolstencroft, NERC/SERC Reprographic Services, p63.

Yorke, H. W.: 1993, in *Massive Stars: Their Lives in the Interstellar Medium*, eds: J. P. Cassinelli & E. B. Churchwell, A. S. P. Conf. Series **35**, p45.

Zheng, X. W., Ho, P. T. P., Reid, M. J., Schneps, M. H.: 1985, Astrophys. J. **293**, 522.

THE MASSIVE HOT CORE ASSOCIATED WITH

G31.41+0.31

C.M. WALMSLEY

Max Planck Institut für Radioastronomie Bonn, Auf dem Hügel 69, D-53121 Bonn, Germany

R.CESARONI

Osservatorio Astrofisico di Arcetri, Largo E. Fermi 5, I-50125 Firenze, Italy

L.OLMI

Arecibo Observatory, PO Box 995, Arecibo, Puerto Rico 00613

and

E.CHURCHWELL and P.HOFNER

Washburn Observatory, University of Wisconsin-Madison, 475 N.Charter St., Madison, WI 53706, USA

Abstract. We have used the IRAM Plateau de Bure Interferometer to observe the ground state and vibrationally excited lines of Methyl Cyanide towards the UCHII region G31.41+0.31. We also obtained a map of the continuum emission at a frequency of 110 GHz. We detect a hot molecular core in the emission both of methyl cyanide and of the 110 GHz continuum. We estimate a temperature of 200 K and a mass of 1000 $M_\odot$ for this compact massive region. We also detect a velocity gradient or shift across the methyl cyanide core whose origin could be due to rotation.

Key words: Molecular Lines – Hot Core – Massive Star Formation

1. G31.41+0.31

Ultra-Compact HII Regions (often abbreviated as UCHII's) serve as a tracer for massive star formation throughout our galaxy. Several recent reviews (*e.g.* Churchwell 1991) summarise our present knowledge of their morphology and physical characteristics. There is some controversy about their lifetime but there seems little doubt that they represent the first easily observable phase in the evolution of an O star. There is also little doubt that they form from extremely dense compact molecular clumps or cores and over the past few years we have been attempting to determine the properties of such molecular cores. This work is described for example in Churchwell *et al.* (1990), Olmi *et al.* (1993), and Cesaroni *et al.* (1994a). One of the results of these studies has been that we have found close to but not coincident with the UCHIIs a number of molecular regions with properties analogous to those of the Orion hot core. These are characterised by much higher temperatures than the molecular gas in their surroundings (temperatures of order 100-200 K as compared to 20-30 K). They are also extremely dense and have masses of at least 10 $M_\odot$.

Astrophysics and Space Science **224**: 173–175, 1995.

© 1995 *Kluwer Academic Publishers.*

One of the most interesting objects of this type which we discovered was the molecular core associated with the UCHII region G31.41+0.31. NH$_3$(4,4) VLA observations by Cesaroni et al. (1994a) showed that there was a very hot ammonia clump (temperature 120 K, ammonia column density 10^{19} cm^{-2}, diameter 0.1 pc) associated with this UCHII region. Moreover Churchwell et al. (1992) found the source to have easily observable lines of methyl cyanide (CH$_3$CN). Methyl cyanide is a symmetric top with several transitions of differing excitation at neighbouring frequencies. It has thus found applications as a "molecular thermometer" and we therefore decided to observe both ground and vibrationally excited lines of this species using the IRAM Plateau de Bure Interferometer at a frequency of 110 GHz (2.7 mm). Our angular resolution was approximately 2 arc seconds which corresponds to 0.08 parsec at the distance of G31.41+0.31 (7.9 kpc). In addition to the maps in the various line transitions, we obtained a map of the continuum emission at 110 GHz .

As we had expected (but in contrast to the results for G34.26 reported by MacDonald in this conference), we found a methyl cyanide clump coincident with the hot core seen in NH$_3$(4,4). The methyl cyanide emission appears slightly extended in our beam implying a core diameter of order 0.04 pc. We detect vibrationally excited levels of methyl cyanide more than 500 K above ground and estimate the gas temperature in the line emitting region to be 200 K.

More surprising was the detection of a 3 mm continuum source of approximately the same size as that seen in CH$_3$CN and which is also coincident with the ammonia hot core. There is no corresponding continuum source seen on our VLA 1.3 cm maps and we conclude that the 3 mm continuum emission is due to heated dust. We derive on this basis a mass of 1000 M$_\odot$ for the G31.41 hot core although this involves "an educated guess" about the 3 mm dust emissivity. However, it is also intriguing that when we map the peak positions determined in different velocity channels and different K-components of methyl cyanide, we see an elongated structure of size approximately 1 arc second. Moreover, when we additionally plot channel velocity against angular offset along the major axis of this structure, we see evidence for a velocity gradient of 400 km s^{-1} pc^{-1}. The interpretation of this gradient (or velocity shift) is unclear but one possibility is rotation. If so, the inferred dynamical mass is of the same order as that inferred for the source of the 3 mm continuum emission. However, there are other possibilities such as outflow and for the moment, we conclude one should keep an open mind. Finally, we note that the results of this study are given in more detail in an article by Cesaroni et al. (1994b).

References

Cesaroni R., Churchwell E., Hofner P., Walmsley C.M., Kurtz S.: 1994a, Astron. & Astrophys. **288**, 903.

Cesaroni R., Olmi L., Walmsley C.M., Churchwell E., Hofner P.: 1994b, Astrophys. J. in press.

Churchwell E., Walmsley C.M., Cesaroni R.: 1990, Astron. & Astrophys. Suppl. **83**, 119.

Churchwell E.: 1991, in *The Physics of Star Formation & Early Stellar Evolution*, eds: C.J. Lada & N. Kylafis, NATO ASI Series.

Churchwell E., Walmsley C.M., Wood D.O.S.: 1992, Astron. & Astrophys. **253**, 541.

Olmi L., Cesaroni R., Walmsley C.M.: 1993, Astron. & Astrophys. **276**, 489.

PLATE II: East tower of the 1894 building. The enclosure at the top houses the **36"** reflector, which was used for spectroscopy until the early 1970s.

STRUCTURE AND CHEMISTRY IN THE HOT
MOLECULAR CORE G34.3+0.15

G.H.MACDONALD and R.J.HABING*

*Electronic Engineering Laboratory, University of Kent, Canterbury, Kent CT2 7NT,
ENGLAND*

and

T.J.MILLAR

Department of Mathematics, UMIST, P.O. Box 88, Manchester M60 1QD, ENGLAND

Abstract. New multifrequency spatial and spectral studies of the hot molecular core
associated with the ultracompact HII region G34.3+0.15 have demonstrated an extremely
rich chemistry in this archetypal hot core and revealed differing spatial structure between
certain species which may be a dynamical effect of chemical evolution. The structure of
the hot core has been studied with the JCMT in the high excitation J=19-18 and J=18-
17 lines of CH_3CN and with the Nobeyama Millimetre Array at 4″ arc resolution in the
J=6-5 transition. Comparison with a VLA $NH_3(3,3)$ map shows a displacement between
peak emission in the two chemical species which is consistent with chemical processing on
a time scale comparable to the dynamical time scale of $\simeq 10^5$ yrs.

A 330-360 GHz spectral survey of the hot core with the JCMT has detected 358 spectral
lines from at least 46 distinct chemical species, including many typical of shocked chemistry
while other species indicate abundances that reflect the chemistry of a previous cold phase.
The first unambiguous detection of ethanol in hot gas has been made. Observations of 14
rotational transitions of this molecule yield a temperature of 125 K and column density
$\simeq 2 \times 10^{15}$ cm^{-2}. This large abundance cannot be made by purely gas-phase processes and
it is concluded that ethanol must have formed by grain surface chemistry.

Key words: molecular processes – ISM:clouds – gas,molecules

1. Introduction

Hot molecular cores in star-forming regions are particularly interesting because their gas-phase chemical compositions are determined by the material
evaporated from the icy mantles of interstellar grains, followed by subsequent reactions in the gas phase. The ultracompact HII region G34.3+0.15
and its associated hot molecular cloud core lie 3.1kpc from the Sun. High
resolution radio continuum maps show that the HII complex comprises two
ultracompact ($\sim 0.3''$) and one compact region (Reid & Ho 1985). The latter
has a prototypical "cometary" morphology, with a compact "head" $\sim 4''$ in
size and a 20″ "tail" trailing to the west (Wood & Churchwell 1989). This
HII region may have acquired its unusual shape either by relative motion
between the young star and the parent cloud (Reid & Ho 1985) or be due
to the HII region undergoing a champagne phase expansion from the cloud

---

* Present address: Canterbury Christ Church College, Canterbury, Kent CT1 1QU

Astrophysics and Space Science **224**: 177–180, 1995.
© *1995 Kluwer Academic Publishers.*

(Garay *et al.* 1986). We have made high resolution observations of line emission from the symmetric top molecule methyl cyanide (CH_3CN) in the hot molecular core of G34.3+0.15 in an attempt to determine its morphology and temperature structure and to investigate its chemical evolution and detailed relationship with the cometary HII region. A 330-360 GHz spectral survey of the hot core has also been made in order to investigate its chemistry and state of excitation.

2. Methyl Cyanide Mapping

The 349.4 GHz J=19-18 and 331.0 GHz J=18-17 lines of CH_3CN were mapped with the JCMT and found to be unresolved at 12" arc resolution. Statistical equilibrium (SE) modelling yielded temperatures of 315±15K and 200±25K respectively for the two transitions. Rotation diagram analysis including lower excitation J=12-11 & J=6-5 data from IRAM (Churchwell *et al.* 1992) & J=6-5 data from Onsala (Bergman & Hjalmarson 1989) gave excellent agreement with the SE modelling results, confirming the expectation that higher J lines probe hotter, denser gas in the cloud core.

Subsequent observations of 110.3 GHz J=6-5 emission of CH_3CN with Nobeyama Millimeter Array (NMA) at 4" arc resolution showed a strong unresolved core with a weak halo extended ~10" NW-SE (Figure 1). Unfortunately, the relatively poor signal/noise ratio in the extended structure seen in the different K components precluded the planned measurement of the detailed temperature structure of the core. However, the CH_3CN halo closely matches the 450μm dust continuum emission (Strong-Jones *et al.* 1991).

Perhaps the most interesting and exciting new result comes from a comparison of the NMA CH_3CN map and the VLA NH_3(3,3) map (Heaton *et al.* 1989). The peak NH_3 emission is offset slightly to the north of the axis of symmetry of the cometary HII region and is coincident with the peak of the 450μm dust emission. Peak CH_3CN emission is found to be displaced 2" NW of the NH_3 peak, an offset much greater than the positional error of either map (Figure 1). Could this morphology arise through a combination of chemical evolution and the relative motion of the young star and the ambient medium? NH_3 molecules form by gas phase reactions during the initial cold accretion of a molecular cloud and freeze out on to the dust grains. Subsequent heating of the dust by a newly formed star returns the ammonia to the gas phase where further chemical processing occurs (Charnley et al. 1992). We have computed the time-dependent abundances of NH_3 and CH_3CN (MacKay, private communication) using a comparable code to Charnley *et al.* for 173 species linked by 1801 reactions using initial abundances constrained by hot core observations and standard dark cloud values (*e.g.* Herbst *et al.* 1989). For T $\sim$ 300K, n(H_2) $\sim 2 \times 10^7$ cm^{-3}, we find that

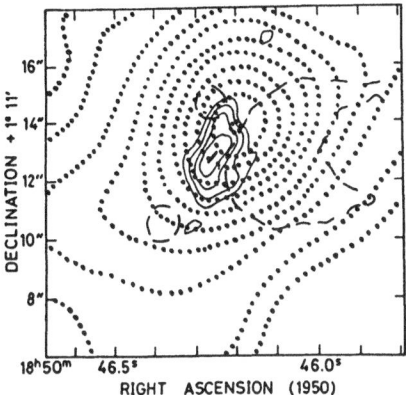

Fig. 1. Comparison of VLA map of $NH_3(3,3)$ emission in the hyperfine components (solid) and NMA map of CH_3CN J=6-5, K=0 emission (dotted). Dashed line indicates the compact "cometary" HII region and two ultracompact components.

the NH_3 abundance falls rapidly by ~ 3 orders of magnitude at an age $\sim$ 10^5 years as the CH_3CN abundance steadily rises. This suggests a scenario in which the NH_3 molecules are evaporated from grains by local heating by a young star embedded in the cometary HII region, swept along by gas flow parallel to the bow shock at the head of the cometary HII region and chemically processed to provide a peak CH_3CN abundance $\sim 10^5$ years later. The 0.04pc displacement of NH_3 and CH_3CN corresponds to a dynamical time $\sim 4 \times 10^4$ years for a relative velocity between the young star and ambient medium of ~ 1 km s^{-1}. Our model is supported by three observational features arising from the present study:

(i) the coincidence of peak NH_3 emission with $450\mu m$ emission from warm dust;

(ii) the relative displacement of NH_3 and CH_3CN peak emission corresponds to the direction and sense of laminar flow around the cometary HII region;

(iii) the commensurate values for the dynamical time for gas flow around the bow shock between the NH_3 and CH_3CN peak positions and the chemical evolutionary time for significant conversion between these two species.

3. 330-360 GHz Spectral Survey

We have made a systematic spectral line survey of the hot core in G34.3+0.15 with JCMT over the 330-360 GHz frequency range. No less than 358 independent lines were detected, of which only 68 could not be identified using available catalogues. The 290 identified lines represent some 46 different species, including isotopomers. The rich chemistry is most probably strongly affected by shock waves driven into the molecular core by the ionisation front and by the relative motion through the ambient medium giving rise to the cometary shape of the HII region. Shock excitation can release grain mantle material into the cloud and drive endothermic reactions in the gas. Observation of SiO indicates that grain cores, not only mantles, are being disrupted. Furthermore, the presence of sulphur-bearing molecules such as CS, SO, SO_2, OCS and HCS^+ is taken as direct evidence of shock chemistry whereas other species such as NH_2D, HCOOD, and HDS indicate abundances that reflect the chemistry of a previous cold phase (Millar 1993).

4. Detection of Ethanol.

Ethanol, C_2H_5OH, has been conclusively detected so far only in SgrB2 (Zuckerman *et al.* 1975) and W51M (Millar *et al.* 1988). In the 330-360 GHz spectral survey of G34.3+0.15, 14 lines were found to correspond to frequencies in a recent extensive laboratory and theoretical study (Sastry *et al.* 1994), all of which are b-type as expected since these lines are ~ 1000 times stronger than a-type. Rotation diagram analysis yields a beam-averaged column density of 2×10^{15} cm^{-2} and rotation temperature of 125K, considerably hotter than in earlier detections. The column density corresponds to a very high fractional abundance of $\sim 4\times10^{-9}$.

5. Conclusions

The NH_3/CH_3CN displacement seen in G34.3+0.15 appears to be due to a combination of dynamics and chemical evolution. The large abundance of ethanol cannot be made purely by gas-phase processes and we conclude that the ethanol must be formed efficiently via surface chemistry.

References

Bergman, P. & Hjalmarson, Å.: 1989 in: *The Physics and Chemistry of Interstellar Molecular Clouds*, eds: G. Winnewisser & J.T. Armstrong, Springer-Verlag, p124.
Charnley, S.B., Tielens, A.G.G.M. & Millar, T.J.: 1992, Astrophys. J. **399**, L71.
Churchwell, E., Walmsley, C.M. & Wood, D.O.S.: 1992, Astron. & Astrophys. **253**, 541.
Garay, G., Rodriguez, L.F. & van Gorkom, J.H.: 1986, Astrophys. J. **309**, 553.
Heaton, B.D., Little, L.T. & Bishop, I.S.: 1989, Astron. & Astrophys. **213**, 148.
Herbst, E., Millar, T.J., Wlodek, S. & Bohme, D.K.: 1989, Astr. & Astrophys. **222**, 205.
Millar, T.J.: 1993 in: *Dust and Chemistry in Astronomy*, eds: T.J. Millar & D.A. Williams, IOP Pub, p249.
Millar, T.J., Olofsson, H., Hjalmarson, Å. & Brown, P.D.: 1988, Astron. & Astrophys. **205**, L5.
Reid, M.J. & Ho, P.T.P.: 1985, Astrophys. J. **288**, L17.
Sastry, K.V.L.N., Pearson, J.C., Herbst, E. & DeLucia, F.C.: 1994, in preparation.
Strong-Jones, F.S., Heaton, B.D. & Little, L.T.: 1991, Astron. & Astrophys. **251**, 263.
Wood, D.O.S. & Churchwell, E.: 1989, Astrophys. J., Suppl. **69**, 831.
Zuckerman, B.E., Turner, B.E., Johnson, D.R., Clark, F.O., Lovas, F.J., Fourikis, N., Palmer, P., Morris, M., Lilley, A.E., Ball, J.A., Gottlieb, C.A., Litvak, M.M. & Penfield, H.: 1975, Astrophys. J. **196**, L99.

OBSERVATIONS AND MODELING OF INFRARED AND MILLIMETER MOLECULAR LINES TOWARDS GL2591

A. NEAL J. EVANS II AND B. JOHN H. LACY

The University of Texas at Austin, Austin, TX 78712-1083, USA

C. JOHN S. CARR

The Ohio State University, Columbus, Ohio 43210-1106, USA

and

D. SHUDONG ZHOU

The University of Illinois, Urbana IL 61801, USA

Abstract. We have observed C_2H_2 and HCN rovibrational transitions near 13 μm in absorption against GL2591. We also have observed rotational transitions at 0.6-3 mm of CS, HCN, H_2CO, and HCO^+. Analysis of the rotational lines, which arise in the extended cloud around the source, shows that no single density model can explain all the data. Models with density and temperature gradients do much better; in particular models with $n(r) \propto r^{-1.5}$ can reproduce the observed pattern of emission line strengths. The abundances show significant depletion compared to models of gas-phase chemistry. The rovibrational data were analyzed in comparison to the absorption line analysis of CO by Mitchell *et al.* (1989). Our data are consistent with the C_2H_2 and HCN absorption arising in the same warm (200 K) and hot (1010 K) components seen in CO, but we see little evidence for the cold (38 K) component seen in CO. The rovibrational lines from higher states ($J \geq 21$) indicate that the hot HCN deviates from LTE, leading to a density of about 3×10^7 cm^{-3}. Comparison of the two data sets shows that the rovibrational absorption of HCN, rather than arising in the extended envelope, must come from a region with a small angular extent. A model in which early-time gas phase abundances are preserved on grain mantles and released at high temperature can explain the data.

Key words: Abundances – Molecular – Infrared

1. Introduction

Millimeter emission spectroscopy and infrared absorption spectroscopy are complementary probes of molecular clouds. Millimeter emission from many species can be mapped across an extended cloud with high spectral resolution, but modest spatial resolution, but some species without permanent dipole moments, such as CH_4 and C_2H_2, cannot be detected. These species, as well as ones with permanent dipoles, can be detected through their rovibrational transitions in the infrared, if substantial column densities lie in front of a source of infrared continuum. Combining these two techniques can lead to new perspectives on cloud structure and abundances.

We apply the two techniques to GL2591, a deeply embedded ($A_V \sim 70$ mag) source with a high velocity outflow. At a distance of 1 kpc, its luminosity is 2×10^4 $L_\odot$. Mitchell *et al.* (1989) detected many CO rovibrational lines and identified three temperature components, cold (38 K), warm (200

Astrophysics and Space Science **224**: 181–184, 1995.
© 1995 *Kluwer Academic Publishers*.

K), and hot (1010 K), with differing velocities.

2. Millimeter Observations and Results

We obtained data on 5 lines of CS, 3 lines of $C^{34}S$, 2 lines of HCN, 1 line of $H^{13}CN$, 2 lines of H_2CO, 1 line of HCO^+, and one line of $H^{13}CO^+$. Data on the $J = 5 - 4$ line of CS was obtained with differing spatial resolutions (11″ and 30″). Observational details are given in Carr *et al.* (1995). Some of the lines were mapped and showed round structure on scales of 30″. Consequently, all modeling assumed spherical symmetry of the envelope. The CS lines were used as the primary constraint on models.

Initial modeling assumed a constant density and temperature, the latter set by the CO observations of Choi *et al.* (1994). An LVG code generated predicted line intensities for a grid of total density ($n = n(H_2) + n(He) + ...$) and column density of the species being observed. The CS data were best fit by $n \sim 5 \times 10^6$ cm^{-3}, but the fit was very poor. It was not possible to fit the $J = 2 - 1$ and $J = 7 - 6$ lines with a single density.

Models with density and temperature gradients were run, using a spherical Monte Carlo code for the excitation and radiative transfer (Choi *et al.* 1995). The code calculated line profiles which could be directly compared to the data. The velocity field was assumed to be microturbulent, with a e^{-1} width of 1.6 km s^{-1}. The temperature was given by $T(r) = 13K r_{pc}^{-0.4}$, where r_1 is the radius from the infrared source in pc, based on modeling of dust continuum emission and the source luminosity. The density was modeled as a power law ($n(r) = n_1 r_{pc}^{-\alpha}$), where n_1 is the density normalized to a radius of 1 pc. The molecular abundance (X) was taken to be constant throughout the envelope. Three values of α were tried (1.0, 1.5, and 2.0). For each α, n_1 and X(CS) were varied to find the best match to the observed lines, including $C^{34}S$. The ratio for CS/$C^{34}S$ was fixed at 20. The best results were obtained with $\alpha = 1.5$. Larger and smaller values for α had trouble matching the observed ratio of CS $J = 5 - 4$ lines in different beams and the ratio of high-J lines to low-J lines. Simulation of spectra at positions away from the peak were also generally fitted better with $\alpha = 1.5$.

For other molecules, n_1 was kept the same as was found from the CS modeling, allowing only X to vary. For HCN and HCO^+, the isotope ratio for C/^{13}C was set to 55, based on studies by Langer and Penzias (1990). The same n_1 gave a reasonable fit for other molecules. The line profiles of the molecules with ^{13}C were well-fitted, but the predicted profiles for the ^{12}C isotopomers were self-reversed, unlike the observations. This discrepancy is well known when modeling opaque lines with microturbulent codes; it can be alleviated if some systematic velocities are present or if the cloud is clumpy and macroturbulent (*e.g.* Park & Hong 1994). The derived abundances depend primarily on the optically thin ^{13}C isotopomers and the assumed isotope ratio.

TABLE I

Abundances (10^{-10}) Relative to H_2 from Millimeter Data

| Species | Observed | Gas-Phase | | With Depletion[a] | |
		HL[b]	LG[c]	10^6 y	10^7 y
CS	4	130	50	6	0.00002
HCN	7	16	300	28	1.8
HCO+	4	95	60	66	0.008
H_2CO	2	33	80	90	0.27

[a] Hasegawa & Herbst (1993), with cosmic ray desorption, at $n = 10^4$ cm^{-3}.
[b] Herbst & Leung (1989).
[c] Langer & Graedel (1989).

The abundances derived from the modeling of the envelope are given in Table 1, along with predictions of various models. The uncertainties are probably about half an order of magnitude. The results are not consistent with any chemical models with only gas-phase chemistry. Instead they suggest depletion over a period exceeding $10^6/n_4$ y, where n_4 is the density in units of 10^4 cm^{-3} at which most of the depletion occurred.

3. Infrared Absorption

The Q branch and seven lines in the R branch of C_2H_2 and five lines in the R branch of HCN were detected. The C_2H_2 lines were modeled assuming LTE since C_2H_2 has no permanent dipole moment, using the temperature components found by Mitchell et al. (1989). Similar abundances (10^{-3} of CO) were found for the warm and hot components, but the abundance in the cold component was less ($\leq 10^{-4}$).

The HCN was modeled with only warm and hot components; an LTE fit could not match the low equivalent widths in levels with $J \geq 21$. Fits using populations from an LVG excitation code provided better fits, with $n \sim 3 \times 10^7$ cm^{-3}. The abundance of HCN derived from the infrared is ~ 400 times that derived from the millimeter lines! In addition, the velocities of the absorption lines (including the CO) do not agree with the millimeter emission lines, and the temperatures required to fit the infrared lines are much higher. All these facts indicate that the molecules seen in absorption come from a small, blue-shifted, hot region, presumably near the infrared source. By requiring that the millimeter emission lines from this region do not exceed those observed, we can set an upper limit of 2–3″ (2000–3000

AU) on the size of the absorbing region.

TABLE II

Abundances (10^{-3}) Relative to CO from Infrared Data

Species	Observed GL2591	Gas-Phase[a] Early Time	Gas + Grain (Active)		
			HH[b]	B90[c]	BCM[d]
C_2H_2	1	5.4	6	9	17
HCN	3	1	280	10	1.3
CH_4	< 15	31	130	5	164

[a] Herbst & Leung (1989) at $t = 10^5$ y, no depletion.
[b] Hasegawa & Herbst (1993), Model N(2100K,CR); sum of abundance in gas and on surface at $t = 10^6$ y.
[c] Brown 1990, Case 3.
[d] Brown et al. 1988, Case A, after grain mantle evaporation.

The abundances (Table II) in the warm and hot gas can be explained if substantial depletion occurred at early times, when the abundances of C_2H_2 and HCN were high. Heating by radiation from the infrared source or by shocks would then liberate these molecules from the grains, restoring the early-time abundances. Models with active chemistry on the grain surfaces tend to overproduce CH_4, which is not seen in GL2591 (Lacy et al. 1991).

Acknowledgements

We acknowledge support from NSF grants AST-9317567, AST-9020292, and AST-8615906. JC acknowledges support from a Columbus Fellowship at The Ohio State University.

References

Brown, P. D.: 1990, Mon. Not. of the RAS **243**, 65.
Brown, P. D., Charnley, S. B. & Millar, T. J.: 1988, Mon. Not. of the RAS **231**, 409.
Carr, J. S., Evans, N. J., II, Lacy, J. H. & Zhou, S.: 1995, Astrophys. J. submitted.
Choi, M. Evans, N. J., II & Jaffe, D.: 1994, Astrophys. J. **417**, 624.
Choi, M. Evans, N. J., II, Gregerson, E. & Wang, Y.: 1995, Astrophys. J. submitted.
Hasegawa, T. I. & Herbst, E.: 1993, Mon. Not. of the RAS **261**, 83.
Herbst, E. & Leung, C. M.: 1989, Astrophys. J., Suppl. **69**, 271.
Lacy, J. H., Carr, J. S., Evans, N. J., II, Baas. F., Achtermann, J. M. & Arens, J. F.: 1991, Astrophys. J. **376**, 556.
Langer, W. D. & Graedel, T. E.: 1989, Astrophys. J., Suppl. **69**, 241.
Langer, W. D. & Penzias, A. A.: 1990, Astrophys. J. **357**, 477.
Mitchell, G. F., Curry, C., Maillard, J-P. & Allen, M.: 1989, Astrophys. J. **341**, 1020.
Park, Y.-S. & Hong, S. S.: 1994, Astron. & Astrophys. in press.

SUBMM OBSERVATIONS OF HERBIG AE/BE SYSTEMS

VINCENT MANNINGS
Department of Physics and Astronomy
University College London
Gower Street
London WC1E 6BT

Abstract. Studies of circumstellar dust and gas in pre-main-sequence systems are in part motivated by a desire to probe possible sites of future or on-going planet formation and, should accretion onto young stars be taking place, to determine whether the observed remnants of embryonic envelopes might yet contribute significantly to the final masses of stars as they evolve towards the main sequence. New 0.35- to 1.3-mm flux measurements have been made for 13 intermediate-mass pre-MS systems in order to estimate total dust+gas masses and to investigate the spatial distribution of circumstellar material. At least two sources may be dominated by compact circumstellar disks; emission from extended envelopes appears to be important for the majority of sources. HD 163296 is an unusual object and warrants further examination.

Key words: pre-main-sequence – disk – envelope

1. Introduction

Pre-main-sequence Herbig Ae/Be stars are the intermediate-mass counterparts to the more numerous solar-type T Tauri stars (see Thé, Pérez & van den Heuvel 1994). They are also, presumably, the evolutionary precursors to the main-sequence Vega-excess sources. As found during the early phases of T Tauri stars, many Ae/Be systems show large near-IR to mm continuum excesses, strongly suggesting thermal emission from circumstellar grains.

IR excesses in the classical, optically visible, T Tauri systems can often be accounted for solely by emission from grains confined to massive circumstellar disks (Beckwith & Sargent 1993, and references therein). Clearly, it would be fascinating to extend this idea to the intermediate-mass region of the pre-MS H-R diagram and then to compare disk masses, accretion rates, disk survival times, temperature distributions, and grain opacity profiles with those of disks around the solar-mass stars. However, the dominant spatial distribution of dust and gas – envelopes versus disks – is in reality unknown for the majority of Herbig Ae/Be systems. Several studies (Berrilli *et al.* 1992; Hillenbrand *et al.* 1992; Natta *et al.* 1993) have tried to disentangle possible flux contributions by disks and envelopes across the spectral energy distribution (SED). Hillenbrand *et al.* conducted a major survey and used near-IR−mm SED morphology to divide most of the catalogued Ae/Be sources into two groups: Group I (possibly disk-dominated) and Group II (envelope-dominated). (Half a dozen sources are placed in a third class,

Astrophysics and Space Science **224**: 185–188, 1995.
© 1995 *Kluwer Academic Publishers.*

Group III, and have SEDs similar to those of classical Be stars.) The aims
of the present work are to fill in the mm/submm portion of the SED for
samples of Group I and Group II systems and to use the new measurements
to probe the dominant sources of long-wavelength emission. The mm measurements have the additional advantage of being (largely) in the optically
thin régime, so that we can make rough estimates of total dust+gas masses.

2. Observations and Analysis

New observations of 0.35- to 1.3-mm continuum fluxes from 13 Herbig Ae/Be
systems have been made using UKT14 at the James Clerk Maxwell Telescope (JCMT, Hawaii). The beamsize was around 20″ FWHM. See Mannings (1994) for the flux densities of each source. The observations were
made during a total of eight nights in 1993 April and December.

Note that a disk of radius $\approx$ 300 AU at a typical distance of 650 pc (for
the present sample) would subtend an angle of $\approx$ 1″ ($\ll$ beam FWHM) and
would therefore appear as a point source.

See Table 1. The optically thin 1.3-mm fluxes have been used to estimate
circumstellar masses (assuming a single dust temperature and a gas:dust
ratio of 100, by mass). Total dust+gas masses are estimated in this way to be
$\sim$ 0.01 to 1 $M_\odot$ so that, should accretion be taking place, the envelopes/disks
may eventually contribute significantly to the final main sequence masses of
their host stars. Note that Group I and Group II sources cannot be distinguished by mass of circumstellar material, nor can we differentiate between
the two classes by the mm/submm spectral index ($F_\nu \propto \nu^\alpha$) listed in the
final column of the Table. A mean value of $\alpha = 3.10 \pm 0.04$ is found, significantly steeper than for disk-dominated T Tauri systems ($\alpha \approx 2$, Beckwith &
Sargent 1991; Mannings & Emerson 1994). Moriarty-Schieven et $al.$ (1994)
also find steep spectral indices (with a distribution peaking around $\alpha = 3$) for
embedded (envelope-dominated) sources in Taurus, which suggests that we
might be dealing with mm/sub-mm emission originating largely in extended
envelopes around most of the present sample of Ae/Be systems.

Although arguments based on disk optical depth suggest that disks are
consistent with circumstellar masses estimated using the new 1.3-mm fluxes,
it is also found (Mannings 1994) that, with the notable exception of HD
163296, the same masses distributed within spherical envelopes matched to
the JCMT beam would lead to V-band extinctions which are consistent
with observed values, both for uniform density and for free-fall envelopes
(Table 1). To decide between the two geometries for the Group I sources,
model SEDs have been computed using a classical disk temperature profile
($T(r) \propto r^{-3/4}$). [The implicit assumption in this exercise is that, for a given
Ae/Be system, all observed excess fluxes from near-IR to mm wavelengths
originate in the same source (see also Weintraub, Kastner & Mahesh 1994;

TABLE I
Herbig Ae/Be systems observed at 0.35-1.3 mm with JCMT/UKT14.

Source	Sp. Type	Env. mass ($M_\odot$)	$A_V(M_{1.3})$ $(A_V(obs))$ Uniform shell $(s=0)$	$A_V(M_{1.3})$ Free-fall shell $(s=3/2)$	mm/sub-mm spectral index $(F_\nu \propto \nu^\alpha)$
GROUP I					
AB Aur	A0	0.01	1.0 (0.4)	0.9	3.45 ± 0.13
CD -42°11721	B7	0.03	2.7 (4.3)	2.6	2.92 ± 0.09
MWC 147	B2	< 0.09	< 0.3 (1.6)	(< 1.0)	—
HD 163296	A0	0.10	9.0 (0.3)	8.6	1.93 ± 0.04
Lk Hα 215	B5	0.11	0.4 (2.1)	1.2	3.47 ± 0.28
MWC 297	09	0.42	4.8 (8.3)	9.9	1.66 ± 0.05
MWC 137	B0	0.73	0.9 (4.5)	4.1	2.94 ± 0.12
MWC 1080	B0	1.14	2.6 (5.3)	9.2	2.97 ± 0.12
GROUP II					
R CrA	F5	0.02	2.7 (0.9)	2.2	3.00 ± 0.10
Elias 1	A6	0.03	2.7 (4.1)	2.6	2.18 ± 0.10
V376 Cas	F0	0.18	1.2 (2.9)	2.9	3.43 ± 0.10
R Mon	B0	0.24	0.9 (4.3)	2.6	2.72 ± 0.15
Lk Hα 198	A5	0.41	2.6 (2.5)	6.6	3.38 ± 0.12

Butner & Natta 1995).] AB Aur, MWC 147 and MWC 297 can be modelled in this way assuming that disks dominate excess near-IR−mm fluxes. AB Aur has also been scanned with *KAO* at 100 μm (Di Francesco *et al.* 1994), with no evidence found for emission extended beyond $\sim$ 2000 AU from the star. However, MWC 297 *is* spatially resolved by *KAO* and it may be that some of the mm/submm flux arises in the same extended region. The new models predict a significant shortfall in 0.35-mm flux for MWC 137, even for a disk which is optically thick at all wavelengths and radii. In this way, envelope emission is found to be probably important/dominant for MWC 137, Lk Hα 215, and MWC 1080.

HD 163296 is an unusual Herbig Ae/Be system. As for MWC 137, Lk Hα 215, and MWC 1080, the submm data cannot be fitted with an SED generated by a classical disk alone. However, HD 163296 is amongst a minority for which IRAS fluxes decline as a power law with increasing wavelength, reminiscent of many classical, disk-dominated T Tauri systems. (The same is true for AB Aur.) A disk with non-standard temperature profile $T \propto r^{-0.62}$ provides an excellent fit to the HD 163296 IRAS and mm/submm fluxes for a disk mass of 0.12 $M_\odot$ and a grain opacity profile $\kappa \propto \lambda^{-0.6}$ (wavelength λ

> 0.25 mm) – note the shallow mm/submm spectral index in Table 1; the opacity index is similar to low values claimed for some T Tauri disks (Beckwith & Sargent 1991; Mannings & Emerson 1994), which might indicate the presence of large grains. Such speculation rests upon the assumption that disk emission dominates excess flux from this source, which has yet to be demonstrated directly. Support for a disk interpretation is, however, offered by single-dish maps of HD 163296 made at 0.8 mm (Mannings 1995), which show that the bright 0.8-mm flux (a) appears to be centred on the star; (b) is not significantly extended with respect to the 14".5-FWHM diffraction-limited beam. The high V-band extinctions predicted for spherical envelopes (Table 1) also suggest a relatively flattened geometry.

The new analysis indicates that, for many Group I Herbig Ae/Be sources, envelopes possibly accompany any disks which might be present. Differences between SEDs of Groups I and II therefore most likely reflect only the relative contributions of the two components. Conclusive statements await a search for compact (disk) emission using mm-wave aperture synthesis – particularly for the nearby sources HD 163296 and AB Aur. Also, the JCMT *SCUBA* array promises sensitive, large-scale submm continuum imaging of extended envelopes in many of these systems.

Acknowledgements

I wish to thank University College London for providing a grant for conference expenses. This research is funded by the Particle Physics and Astronomy Research Council.

References

Beckwith, S.V.W. & Sargent, A.I.: 1991, Astrophys. J. **381**, 250.
Beckwith, S.V.W. & Sargent, A.I.: 1993, in *Protostars and Planets III*, eds: E.H. Levy & J.I. Lunine, University of Arizona Press, p521.
Berrilli, F. *et al.* : 1992, Astrophys. J. **398**, 254.
Butner, H.M. & Natta, A.: 1995, Astrophys. J. to appear in Feb. 10 issue.
Di Francesco, J., *et al.* : 1994, Astrophys. J. **432**, 710.
Hillenbrand L.A., *et al.* : 1992, Astrophys. J. **397**, 613.
Mannings, V. & Emerson, J. P.: 1994, Mon. Not. of the RAS **267**, 361.
Mannings, V.: 1994, Mon. Not. of the RAS to appear in December 1 issue.
Mannings, V.: 1995, *HD 163296: A Disk-Dominated Herbig Ae/Be System?* in preparation.
Moriarty-Schieven, G.H., *et al.* : 1994, Astrophys. J. in press.
Natta, A., *et al.* : 1993, Astrophys. J. **406**, 674.
Thé, P.S., Pérez, M.R. & van den Heuvel, E.P.J. (eds.): 1994, *The Nature and Evolutionary Status of Herbig Ae/Be Stars*, ASP Conference Series, Vol. **62**.
Weintraub, D.A., Kastner, J.H. & Mahesh, A.: 1994, Astrophys. J. **420**, L87

MILLIMETRE AND SUBMILLIMETRE OBSERVATIONS

OF HIGH-MASS STAR FORMING CORES

R. ZYLKA
Max-Planck-Institut für Radioastronomie
Auf dem Hügel 69
53121 Bonn, Germany

Abstract. I summarize recent millimetre and submillimetre observations of cloud cores where massive star formation is currently taking place. The first systematic continuum surveys in this wavelength range obtained with single dish telescopes and high-resolution data of NGC 2024 and W51A obtained with the Plateau de Bure interferometer are presented in more detail. Also given is a discussion of observing methods and some of the difficulties involved with the observations.

Key words: interstellar medium: clouds – dust – stars: formation

1. Introduction

The knowledge of the density and temperature structure of molecular clouds is essential for understanding the formation of stars. To derive a detailed picture of the gravitational collapse, high resolution data are necessary covering the whole range of wavelengths, at which the collapsing object is likely to emit. Currently the resolution of the (sub)millimetre observations is a few times higher than of the other relevant data. Large single dish telescopes operating down to $350\,\mu$m complemented by the millimetre interferometers with arc-second resolution opened new possibilities in investigating small-scale structure of molecular clouds. Many high-mass star forming regions were investigated in the millimetre and submillimetre wavelength range during the last years. The picture of star formation that emerges from these observations was recently summarized by Mezger (1994).

After a brief discussion of the observing methods and the problems specific to the submillimetre regime, I will review recent data of high-mass star forming cores obtained with the large single dish telescopes (CSO, JCMT, IRAM 30-m and SEST) and the Plateau de Bure (PdB) interferometer.

2. Mapping Techniques in Millimetre and Submillimetre Range

The current standard for continuum mapping is the double beam scanning technique (Emerson *et al.* 1979), also called on-the-fly (OTF) method. The angular distance between the ON and the OFF beam is usually a few times the resolution of the telescope. The OTF mapping rejects very effectively the atmospherical emission and has small observational overheads. Since the differential signal of the intensity distribution is measured, subsequent

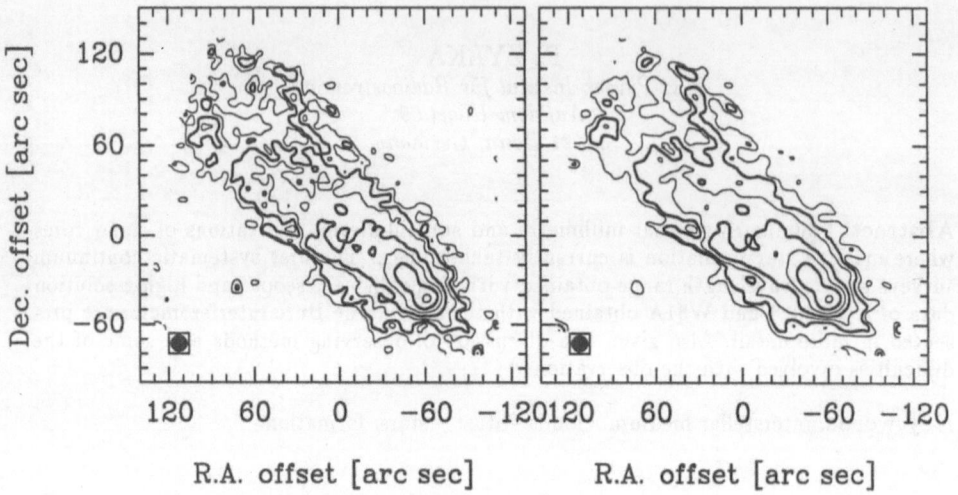

Fig. 1.   Result of a 'sky noise' filtering (Zylka in preparation) of the Sgr C data obtained
with the 7-channel MPIfR bolometer array. Left: not filtered data, rms ~ 35 mJy/12″beam;
right: filtered data, rms ~ 28 mJy/12″beam. Contour levels are the same in both images
and start at 100 mJy/12″beam.

restoration is required in order to obtain the equivalent of a single beam
image. In addition to the restoration technique of Emerson *et al.* the double-
beam Maximum Entropy Method (DBMEM, Richer 1992) or mask-and-shift
method (Zylka, in preparation) can be used. Using sensitive bolometers, the
integration time/pixel is usually on the order of 1 second.

New possibilities are opening currently in the submillimetre regime with
bolometer arrays: different mapping modes can be used, larger fields mapped.
As a result of the anomalous refraction (Altenhoff *et al.* 1987, Zylka *et al.*
1992, Zylka *et al.* 1994) an image obtained with an N-channel array is always
better than the average of N independent images. Further improvement can
be obtained using sky-noise suppressing algorithms (see Fig. 1, Sievers *et al.*
1994).

In spectroscopy mapping is performed usually in the ON-OFF mode, where
the change between the ON and the OFF position is obtained by moving the
telescope or a chopper. Integration times are much longer than in continuum
(usually $\geq$ 20 sec). The advantage of this method is that the OFF position
can be chosen to be free of emission, but its observational overheads are
very large. Different mapping modes are the single and double beam scanning, with the same advantages and problems as in continuum. With large
heterodyne arrays similar possibilities are arising as in continuum.

The calibration accuracy is comparable in spectroscopy and in continuum.
It is very much wavelength dependent. Usually errors are $\leq$ 15% at 1.3 mm

and $\geq 40\%$ at 0.35 mm. The largest uncertainty introduces the estimation of the atmospheric transparency, although the contribution of the telescope error beam may sometimes be even more important. Because the bandwidth of the bolometers used for continuum observations is large (≥ 50 GHz), the shift of the effective wavelength due to the different spectral energy distributions of the calibrator and the target source and effects due to (time variable) optical depth of the atmosphere have also be taken into account.

3. Examples of Single-Dish Studies

During the last few years continuum maps of high-mass star forming regions have been obtained with the highest resolution possible in the (sub) millimetre range. Mezger *et al.* (1990) investigated the OMC1 and OMC2 regions in the Orion A molecular cloud at 1.3 mm with 11″ resolution. Sievers *et al.* (1991) observed the W49A and W51A GMC and giant H II region complexes in the wavelength range from 3 down to 0.87 mm with 8″ resolution at the shortest wavelength. With the full spectral coverage possible from the ground (3 to 0.35 mm), Hobson *et al.* (1993) observed the M17 SW cloud core (Fig. 2). The physical state of high-mass star forming GMCs in spiral arms, in the interarm region, and in the Galactic Centre has been compared by Gordon *et al.* (1993) and Mezger (1994).

Recently a few (sub)millimetre surveys of high-mass star forming regions were performed. Lis and Carlstrom (1994) conducted an unbiased 0.8 mm survey in the Galactic Centre region with the CSO. At 1.3 mm, surveys with the IRAM 30-m and the SEST of the Orion B molecular cloud, FIR-strong cores and the Galactic Centre region have been performed by Launhardt *et al.* (1994), Mooney *et al.* (1994) and Zylka *et al.* (in preparation), respectively.

3.1 THE 1.3 MM SURVEY OF HIGH-MASS STAR FORMING CORES

In an on-going investigation of the morphology of high-mass star forming cloud cores, Mooney *et al.* (1994) selected a subset of 51 IR-strong and 27 IR-quiet GMCs. According to a qualitative classification, IR-strong clouds are those whose molecular emission is associated with strong, well-defined emission in the IRAS FIR bands, while IR-quiet clouds are those associated with relatively weak FIR emission, not visible directly as a maximum in the IRAS images. The total IR luminosities of the selected objects range from $\sim 10^{4.2-7.3}$ for the IR-strong and $\sim 10^{2.2-5.3}$ for the IR-quiet clouds. Approximately 3/4 of IR-strong and about half of the IR-quiet clouds are located in or near Galactic spiral features.

The molecular line and dust emission maps of the IR-strong cloud SRBY 162 are shown in Fig. 3. This figure illustrates the basic selection process and observing strategy applied by Mooney *et al.* (1994). Using the FCRAO

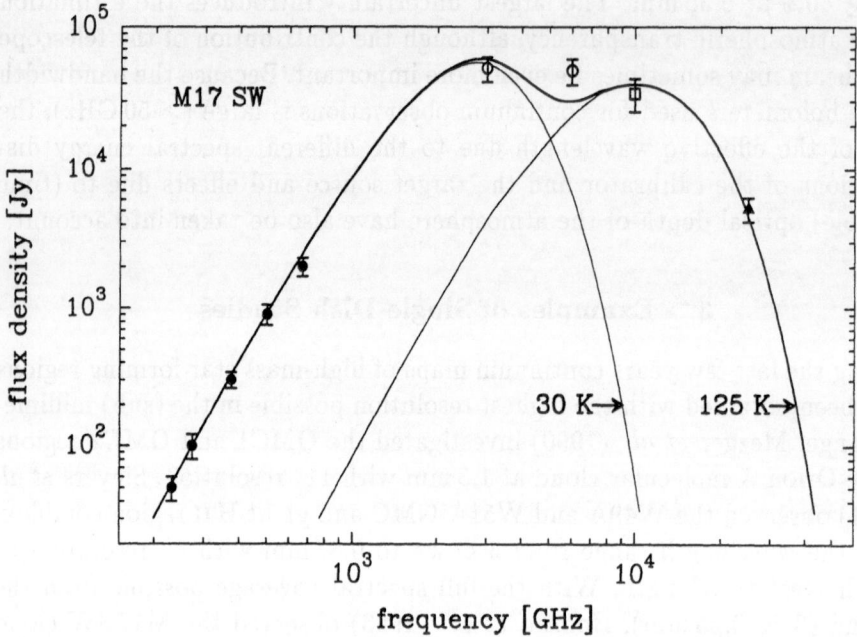

Fig. 2. The continuum spectrum of the of M17 SW complex with spectral fits indicating the range of dust temperatures present in this region. The filled circles correspond to the $1.3 - 0.45$ mm fluxes of Hobson *et al.* (1993), the open circles are the 100- and 50 μm fluxes of Meixner *et al.* (1992), the open square is the 30 μm flux obtained by Gatley *et al.* (1979) and the filled triangle is the flux derived by Hobson and Ward-Thompson (1994) by a MEM reconstruction of IRAS 12μm data. Because at wavelengths shortwards of 0.35mm no data with resolution comparable to that of the the mm- and submm measurements are available, integrated flux densities (integrated over regions considerably larger than the compact condensations observed) have to be used to derive the average physical parameters (see Hobson *et al.* (1993) for details).

CS $(2-1)$ and IRAS 100 μm maps for reference, maps of the 1.3 mm continuum were first measured with the SEST 15-m antenna (middle panel). The SEST maps were then used as guides for higher-resolution follow-up continuum observations with the IRAM 30-m telescope (lower panel). Altogether, 14 IR-strong and 4 IR-quiet clouds were surveyed by Mooney *et al.* (1994). For a few of the cloud cores, the CS $(2-1)$ maps were also used as guides to obtain high-resolution CS $(3-2)$ maps with the IRAM 30-m antenna (lower panel left). The CS $(3-2)$ integrated emission exhibits a striking morphological similarity to the high-resolution IRAM 30-m 1.3 mm continuum emission, demonstrating that the dust continuum indeed traces the high-density regions of molecular cloud cores. This figure also illustrates how, with increasing angular resolution, the structure of high-mass star forming regions separates into continually smaller and denser clumps. Explicitly, in

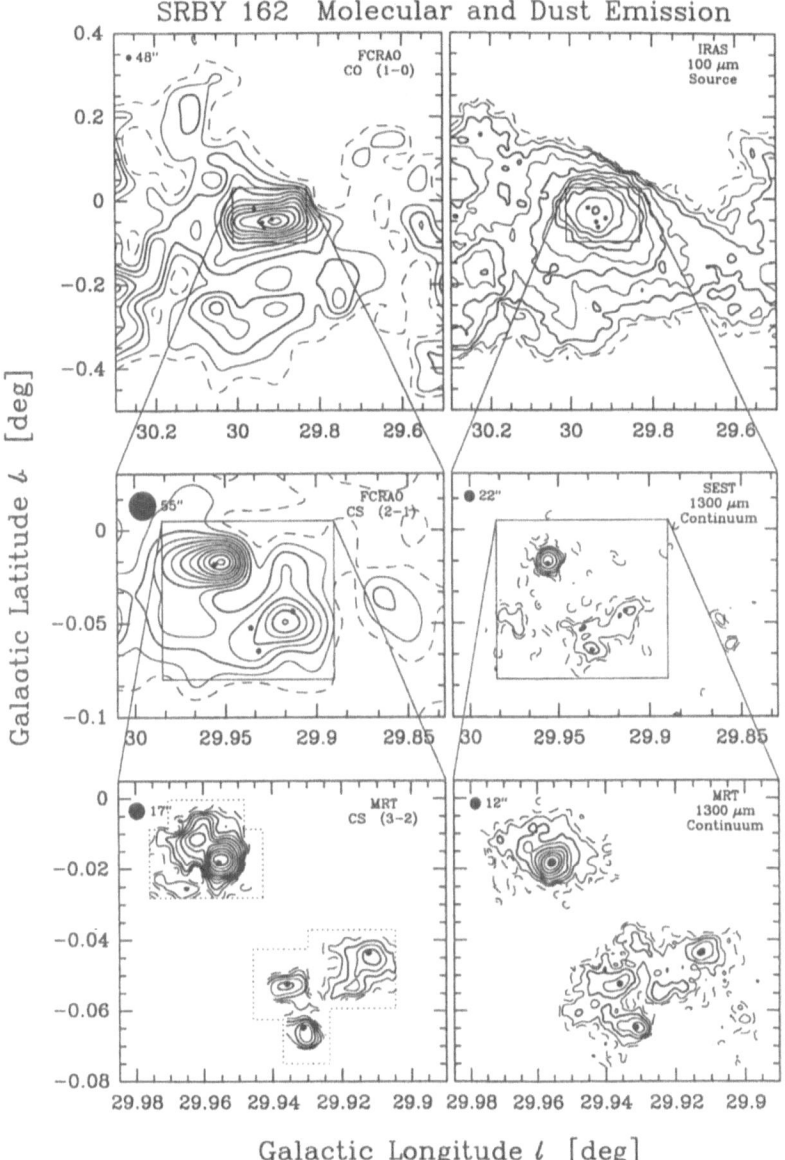

Fig. 3. A comparison of low- to high-resolution molecular line emission (left panels) with dust emission (right panels) from the IR-strong cloud SRBY 162 (Mooney *el al.* 1994). Low contours are dashed lines. Top panels: The CO (1 – 0) integrated emission contoured from a low of 48 with a spacing of 15 $K \cdot km \cdot s^{-1}$; the IRAS 100 μm surface brightness contoured from a low of 275 MJy/str with a logarithmic spacing equal to a factor of $\sqrt{2}$ the previous level. Middle panels: The CS (2 – 1) integrated emission contoured from a low of 1.6 with a spacing of 1.4 $K \cdot km \cdot s^{-1}$; the SEST 1.3 mm continuum emission contoured from a low of 300 mJy with a logarithmic contour spacing equal to a factor of $10^{0.2}$ the previous level. Bottom panels: The CS (3 – 2) integrated emission contoured from a low of 5.0 $K \cdot km \cdot s^{-1}$ with a logarithmic spacing equal to a factor of $10^{0.1}$ the previous level; the IRAM 30-m 1.3 mm continuum emission contoured from a low of 75 mJy/12″beam with a logarithmic contour spacing equal to a factor of $10^{0.2}$ the previous level. For reference, the positions of the compact dust components in the IRAM 30-m continuum map are marked with solid circles.

a sequence of decreasing size and increasing morphological structure and density, the core of the GMC as traced by the low-resolution CO emission ($n \geq 10^2$ cm^{-3}) resolves into a multicomponent moderate-resolution CS core region ($n \geq 10^3$ cm^{-3}), which separates further into extended dust continuum sources ($n \geq 10^4$ cm^{-3}, $N_H \geq 10^{23}$ cm^{-3}), within which compact dust components ($n \geq 10^5$ cm^{-3}) are situated.

Mooney et al. (1994) found that all of the IR-strong dust continuum sources exhibit an extended morphology within which compact components are embedded, providing multiple sites of high-mass star formation (Fig. 3). Because they are associated with large extended H II regions, in most of the CS core regions the current star formation appears to be secondary. Very little evidence for purely first generation high-mass star formation was found, which may indicate that once the first generation occurs, second generation star formation proceeds rapidly. Only in G38.92/38.96 and possibly in G35.20 Mooney et al. (1994) find indications suggesting that the current star formation there is first generation. The overall morphology of most of the dust continuum sources within the CS core regions reveals compact cores adjacent to more diffuse structure. The secondary star formation in sources with compact morphologies may be at a very early stage.

The physical parameters of the parental clouds, the extended dust sources and the dense compact cores identified by Mooney et al. (1994) are summarized in Table 1. On the average, about 40% of the total mass is contained in the compact components.

TABLE I

Physical parameters of the sources identified by Mooney et al. (1994)

component	mass [M$_\odot$]	vol. density [cm^{-3}]	col. density [cm^{-2}]	size [pc]
parental clouds	$10^{5.0-6.6}$	200−1200	$10^{22.5-23.0}$	20−100
extended sources	$10^{2.8-5.0}$	$10^{4.3-5.0}$	$10^{23.5-23.8}$	1.2−7.3
compact cores	$10^{1.8-4.3}$	$10^{4.7-6.5}$	$10^{23.0-25.0}$	0.1−1.8

The radial density profile $n(r) = n_0(r/r_0)^{-\alpha}$, where n_0 is the density at the radius of the cloud edge r_0, calculated for the IR-strong clouds from the effective diameters and average densities of the parental GMC, the CS core region, the extended dust sources and the compact cores, is $\alpha \leq 1.7 - 1.8$.

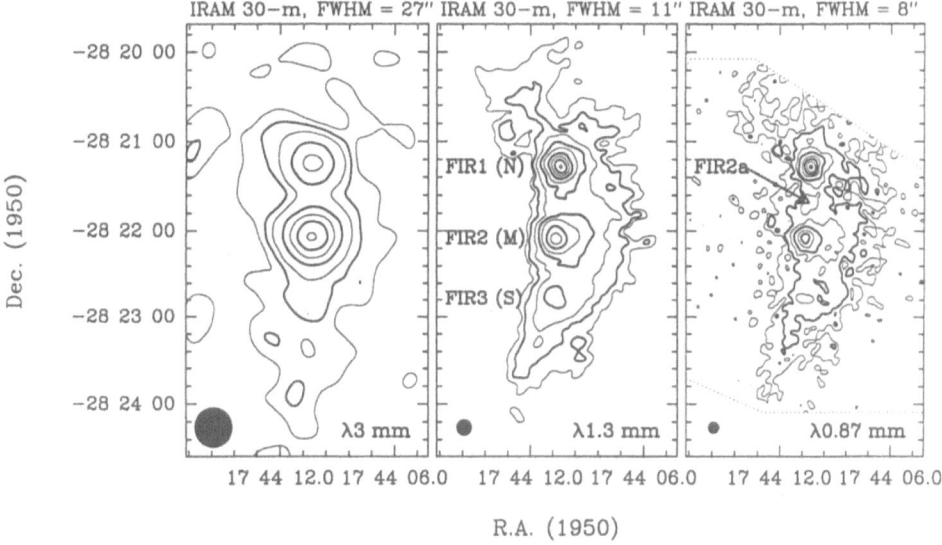

Fig. 4.   Sgr B2 observed with the IRAM 30-m by Salter *et al.* (1989, $\lambda = 3.2$ mm) and
Gordon *et al.* (1993, $\lambda = 1.3$ and 0.87 mm). The contours are: at 3.2 mm 5, 10, and 18
to 98% in steps of 20% of the maximum of 2468 mK/27″beam; at 1.3 mm 1.5, 2.5, 5, 10,
and 18 to 98% in steps of 20% of the maximum of 55 Jy/11″beam; at 0.87 mm 2.5, 5, 10,
and 18 to 98% in steps of 20% of the maximum emission of 177 Jy/8″beam. The three
compact components are identified in the middle pannel. The beam areas are shown in
the lower left hand corner of each map.

3.2   Surveys of the Central 200 pc of the Galaxy

In the Galactic Centre, Lis and Carlstrom (1994) conducted an unbiased survey of the 0.8 mm continuum emission from the central $1.5° \times 0.2°$ region
with the 10-m CSO telescope (HPBW = 25″). They found a number of compact dust condensations forming an elongated ridge of emission stretching
from the radio Arc ($l = 0.12°$) to Sgr B2 ($l = 0.7°$). The spatial correlation between these cores and the compact far-infrared sources/compact H II
regions is poor. This implies that many of the dust condensations are heated
by a diffuse external radiation field rather than by embedded young stars.
Further investigation of this ridge by Lis *et al.* (1994) confirmed this suggestion. Toward the known high-mass star forming regions, such as Sgr B2 and
Sgr C, the large-scale distribution of the dust continuum emission correlates
well with the distribution of the radio continuum and far-infrared emission.
Using the IRAM 30-m telescope and SEST Gordon *et al.* (1993) observed
Sgr B2. Zylka *et al.* (1990) and Zylka *et al.* (in preparation) mapped the
central $\sim 20' \times 15'$ of the Galactic Centre, the main star forming regions
Sgr B1, Sgr C (Fig. 1) and Sgr D, as well as some previously unknown cores.

The results of these surveys further confirm that the current high-mass star formation activity within the central region is considerably lower than in the spiral arms.

The maps of Sgr B2 at 3.2 mm, 1.3 mm and 0.87 mm of Gordon *et al.* (1993) show three compact ($\leq$ 0.4pc) components of dust emission embedded in an extensive (4pc × 11pc) low-density envelope (Fig. 4). Combining the flux densities from the SEST and IRAM 30-m observations with other measurements at higher frequencies Gordon *et al.* (1993) constructed the spectral energy distribution for the entire Sgr B2 complex and its three principal components. Based on a spatial and spectral decomposition, they determined the physical parameters of the four components. The dust temperature of the three compact sources is $\sim$ 45 − 60K and that of the extended envelope $\sim$ 19K. The extended envelope contains the bulk of the mass ($\sim$ 95% of the total $M_H \sim 8\,10^5\ M_\odot$), whereas the compact star-forming cores contribute more than half of the luminosity. Comparison with high-resolution IRAS 60 and 100 μm images shows that the dust within the envelope of the Sgr B2 complex observed at millimetre and submillimetre wavelengths is too cold to be detected in the IRAS data, which mainly traces the warmer dust in the outer skin of GMCs. Similar studies of Sgr B2, Sgr C and Sgr D cores have been performed by Lis *et al.* (1991). The luminosity/mass ratio observed in the Galactic Centre sources is an order of magnitude lower than in the star forming regions W49A and W51A located within spiral arms.

4. Plateau de Bure Observations of NGC 2024 and W51A

The NGC 2024 FIR region contains seven dense dust condensations detected by bolometric observations (Mezger *et al.* 1988). Their identification as isothermal protostars had been challenged by a number of lower resolution molecular line studies claiming that there is a large amount of warm dust, while 12″ resolution measurements of $C^{18}O$ (2 − 1) by Mauersberger *et al.* (1992) have suggested that most molecules are frozen out onto cold dust grains. To settle this question, Wiesemeyer *et al.* (1994) used the high resolution (3″) of the Plateau de Bure (PdB) interferometer to simultaneously map the emission of the $C^{34}S$ (2 − 1) line and the 3 mm continuum from the most prominent condensations FIR5 and FIR6. Both condensations were detected. They are even more compact than previously thought. From a comparison with the VLA continuum data, they found the 3 mm continuum to be emitted by dust. VLA NH_3 maps of Mauersberger *et al.* (1992) and Ho *et al.* (1993) show that these condensations are fairly cold ($T_K \leq 40\,K$). The $C^{34}S$ emission is much less than expected from the strength of the dust emission and from the resulting H_2 column density. The molecular gas is anticorrelated with the dust emission, which may indicate that in NGC 2024 polar molecules are frozen onto dust grains. The fact that FIR5 and FIR6 are sources of molecular outflows suggests that this process begins in an early phase of (proto-)stellar evolution.

In the W51A region three sub-condensations associated with the H II regions W51d, W51e1/e2 and with W51FIR3 were observed by Uchida et $al.$ (1994). In all three regions they found ultra-high density clumps ($10^6 - 10^9$ cm^{-3}) with linear sizes on the order of 5×10^{17} cm. Their PdB data show that in W51e1 the 3 mm continuum emission is also emitted by dust, but the C^{34}S line emission is correlated with the dust continuum emission. The reason for this may be the much higher temperature of the detected sub-cores ($T_K > 100$ K), restraining the physical parameters at which the depletion of molecules begins. Based on the observed 3 mm continuum flux, they determined a total clump mass of 3×10^3 M$_\odot$.

5. Summary

The recent data obtained with large single-dish telescopes suggest that the precursors of high-mass stars appear to be cloud cores with hydrogen masses $M_H \sim 10^{\,3.1 - 4.6}$ M$_\odot$, column densities $N_H \sim 10^{\,23.3 - 25.6}$ cm^{-2}, and volume densities $n_H \sim 10^{\,5.2 - 7.2}$ cm^{-3}. The upper limits refer to W49A and W51A, the lower limit to OMC1. Masses of the parental clouds are similar – a few 10^5 M$_\odot$ – but the volume densities of GMCs in spiral arms and in the Galactic Centre are ~ 100 times higher than in the Orion A cloud, which represents a typical interarm GMC. Massive star formation activity – as measured by the L_{IR}/M_H ratio – is highest in W49A and W51A, about an order of magnitude lower in Sgr B2, and another order of magnitude lower in the Orion A cloud. The arc-seconds resolution PdB data further confirm that under the extreme conditions at which star formation takes place ($n_H \geq 10^6$ cm^{-3} and $T \leq 20$K), molecules such as CO, CS and NH$_3$-appear to freeze out and form mantles around dust grains and therefore are no longer reliable tracers of hydrogen column densities Dust emission remains optically thin up to column densities of $N_H \sim 10^{26}$ cm^{-2} and, according to our current knowledge, it is a better tracer of high density condensations.

Acknowledgements

Drs. R. Güsten, K. Uchida and H. Wiesemeyer are kindly thanked for letting me quote their results on NGC 2024 and W51A prior to publication and Dr. M.P. Hobson for sending me the M17 data. Drs. P. Cox, W. Duschl, P.G. Mezger, T. Mooney, K. Uchida and D. Ward-Thompson are acknowledged for their comments. Finally, it is a pleasure to warmly thank the organizers of this conference.

References

Altenhoff W.J., Baars J.W.M., Downes D. & Wink J.-E.: 1987, Astron. & Astrophys. **184**, 381.

Emerson D.T., Klein U. & Haslam C.G.T.: 1979, Astron. & Astrophys. **43**, 159.

Gatley I., Becklin E.E., Sellgren K. & Werner M.W.: 1979, Astrophys. J. **233**, 575.

Gordon M.A., Berkermann U., Mezger P.G., Zylka R., Haslam C.G.T., Kreysa E., Sievers A.W. & Lemke R.: 1993, Astron. & Astrophys. **280**, 208.

Ho P.T.P, Yun-Lou P., Torrelis J.M., Gómez J.F., Rodríguez L.F. & Cantó J.: 1993, Astrophys. J. **408**, 565.

Hobson M.P., Padman R., Scott P.F., Prestage R.M. & Ward-Thompson D.: 1993, Mon. Not. of the RAS **264**, 1025.

Hobson M.P. & Ward-Thompson D.: 1994, Mon. Not. of the RAS **267**, 141.

Launhardt *et al.* : 1994, in preparation.

Lis. D.C., Carlstrom J.E. & Keene J.: 1991, Astrophys. J. **380**, 429.

Lis. D.C. & Carlstrom J.E.: 1994, Astrophys. J. **424**, 189.

Lis. D.C., Menten K.M., Serabyn E. & Zylka R.: 1994, Astrophys. J. **423**, L39.

Mauersberger R., Wilson T.L., Mezger P.G., Gaume R. & Johnston K.J.: 1992, Astron. & Astrophys. **256**, 651.

Meixner M., Hass M.R., Tielens A.G.G.M., Erickson E.F. & Werner M.: 1992, Astrophys. J. **390**, 499.

Mezger P.G., Chini R., Kreysa E., Wink J.-E. & Salter C.J.: 1988, Astron. & Astrophys. **191**, 44.

Mezger P.G., Wink J.-E. & Zylka R.: 1990, Astron. & Astrophys. **228**, 95.

Mezger P.G.: 1994, Astrophys. & Space Sci. **212**, 197.

Mooney T., Sievers A.W., Mezger P.G., Solomon P.M., Kreysa E., Haslam C.G.T. & Lemke R.: 1994, Astron. & Astrophys. in press.

Richer J.S.: 1992, Mon. Not. of the RAS **254**, 156.

Salter C.J., Emerson D.T., Steppe H. & Thum C.: 1989, Astron. & Astrophys. **225**, 167.

Sievers A.W., Mezger P.G., Gordon M.A., Kreysa E., Haslam C.G.T. & Lemke R.: 1991, Astron. & Astrophys. **251**, 231.

Sievers A.W., Reuter H.-P. & Greve A.: 1994, IRAM Newsletter **17**, 5.

Uchida K.I., Wiesemeyer H. & Güsten R.: 1994, Astron. & Astrophys. to be submitted.

Wiesemeyer H., Güsten R., Wink J.-E. & Yorke H.W.: 1994, Astron. & Astrophys. to be submitted.

Zylka R., Mezger P.G. & Wink J.-E.: 1990, Astron. & Astrophys. **234**, 133.

Zylka R., Mezger P.G. & Lesch H.: 1992, Astron. & Astrophys. **261**, 119.

Zylka R., Mezger P.G., Ward-Thompson D., Duschl W. & Lesch H.: 1994, Astron. & Astrophys. in press.

AN IRAS ATLAS OF YSOS IN NEARBY DARK CLOUDS

D.O.S. WOOD

National Radio Astronomy Observatory, P.O. Box O, Socorro NM, 87801 USA

and

P.C. MYERS

Harvard-Smithsonian Center for Astrophysics, 60 Garden Street, Cambridge MA, 02138 USA

Abstract. We are using the new *IRAS* Sky Survey Atlas to construct a catalog of Galactic dark clouds and their associated Young Stellar Objects (YSOs). From *IRAS* 100 and 60μm data we calculate images of the 100μm optical depth distribution in the clouds. Using the *IRAS* Faint Source Catalog we identify sources which have colors indicative of YSOs and have positions that lie inside cloud boundaries in the optical depth images. Here we present some preliminary results of applying this analysis to the Chamaeleon dark cloud complex.

Key words: Infrared – Dust – Dark Clouds – Young Stellar Objects

1. *IRAS* Images of Dark Clouds

We have begun a project, based on the new *IRAS* Sky Survey Atlas (ISSA), to construct *IRAS* 100μm optical depth images of dark clouds and catalog their physical properties. Our goal is to construct an all sky catalog of dark cloud structures in order to obtain better statistics on their physical properties and associated star formation. Here we present one aspect of this work, identifying candidate young stellar objects selected from the Faint Source Catalog (FSC) which have projected positions inside a cloud boundary.

Our method for mapping dark clouds structures with *IRAS* is presented in detail in Wood, Myers & Daugherty (1994), hereafter WMD. WMD demonstrated, as did Snell *et al.* (1989) and Langer *et al.* (1989), that *IRAS* 100μm optical depth images are at least as good a tracer of dark cloud structures as CO. Like CO, *IRAS* 100μm images trace the warm surfaces of clouds, but unlike CO, *IRAS* 100μm optical depth does not reach unity until an A$_V$ of ~2500 mag. This makes *IRAS* an excellent probe of cloud structures except where they are confused along the line of sight (within ~3° of the Galactic plane in the inner Galaxy). *IRAS* also has all sky coverage, ~1 arcminute resolution and can image structures ranging from as little as a tenth to as much as several hundred magnitudes in A$_V$.

A major difference between the image processing of the present work and that of WMD is that we now use images from the ISSA which have had zodiacal light subtracted using a physical model of dust in the solar system. WMD used a simple 'flat-fielding' technique to remove zodiacal light. The better zodiacal light subtraction and destriping of the ISSA produces better results than the previous work. Figure 1 shows an example of the 100μm

Astrophysics and Space Science **224**: 199–202, 1995.

D. WOOD AND P. MYERS

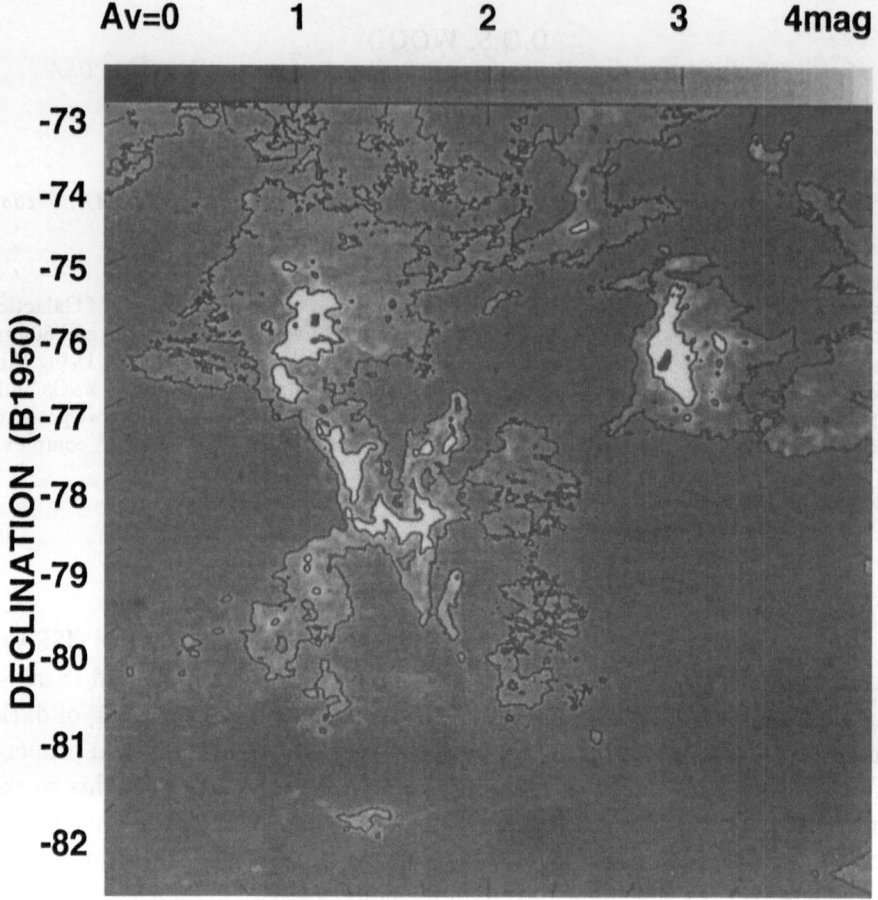

Fig. 1.   *IRAS* 100μm optical depth image of the Chamaeleon I, II and III dark cloud complex. Contours are at ∼2 and 4 mag Av.

optical depth images we have produced. The field of view is 12.5° × 12.5° and shows the remarkable structures of the Chamaeleon I, II and III dark cloud complex. Although the structure of Chamaeleon I and II were known previously, the III cloud is not as well studied. We also image the structures of several previously unknown clouds and cores in the region.

2. Identifying YSOs

We use the selection criteria developed by WMD to identify candidate YSOs. A FSC source is a YSO candidate if its 25μm flux is greater than its 12μm flux and it has a detection at 25 or 60μm with a detection in at least one

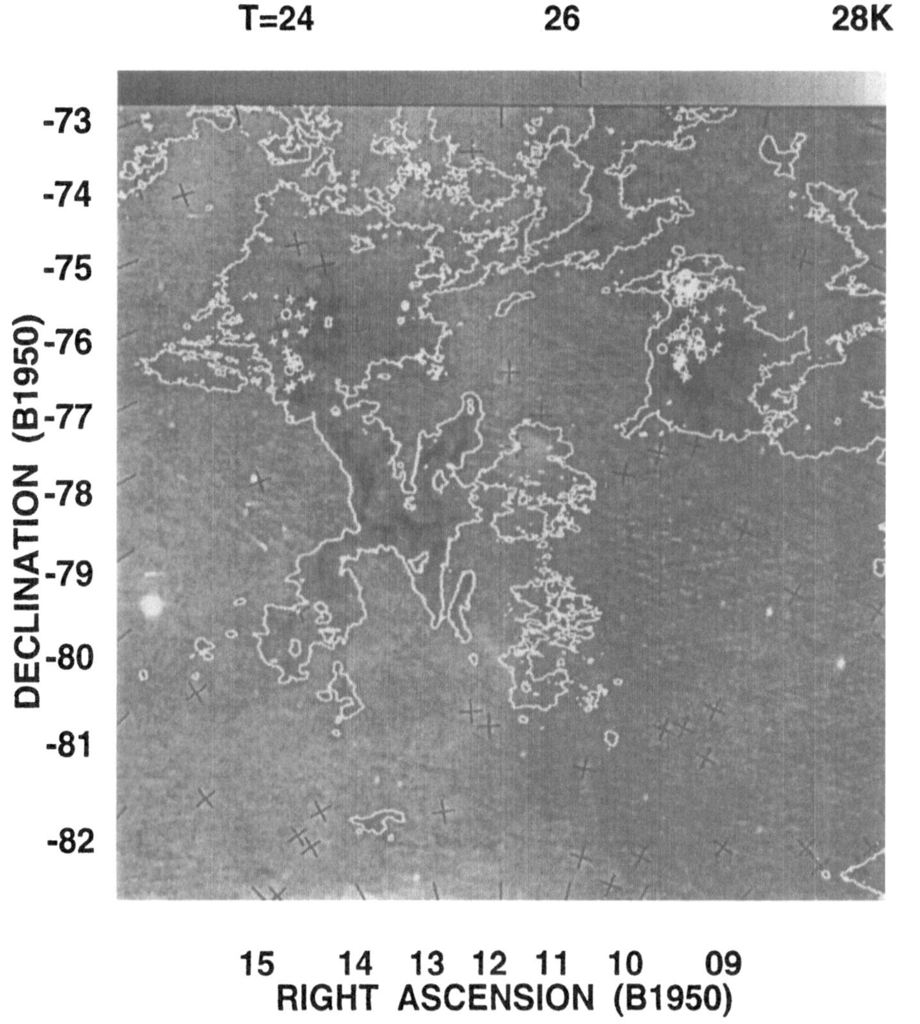

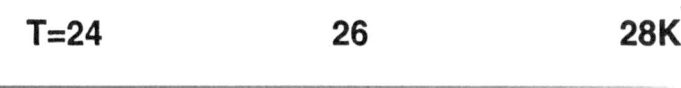

Fig. 2. *IRAS* 100/60μm color temperature image. White circles mark the positions of T Tauri stars from Hartigan (1993); white crosses mark pre-main-sequence stars from Gauvin & Strom (1992). The candidate YSOs selected from the FSC are marked with a black cross. A white contour at 2 mag Av shows the boundary of the cloud.

other band. In addition, we excluded sources in any catalog of stellar objects, planetary nebulae, or galaxies. Applying these criteria to the Chamaeleon region shown in Figure 1, we select 82 sources from the FSC. They are plotted as black crosses in Figure 2. White crosses and circles in Figure 2 mark the positions of known YSOs taken from Hartigan (1993) and Gauvin & Strom (1992).

Of the 82 sources selected from the FSC, 49 are projected inside a 2 A_v contour of a cloud boundary. Twelve of these candidates, which appear to be associated with the cloud and have colors indicative of embedded objects, do not appear in the previous YSO surveys. They are more distributed throughout the clouds than the young clusters studied by Hartigan (1993) and Gauvin & Strom (1992). This is due, in part, to the more limited regions surveyed by these authors. Most of our candidate YSOs are located in less well studied regions such as Chamaeleon III and the clouds to the west and north of Chamaeleon II. We plan to obtain optical spectroscopy to confirm whether they are YSOs or not.

3.   Future Work

We have developed automated techniques to identify and catalog cloud structures in 100μm optical depth maps. This system also searches the FSC for sources with colors typical of YSOs which are projected inside a cloud contour. Our goal is to construct an Atlas of Dark Clouds and their associated YSOs. This work is currently in progress. Although it is clear from Figure 2 that our Atlas of YSOs will not be a complete survey of the embedded population, largely due to sensitivity and confusion limits in the FSC, we will find many of the YSOs associated with dark clouds. More important, our survey will have all-sky coverage (except for the regions within $\sim 3°$ of the Galactic plane in the inner Galaxy) and as such will be a valuable resource for future investigation.

4.   Acknowledgements

This research was supported in part by NASA contract NAGW-3401 of the Astronomical Data Program.

References

Gauvin, L.S. & Strom, K.M.: 1992, Astrophys. J. **385**, 217.
Hartigan, P.: 1993, Astron. J. **105**, 1511.
Langer, W. D. *et al.* : 1989, Astrophys. J. **337**, 378.
Snell, R. L., Heyer, M. H., & Schloerb, F. P.: 1989, Astrophys. J. **337**, 738.
Wood, D. O. S., Myers, P. C. & Daugherty, D. A.: 1994, Astrophys. J., Suppl. **97**, in press
    (WMD).

A SELF-CONSISTENT MODEL FOR THE S140/L1204
REGION

NIGEL R. MINCHIN

*Department of Physics, Queen Mary and Westfield College, University of London, Mile
End Road, London E1 3NS, U.K.*

DEREK WARD-THOMPSON

Royal Observatory, Blackford Hill, Edinburgh EH9 3HJ, U.K.

and

GLENN J. WHITE

*Department of Physics, Queen Mary and Westfield College, University of London, Mile
End Road, London E1 3NS, U.K.*

Abstract. Comparison of submillimetre continuum observations of the L1204/S140 complex with previous high resolution CS, NH₃ and CI observations provides evidence that, for the first time, demonstrates the PDR and outflow are intimately linked. The only scenario that is able to explain all of the available molecular and atomic emission line data and our submillimetre continuum data, is one in which the outflow has expanded towards the edge of the molecular cloud and the edge of the blueshifted outflow lobe is now bounded by the expanding HII region. The NH₃ and continuum emission emanate from the inner edge of the outflow lobe, shielded from the external UV field.

Key words: Stars: early-type – interstellar medium: HII regions – jets and outflows

1. Introduction

The L1204 molecular cloud/S140 HII region complex is one of the most fascinating and complex regions for observing the interaction of an embedded molecular outflow and a PDR with their parent molecular cloud. At the southwestern extremity of the L1204 molecular cloud is an edge-illuminated PDR (*e.g.* Blair *et al.* 1978; Evans *et al.* 1987; White & Padman 1991), adjacent to the S140 HII region, powered by the nearby B0 star HD 211880. The complex lies at a distance of 910 pc (Crampton & Fisher 1974). Only ∼70 arcsec to the northeast of the PDR is an embedded cluster of three infrared sources (*e.g.* Beichman *et al.* 1979) which lie at the centre of a high-velocity molecular outflow (*e.g.* Bally & Lada 1983; Snell *et al.* 1984), and which have been identified as three embedded stars of spectral type B1.5– B2 (Evans *et al.* 1989). The outflow axis is along the southeast-northwest direction, parallel to the PDR. The blue and redshifted outflow lobes are separated by ∼35 arcsec, with a high degree of overlap, implying the outflow axis is directed close to the line of sight.

Minchin, White & Padman (1993–hereafter Paper I) recently compared high-resolution (∼10 arcsec), single velocity channel, [CI] $^3P_1 \rightarrow {}^3P_0$ observations of White & Padman (1991) to various ^{12}CO and ^{13}CO emission line

Astrophysics and Space Science **224**: 203–209, 1995.
© 1995 *Kluwer Academic Publishers.*

maps. The CI emission is mainly confined to a clumpy, elongated, ridge-like feature adjacent to the edge of the molecular cloud and coincident with a similar feature seen in ^{12}CO emission. The coincidence of the CI and CO features is interpreted as evidence that the molecular material is highly clumpy. This allows the UV radiation to penetrate into the cloud before dissociating carbon-bearing molecules, producing CI at the clump edges. There is a second region of intense CI emission, extending to the northeast *away from* the PDR, towards the centre of the cloud. This is contradictory to PDR models, and has led to speculation that it is produced by the radiation field from the embedded infrared cluster (Hayashi & Murata 1992; Paper I).

Recent multi-channel observations of the [CI] $^3P_1$–$^3P_0$ line (also at ∼10 arcsec) not only confirm that CI emission emanates from the PDR adjacent to the HII region/molecular cloud interface, but also reveals intense CI emission towards the outflow source, mainly at velocities blue and redshifted relative to the ambient cloud, not covered by the White & Padman observations (Minchin *et al.* 1994a – hereafter Paper II). There is also an arc of blue and redshifted emission extending from the peak for at least 30 arcsec to the south, which is adjacent to a similar feature observed in blueshifted C^{17}O J=3→2 emission. This implies the CI emission originates from the inner edge of the blueshifted molecular outflow wall. It is argued that the most plausible mechanism for producing the CI emission from the outflow region is the effect of shocks on the physical and chemical processes at the interface between the stellar wind and the blueshifted outflow cavity wall. A more extensive C^{18}O J=3→2 emission line map confirms the arc feature observed in CI and C^{17}O J=3→2 emission, but also reveals a second, similar arc feature extending towards the east (Minchin, White & Ward-Thompson 1994b–hereafter Paper III). It is thus an example of the classic ''tuning fork'' morphology, where emission at the ambient cloud velocity is tracing the outflow cavity wall of the blueshifted lobe.

2.   The morphology of the S140/L1204 region

2.1   COMPARISON OF CONTINUUM OBSERVATIONS WITH RECENT ATOMIC AND MOLECULAR LINE DATA

Figure 1 shows the contours of 800 μm continuum emission overlayed upon greyscale images of the CS J = 1→0 (Hayashi and Murata 1992), NH$_3$ (1,1) (Zhou *et al.* 1993) and CI $^3P_1$–$^3P_0$ (White & Padman 1991) line emission. The 800 μm continuum emission map was obtained using the UKT14 bolometer on the JCMT in April 1993.

Before discussing each figure in detail, let us point out some general similarities. Figures 1a–c show there to be a prominent, elongated ridge of emission extending from south of the infrared sources (shown on each of the figures) to the northeast – designated the eastern ridge. There is

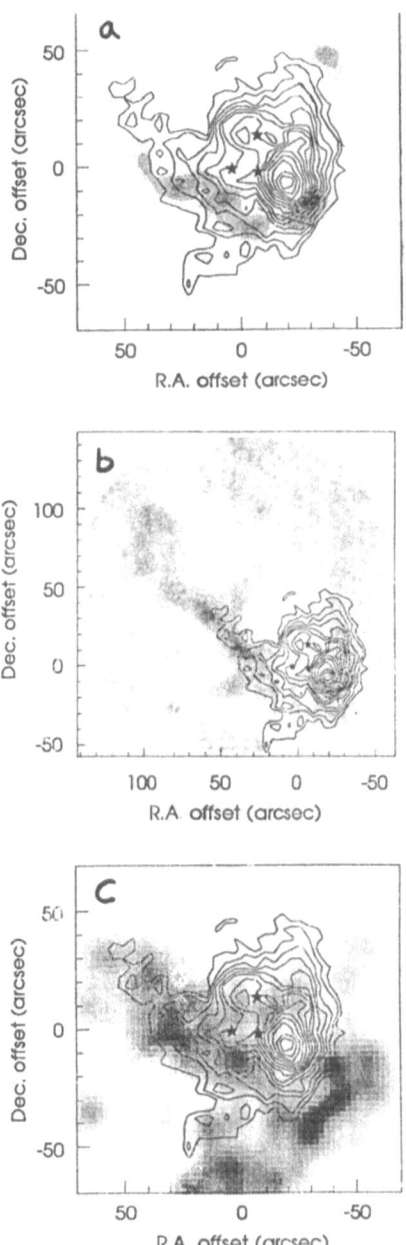

Fig. 1. Contours of the 800 μm continuum emission overlayed upon greyscale images of (a) the integrated intensity of the CS J=1→0 line emission – from Hayashi and Murata (1992) (b) the integrated intensity of the NH$_3$ (1,1) line emission – from Zhou *et al.* (1993) (c) the CI $^3P_1-^3P_0$ line emission in a single velocity channel – from White & Padman (1991).

a second, though less prominent and extensive, ridge extending from the southwest of the infrared sources to the northwest – designated the western ridge. This feature is particularly bright in CS emission, but is also clearly observed in NH_3 and submm continuum emission (at all three of the observed wavelengths), but is not seen in CI emission.

The CS J=1→0 line emission (Fig 1a) is resolved into a clumpy, horseshoe-shaped structure, with a diameter of ~0.3 pc, width of ~0.05 pc, a total mass of 60 $M_\odot$, and individual clump masses of 0.3-8.5 $M_\odot$ (Hayashi & Murata 1992). The CS map has an effective resolution of 10 arcsec, slightly higher than the 14 arcsec resolution of the 800 μm observations. The position of peak continuum emission (SMM1) is offset by ~12 arcsec to the northeast of the main peak of CS emission, designated peak G by Hayashi & Murata, with a derived mass of 8.5 $M_\odot$ at position RA (1950) = 22^h 17^m $37^s.78$, Dec. (1950) = $+63°$ $03'$ $28''.9$. Both the CS and continuum peaks lie along a line directed northeast-southwest that also points towards the external illuminating star, HD211880, and the embedded IR sources. Both the eastern and western CS ridge features appear to be located 10–20 arcsec outside of the more compact and centrally peaked continuum emission.

The NH_3 (1,1) emission (Fig. 1b) also delineates a clumpy horseshoe morphology at the high resolution of the Zhou *et al.* observations (5 arcsec FWHM). Although the basic morphology is similar to the CS J=1→0 line, the eastern ridge is more elongated (~2.5 arcmin compared to ~1.5 arcmin), but the western ridge is noticeably less prominent and extended. The comparison between continuum and NH_3 (1,1) emission is better than for the CS or CI lines, with both eastern and western ridges closely coincident. The main NH_3 peak, designated clump 3 by Zhou *et al.* is at position RA (1950) = 22^h 17^m $38^s.9$, Dec. (1950) = $+63°$ $03'$ $32''$, between the continuum and CS peaks and along the same line that also contains the external illuminating star and embedded infrared sources. The closer positional agreement between the NH_3 and continuum peaks may imply that NH_3 is enhanced by evaporation from heated dust grains, or could simply be due to the fact that CS is more difficult to excite than NH_3.

The CI $^3P_1-^3P_0$ line emission (Fig. 1c) is the integrated intensity in a 1.3 km s^{-1} channel centred at -8 km s^{-1} (White & Padman 1991) and thus traces emission at the ambient velocity of the molecular cloud. There is intense CI emission from the eastern ridge which, like the NH_3 emission, is closely coincident with the equivalent continuum emission feature. Conversely, the western ridge is not seen in CI emission. There is, however, a clumpy, elongated ridge-like feature adjacent to the southwestern edge of the molecular cloud. This is the position of the edge-on PDR and has been discussed in detail elsewhere (*e.g.* Papers I, II and III and references therein). The lack of significant continuum, CS or NH_3 emission from the PDR is in accord with models of both homogeneous and clumpy molecular clouds (*e.g.* Tielens & Hollenbach 1985; Stutzki *et al.* 1988).

Full velocity coverage of the CI $^3P_1$–$^3P_0$ line, for a limited part of the outflow/PDR regions, was presented in Paper II. This agrees closely with the White & Padman (1991) single channel image, but does not extend across the region of the eastern ridge. It should be pointed out that there *is* considerable CI emission from positions towards the infrared sources, but this is produced within the outflow and is thus observed at blue and redshifted velocities. Hence it was not observed in the single channel observations of White & Padman.

Clearly the S140 region is extremely complex, with certain aspects of the observations seemingly contradictory. In particular, how can CI emission from the eastern ridge feature be explained? The observation of CI emission from the PDR at the southwestern edge of the cloud is as expected from clumpy cloud models (Papers I and II), but the presence of an elongated CI feature that is almost orthogonal to the plane of the PDR is puzzling. The fact that the northern extension to the blueshifted CO emission (Paper I) lies even further within the eastern ridge feature may suggest that this is the limb-brightened edge to the northern extension of the blueshifted outflow lobe. As CS is a high density tracer this would seemingly agree with the scenario of a dense swept-up shell of material at the outer edge of an outflow cavity wall.

We also know that the CI emission is coincident with the NH_3 and continuum emission along the eastern ridge, *all* of which are offset from the CS emission. This contradicts PDR models, and is in marked contrast to observations of the PDR at the southwestern edge of the molecular cloud. As discussed in Papers I, II and III, at this position the only significant observed emission is from CI and ^{12}CO (due to effective self-shielding and clumpy structure) lines. Significant CS, NH_3 and submm continuum emission is only observed behind the PDR, at the positions of clump G, clump 3 and SMM1 respectively. The CS emission peak is observed to be *closer* towards the PDR than either the NH_3 or continuum emission peaks, quite the opposite of the situation along the eastern ridge.

We must also consider the kinematics of the region. The multi-channel CI $^3P_1$–$^3P_0$ line map of the central region of S140 shows there to be both blue and redshifted CI emission produced at the inner edge of the blueshifted molecular outflow cavity wall, as traced by blueshifted $C^{17}O$ emission (Paper II). The situation along the eastern ridge is quite different. Here the CO emission sits inside the CI ridge feature and the observed CI emission is at the velocity of the ambient cloud. The obvious implication is that the CI emission *cannot be produced within the molecular outflow*. Alternatively, the fact that CI emission from eastern ridge is observed at the same velocity as the ambient cloud, and is almost as intense as the CI emission from the PDR at the southwestern edge of S140, implies it is produced in the same

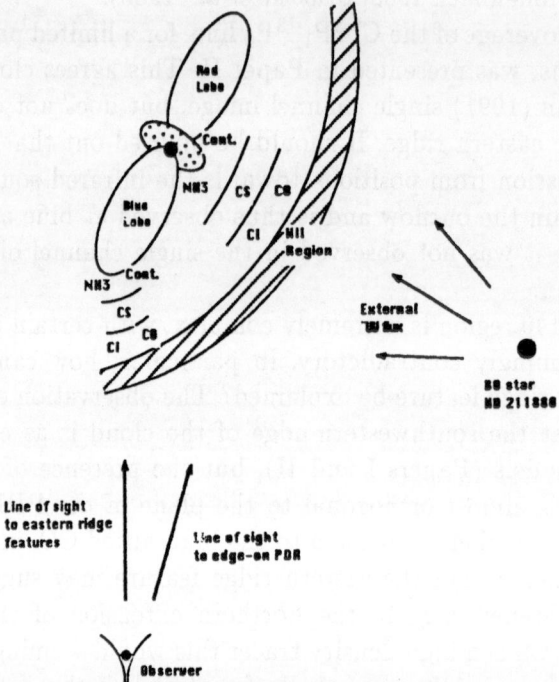

Fig. 2. Schematic representation of the S140/L1204 region showing the plane that contains the observer, the external illuminating star HD 211880, the HII region/molecular cloud interface and the embedded molecular outflow. The outflow has expanded towards the edge of the molecular cloud such that the edge of the blueshifted outflow lobe is bounded by the expanding HII region. The outside edge is therefore an externally illuminated PDR. Thus the CI emission emanates from the outer edge of the cloud (the PDR), with the CS emission tracing the compressed high density gas, effectively sandwiched between the expanding outflow and PDR regions. The NH_3 and continuum emission emanates from the inner edge of the outflow lobe, shielded from the external UV field. The CI, CS, NH_3 and continuum emission peaks are along the same observers line of sight and thus appear overlapping. Towards the southwestern rim of the molecular cloud the geometry of the PDR is such that it is edge-on when viewed along the observers line of sight and thus the CI, CS, NH_3 and continuum emission layers are spatially separate and distinct

manner.

2.2 A SELF-CONSISTENT MODEL FOR THE L1204/S140 REGION

The only morphology for the region that is consistent with all the molecular, atomic and submm continuum data is as follows. The eastern ridge feature is the dense, clumpy edge of the blueshifted outflow lobe that is closest to the observer along the line of sight. This naturally explains the overlapping CS, NH_3, CI and continuum features. The outflow has expanded towards the edge of the molecular cloud such that the edge of the blueshifted outflow lobe is bounded by the expanding HII region. The outside edge is therefore an externally illuminated PDR. Thus the CI emission emanates from the outer edge of the cloud (the PDR), with the CS emission tracing the compressed high density gas, effectively sandwiched between the expanding outflow and PDR regions. The NH_3 and continuum emission emanates from the inner edge of the outflow lobe, shielded from the external UV field. A schematic of the region is shown in figure 2.

This is also consistent with what we observe from the edge-on PDR at the southwestern rim of the cloud. Here the geometry allows us to observe the different molecular, atomic and continuum emission "layers". The CS, NH_3, CI and continuum peaks are *all* along a line that also connects IRS1 and the external illuminating source. The CI emission peak on the PDR is bounded by the CS peak, with the NH_3 and continuum peaks closest to the centre of the cloud. The close proximity of the CI and CS peaks and the fact that the CS emission along the western ridge is located 10-20 arcsec outside the more centrally peaked continuum emission may imply that, here too, part of the outflow structure is bounded by the expanding HII region, producing a PDR on the outer edge. Velocity channel maps of CS $J = 1{\rightarrow}0$ from Hayashi & Murata (1992), CS $J = 3{\rightarrow}2$ from Stutzki *et al.* (1994) and NH_3 (1,1) from Zhou *et al.* (1993), imply that emission along the eastern and western ridges is mainly observed at close to the ambient velocity of the molecular cloud ($\sim$ -8 km s^{-1}), but that many clumps are observed at either blue or redshifted velocities. This is consistent with the scenario of the dense outflow wall compressed from two directions.

References

Bally J. & Lada C.: 1983, Astrophys. J. **265**, 824.
Beichman C. A., Becklin E. E. & Wynn-Williams C. G.: 1979, Astrophys. J. **232**, L47.
Blair G. N., Evans N. J., Vanden Bout P. A. & Peters W. L.: 1978, Astrophys. J. **219**, 896.
Crampton D. & Fisher W. A.: 1974, *Pub. Dom. Astrophy. Obs.* 14, 12.
Evans N. J., Kutner M. L. & Mundy L. G.: 1987, Astrophys. J. **323**, 145.
Evans N. J., Mundy L. G., Kutner M. L. & DePoy D. L.: 1989, Astrophys. J. **346**, 212.
Hayashi M. & Murata Y.: 1992, Publ. of the ASJ **377**, 391.
Minchin N. R., White G. J. & Padman R.: 1993, Astron. & Astrophys. **277**, 595 (**Paper I**).
Minchin N. R., White G. J., Stutzki J. & Krause D.: 1994a, Astron. & Astrophys. in press (**Paper II**).
Minchin N. R., White G. J. & Ward-Thompson D.: 1994b, Astron. & Astrophys. submitted (**Paper III**).
Snell R. L., Scoville N. Z., Sanders D. B. & Erickson N. R.: 1984, Astrophys. J. **284**, 176.
Stutzki J., Stacey G. J., Genzel R., Harris A. I., Jaffe D. T. & Lugten J. B.: 1988, Astrophys. J. **332**, 379.
Tielens A. G. G. M. & Hollenbach D. J.: 1985, Astrophys. J. **291**, 722.
White G. J. & Padman R.: 1991, *Nature* **354**, 511.
Zhou S., Evans N. J., Mundy L. G. & Kutner M. L.: 1993, Astrophys. J. **417**, 613.

PLATE III: Entrance to the Observatory grounds.

PERIODICAL LIGHT VARIATIONS OF T TAURI STARS
AS A RESULT OF DISK ACCRETION

S.A.LAMZIN

Sternberg Astronomical Institute, Moscow V-234, 119899 Russia

Abstract. I argue that temperatures of spots, responsible for observed periodical light variations of T Tauri stars (TTS), are not known with reliable accuracy to discriminate between chromospheric and accretion theories of TTS 's phenomenon. The hypothesis is set up that spots on classical TTS (CTTS) are due to heating of stellar surface by radiation from a collisional accretion shock, whereas spots on weak line TTS (WTTS), at least in some cases, are connected with a collisionless accretion shock rather than chromospheric activity. Possible scenarios of WTTS interaction with circumstellar matter are discussed.

Key words: Pre-main sequence stars – Spots – Accretion – Shock waves

Periodical light variations ($P \sim 3^d$), which are interpreted as a result of rotational modulation of surface temperature's inhomogenities (spots), have been detected in a number of WTTS and CTTS. According to present day paradigm, activity in CTTS is caused by disk accretion of circumstellar matter, whereas in the case of WTTS one observes enhanced chromospheric activity analogous to solar activity (Bertout 1989). Spot temperatures T_s are considered one of the most important arguments in favour of this viewpoint: it was found that WTTS spots are cold ($T_s < T_{eff}$), while CTTS spots are most often hotter than the surrounding photosphere. I suppose however that it is not so obvious because the theoretical models used up to now to derive temperature of TTS spots are too crude.

Observed profiles of C IV 1548 Å, He II 4686 Å and some other lines indicate that hot ($T > 10^4$ K) regions of CTTS are connected with accretion shock rather than boundary layer (Gomez de Castro *et al.* 1994). I suppose therefore that the global dipole-like magnetic field of a CTTS disrupts the innermost part of the accretion disk and channels gas flow onto the star. Infalling matter is sharply decelerated in strong shocks near the stellar surface and is heated up to $T_{max} \sim 10^6$ K. Then the gas cools via volume emission of UV and X-ray photons down to some thousands of degrees. Gas velocity decreases, density increases and the post-shock region steeply transits into the upper layers of stellar atmosphere. Half the quanta from the post-shock region moves inward and is absorbed in the upper layers of stellar atmosphere. Large optical depths of H and He resonant lines as well as high gas density are reasons to assume that continuum rather than line emission plays a dominant role in the energy balance of this region. Thus I suppose that CTTS excess (veiling) continuum originates in a transition region between the post-shock zone and the upper layers of stellar atmosphere. It is this region I identify with CTTS "spots", because the pre-shock zone is apparently transparent to continuum emission longward from the Lyman

Astrophysics and Space Science **224**: 211–214, 1995.

limit (Lamzin 1994).

Let us introduce a system of spherical coordinates with its origin at the center of the star, the $\varphi = 0$ plane of which passes through the line connecting the stellar center with the Earth. If rotational and magnetic axes of the star do not coincide, such that the angle α between them is not too large, then accretion zones at the stellar surface form two ring-like hot spots ($\theta_1 \leq \theta \leq \theta_2$ and $\pi - \theta_2 \leq \theta \leq \pi - \theta_1$), axially symmetric relative to the magnetic axis ($\theta = 0$). The angle β between the stellar magnetic axis and the direction to the Earth depends on the rotational phase ϕ : $\cos\beta = \sin\alpha \sin i \cos(2\pi\phi) + \cos\alpha \cos i$, where i – the inclinational angle of the stellar rotational axis. This is why the projected area of the spots changes with time producing light variations (Giovannelli et al. 1991, Kónigl 1991). For an arbitrary point on the stellar surface with coordinates (θ, φ), the angle γ between the local normal and the direction to the Earth can be expressed as follows: $\cos\gamma = \sin\beta \sin\theta \cos\varphi + \cos\beta \cos\theta$.

Let $I^*(\lambda, \gamma)$ and $I^s(\lambda, \gamma)$ be monochromatic specific intensities of both stellar and spot continuum emission respectively. If d – the distance to the star, R_* – its radius and $C = \pi R_*^2/d^2$, one can write for the monochromatic flux near the Earth:

$$\mathcal{F}_\lambda = C \left[\int_0^\pi \sin\theta d\theta \int_0^{2\pi} I_\lambda^* \cos\gamma d\varphi + \int\!\!\!\int_{2\,spots} (I_\lambda^s - I_\lambda^*) \cos\gamma \sin\theta d\theta d\varphi \right] \quad (1)$$

It is convinient to write I^* in the form $I^* = B_\lambda(1 - \varepsilon_\lambda + \varepsilon_\lambda \cos\gamma)$, where B_λ – Planck function for stellar T_{eff} – see e.g. Bouvier et al. (1986, 1993). One should note however that:

1) the Spectral energy distribution of late-type subgiants differs very strongly from a black-body shortward of 5000 Å. For example $U - B = 0.39$ for a T=4000 K black-body, while for K7 subgiants $U - B \simeq 1.8$ (Straizys 1977).

2) a limb darkening law $I(\gamma)$ written above is not valid for U and B filters in the case of stars with $T_{eff} < 4000$ K, because it predicts $I < 0$ for $\gamma \sim \pi/2$ (Al-Naimy 1978).

These are serious enough reasons by themselves to doubt results of spot temperature determination bearing in mind that almost all TTS have spectral class later K0. But the situation is much more uncertain with radiation from the spots. To calculate the spectrum and "limb darkening" law $I_\lambda^s(\gamma)$ of spot continuum emission ab initio it is necessary to calculate the structure of the stellar atmosphere with non-zero gas density at the upper boundary heated by an external flux of UV and X-ray radiation. This problem has not been solved up to now, so different approximations have to be used for $I^s(\lambda, \gamma)$.

Herbst & Koret (1988) have supposed $I_\lambda(\gamma) = B_\lambda = const(\gamma)$. In this case one underestimates the spot temperature if $T_s < T_*$ and overestimates it otherwise. Bouvier et al. (1993) have assumed that $\varepsilon_* = \varepsilon_s$. No specific geometry for spots were assumed in this work, but the method which has been used to determine the spot temperature is less than perfect from a mathematical viewpoint. I have tried to reproduce their observational data using the same approximation, but have used the geometry described above. In such a case Eq.(1) gives:

$$\mathcal{F}_\lambda = CI_\lambda^*(0) \left\{ 1 - \frac{\varepsilon}{3} + Q_\lambda \left[\frac{2}{\pi}(1 - \varepsilon)(\Psi_1 - \Psi_2) + \varepsilon(\Phi_1 - \Phi_2) \right] \right\}, \quad (2)$$

where $(k = 1, 2) \, Q_\lambda = I_\lambda^s/I_\lambda^* - 1$, $\Phi_k = \sin^2 \beta \cos \theta_k + (\cos^2 \beta - 1/3) \cos^3 \theta_k$,

$$\Psi_k = \begin{cases} \frac{\pi}{2}(1 - \sin^2 \theta_k \cos \beta) & \theta_k \le \frac{\pi}{2} - \beta \\ \sin^{-1}\left(\frac{\cos \theta_k}{\sin \beta}\right) - \sin^2 \theta_k \cos \beta \tan^{-1}\left(\frac{\cos^2 \theta_k \cos \beta}{y_k}\right) + y_k, & \text{otherwise} \end{cases}$$

such that $y_k = (\sin^2 \beta - \cos^2 \theta_k)^{1/2} \cos \theta_k$.

I have succeeded, for example, to reproduce observed UBVRI light curves of LkCa-19 ($T_{eff} = 5240$ K) with the same accuracy as Bouvier et al.(1993) adopting $\alpha = 10°$, $i = 30°$, $\theta_1 = 20°$, $\theta_2 = 45°$ for "hot" spot ($T_s = 7680$ K), whereas they have found that the spot(s) should be "cold": $T_s = 4770$ K. Bouvier et al. (1986) assume that DN Tau's spot has a circular form and have found in the $\varepsilon_s = \varepsilon_*$ approximation $T_s < 3200$ K, while I have found from Eq.(2) and the same observational data $T_s = 6140$ K ($\alpha = 23°$, $i = 20°$, $\theta_1 = 34°$, $\theta_2 = 46°$).

I'd like to stress that one should not consider my parameters for LkCa-19 and DN Tau as reliable ones: I present them as counter-example only to demonstrate the uncertainties of T_s determination. But these stars are WTTS, so one can propose that at least in some cases WTTS spots can be hot. Strassmeir et al. (1994) have found from Doppler imaging that WTTS V410 Tau has both cold and hot spots. Hot WTTS's spots can be analogous to solar plages, but it is interesting to check if they are connected with accretion as in the case of CTTS, because circumstellar matter is also observed in the vicinity of some WTTS (Osterloh & Beckwith 1994). Further, it is known that some WTTS e.g. DoAr21 were recently CTTS (Herbig & Bell 1988).

Let us assume that the accretion rate onto a CTTS with given magnetic field strength has decreased due to some reasons. If Alfvenic radius r_A ($\simeq$ inner radius of the disk) becomes larger than the co-rotation radius r_C, then the magnetic field will throw back the plasma from the star and stationary accretion will stop. I suppose that observational manifestations of this so-called "propeller" regime would not be very striking. The case $r_A \sim 10 R_* <$

r_C seems much more interesting. One can expect that the near infrared excess in this case would be negligible similar to the emission line strength. But decreasing the infall gas density should change the situation not only qualitatively but quantitatively.

It can be shown that for a typical CTTS the pre-shock gas density $\sim 10^{12}$ cm^{-3} (Lamzin 1994). Let us compare the mean free path of protons (in plasma) relative to binary collisions with their Larmor radius in a global stellar magnetic field (Zeldovich & Raizer 1967):

$$\frac{r_L}{l_{pp}} \sim \left(\frac{N_0}{3 \times 10^{12} \text{ cm}^{-3}} \right) \left(\frac{10^4}{T \text{ K}} \right)^{3/2} \left(\frac{500}{H \text{ G}} \right).$$

It follows from this relation that in the case of CTTS an accretion shock is still collisional, but if the infall gas density decreases by ~ 10 times the shock transforms into a collisionless one.

Collisionless shocks may be the source of relativistic particles responsible for the observed nonthermal radio emission from WTTS (Philips *et al.* 1991, White *et al.* 1992, Skinner 1993) and their hard (up to 8 KeV), constant during hours X-ray emission (Koyama *et al.* 1994). According to preliminary estimations by E.Berejko (private communication) one can expect acceleration of protons up to energies ~ 1 GeV. It would be very interesting to check if interaction of high energy protons with dust grains of the primordial Solar Nebula can explain isotopic anomalies found in meteorites. A collisionalless hypothesis appears so attractive that I recommend to investigate it in more detail.

References

Al-Naimy H.M.: 1978, Astrophys. & Space Sci. **53**, 181.
Bertout C.: 1989, Ann. Rev. of Astron. & Astrophys. **27**, 351.
Bouvier J., Bertout C. & Bouchet P.: 1986, Astron. & Astrophys. **158**, 149.
Bouvier J., Cabrit S., Fernandez M. *et al.* : 1993, Astron. & Astrophys. **272**, 176.
Giovannelli F.,Rossi C.,Errico L. *et al.* : 1991, in *Angular Momentum Evolution of Young Stars*, eds: S. Catalano & J.R. Stauffer, Kluwer Acad. Publ.,Dordrecht, p97.
Gomez de Castro A., Lamzin S.A. & Shatskyi N.I.: 1994, Astronomy Reports **38**, 536.
Herbig G.II. & Dell K.R.: 1988, Lick Observatory Bull. **1111**.
Herbst W. & Koret D.L.: 1988, Astron. J. **96**, 1949.
Kónigl A.: 1991, Astrophys. J. **370**, L39.
Koyama K.,Maeda Y.,Ozaki M. *et al.* : 1994, Publ. of the ASJ **46**, L125.
Lamzin S.A.: 1994, Astron. & Astrophys. submitted.
Osterloch M. & Beckwith S.: 1994, Astrophys. J. in press.
Philips R.W., Lonsdale C.J. & Feigelson E.D.: 1991, Astrophys. J. **382**, 261.
Skinner S.L.: 1993, Astrophys. J. **408**, 660.
Straizys V.L.: 1977, *Multicolor stellar photometry*, Mokslas publishers: Vilnus.
Strassmeier K.G., Welty A.D. & Rice J.B.: 1994, Astron. & Astrophys. **285**, L17.
White S.M., Pallavicini R. & Kundu M.R.: 1992, Astron. & Astrophys. **259**, 149.
Zeldovich Ya.B. & Raizer Yu.P.: 1967, *Physics of Shock Waves and High Temperature Hydrodynamic Phenomena*, Academic Press, New York & London.

SESSION D:

DUST & GAS; COMPOSITION & CHEMISTRY

MOLECULES AND DUST IN CIRCUMSTELLAR SHELLS

W. W. DULEY

University of Waterloo, Waterloo, Ontario, Canada N2L 3G1

Abstract. Circumstellar shells provide a unique environment for the study of dust formation and the relation of dust composition to specific atomic and molecular components.
As a specific example, the formation of carbonaceous dust is discussed in relation to the
presence of polycyclic aromatic hydrocarbon molecules and their survival in the interstellar medium. Some conclusions will be drawn concerning the composition of carbonaceous
dust in circumstellar sources and that in the diffuse interstellar medium.

1. Introduction

The infrared spectrum of circumstellar shells of evolved stars is typically characterized by continuum emission upon which are often superimposed
emission and absorption features associated with a variety of chemical species.
Oxygen-rich M stars and planetary nebulae show the 9.7 and 18.5 μm spectral bands of silicates, which carbon stars often exhibit emission by SiC
particles at $\lambda \simeq 11.5\mu$m. The spectra of post AGB stars and many carbon-
rich planetary nebula show the emission features at 3.3, 6.2, 7.7, 8.6 and
11.3 μm associated with polycyclic aromatic hydrocarbon (PAH) molecules
superimposed on a continuum that arises due to thermal emission from small
particles of amorphous carbon. The growth of these particles is undoubtedly related to chemical species present in the extended atmospheres of AGB
stars and so it becomes of interest to examine any connection that might
exist between PAH molecules and this dust. Specifically, a) do these carbon grains derive from the condensation of PAH molecules? b) are PAH
molecules ejected in significant quantities in this way into the interstellar
medium? and c) what is the relation between PAH species observed in planetary nebulae and photodissociation regions and PAH molecules, in these
atmospheres?

2. Molecules and Dust in Carbon Shells

Many carbon rich AGB and post-AGB stars emit infrared continua that can
be identified with emission from small amorphous carbon grains (Martin
& Rogers 1987, Sopka *et al.* 1985, Waters *et al.* 1989). Graphite grains are
not observed and PAH absorption/emission is seen only in post-AGB objects
such as HR 4049, and then only weakly (Waters *et al.* 1981). PAH emission
is, however, detected in the late stages of the evolution of carbon rich novae
such as Nova Cen 1986 (Smith *et al.* 1993) and in fully evolved objects such
as the planetary nebula NGC 7027 (Russell *et al.* 1977).

Astrophysics and Space Science **224**: 217–221, 1995.
© *1995 Kluwer Academic Publishers.*

While significant amounts of PAH do not seem to be present in post-AGB objects extending to CRL 618 a proto-planetary nebula (Lequeux *et al.* 1990) an absorption feature centered near 3.4 μm and identified with hydrogenated amorphous carbon (HAC) is observed in this object and suggests the presence of an appreciable quantity of a low temperature form of HAC (Duley 1994). This suggests that the carbon dust in late carbon stars such as IRC+10216 is a highly unsaturated form of amorphous carbon, but that this material evolves with time into HAC. It is significant, in this regard, that a weak extended red emission (ERE) attributable to HAC luminescence (Duley 1985) has been recently detected in the spectrum of the nova QV Vul (Scott *et al.* 1994).

3. Chemistry in Carbon Shells

The presence of PAH molecules in evolved post AGB objects has led to the suggestion that these molecules are formed in carbon rich stellar atmospheres of AGB stars and are then ejected with dust into the interstellar medium (see Allamandola *et al.* 1989). The formation of graphite grains by condensation of atomic and molecular carbon species in carbon stars has been suggested for some time (Tsuji 1964) however the role of PAH molecules in the formation of carbon dust has only recently been investigated (Keller 1987, Gail & Sedlmayr 1985, Frenklach & Feigelson 1989, Cadwell *et al.* 1994). PAH molecules form by reactions involving C_2H and C_2H_2 as primary chemical species leading to benzyne, a six membered ring, as the initial ring compound. Subsequent reactions proceed to yield phenyl-acetylene and then naphthalene radicals. The overall effect is to convert acetylene to PAH.

A comprehensive study involving over 100 reactions and 40 chemical species by Frenklach & Feigelson (1989) has explored possible regimes for PAH condensation in carbon atmospheres under AGB conditions. They find that the absolute quantities of PAH are sensitive to many parameters and that PAH production is important only over a narrow temperature range (900-1100 K) in a limited range of objects. As a result, they suggest that the homogeneous nucleation of PAH molecules to form a carbon solid in these atmospheres is unlikely. Heterogeneous nucleation on SiC particles is, however, favoured and may lead to the ejection of SiC grains coated with an amorphous carbon containing PAH groups (Cadwell *et al.* 1994). This material is likely a highly unsaturated form of amorphous carbon.

4. Carbon Dust in Post AGB Objects

While the ingredients for carbon dust in post AGB shells and planetary nebulae have been ejected, in some part, from the envelope of a precursor AGB star, it is likely that this material is significantly modified in protoplanetary nebulae. This modification will occur as gas and dust are exposed to UV radiation in the nebula. The effect of hard UV radiation on amorphous carbon is to produce partial graphitization (Jones *et al.* 1990) and a reduction in band gap energy but exposure to H atom fluxes can reverse this trend, leading to rehydrogenation (Jones *et al.* 1990, Furton & Witt 1993). Rehydrogenation of amorphous carbon by exposure to H atoms in the presence of UV radiation has been shown to produce HAC in laboratory experiments (Furton & Witt 1993). This rehydrogenated material will exhibit 3.4 μm absorption and has been shown to emit an ERE-like luminescence. The observation of a 3.4 μm CH$_n$ absorption band in the pre-planetary nebula CRL 618 (Lequeux *et al.* 1990) and ERE emission in a later stage of dust evolution in nova QV Vul (Scott *et al.* 1994) can then be understood as the rehydrogenation of amorphous carbon dust in the presence of UV radiation and H atoms, and suggests that the normal evolution of carbon dust may be from amorphous carbon in AGB stars to HAC in pre-planetary nebulae. Indeed, the appearance of HAC in absorption at 3.4 μm may signal the onset of formation of a planetary nebula.

5.   PAH Molecules in Planetary Nebulae and Interstellar Clouds

Under more vigorous conditions of rehydrogenation and in the presence of shock, HAC grains will be dissociated leading to the evolution of atomic and molecular fragments (Duley & Jones 1990, Duley 1993). These fragments will contain some of the structural elements of HAC which have been shown to include clusters of C$_6$ rings (Robertson & O'Reilley 1987). Excitation of these molecules either during grain disruption, or subsequently in the gas phase will yield the usual discrete IR emission features attributed to PAH. It is important to note that these molecules as such do not have to be present in dust during condensation in order to be liberated on dissolution of HAC. Scanning tunneling microscope studies of HAC (*e.g.* Cho *et al.* 1992) show that these structural elements are commonly present in HAC prepared by vapour deposition using a variety of gaseous precursors.

Liberation of PAH molecules from HAC by shocks may, in fact, constitute a primary source of interstellar PAH. As HAC can be deposited by accretion of C and C$^+$ on silicate grains in interstellar clouds (Duley & Williams 1988), shocking of coated grains can be an efficient production route for PAH. The rate of formation of PAH is given by

$$R_{PAH} = \frac{S_c n_c < n_g \pi a^2 > v_c \gamma}{N_c} \; cm^{-3} sec^{-1} \tag{1}$$

where S_c is the sticking efficiency for C, C$^+$ on dust, n_c is the density of C, C$^+$, $< n_g \pi a^2 >$ is the integrated area of dust grains, v_c is the speed of a C atom, γ is the fractional amount of carbon on dust in the form of PAH molecules each containing N_c carbon atoms. With $< n_g \pi a^2 > = 2.1 \times 10^{-21}$ n, T = gas kinetic temperature = 56 K, $n_c = 10^{-4}$ n, n = 300 cm^{-3}, $S_c =$ 1.6 x 10^{-3} (Duley & Smith 1994), $\gamma = 0.3$, $N_c = 14$ (*i.e.* as for anthracene)

$$R_{PAH} = 1.2 \times 10^{-18} \frac{\gamma}{N_c} = 2.6 \times 10^{-20} cm^{-3} sec^{-1} \qquad (2)$$

Some fraction, f, of this material is released by a shock after some time interval, t_s. Then

$$n_{PAH} = 2.6 \times 10^{-20} f t_s cm^{-3} \qquad (3)$$

With $f = 0.1$ and $t_s \simeq 10^{15}$ sec, $n_{PAH} = 2.6$ x 10^{-6}cm^{-3} or $n_{PAH}/n = 8.7$ x 10^{-9}. This ejected PAH would recondense on grains over a timescale $\tau = 1/< n_g \pi a^2 > \nu_{PAH} \simeq 5x10^{17}/n$ sec = 1.6 x 10^{15} sec for n = 300 cm^{-3}. This suggests that a significant concentration of gaseous PAH can be maintained by shocking of HAC grains in diffuse clouds if $t_s \leq 50My$.

6. Conclusions

It can be concluded from chemical equilibrium calculations that PAH molecules are not generally abundant components in most carbon stars and do not therefore lead directly to dust formation. The dust in carbon stars is then more likely an unsaturated amorphous carbon. Fits to the IR emissivity of carbon stars support this conclusion. On ejection from AGB stars, this amorphous carbon dust is rehydrogenated to HAC by reaction with hydrogen during the pre-nebula phase. This dust exhibits a 3.4 μm CH$_n$ absorption band similar to that observed in Galactic center sources. It should also show an ERE emission. Further hydrogenation and shocking in the planetary nebula phase leads to the dissolution of the HAC dust and to the liberation of PAH molecules which are excited to yield the UIR features. It is shown that a similar process may occur in general in interstellar clouds as HAC, created by accretion of C and C$^+$ on silicate cores is dissociated in shocks.

Acknowledgements

This research was supported by grants from the NSERC of Canada.

References

Allamandola, L.J., Tielens, A.G.G.M. & Barker, J.R.: 1989, Astrophys. J., Suppl. **71**, 733.
Cadwell, B.J., Wang, H. Feigelson, E.D. & Frenklach, M.: 1994, Astrophys. J. in press.
Cho, N.H., Veirs, D.K., Ager, J.W., Rubin, M.D., Hopper, C.B. & Bogy, D.B.: 1992, J. Appl. Phys. **71**, 2243.
Duley, W.W.: 1985, Mon. Not. of the RAS **215**, 259.

Duley, W.W.: 1993, in *Dust and Chemistry in Astronomy*, eds: T.J.Millar & D.A. Williams, Bristol: IOP Publ.

Duley, W.W.: 1994, Astrophys. J. **430**, L133.

Duley, W.W. & Jones, A.P.: 1990, Astrophys. J. **351**, L49.

Duley, W.W. & Smith, R.G.: 1994, Astrophys. J. submitted.

Duley, W.W. & Williams, D.A.: 1988, Mon. Not. of the RAS **230**, 1P.

Frenklach, M. & Feigelson, E.D.: 1989, Astrophys. J. **341**, 372.

Furton, D.G. & Witt, A.N.: 1993, Astrophys. J. **415**, L51.

Gail, H.P. & Sedlmayer, E.: 1985, Astron. & Astrophys. **148**, 183.

Jones, A.P., Duley, W.W. & Williams, D.A.: 1990, Quart. J. of the RAS **31**, 567.

Keller, R.: 1987, in *Polycyclic Aromatic Hydrocarbons and Astrophysics*, eds: A. Leger, L. d'Hendecourt & N. Boccara, Dordrecht, Reidel, p3.

Lequeux, J. & Jourdain de Muizon, M.: 1990, Astron. & Astrophys. **240**, L19.

Martin, P.G. & Rogers, C.: 1987, Astrophys. J. **322**, 274.

Robertson, J. & O'Reilly, E.P.: 1987, Phys. Rev. B: Solid State **35**, 2946.

Russell, R.W., Soifer, B.T. & Willner, S.P.: 1977, Astrophys. J. **217**, L149.

Scott, A.D., Evans, A. & Rawlings, J.M.C.: 1994, Mon. Not. of the RAS **269**, L21.

Smith, C.H., Aitken, D.K. & Roche, P.F.: 1993, preprint.

Sopka, R.J., Hildebrand, R., Jaffe, D.T., Gatley, I., Roellig, T., Werner, M., Jura, M. & Zuckerman, B.: 1985, Astrophys. J. **294**, 242.

Tsuji, T.: 1964, *Ann. Tokyo Astron. Obs., 2nd Series* **9**, 1.

Waters, L.B.F.M., Lamers, H.J., Snow, T.P., Mathlener, E., Trams,N.R., van Hoof, P.A.M., Waelkens, C., Se ab, C.G. & Stanga, R.: 1989, Astron. & Astrophys. **211**, 208.

PLATE IV: The original Observatory on Calton Hill in the City of Edinburgh.

CIRCUMSTELLAR DUST AND CIRCUMSTELLAR
EXTINCTION CURVES

N.V. VOSHCHINNIKOV
Astronomical Institute, St. Petersburg University, St. Petersburg, 198904 Russia

and

F.J. MOLSTER and P.S. THÉ
Astronomical Institute "Anton Pannekoek", University of Amsterdam, Amsterdam,
The Netherlands

Abstract. We have calculated the circumstellar extinction curves produced by dust grains which absorb and scatter the stellar radiation in the shells of pre-main-sequence stars. A Monte Carlo method was used to model the radiative transfer in non-spherical shells. The dependence on the particle size distribution and the dust shell parameters has been examined.

The application of the theoretical results to explain the extinction and polarization of the Herbig Be star HD 45677 shows that the dust shell is not disk-like and that very small grains are absent in it.

Key words: circumstellar matter – stars: pre-main-sequence – dust – extinction

1. Introduction

Many stars are surrounded by dust shells. They are primarily revealed by the IR excesses and variable intrinsic polarization. Another observational feature detected from the UV observations is the significant deviation of the extinction curves for hot young stars from the average interstellar extinction law which was treated as anomalous extinction (Steenman & Thé, 1991). It was observed for many Herbig Ae/Be stars (*e.g.* Sitko *et al.* 1981 and Thé *et al.* 1985) which are sufficiently bright in the UV and do not have numerous molecular bands as in late-type stars.

We consider the circumstellar (CS) extinction taking into account not only the obscured stellar radiation but also the light scattered by dust in the shell. A more detailed analysis will be published elsewhere (Voshchinnikov *et al.* 1994).

2. Model

We assume that the star lies in the center of a homogeneous oblate spheroidal shell with a central cavity free of dust and calculate the Stokes vector of the produced radiation. Its first parameter $[I(\lambda)]$ is the sum of the stellar brightness obscured by dust in the line of sight $[I_*(\lambda) \cdot e^{-\tau_i^{\text{ext}}(\lambda)}]$ and the scattered radiation $[I_{\text{sca}}(\lambda)]$. It may be used to define the curves of "the

Astrophysics and Space Science **224**: 223–226, 1995.
© 1995 *Kluwer Academic Publishers.*

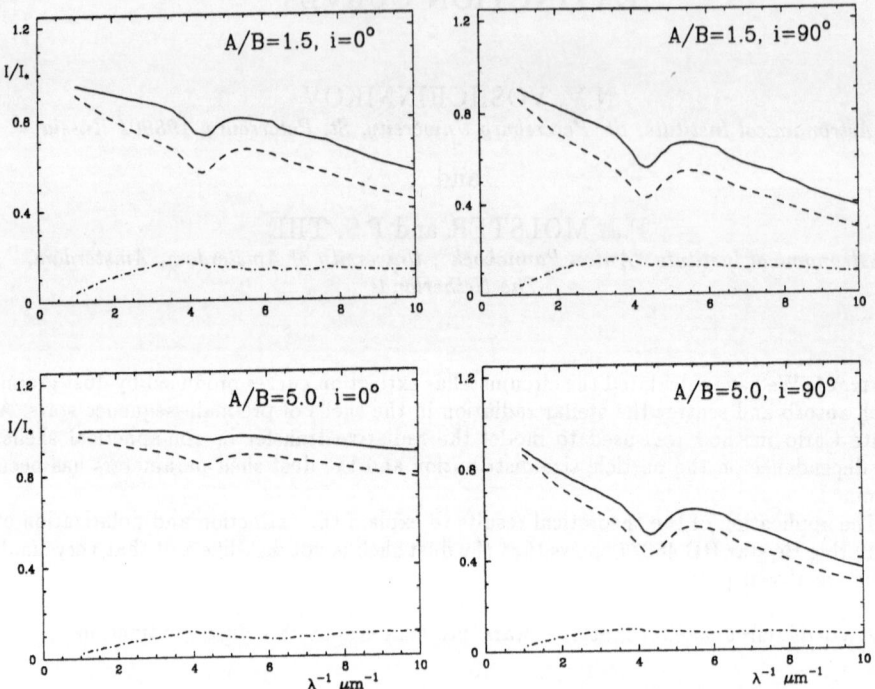

Fig. 1. Wavelength dependencies of the normalized intensity (Mie scattering, standard MRN mixture: $a_- = 0.005$ μm, $a_+ = 0.25$ μm, $q = 3.5$, $n_{Si}/n_C = 1.07$, $\tau_{90}^{ext}(0.55\,\mu\text{m}) = 0.3$). Solid lines: total radiation, short-dashed lines: scattered radiation, dot-dashed lines: obscured stellar radiation.

circumstellar extinction"

$$A^{cs}(\lambda) = -2.5 \log\left[I(\lambda)/I_*(\lambda)\right] = -2.5 \log\left[e^{-\tau_i^{ext}(\lambda)} + I_{sca}(\lambda)/I_*(\lambda)\right]. \quad (1)$$

Here, $\tau_i^{ext}(\lambda)$ is the optical path of the stellar radiation inside the shell in the direction of an observer.

We considered shells with various aspect ratios (A/B), viewing angles (i), and the optical thicknesses along the main axis $[\tau_{90}^{ext}(\lambda_0)]$. The properties of the CS dust were modeled by changing the parameters of the silicate-graphite mixtures of grains (MRN mixture; Mathis et al. 1977): the lower (a_-) and upper (a_+) size cutoffs, the index of the power-law size distribution (q), and the ratio of the number densities of silicate and graphite grains (n_{Si}/n_C).

The detailed description of the model is presented in the paper of Voshchinnikov & Karjukin (1994).

3. Some Results

Numerous calculations have been made to understand how the scattered radiation alters the CS extinction. Fig. 1 – 3 show the intensity and extinc-

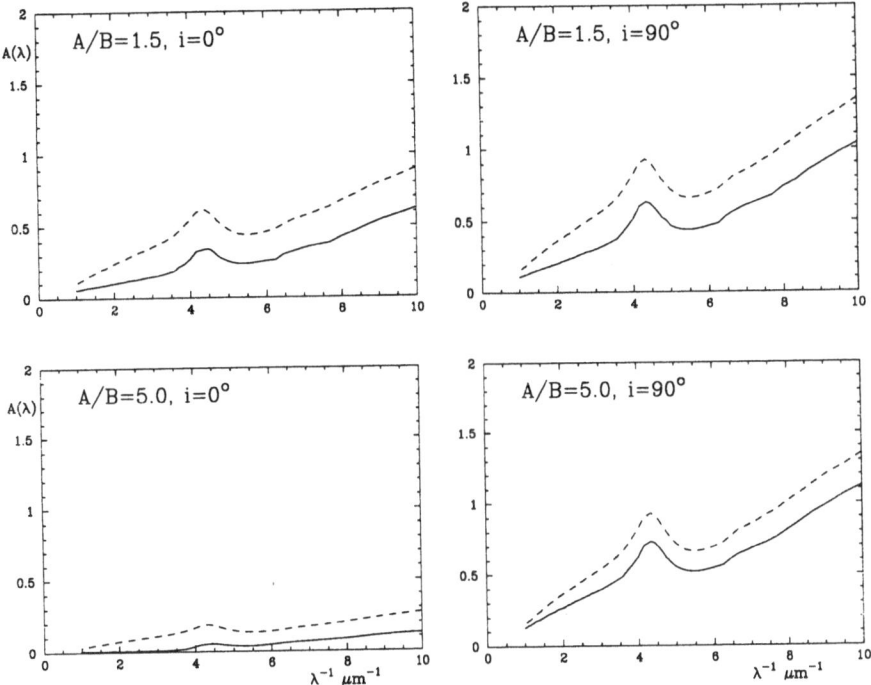

Fig. 2.   Wavelength dependencies of the non-normalized extinction for the same model as in Fig. 1. Solid lines: "circumstellar" extinction (with the scattered radiation), dashed lines: "interstellar" extinction (without the scattered radiation).

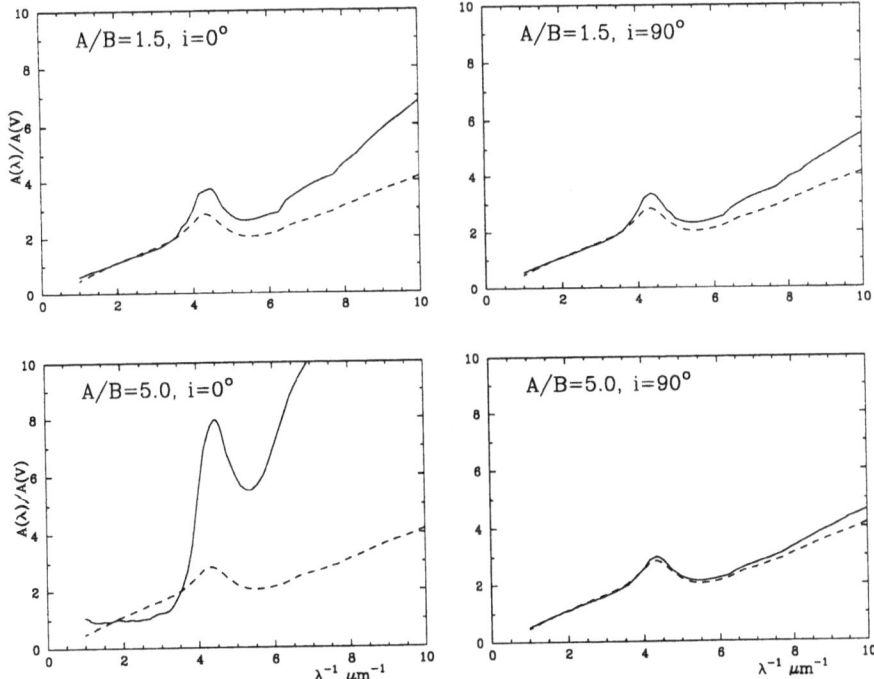

Fig. 3.   The same as in Fig. 2, but now for the normalized extinction.

tion curves for four different cases of the shell shape and viewing angle. It is seen (Fig. 2) that the scattered radiation always dilutes the non-normalized extinction. Its influence on the normalized circumstellar extinction curves $A^{CS}(\lambda)/A^{CS}(V)$ (see Fig. 3) is maximal for shells seen pole-on.

It should be noted that the observations of the extinction in the UV (especially at $\lambda < 1600$ Å) as well as of the intrinsic polarization are very important to constrain the parameters of dust grain models.

We used the results of our modelling to interpret the observations of the pre-main-sequence star HD 45677 (B2: IV [e]) possessing a circumstellar dust shell (Grady *et al.* 1994). Its extinction curve shows a small 2200 Å-feature and a flat extinction in the far-UV (Sitko *et al.* 1981). Besides this, polarization reversal was discovered near $\lambda \approx 2400$ Å (Schulte-Ladbeck *et al.* 1992). The main observational features can be explained if we choose a model with the following parameters — dust grains: $a_- = 0.05$ μm, $a_+ = 0.25$ μm, $q = 5.0$, $n_{Si}/n_C = 1.07$; dust shell: $A/B = 2$, $\tau_{90}^{ext}(0.15\,\mu m) = 2.1$, $i = 90°$, i.e. oriented edge-on. The main discrepancy found between theory and observations concerns the polarization at $\lambda < 2000$ Å. Apparently, it may be reduced by varying the grain size distribution.

4. Acknowledgements

We thank Carol Grady for useful comments and Mario van den Ancker for helping us with the derivation of the extinction curve of the star HD 45677. N.V. would like to thank the ESO C&EE Programme (grant A-01-093), the "Nederlandse Organisatie voor Wetenschappelijk Onderzoek" and the "Leids Kerkhoven Bosscha Fonds" for financial support.

References

Grady, C.A., Pérez, M.R. & Thé, P.S.: 1994, in *The Nature and Evolutionary Status of Herbig Ae/Be Stars*, eds: P.S. Thé *et al.* , ASP Conference Series V. **62**, p409.
Mathis, J.S., Rumpl, W. & Nordsieck, K.H.: 1977, Astrophys. J. **217**, 425.
Schulte-Ladbeck, R.E. *et al.* : 1992, Astrophys. J. **401**, L105.
Sitko, M.L., Savage, B.D. & Meade, M.R.: 1981, Astrophys. J. **246**, 161.
Steenman, H. & Thé, P.S.: 1991, Astrophys. & Space Sci. **184**, 9.
Thé, P.S., Felenbok, P., Cuypers, H. & Tjin A Djie, H.R.E.: 1985, Astron. & Astrophys. **149**, 429.
Voshchinnikov, N.V. & Karjukin, V.V.: 1994, Astron. & Astrophys. **288**, 883.
Voshchinnikov, N.V., Molster, F.J. & Thé, P.S.: 1994, Astron. & Astrophys. in preparation.

THE FIR EMISSION OF DUST PARTICLES
AROUND C-RICH IRAS SOURCES

A. BLANCO, A. BORGHESI, S. FONTI and V. OROFINO

Department of Physics, University of Lecce, C.P. 193, 73100 Lecce, Italy

Abstract. The spectra of 26 IRAS sources have been fitted using experimental optical data obtained in our laboratory for different types of carbon and silicon carbide (SiC) submicronic particles. Laboratory data are the main input for a radiative transfer model, which, taking into account several important parameters of the circumstellar matter and of the central star, evaluates the temperature gradient across the envelope and hence the emission best-fit spectrum. In this way it has been possible to fit, with good accuracy, most of the examined IRAS spectra in the range 8 – 100 μm. The sources have been chosen among those showing SiC feature around 11 μm and the best fits have been obtained by carefully assessing the amount and the type (α & β) of SiC particles present in the grain mixtures.

1. Introduction

In a previous paper (Blanco *et al.* 1994 – hereinafter Paper I), we have fitted the spectra of eight IRAS sources chosen among the brightest carbon stars and exhibiting a prominent circumstellar emission feature at 11.3 μm, generally attributed to silicon carbide (SiC) dust grains. We obtained good fits, between 8 and 20 μm, using the optical properties of a mixture of amorphous carbon (AC) and SiC submicronic grains, produced and characterized in our laboratory. Such grains are widely accepted as the main constituents of the solid component in the envelopes of carbon stars (Rowan-Robinson and Harris 1983; Baron *et al.* 1987). The extinction laboratory data were used as inputs for a simple theoretical model which simulated the dust envelope by means of a homogeneous and isothermal shell with a spherical geometry. The main result was the clear possibility to discriminate between the two different crystalline forms of SiC and hence between the sources containing only α-SiC and those containing both α-SiC and β-SiC. We also attempted an evolutionary interpretation of such diversity. The main problem was that the isothermal model prevented us from fitting the whole IRAS spectrum up to 100 μm. Moreover the number of sources was too small to allow any meaningful statistical analysis. We present here an analysis extended to a set of 26 IRAS sources, made by means of an improved model, in which the use of the Leung-Spagna radiative transfer code (Egan *et al.* 1988) allowed us to calculate the temperature gradient across the shell for each component of the mixture separately.

Astrophysics and Space Science **224**: 227–231, 1995.

© *1995 Kluwer Academic Publishers.*

2. Model & Results

Our approach in modeling the spectra of C-rich IRAS sources is based on the use of the experimental extinction cross-sections of cosmic dust analogues directly measured in our laboratory. Such data are fed into an appropriate radiation transfer model to simulate the observed spectra. In the present work we use a three component mixture of AC, α-SiC and β-SiC submicronic grain, which has been experimentally tested (Orofino *et al.* 1991) and already successfully used in Paper I. The model used in that occasion needed as further input parameters the circumstellar absorption at 5 μm, the star temperature, the ratio between the external radius of the shell and the radius of the star, the ratio between the outer and inner radii of the shell and the dust shell temperature. Since the last parameter can be regarded only as an equivalent temperature the model can give acceptable results only for a narrow wavelength range. For this reason, in Paper I, we had to limit our analysis to the region of the SiC feature (8 – 20 μm), ignoring the IRAS photometric data at 25, 60 and 100 μm.

In this work we have used an improved radiation transfer model, based on the Leung-Spagna radiative transfer code, which allowed us to determine the temperature in each point of the shell for each component of the grain mixture. The model allows also to change the exponent of the density law, but we assumed here $\rho \propto R^{-2}$, which is widely accepted for carbonaceous grains (Kömpe *et al.* 1994). Therefore the input parameters for the model used in this work are: the star temperature T_* , the optical depth τ at 5500 Å, the ratios R_I/R_* and R_E/R_I , where R_* is the stellar radius while R_I and R_E are the shell internal and external radius respectively. We used of course as further free parameters the relative amount of each component of the mixture, evaluated according to the expected elemental abundances in the envelopes of carbon stars (Orofino *et al.* 1991). It is worthwhile to note that the values of all the input parameters have been carefully chosen, taking into account the possible relations that might exist among them and all the relevant information available in literature.

At present we have extended the original sample of 8 sources, studied in Paper I, up to 26 objects. Such number is certainly not high enough to perform a thorough analysis of the type of those already made by Baron *et al.* (1987), Willems (1988) and Chan & Kwok (1990). We feel, however, that it is important to assess a definite procedure, which, taking into account all the relevant information, could considerably reduce the uncertainties on each of the input parameters. At this stage therefore the number of sources can increase only very slowly, since each object has to be carefully examined individually.

The list of the 26 IRAS objects is given in Table 1, together with the values of the best-fit parameters used for each source. We want to stress that, whenever possible, the values of the best-fit parameters have been chosen according to the available information. As an example the star temperatures have been selected close to the values given by Bergeat *et al.* (1976), while the optical depths have been checked against the existence of an optical counterpart. Moreover the values of the stellar radii have been evaluated according to the empirical relation given by Bergeat *et al.* (1978).

TABLE I

IRAS sources and values of the best-fit parameters.

IRAS name	T_*(K)	τ	$R_*/R_\odot$	R_I/R_*	R_E/R_I	AC(%)	α-SiC(%)	β-SiC(%)
00247+6922	1800	3	411	3	10^3	92	8	
02152+2822	1200	10	2600	3	10^3	85	15	
02270-2619	2300	0.5	232	3	10^3	85	15	
03374+6229	2300	0.5	232	6	10^4	75	15	10
04307+6210	2000	1.5	254	3	10^3	80	12	8
04573-1452	1800	1	411	3	10^3	75	15	10
05028+0106	2500	0.5	310	4	10^4	85	15	
05377+1346	1600	1	702	3	10^2	75	15	10
05426+2040	2500	0.5	310	4	10^4	80	20	
06331+3829	2500	0.5	310	4	10^4	90	10	
06342+0328	1300	0.8	1805	3	10^3	85	15	
07065-7256	2000	0.6	254	3	10^3	75	15	10
07098-2012	1600	1.7	702	3	10^3	90	10	
08050-2838	2400	5	269	3	10^3	92	8	
08073-3608	1600	1	702	3	10^3	80	20	
08340-3357	2000	8	254	3	10^3	92	8	
08416-2525	1800	0.8	411	3	10^3	75	15	10
09521-7508	1600	1.5	702	3	10^3	90	10	
09533-4120	1600	0.5	702	3	10^3	75	25	
10154-4950	2000	0.8	254	5	10^4	75	15	10
10249-2517	2000	0.8	254	5	10^4	75	15	10
11308-1020	1600	0.5	702	3	10^3	75	15	10
12427+4542	2400	0.4	269	4	10^4	90	10	
12447+0425	2400	0.5	269	3	10^3	80	15	5
15082-4808	1600	8	702	3	10^2	90	10	
17556+5813	1800	0.8	411	3	10^3	75	15	10

A typical fit is shown in Figure 1 for IRAS 08416-2525. The result is quite good, but it is interesting to note that there is a clear discrepancy in the long wavelength wing of the SiC feature; the different shape of the emission band could be interpreted in terms of different grain size distribution, giving

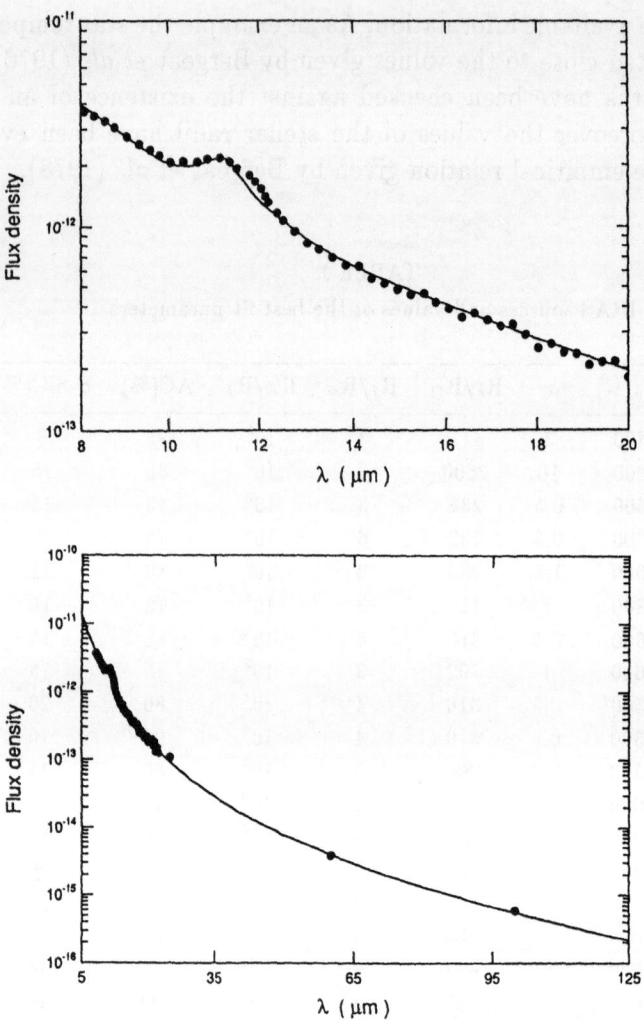

Fig. 1.   Best fit curve (solid line) of the observed spectrum of IRAS 08416-2525 (dots); in the lower box is shown the whole IRAS spectral range, while in the upper box the region around the 11.3 μm feature is presented in more detail.

interesting clues for further laboratory and theoretical work.

3.   Conclusions

The agreement between the observations and the spectra calculated from laboratory data is, in most cases, rather good and allows to state clearly which sources are fitted using a mixture of AC and α-SiC only and which of them require also the presence of β-SiC.

Unfortunately no definite conclusion can be drawn about a correlation between such diversity and any other physical property of the selected sample. At present therefore, it is impossible to discriminate between the scenario we proposed in Paper I and that suggested by Chan & Kwok (1990), concerning the correlation between the optical depth and the presence of α- and/or β-SiC. However, it is worthwhile to check the possible existence of such a correlation, since it could be of help in understanding the evolution of the sources and the history of the circumstellar grains.

To achieve this result it would be necessary to increase the number of objects examined. Each of them however has to be carefully checked in order to collect all the available relevant information useful to assess as precisely as possible the acceptable values of the input parameters.

New observations of improved quality (UKIRT, ISO) can also be of great help, since at present resolution and S/N ratio of the observed spectra are worse than those achieved for the laboratory data. On the other hand experimental work is needed to account for other possible components of the dust mixture and to better reproduce the real grain size distribution.

References

Baron Y., de Muizon M., Popoular R. & Pgouri B.: 1987, Astron. & Astrophys. **186**, 271.

Bergeat J., Lunel M., Sibille F. & Lefevre J.: 1976, Astron. & Astrophys. **52**, 263.

Bergeat J., Sibille F. & Lunel M.: 1978, Astron. & Astrophys. **64**, 423.

Blanco A., Borghesi A., Fonti S. & Orofino V.: 1994, Astron. & Astrophys. **283**, 561 **(Paper I)**.

Chan S. J. & Kwok S.: 1990, Astron. & Astrophys. **237**, 354.

Egan M. P., Leung C. M. & Spagna G. F. Jr.: 1988, Comput. Phys. Commun. **48**, 271.

Kömpe C., Gürtler J. & Henning Th.: 1994, this conference.

Orofino V., Mennella V., Blanco A., Bussoletti E., Colangeli L. & Fonti S.: 1991, Astron. & Astrophys. **252**, 315.

Rowan-Robinson M. & Harris S.: 1983, Mon. Not. of the RAS **202**, 797.

Willems F. J.: 1988, Astron. & Astrophys. **203**, 51.

Willems F. J.: 1988, Astron. & Astrophys. **203**, 65.

PLATE V: One of the treasures of the Crawford Library. this illustration is a stylised representation of the construction and instrumentation of Tycho Brahe's observatory, *Uraniborg*, on the island of Hven, Denmark. In the foreground is his great Mural Quadrant with which he practised positional astronomy to remarkable standards of accuracy. Tycho is the figure to the right of centre.

THE INTERPRETATION OF OBSERVATIONS OF ICE AND SILICATE IR FEATURES AND POLARIZATION ACROSS IT IN THE SPECTRA OF PROTOSTELLAR OBJECTS BN AND AFGL 2591

ALEX E. IL'IN

Pulkovo Observatory, St.Petersburg, 196140, RUSSIA

Abstract. The absorption features of ice at 3 μm and of silicate at 10 and 20 μm as well as the linear polarization across it have been calculated. The interpretations of data for protostellar objects BN and AFGL 2591 are made.

The model of partially aligned spinning spheroidal grains with the Purcell's suprathermal alignment mechanism and the power law size distribution are considered. Core-mantle, porous and composite particles are investigated in the Rayleigh approximation. In the case of composite and porous grains the effective refractive indices are computed with the approximate Bruggeman rule.

The influence of grain chemical composition, elongation and structure has been investigated. The distinctions in polarization between composite and core-mantle models are found. The mixing of grain materials smooths the individual spectral features of chemical components. The adding of graphite causes the shift of the 10 μm peak position to shorter wavelengths. When the fraction of graphite increases, the peak at 20 μm decreases and shifts to shorter wavelengths. The increase of elongation reduces the negative polarization at the 3 μm feature, shifts the 10 μm and 20 μm peak positions to longer wavelengths, and raises the strength of the 20 μm band. The porosity leads to similar effects.

It was found that the studied grain models are in good agreement with observational data for BN and AFGL 2591 objects. The absence of polarization excess near 3 μm for AFGL 2591 is attributable to a decrease in elongation of large grains as a result of coagulation. The attention is attracted to the problem of real distribution of the interstellar dust grains over the elongation parameter. The determination this distribution function and the study of its evolution in the processes of accretion and coagulation are necessary.

SESSION E:

CHEMICAL EVOLUTION
OF
CIRCUMSTELLAR MATTER
SURROUNDING
YOUNG STELLAR OBJECTS

CHEMICAL EVOLUTION OF CIRCUMSTELLAR MATTER
AROUND YOUNG STELLAR OBJECTS

EWINE F. VAN DISHOECK
Leiden Observatory, P.O. Box 9513
2300 RA Leiden, The Netherlands

and

GEOFFREY A. BLAKE
California Institute of Technology 170-25
Pasadena, CA 91125, USA

Abstract. Recent observational studies of the chemical composition of circumstellar matter around both high– and low–mass young stellar objects are reviewed. The molecular abundances are found to be a strong function of evolutionary state, but not of system mass or luminosity. The data are discussed with reference to recent theoretical models.

Key words: Molecular abundances – young stellar objects – (sub)millimeter observations

1. Introduction

The birth of stars occurs deep inside molecular clouds and is accompanied by enormous changes in the physical parameters of the gas and dust (see Shu *et al.* 1993 for an overview). In the earliest stages of collapse, the density increases from typically 10^4 cm^{-3} in the interstellar cloud core to $> 10^8$ cm^{-3} within 100 AU of the young stellar object. At a somewhat later stage, radiation emerges from the newly–formed star and bipolar outflow(s) create shocks which can result locally in very high temperatures. Because the chemical composition of the gas and dust must be affected by these changes in temperature, density, and radiation field, molecular abundances can potentially serve as very sensitive probes of the evolutionary stage of the objects. Moreover, the molecular observations provide excellent diagnostics of the physical parameters and the kinematics of processes such as infall, outflow or rotation in a circumstellar disk or envelope. Conversely, the chemistry itself can influence the physical evolution, since molecules are the main coolants of the gas. Also, the fractional ionization controls the interactions of both ions and neutrals with any magnetic fields present, and thus the rates of collapse. In this chapter, a summary will be given of the latest theoretical and observational studies of the chemistry around both high–mass and low–mass young stellar objects (YSOs). More detailed overviews have been presented in van Dishoeck *et al.* (1993) and Hartquist *et al.* (1993).

Astrophysics and Space Science **224**: 237–249, 1995.

2.   Observations

Although molecules in high–mass star–forming regions such as Orion and SgrB2 have been observed for more than two decades, a systematic study of the chemical modifications induced by both high– and lower–mass star formation has been undertaken only in the last few years. The recent work has been stimulated by the availability of new large aperture (sub–)millimeter telescopes such as the JCMT 15m and CSO 10.4m equipped with sensitive SIS receivers. Because of the higher frequencies, the beam sizes are typically only 15″–20″, corresponding to ~3000 AU in the nearest molecular clouds such as Taurus and Ophiuchus. This size is comparable to that of the primitive solar nebula. Moreover, the submillimeter observations probe primarily the higher density $n(H_2) > 10^5$ cm^{-3}, warmer $(T > 20$ K$)$ gas intimately associated with YSOs. In contrast, earlier millimeter observations had typical beam sizes of 60″ or more, and were mostly sensitive to the colder $(T \approx 10$ K$)$, lower density $(n \approx 10^4$ cm$^{-3})$ cloud core. Care should therefore be taken in comparing abundances taken with different telescopes at dissimilar frequencies, since they may not probe the same environment.

The earliest stages of star formation are accompanied by hundreds to thousands of magnitudes of extinction, so that observations are only possible at (sub–)millimeter and radio wavelengths. However, at later stages the near–infrared radiation becomes stronger (Lada 1991) and detection of the infrared source may be aided by other observable features such as the orientation of the outflow. In this phase, high resolution infrared absorption (and occasionally emission) line observations toward YSOs can be obtained with new cryogenic echelle spectrometers such as those available on UKIRT and the IRTF. Spectra of, *e.g.* the CO molecule and its isotope provide complementary information on the physical parameters along a pencil beam line of sight (Mitchell *et al.* 1990). The infrared technique also allows detection of molecules such as CH$_4$ and C$_2$H$_2$ which cannot be observed by their rotational lines at millimeter wavelengths (Lacy *et al.* 1989, 1991). Most importantly, it permits observations of solid state features due to molecules in icy grain mantles (Whittet 1993), thereby complementing the (sub–)millimeter data on the gas phase composition.

3.   Chemical Scenario and Model Results

3.1 SCENARIO

Several chemically distinct stages can be distinguished during the star formation process. Before the onset of collapse, the chemistry in the quiescent parental cloud core is thought to be dominated by gas–phase ion–molecule reactions, which proceed rapidly at the low temperatures of 10–20 K. This chemistry has been studied in detail by, *e.g.* Herbst & Leung (1989) and

Millar *et al.* (1991b). During the subsequent collapse phase, the densities increase from $\sim 10^4$ cm^{-3} to $\sim 10^8$ cm^{-3} or more, whereas the temperature remains low because molecules are such effective coolants. As a result, the time scale for molecular collisions with grain surfaces becomes shorter than the free–fall time scale, so that most molecules will condense onto the grains (Walmsley 1992). Once on the grain, surface reactions and photochemical reactions in icy mantles can modify the composition and increase the chemical complexity (Tielens & Allamandola 1987; Greenberg *et al.* 1993).

After the star has formed, its radiation can heat the gas, which induces higher temperatures and drives photochemical reactions. More importantly, the temperature of the grains also increases, resulting in the evaporation of adsorbed molecules, probably in a sequence according to their sublimation temperatures. For example, non–polar molecules such as N_2, CO and CH_4 have sublimation temperatures of 20–30 K, whereas polar species such as HCN, CH_3OH and H_2O do not evaporate until 95–150 K (Mumma *et al.* 1993). Although non–polar molecules such as CO embedded in an H_2O–ice matrix will be somewhat stronger bound, they can still sublime at grain temperatures around 50 K. Another, probably more efficient, way of removing the icy grain mantles is in shocks created by the interaction of bipolar outflow(s) with the surrounding envelope. If such shocks are sufficiently powerful, the refractory grain cores can be (partly) destroyed as well, resulting in an increase of, *e.g.* the gas phase Si abundance. The shocks also create localized regions of very high temperature ($T \approx 1000 - 2000$ K), in which endothermic reactions leading to OH, H_2O and H_2S, for example, can occur.

The first attempt to model these different chemical stages was made by Brown, Charnley & Millar (1988). They started from an entirely atomic cloud (except for H_2) at $n_H \approx 3000$ cm^{-3}, and followed the chemistry in the subsequent free–fall collapse. Atomic species such as C, O and N are gradually transformed into CO, O_2, H_2O and N_2 through gas phase reactions, but since the collapse phase lasts $\lesssim 10^6$ yr, the conversion to molecules does not reach completeness. At $\sim 10^6$ yr, the depletion time scale becomes so short that virtually all species (except those consisting of H, He and their isotopes) freeze out abruptly onto the grains. The subsequent grain surface reactions are the most uncertain part of the model. Brown *et al.* use the simplest assumption that the primary result of these reactions is hydrogenation of the remaining atomic C, O and N to CH_4, H_2O and NH_3. The chemical calculation is then started again once some heating event associated with the newly–formed YSO releases all grain mantle species instantaneously back into the gas phase. The ensuing gas phase chemistry recycles these molecules to their normal quiescent "equilibrium" values on a time scale of $\sim 10^5$ yr.

3.2 RECENT MODELS

In recent years, the various chemical phases have been studied individually in more detail. In particular, Rawlings et al. (1992) have performed a comprehensive study of the chemistry during the collapse phase, using a more realistic dynamical model in which an isothermal sphere with a r^{-2} density gradient is formed in $\sim 10^6$ yr, followed by a self–similar collapse. Molecular abundances as functions of time and position in the collapsing envelope have been presented, and the resulting line profiles computed. One difference compared with the earlier work is that not all species deplete simultaneously onto the grains; the abundances of selected molecules such as HCO^+, OH and SO can temporarily *increase*. Thus, the profiles of such species are expected to be wider than those of molecules like NH_3, which accrete rapidly. This transient increase is linked to species such as H_3^+, whose abundance greatly increases when the gas–phase abundances of its dominant removal partners —CO and O— drop.

Grain surface chemistry has received recent attention through the work of Hasagawa & Herbst (1993a, b) and Willacy & Williams (1993). The former authors point out that tunneling reactions with H_2 may be more important than previously thought, which tend to quench complex molecule formation on dust particles. Breukers (1991) and Shalabiea & Greenberg (1994) have modeled in more detail the chemistry in icy mantles by extending the earlier work of d'Hendecourt et al. (1985). In their work, photochemical reactions between solid H_2O and CO rather than grain surface reactions lead to the production of H_2CO and CH_3OH. The latter species is observed in substantial amounts in the solid state (Allamandola et al. 1992), whereas solid H_2CO has recently been tentatively identified (Schutte et al. 1994). Laboratory experiments of surface binding energies of various species have been performed by Sandford & Allamandola (1993).

Blake et al. (1987) suggested that injection of molecules, in particular H_2O, from grain surfaces can significantly affect the gas phase chemistry and result in the complex organic species observed in the Orion hot core. After solid CH_3OH proved to be an abundant ice mantle constituent from observations, it rapidly became apparent that injection of CH_3OH is much more effective in creating observed species like CH_3OCH_3 and $HCOOCH_3$ than injection of H_2O (Millar et al. 1991a; Charnley et al. 1992; Caselli et al. 1993). Also, injection of e.g. NH_3 is found to lead to enhanced production of HCN, HC_3N and CH_3CN through subsequent gas phase reactions. As Figure 1 shows, the time scale for production of these species is $\sim 10^4 - 10^5$ yr, after which the abundances drop rapidly back to their quiescent values. Thus, the abundances of these complex organic species can potentially serve as sensitive "clocks" of the evolutionary state of the YSO.

In summary, the collapse phase is expected to be characterized by large

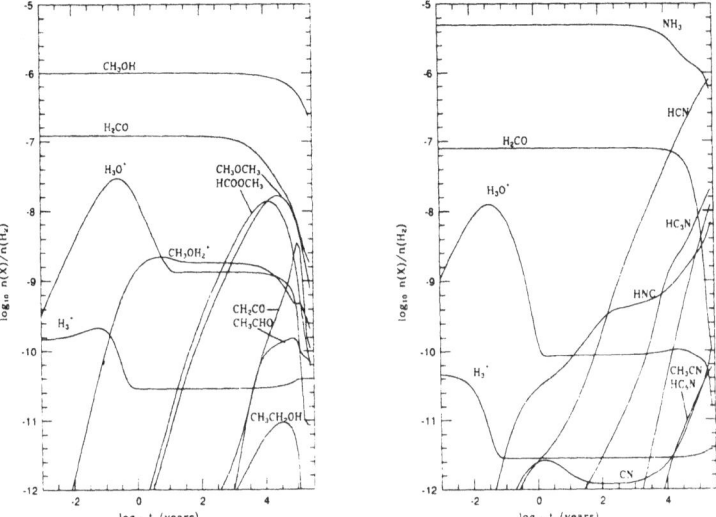

Fig. 1.   Models of the gas phase chemical evolution in the Orion Compact Ridge (left) and Hot Core (right) after injection of molecules from the grain mantles at $t=0$ (Charnley *et al.* 1992).

depletions of molecules from the gas phase. Grain surface reactions modify the chemical composition, and lead to more hydrogenated and/or more complex species depending on the local conditions. In the later "hot core" phase, molecules in the icy mantles are liberated back into the gas phase, and drive a rapidly evolving chemistry through ion–molecule and neutral–neutral reactions, resulting in the creation of more complex organic species. Shocks associated with outflowing matter can enhance Si– and S–bearing molecules such as SiO and SO_2; the return of only a small fraction ($\sim 1\%$) of the cosmic abundance of these elements leads to easily observable effects.

3.3 OPEN QUESTIONS

The main uncertainties in the current models lie in the details of the grain surface chemistry. Which molecules are formed on the grains, and which are "second generation" species produced in the gas phase by "hot core-like" chemistry? Is the injection gradual or instantaneous? How does the grain surface chemistry depend on the composition of the gas? Are there chemically different layers? Current models suggest that if sufficient atomic H, O, C, N and CO are present, H_2, CH_4, NH_3, H_2CO and CH_3OH are readily produced. On the other hand, little atomic H likely leads to solid O_2, CO_2 and complex cyanogens (Tielens 1989; Caselli *et al.* 1993). Future observations of the composition of ices, especially with the ISO satellite, should provide much better constraints on such models.

A second question is whether the "hot core–like" models can also explain the large abundances of Si– and S–bearing species such as SiO, SO and SO_2 that are commonly observed in YSO surroundings. Grain surface reactions are likely to produce H_2S, which, when released back in the gas phase, can lead to S, SO and SO_2 through reactions with H, OH and O_2 (Charnley 1993, in Millar 1993). Similarly, MacKay (this volume) has suggested that the release of SiH_4 from the grain surface can lead to SiO.

The origin of the large abundances of deuterated species is also still puzzling. In one scenario, these species are formed in the gas phase through fractionation reactions in the cold parental cloud core, and are subsequently frozen out onto grains (Walmsley *et al.* 1987). Alternatively, their abundances can be enhanced through surface reactions in grain mantles with atomic D, which is more abundant in the gas phase than atomic H in a high density cloud (Tielens 1983).

4.   High–Mass Star–Forming Regions

4.1   Orion/KL vs. Orion S

The best studied region of high–mass star formation is the Orion/KL core, centered on the luminous infrared source IRc2 ($\sim 10^5 L_\odot$). Many molecular line observations have been performed toward this source, the most detailed of which are the OVRO survey in the 230 GHz window (Blake *et al.* 1987), the CSO survey in the 345 GHz window (Groesbeck *et al.* 1994) and the partial JCMT survey at 345 GHz (Sutton *et al.* 1994; see also references cited). The main advantage of line surveys over selected frequency settings is that several lines of each molecule and its isotopes are detected, which allows independent determination of the physical conditions, molecular abundances and optical depth effects. The 230 GHz survey showed that the above mentioned chemical effects are clearly present and that at least four physically and chemically different components can be distinguished within the 30″ beam (0.07 pc). The first is the Orion ridge, which has narrow lines and contains the quiescent parental cloud material. Its chemical composition is dominated by simple molecules such as HCO^+, CS and unsaturated radicals like CN and C_2H. The second component is the hot core, which appears to be a warm clump of gas heated by IRc2 with larger line widths. It contains abundant cyanogen species such as HCN and CH_3CN, as well as the deuterated species HDO, NH_2D and even D_2CO. The third component, the compact ridge, also has "hot core–like" characteristics and shows the complex oxygen–containing organic species mentioned above more prominently, especially CH_3OH, CH_3OCH_3, and $HCOOCH_3$. This clump of warm gas is located only 10″ away from the hot core, and may represent the interface of the outflow with the ridge. The fourth component, the outflow itself, can be clearly distinguished through very broad line profiles of species such as SiO,

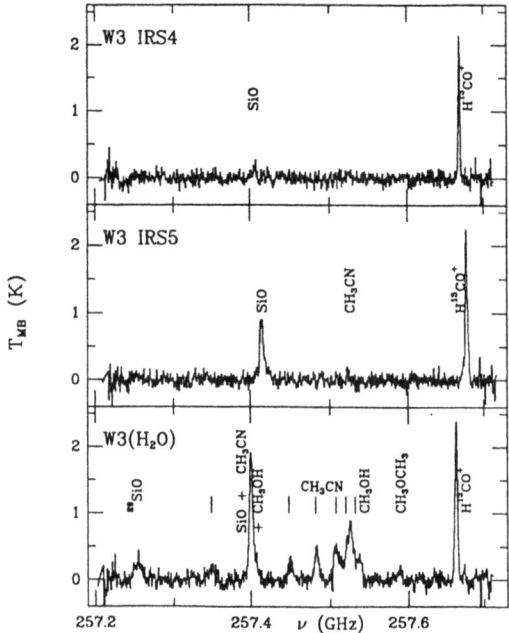

Fig. 2. Spectra around 257 GHz obtained with the JCMT toward W3 IRS4, W3 IRS5 and W3 (H₂O) (Helmich *et al.* 1994).

SO, SO_2 and H_2S. The distribution of molecules over the various components has recently been imaged directly through interferometer observations which can separate the hot core and compact ridge components (*e.g.* Plambeck & Wright 1988; Minh *et al.* 1993).

To what extent are these characteristics found in other regions? The Orion ridge contains another, equally massive clump only 1.5′ S of IRc2. This object also shows obvious signposts of star formation activity through a very well collimated outflow (*e.g.* Schmid–Burgk *et al.* 1990). A CSO 345 GHz line survey of this object has been performed by Groesbeck *et al.* (1994). Compared with Orion/KL, Orion–S shows dramatically fewer lines, especially of S–bearing species such as SO_2 and complex organics such as CH_3OH and CH_3OCH_3. Interferometer observations by McMullin, Mundy & Blake (1993) confirm this picture. Simple species such as $H^{13}CO^+$ in the ridge material are still prominent, but SO_2 is virtually absent and CH_3OH very weak. Some SiO is seen, which appears to be associated with the outflow. The authors conclude that Orion–S lies at an earlier evolutionary stage than Orion–KL, in which an outflow has already developed, but in which few complex molecules are yet found.

4.2   W3

Another good example of evolutionary effects on the chemical composition of the gas is formed by the JCMT 230/345 GHz survey of three YSO's in the W3 giant molecular cloud: W3 IRS4, W3 IRS5 and W3 (H_2O) (Helmich *et al.* 1994). Spectra of these three objects are presented in Figure 2: it is clear that simple species such as $H^{13}CO^+$ are strong in W3 IRS4, whereas SiO and SO_2 are prominent in W3 IRS5. Complex organics such as CH_3OH and CH_3CN are abundant in W3 (H_2O). These data have led to a chemical sequence, in which the earliest phase is formed by W3 IRS5. Here the interaction of the strong outflow with the surroundings has produced SiO and SO_2, but most other species are still sitting on the grains. Since IRS5 is a very strong infrared source, the composition of the icy mantles along the line of sight can actually be determined in this case. Indeed, it is found that the abundance of solid CH_3OH seen by Allamandola *et al.* (1992) is a factor of 10^4 larger than that of gas–phase CH_3OH. Thus, W3 IRS5 appears to be just prior to the "hot core" phase; only a few $\times 10^4$ yr may be sufficient to remove most species from the grains, since the gas phase temperature has already reached $\sim$100 K. W3 (H_2O), located close to W3 (OH) and associated with prominent H_2O maser emission, is somewhat further developed and shows already all the characteristics of a "hot core–like" region. The abundance of gas–phase CH_3OH is at least two orders of magnitude larger than that in W3 IRS5, and has led to complex organics such as CH_3OCH_3 and $HCOOCH_3$. Finally, W3 IRS4 appears to be the oldest of the three objects, in which the H II region has already broken free of the molecular cloud and in which the chemistry has returned to its more quiescent gas phase values.

Other examples of "hot core–type" objects which show characteristic emission of complex molecules are SgrB2 (*e.g.* Turner 1991; Sutton *et al.* 1991), G34.3 (MacDonald, this volume), and W51. An interesting object for which complementary infrared absorption lines have recently been performed is GL 2591 (Carr *et al.* 1994; Evans, this volume).

5.   Low Mass Star–Forming Regions

In order to investigate the sensitivity of the chemical modifications to the YSO luminosity or mass, a number of regions which are suspected to be in the process of forming $\sim$solar mass stars have recently been surveyed.

5.1   NGC 1333 IRAS4

NGC 1333 IRAS 4 is an extreme class I YSO embedded in the L1450 molecular cloud ($d \approx 350$ pc). Its submillimeter continuum emission is very strong and reveals a wide binary system with masses of $\sim$4 and 3 $M_\odot$, respectively, and a projected separation of $\sim$30'' (Sandell *et al.* 1991). The very well collimated molecular outflow suggests a dynamical age of only a few thousand yr. Blake *et al.* (1994b) have searched for several molecules toward IRAS 4, but find only weak emission. Most profiles can be decomposed into two components: a narrow "core" feature and a broader "wing" component. If the submillimeter dust continuum is used to estimate the H_2 column density of the core component, *all* observed molecules, including CO, are depleted by factors of at least 20–50 compared with normal interstellar clouds.

In contrast, the abundances of species such as SiO, HCN and CH_3OH are enhanced by several orders of magnitude in the high velocity "wing" component, and are comparable to those in Orion/KL, except for SO_2. Such large abundances are virtually impossible to reproduce with pure gas–phase models, and are therefore thought to result from the desorption of grain mantles. Because their excitation temperatures are low, $T_{ex} \approx 20 - 40$ K, grain–grain collisions induced by velocity shear zones surrounding the outflow may be responsible. Only SiO appears to be kinematically well removed from the bulk of the core velocity, and likely arises from directly shocked material.

5.2   IRAS 16293 –2422

IRAS 16293 –2422 is also a young binary system with a modest luminosity of $\sim$27 $L_\odot$ located in Ophiuchus (d=160 pc). Two millimeter dust continuum sources are found, each of total mass $\sim$0.5 $M_\odot$ and separated by only 5″ (840 AU). Sensitive spectral line surveys by Blake *et al.* (1994a) and van Dishoeck *et al.* (1995) using the JCMT and CSO show that IRAS 16293 is a much richer source of molecular emission than NGC 1333 IRAS 4. The analysis of the excitation of the various species reveals that at least 3 different components are present in the 20″ (3000 AU) beam. In contrast with NGC 1333 IRAS4, however, these components cannot be clearly separated kinematically due to the orientation of the system.

The first component has an effective size of only 3″ (500 AU) with temperatures in excess of 100 K, and densities of order 10^7 cm^{-3}. It has large abundances of SiO, SO_2, CH_3CN and CH_3OH, and likely represents the interaction of the outflow(s) with the surrounding envelope. Indeed, interferometer observations of SO reveal a strong concentration around the southern of the two YSO's, from which the outflow(s) emanate (Mundy *et al.* 1992). This component therefore appears similar to the "wing" component seen in NGC 1333, except that the excitation temperature of the organic molecules is higher and SO_2 much stronger. The abundances in IRAS 16293 approach those of the "plateau" and "hot core" features seen in Orion/KL.

The second component has a size of $\sim$ 10″ (1500 AU) and represents the accreting envelope, which provides the bulk of the (sub–)millimeter continuum emission. The kinetic temperature of $\sim$40 K is close to the dust temperature. It is prominent in species such as HCO^+, CS and H_2CO, but molecules like NH_3 may be depleted by up to an order of magnitude (Mundy *et al.* 1992). The third component is at least 20″ in extent (>3000 AU) and appears to be colder ($T \approx 20$ K) and less dense than the other regions. It shows high abundances of radicals and unsaturated species such as CN, C_2H and C_3H_2, and probably represents the outer envelope or cloud core. It can be considered the equivalent of the "ridge" component in Orion/KL.

5.3 OTHER REGIONS

The results for the NGC 1333 IRAS 4 "core" component provide some of the strongest evidence to date that molecules are indeed severely frozen out onto grains in the densest parts of the accreting envelope. Additional cases for which significant depletions have been suggested are NGC 2024 FIR5 (Mauersberger *et al.* 1992), IRAS 05338–0624 (McMullin, Mundy & Blake 1994b), and Serpens FIRS1 (McMullin *et al.* 1994a), which are all at a very early stage of evolution (see also Mundy, this volume). In addition, low abundances of species other than CO have been found in circumstellar envelopes at later stages, in particular for CS in HL Tau (Blake, van Dishoeck & Sargent 1992) and T Tau (van Langevelde, van Dishoeck & Blake 1995).

The kinematic association of SiO with the outflow component is beautifully demonstrated through interferometer observations of L1448 (Guilloteau *et al.* 1992). Additional single–dish evidence is found for B1 and Cep A (Martín–Pintado *et al.* 1992), L1157 (Mikami *et al.* 1992), NGC 2071 (Masson & Chernin 1993) and NGC 1333 IRAS 2 (Sandell *et al.* 1994).

6. Concluding Remarks

The above examples demonstrate that the chemical composition of the circumstellar matter around both high–mass and low–mass young stellar objects depends sensitively on the evolutionary state of the region. Indeed, comparison of submillimeter surveys of various objects demonstrates that the "hot core" phase characteristic of Orion/KL is by far the spectroscopically most prolific phase. The data also show that at least qualitatively, there appear to be no significant differences of low–mass objects compared with high–mass cases, over luminosities ranging from 10 to 10^5 $L_\odot$: components similar to the Orion ridge, hot core/compact ridge and plateau are also found in low mass objects. Quantitative comparison is complicated by the fact that several physically different regimes are usually present within the single dish beams, which complicates a reliable determination of the abundances in each of the individual components. Higher spatial resolution data using interferometers will be essential to disentangle the various regions.

In general, the chemical scenario sketched in §3 seems to be supported by observations, but more detailed chemical modeling and laboratory work is sorely needed. As mentioned above, one of the principal uncertainties is the extent of the chemical processing on the grains. Also, the mechanisms for returning molecules to the gas phase warrant further research. Is gentle heating due to radiation from the YSO dominant, or is it mostly due to shocks created by the interaction of the outflow with the surrounding envelope? The dominant mechanism may well depend on the luminosity of the system, and on the species considered. For example, there is good kinematical evidence that SiO is formed in the region of "prompt" interaction, whereas other species may be returned mostly in the "steady–state" turbulent regions around the outflow (see Masson, this volume).

Future observational work should involve more systematic single dish and interferometer studies of the molecular composition of the circumstellar matter surrounding YSOs of various luminosities and age. Combination with complementary high–resolution ground–based and ISO infrared observations of both the gas and the dust, whenever feasible, will be essential in testing the complete picture of the chemical evolution of these objects.

This work was supported by a PIONIER grant from the Netherlands Organization of Scientific Research (NWO), and by NASA grant NAGW–2297 as well as the David and Lucille Packard Foundation.

References

Allamandola, L.J., Sandford, S.A., Tielens, A.G.G.M. & Herbst, T.M.: 1992, Astrophys. J. **399**, 134.

Blake, G.A., Sutton, E.C., Masson, C.R. & Phillips, T.G.: 1987, Astrophys. J. **315**, 621.

Blake, G.A., van Dishoeck, E.F. & Sargent, A.: 1992, Astrophys. J. **391**, L99.

Blake, G.A., van Dishoeck, E.F., Jansen, D.J., Groesbeck, T. & Mundy, L.G.: 1994a, Astrophys. J. **428**, 680.

Blake, G.A., Sandell, G., van Dishoeck, E.F., Groesbeck, T.D., Mundy, L.G. & Aspin, C.: 1994b, Astrophys. J. in press.

Breukers, R.: 1991, Ph.D. Thesis, University of Leiden.

Brown, P.D., Charnley, S.B. & Millar, T.J.: 1988, Mon. Not. of the RAS **231**, 409.

Carr, J.S., Evans, N.J., Lacy, J.L. & Zhou, S.: 1994, Astrophys. J. in press.

Caselli, P., Hasegawa, T.I. & Herbst, E.: 1993, Astrophys. J. **408**, 548.

Charnley, S.B., Tielens, A.G.G.M. & Millar, T.J.: 1992, Astrophys. J. **399**, L71.

d'Hendecourt, L.B., Allamandola, L.J. & Greenberg, J.M.: 1985, Astron. & Astrophys. **152**, 130.

Greenberg, J.M., Mendoza–Gómez, C.X., de Groot, M.S. & Breukers, R.: 1993 in: *Dust and Chemistry in Astronomy*, eds: T.J. Millar & D.A. Williams, IOP: Bristol, p271.

Groesbeck, T.D. *et al.* : 1994, in preparation.

Guilloteau, S., Bachiller, R., Fuente, A. & Lucas, R.: 1992, Astron. & Astrophys. **265**, L49.

Hartquist, T.W., Rawlings, J.M.C., Williams, D.A. & Dalgarno, A.: 1993, Quart. J. of the RAS **34**, 213.

Hasegawa, T.I. & Herbst, E.: 1993a, Mon. Not. of the RAS **261**, 83.

Hasegawa, T.I. & Herbst, E.: 1993b, Mon. Not. of the RAS **263**, 589.

Helmich, F.P., Jansen, D.J., de Graauw, T., Groesbeck, T. & van Dishoeck, E.F.: 1994, Astron. & Astrophys. **283**, 626.

Herbst, E. & Leung, C.M.: 1989, Astrophys. J., Suppl. **69**, 271.

Lacy, J.H., Evans, N.J., Achtermann, J.M., Bruce, D.E., Arens, J.F. & Carr, J.S.: 1989, Astrophys. J. **342**, L43.

Lacy, J.H., Carr, J.S., Evans, N.J., Baas, F., Achtermann, J.M. & Arens, J.F.: 1991, Astrophys. J. **376**, 556.

Lada, C.J.: 1991 in: *The Physics of Star Formation and Early Stellar Evolution*, eds: C.J. Lada & N.D. Kylafis, Kluwer: Dordrecht, p329.

Martín-Pintado, J., Bachiller, R. & Fuente, A.: 1992, Astron. & Astrophys. **254**, 315.

Masson, C.R. & Chernin, L.M.: 1993, Astrophys. J. **414**, 230.

Mauersberger, R., Wilson, T.L., Metzger, P.G., Gaume, R. & Johnston, K.J.: 1992, Astron. & Astrophys. **256**, 640.

McMullin, J.P., Mundy, L.G. & Blake, G.A.: 1993, Astrophys. J. **405**, 599.

McMullin, J.P., Mundy, L.G., Wilking, B.A., Hezel, T. & Blake, G.A.: 1994a, Astrophys. J. **424**, 222.

McMullin, J.P., Mundy, L.G. & Blake, G.A.: 1994b, Astrophys. J. in press.

Mikami, H., Umemoto, T., Yamamoto, S. & Saito, S.: 1992, Astrophys. J. **392**, L87.

Millar, T.J.: 1993 in: *Dust and Chemistry in Astronomy*, eds: T.J. Millar & D.A. Williams, IOP: Bristol, p249.

Millar, T.J., Herbst, E. & Charnley, S.B.: 1991a, Astrophys. J. **369**, 147.

Millar, T.J., Rawlings, J.M.C., Bennett, A., Brown, P.D. & Charnley, S.B.: 1991b, Astron. & Astrophys. Suppl. **87**, 585.

Minh, Y.C., Ohishi, M., Roh, D.G., Ishiguro, M. & Irvine, W.M.: 1993, Astrophys. J. **411**, 773.

Mitchell, G.F., Maillard, J.-P., Allen, M., Beer, R. & Belcourt, K.: 1990, Astrophys. J. **363**, 554.

Mumma, M.J., Weissman, P.R. & Stern, S.A.: 1993 in: *Protostars & Planets III*, eds: E.H. Levy & J. Lunine, Univ. Arizona: Tucson, p1177.

Mundy, L.G., Wootten, H.A., Wilking, B.A., Blake, G.A. & Sargent, A.I.: 1992, Astrophys. J. **385**, 306.

Plambeck, R.L. & Wright, M.C.H.: 1988 in: *Molecular Clouds in the Milky Way and External Galaxies*, eds: R.L. Dickman, *et al.*, Springer: Berlin, p182.

Rawlings, J.M.C., Hartquist, T.W., Menten, K.M. & Williams, D.A.: 1992, Mon. Not. of the RAS **255**, 471.

Sandell, G., Aspin, C.A., Duncan, W.D., Russell, A.P.G. & Robson, E.I.: 1991, Astrophys. J. **376**, L17.

Sandell, G., Knee, L.B.G., Aspin, C.A., Robson, E.I. & Russell, A.P.G.: 1994, Astron. & Astrophys. **285**, L1.

Sandford, S.A. & Allamandola, L.J.: 1993, Astrophys. J. **417**, 815.

Schmid-Burgk, J., Güsten, R., Mauersberger, R., Schulz, A. & Wilson, T.L.: 1990, Astrophys. J. **362**, L25.

Schutte, W.A., Gerakines, P.A., van Dishoeck, E.F., Greenberg, J.M. & Geballe, T.R.: 1994, to appear in: *Physical Chemistry of Molecules and Grains in Space*, Amer. Inst. Phys.

Shalabiea, O. & Greenberg, J.M.: 1994, Astron. & Astrophys. **290**, 266.

Shu, F.H., Najita, J., Galli, D., Ostriker, E. & Lizano, S.: 1993 in: *Protostars & Planets III*, eds: E.H. Levy & J. Lunine, Univ. Arizona: Tucson, p3.

Sutton, E.C., Peng, R., Danchi, W.C., Jaminet, P.A., Sandell, G. & Russell, A.P.G.: 1994, Astrophys. J., Suppl. in press.

Sutton, E.C., Jaminet, P.A., Danchi, W.C. & Blake, G.A.: 1991, Astrophys. J., Suppl. **77**, 255.

Tielens, A.G.G.M.: 1983, Astron. & Astrophys. **119**, 177.

Tielens, A.G.G.M.: 1989 in: *Interstellar Dust*, eds: L.J. Allamandola & A.G.G.M. Tielens, Kluwer: Dordrecht, p239.

Tielens, A.G.G.M. & Allamandola, L.J.: 1987 in: *Interstellar Processes*, eds: D. Hollenbach & H.A. Thronson, Kluwer: Dordrecht, p333.

Turner, B.E.: 1991, Astrophys. J., Suppl. **76**, 617.

van Dishoeck, E.F., Blake, G.A., Draine, B.T. & Lunine, J.I.: 1993 in: *Protostars & Planets III*, eds: E.H. Levy & J. Lunine, Univ. Arizona: Tucson, p163.

van Dishoeck, E.F., Blake, G.A., Jansen, D.J. & Groesbeck, T.: 1995, Astrophys. J. submitted.

van Langevelde, H.J., van Dishoeck, E.F. & Blake, G.A.: 1995, in preparation.

Walmsley, C.M.: 1992 in: *Chemistry and Spectroscopy of Interstellar Molecules*, eds: D.K. Bohme, *et al.* , Univ. of Tokyo: Tokyo, p267.

Walmsley, C.M., Hermsen, W., Henkel, C., Mauersberger, R. & Wilson, T.L.: 1987, Astron. & Astrophys. **172**, 311.

Whittet, D.C.B.: 1993 in: *Dust and Chemistry in Astronomy*, eds: T.J. Millar & D.A. Williams, IOP: Bristol, p9.

Willacy, K. & Williams, D.A.: 1993, Mon. Not. of the RAS **260**, 635.

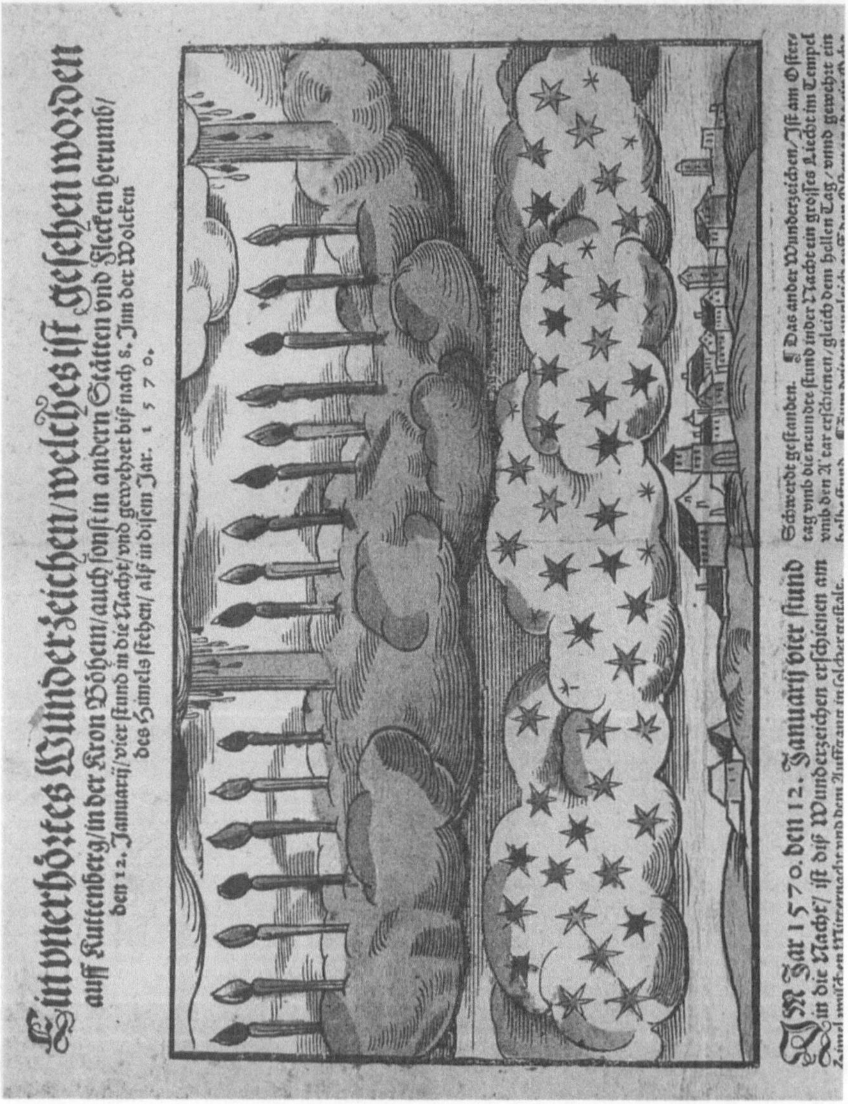

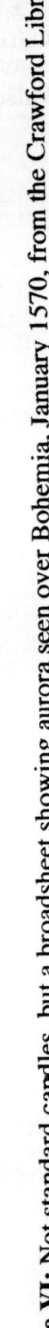

Plate VI: Not standard candles, but a broadsheet showing aurora seen over Bohemia, January 1570, from the Crawford Library.

HOT CORE CHEMISTRY

S.B. CHARNLEY

Astronomy Department, University of California at Berkeley
and Space Science Division, NASA Ames Research Center

Abstract. Recent theoretical work on the chemistry in regions of massive star formation is summarised.

Key words: ISM: molecules – stars: formation

1. Introduction

During formation of a dense core and during isothermal collapse to form a protostar, molecular accretion on to cold dust grains forms icy mantles containing the products of both gas and solid phase chemistries. Heating of their surrounding dust by newly-formed protostars will evaporate the volatile ices and this outgassing drives an active chemistry in the hot gas. Hot cores are dense, warm regions associated with young protostars and are understood to be such regions where evaporation of icy grain mantles has occurred (Charnley 1994).

Observations of several star-forming regions demonstrate that hot core chemistry is extremely rich and that strong chemical differentiation apparently exists between individual cores (Blake *et al.* 1987; Sutton *et al.* 1991, 1994; Turner 1991; Ziurys & McGonagle 1993) The molecular inventory of hot cores is typified by large abundances of fully-saturated molecules such as water and ammonia, as well as more complex organic molecules including methanol, ethanol, dimethyl ether, methyl formate, ketene, formaldehyde, acetaldehyde, formic acid and several nitriles. Many sulphur-bearing molecules are also present. Theoretical models of these sources have shown that with gas phase chemistry and accretion in the cold gas, during collapse, simple grain surface reactions can indeed qualitatively account for some aspects of their chemistry (Brown *et al.* 1988), but also that most of the observed molecules are in fact synthesised in the hot gas from the mantle constituents (Charnley *et al.* 1992).

Simple mantle molecules are precursors of large interstellar molecules, and so hot cores could contain some of the most complex molecules in the interstellar medium. This contribution summarises the organic chemistry expected in hot cores, as well some recent results concerning the molecular chemistry of second-row elements.

2. Organic Chemistry

When mantles are removed a nonequilibrium chemistry takes place that is characterised by their composition. Methanol-rich ices lead to high abundances of dimethyl ether $((CH_3)_2O)$ and methyl formate $(HCOOCH_3)$, whereas ammonia-rich ices lead to several molecules containing C-N bonds (Charnley et al. 1992; Charnley & Tielens 1992). Ethanol cannot be formed efficiently in hot gas and a solid state origin is required. Self-protonation of alcohols leads to pure ethers and ion-molecule reactions between different alcohols produces mixed ethers (Karpas & Mautner 1989). The ion-molecule chemistry driven by ethanol produces diethyl ether, methyl ethyl ether and ethyl formate (Charnley et al. 1995). The presence of methanol and ethanol in mantles suggests that alkyl cation transfer reactions from their protonated forms may be the key process to complexity in hot cores and that the alcohol and nitrogen chemistries may be interconnected (Charnley 1995a). Specifically, CH_3^+ transfer to acetaldehyde and HCN could lead to acetone and provide an additional source of methyl cyanide. Ethyl cyanide is widely observed in star-forming regions and the first step in its synthesis may be $C_2H_5^+$ transfer to HCN. If these reactions are indeed of primary importance then this scenario would yield further insight into surface reaction pathways (Charnley & Tielens 1994).

3. Sulphur and Silicon Chemistry

Hydrogen sulphide is observed to be enhanced in star-forming regions with estimated abundances that are much higher than those measured for dark clouds (Minh et al. 1991). This suggests that some H_2S has been released from grain mantles. At gas temperatures in the range 100-300K, H_2S will react with atomic hydrogen to form SH; this molecule then reacts with H to produce free sulphur atoms since their reaction with H_2 is inhibited at these temperatures.

Calculations show that this sequence drives the sulphur chemistry in hot cores and produces all the observed S-bearing species (Charnley 1995b) Almost all the original H_2S is converted to SO and SO_2. The predicted abundance ratios are in reasonable agreement with existing observations and may allow hot core sulphur chemistry to be distinguished from that of shock waves. Mackay (1995) has shown that silane (SiH_4) from grains can produce SiO in hot cores and that, like the sulphur, the silicon chemistry is sensitive to the presence of molecular oxygen in the mantles.

4. Phosphorus Chemistry

Hydrogenation on grain surfaces suggests that the major repository of solid phosphorus is phosphine (PH_3). The nondetection of PH_3 in dark clouds is in agreement with ion-molecule chemical theory but its absence in dense, warm regions presents a problem. Turner *et al.* (1990) proposed that either phosphorus does not hydrogenate on grains, or that phosphine has an anomalously high binding energy, or that phosphine is destroyed when evaporated. Models of hot core phosphorus chemistry (Charnley & Millar 1994) show that, in warm gas, P atoms are rapidly broken out of phosphine by a sequence of H atom reactions. The original PH_3 is ultimately converted to atomic P and to PN (via PO), the only phosphorus-bearing molecule detected in molecular clouds.

5. The Origin of Hot Core Molecules

In the context of hot core theory it is possible to group those molecules observed (and expected) to be present according to their formation environment: either in cold gas, or on grain surfaces, or in the hot gas (Charnley & Tielens 1994). Deuterated molecules principally form in the first two environments (Brown & Millar 1989) and some (like HCN and CH_3CN) may have significant contributions from more than one. The three classifications are:

5.1 MOLECULES FORMED IN COLD GAS

CO, C_2H_2, C_2H_4, HCN, N_2, O_2, CS.

5.2 MOLECULES FORMED ON GRAIN SURFACES

H_2O, CH_4, NH_3, H_2S, PH_3 SiH_4.
CH_3OH, H_2CO, C_2H_5OH, $HCOOH$, CH_3CHO, NH_2CHO.
C_2H_4, C_2H_6, H_2CS, OCS.

5.3 MOLECULES FORMED IN HOT GAS

$(CH_3)_2O$, $HCOOCH_3$, CH_2CO, CH_3CHO, $(CH_3)_2CO$,
$(C_2H_5)_2O$, $CH_3OC_2H_5$, $HCOOC_2H_5$.
HCN, CH_3CN, C_2H_3CN, C_2H_5CN, HC_3N, CH_3NH_2, CH_2NH.
SiO, SO, SO_2, PN, PO.

6. Summary

Hot core sources provide a natural laboratory in which to study the products of grain surface reactions and the chemistry of warm, dense gas following mantle evaporation. The gas phase chemistry around forming *low-mass* protostars will be determined by the material that is evaporated from the dust (van Dishoeck 1995). The density and velocity structure of the infalling envelope, as well as the accretion luminosity, depend on the mass accretion rate (*e.g.* Adams & Shu 1985). The spatial and temporal distribution of dust temperature and density around such an object can therefore be related to the infall physics. As mantle evaporation is rapid and the gas phase chemistry is highly transient, it may be possible to relate certain chemical timescales to the dynamical ones. Detailed modelling, currently underway, of this dynamical-chemical scenario may lead to the identification of a 'molecular clock' for the star-formation process. This would, in principle, allow the state of star-forming cores to be inferred from their observed molecular composition.

References

Adams, F.C. & Shu, F.H.: 1985, Astrophys. J. **296**, 655.
Blake, G.A., Sutton, E.C., Masson, C.R. & Phillips, T.G.: 1987, Astrophys. J. **315**, 621.
Brown, P.D., Charnley, S.B. & Millar, T.J.: 1988, Mon. Not. of the RAS **231**, 409.
Brown, P.D. & Millar, T.J.: 1989, Mon. Not. of the RAS **237**, 661.
Charnley, S.B.: 1994 in: *Molecules and Grains in Space*, ed: I. Nenner, AIP Conference Proceedings No. 312, American Institue of Physics, p155.
Charnley, S.B.: 1995a in: *The Cosmic Dust Connection*, eds. J.M. Greenberg & V. Pirronello, Kluwer Academic Publishers, in press.
Charnley, S.B.: 1995b, Astrophys. J. in press.
Charnley, S.B., Kress, M.E., Tielens, A.G.G.M. & Millar, T.J.: 1995, Astrophys. J. in press.
Charnley, S.B. & Millar, T.J.: 1994, Mon. Not. of the RAS **270**, 570.
Charnley, S.B., Tielens, A.G.G.M. & Millar, T.J.: 1992, Astrophys. J. **399**, L71.
Charnley, S.B. & Tielens, A.G.G.M.: 1992 in: *The Astrochemistry of Cosmic Phenomena*, IAU Symposium No. 150, ed: P.D. Singh, Kluwer Academic Publishers, p317.
Charnley, S.B. & Tielens, A.G.G.M.,: 1994, in preparation.
Karpas, Z. & Mautner, M.: 1989, J.Phys.Chem. **93**, 1859.
Mackay, D.: 1995, these proceedings.
Minh, Y.C., Irvine, W.M. & McGonagle, D.: 1991, Astrophys. J. **366**, 192.
Sutton, E.C., Jaminet, P.A., Danchi, W.C. & Blake, G.A.: 1991, Astrophys. J., Suppl. **77**, 255.
Sutton, E.C., *et al.* : 1994, preprint.
Turner, B.E.: 1991, Astrophys. J., Suppl. **76**, 617.
Turner, B.E., *et al.* : 1990, Astrophys. J. **365**, 569.
van Dishoeck, E.F.: 1995, these proceedings.
Ziurys, L.M. & McGonagle, D.: 1993, Astrophys. J., Suppl. **85**, 155.

MOLECULAR GAS AROUND SYMBIOTIC MIRAS:

MASERS IN R AQUARII AND H1−36

R. J. IVISON
Royal Observatories, Blackford Hill, Edinburgh EH9 3HJ.

E. R. SEAQUIST
Dept of Astronomy, University of Toronto, 60 St George St., Toronto M5S 1A7.

and

P. J. HALL
Australia Telescope National Facility, CSIRO, PO Box 76, Epping NSW 2121.

Abstract. The concept of studying symbiotic Miras via their molecular emission holds the promise of an improved understanding of the circumbinary environment in symbiotics, but following the largely ineffective maser surveys conducted during the 1970s and 80s there has been a widespread misconception that this avenue of study is a dead end. Here, we present spectral line and continuum maps of R Aqr and H1−36, two objects in which OH and H_2O masers were detected during a recent survey of symbiotic Miras.

Key words: masers – R Aquarii – H1−36 – AGB stars

1. Introduction

In isolated Miras, mass loss generates a dust-rich circumstellar shell (CS) which protects against photodissociation by interstellar UV and thereby harbours molecular gas. Masers are a common consequence of this process. Since D(usty)-type symbiotic binaries contain AGB stars, it might be expected that they too would support masers. Once found, we could employ lines such as OH, SiO and H_2O to probe the morphology and dynamics of the neutral gas, the position and radial velocity of the AGB component and the terminal velocity of its wind.

Until recently, there was an important difference between the behaviour of isolated Miras and that of their cousins in the symbiotic binaries: most field Miras are associated with OH masers, whereas none of the 16 symbiotic Miras had shown any strong evidence of 1612-MHz OH emission, nor of main-line OH emission (see Ivison, Seaquist & Hall 1994).

The same story applied to water masers: five symbiotic Miras had been searched for emission from the H_2O $6_{16} \rightarrow 5_{23}$ transition at 22 GHz, and none were detected. Only two symbiotics have been associated with maser emission of any kind: H1−36 and R Aqr, which support SiO masers (Allen *et al.* 1989; Hall *et al.* 1990).

Given all the potential information stored in maser lines, we ignored the bad press concerning molecular studies of symbiotics and undertook a large maser-line survey of symbiotic Miras; one carefully planned to reach the

Astrophysics and Space Science 224: 255–258, 1995.

TABLE I

Details of the VLA observations.

Source	Configuration & Date (1994)	Line	Velocity resolution	Beamsize ($''$) and PA ($°$)	Spectral rms (mJy)
H1−36	A − May 01	OH	$0.34\,\mathrm{km\,s^{-1}}$	2.16×0.62, 172	4.4
	A − May 01	H_2O	$1.59\,\mathrm{km\,s^{-1}}$	0.29×0.06, 22	5.8
R Aqr	A − May 02	OH	$1.36\,\mathrm{km\,s^{-1}}$	1.39×0.82, 169	2.3
	BnA − May 20	H_2O	$0.40\,\mathrm{km\,s^{-1}}$	0.22×0.11, 55	4.2

same luminosity level as the surveys of isolated Miras (Seaquist & Ivison, *in preparation*). We shall avoid a detailed description of the survey here, and say only that data has been obtained for most of the symbiotic Miras, covering the transitions at 1612 MHz (OH, using the VLA and ATCA), 22 GHz (H_2O, using the VLA), 86 GHz (SiO, using the OVRO millimetre array and SEST) and, finally, for a handful of stars, 321 GHz (H_2O, using the JCMT).

OH and H_2O masers were discovered in both R Aqr and H1−36 and were described in detail by Ivison *et al.* (1994). The original survey observations were designed to aid detection (large bandwidths, compact arrays) rather than to probe masers with high spatial and spectral resolution. It was impossible, for example, to rule out a thermal origin for the line emission. Here we present some follow-up observations of R Aqr and H1−36, using the VLA.

2. Observations and Discussion

Details of the spectral-line observations are given in Table 1. For OH, we recorded two bands, centred at 1612.231 and 1667.359 MHz; the mode chosen for H_2O gave one band centred at 22235.08 MHz. Some continuum observations were also obtained. The H1−36 OH data were self-calibrated using channels containing line emission.

In H1−36, the 1612-MHz OH line centroid ($v_{\mathrm{lsr}} = -124.5\,\mathrm{km\,s^{-1}}$) is displaced by $-5.5\,\mathrm{km\,s^{-1}}$ from the mainline centroid, though a faint red wing stretches to the centre of the mainline (see Figure 1). The full widths are $8\,\mathrm{km\,s^{-1}}$ for both of the lines, whilst the FWHM are 4 and $6\,\mathrm{km\,s^{-1}}$ at 1612 and 1667 MHz, respectively. The velocity of the strongest peak at 1612 MHz corresponds with the 1993 May and July values (to within the errors). This is also true of the H_2O line (which is at only 25% of its 1993 intensity).

If the observed 1612-MHz peak is the blueward component of the usual OH pair, then we can estimate a terminal wind velocity of $10\,\mathrm{km\,s^{-1}}$ from its displacement from H1−36's SiO and H92α lines; furthermore, if the unseen redward peak has been obscured by optically thick, ionized gas (an idea first

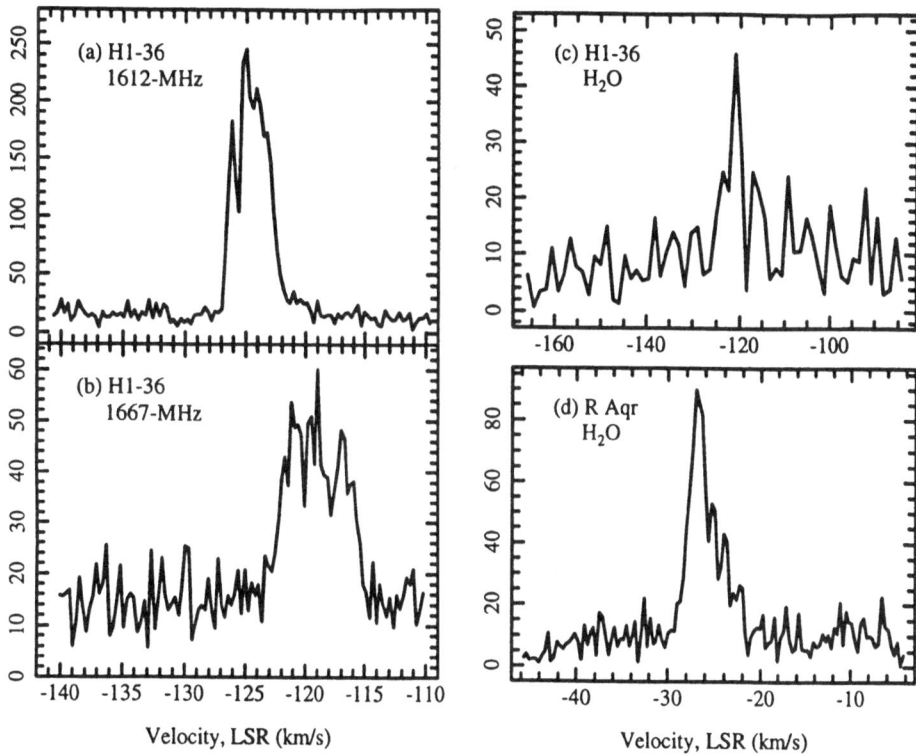

Fig. 1.   Vector averaged cross-power spectra. F_ν in mJy.

proposed by Allen *et al.* 1989) then $\tau \geq 3.6$ at 1612 MHz.

In R Aqr, the H_2O line has a full width of $8 \, km \, s^{-1}$. The FWHM is around $4 \, km \, s^{-1}$, broader than in 1993. The strongest peak is at $-27.0 \, km \, s^{-1}$ which is a very interesting result as there is no corresponding emission in the 1993 spectrum and a possible cause is the orbital motion of the Mira component. Continued monitoring may at last yield a reliable period for this system. The OH spectra are not shown as no line is visible: the limits (for linewidths of $5 \, km \, s^{-1}$) are $3\sigma < 16 \, mJy \, km \, s^{-1}$ at both 1612 and 1667 MHz .

The spatial resolution of our data allows us to limit T_b to $> 9 \times 10^3$, $> 3 \times 10^3$ and $> 1.2 \times 10^4$ K for OH and H_2O in H1−36 and H_2O in R Aqr. The emission mechanism is undoubtedly non-thermal.

Figure 2 shows the positions of the masers with respect to continuum emission from the ionized gas in both systems. It is evident that the masers are offset from the continuum centroid in H1−36. The 0.5-arcsec offset corresponds to several thousand AU in the plane of the sky. Incidentally, the ionized gas is unresolved by our 0.73×0.25-arcsec beam.

In R Aqr, the H_2O maser is coincident (< 0.05 arcsec or < 13 AU) with the centroid of the 22-GHz continuum map. This suggests that the ionized

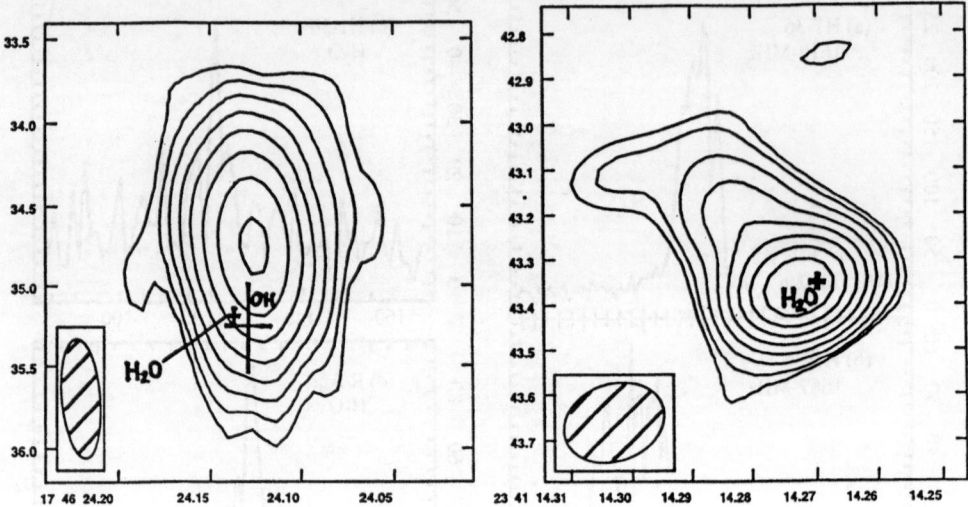

Fig. 2. A-configuration continuum maps of H1−36 (left, 8.4 GHz, 1991 September) and R Aqr (right, 22 GHz, 1994 May), with the positions of the masers marked and labelled.

gas and the maser are virtually co-spatial, which can be understood if the masing gas is offset in the line of sight, or if it is shielded by the Mira from the hot star.

3. Concluding Remarks

It is becoming clear that masers are a useful tool with which we can probe the morphology and dynamics of the dusty, neutral material in symbiotic Miras and relate it to that of the ionized gas. As a bonus, our radial velocity measurements of the R Aqr H_2O maser may eventually betray its hitherto elusive binary period. S2 VLBI observations of H1−36 are planned.

Acknowledgements

This work was supported by an operating grant to ERS from the Natural Sciences and Engineering Research Council of Canada.

References

Allen, D.A., Hall, P.J., Norris, R.P., Troup, E.R., Wark, R.M., Wright, A.E.: 1989, Mon. Not. of the RAS **236**, 363.
Hall, P.J., Wright, A.E., Troup, E.R., Wark, R.M., Allen, D.A.: 1990, Mon. Not. of the RAS **247**, 549.
Ivison, R.J., Seaquist, E.R. and Hall, P.J.: 1994, Mon. Not. of the RAS **269**, 218.

SESSION F:

LATE EVOLUTIONARY
STAGES
OF
HIGH MASS STARS

CIRCUMSTELLAR NEBULAE AROUND OFPE/WN9 STARS

A. NOTA, A.PASQUALI *, L.DRISSEN, C.LEITHERER and C.ROBERT
Space Telescope Science Institute
3700 San Martin Drive, Baltimore, MD.

and

W. SCHMUTZ
Institut für Astronomie, ETH-Zentrum,
Zürich, Switzerland

Abstract. New high resolution echelle observations of Ofpe/WN9 stars in the Large Magellanic Cloud have been obtained in the wavelength region 4000 – 8000 Å. We find that five Ofpe/WN9 stars display in their spectra nebular emission lines [NII], [SII] and also [OIII], previously unreported for BE381 and S119. At least in these two cases we can conclude that the stars are surrounded by an associated nebula, thus strengthening the relationship between Ofpe/WN9 stars and LBVs.

1. Introduction

Ofpe/WN9 transition stars are a crucial component in the population of stars in the uppermost part of the HR Diagram. Walborn (1982) introduced the classification, indicating that the stars could neither be classified as pure WNL stars, nor as pure Of stars. Nebular lines from the surrounding gas hinted at enhanced nitrogen abundances, consistent with the very evolved nature of these objects. Bohannan & Walborn (1989) summarized the properties of all Ofpe/WN9 stars in the Large Magellanic Cloud (LMC). The significance of Ofpe/WN9 stars was recognized when Stahl *et al.* (1983) discovered the transformation of one Ofpe/WN9 star into an LBV on a timescale of less than a few years. *Vice versa*, Stahl (1986) demonstrated that the well-established LBV AG Carinae, when in a high-temperature state, displays all spectral characteristics of Ofpe/WN9 stars. Therefore it seems likely that there is a close relationship between the entire class of LBVs and Ofpe/WN9 stars. This relationship has been further explored by Smith, Crowther & Prinja (1994) who propose that some LBVs are the extension of the WN sequence towards later spectral types, hence unifying the classes of WNL, LBV and Ofpe/WN9 stars. Many LBVs and WNLs are known to be surrounded by circumstellar shells. Our finding that some Ofpe/WN9 stars are also surrounded by circumstellar shells could further strengthen the connection.

---

* Osservatorio di Arcetri, Firenze

Astrophysics and Space Science **224**: 261–265, 1995.
© 1995 *Kluwer Academic Publishers.*

2. Observations and Results

We observed our sample of Ofpe/WN9 stars with the high resolution echelle spectrograph EMMI, coupled to the NTT at ESO, La Silla, on the nights of September 17 – 20, 1991. The data were reduced with standard procedures and the extraction of the echelle orders was performed using the IRAF task DOECSLIT.

Five of the Ofpe/WN9 stars observed show nebular emission in our echelle spectra, namely S119, R99, R84, BE381 and HDE269927c. Walborn (1982) already discovered the presence of forbidden [N II] lines at 6548 Å and 6583 Å in the spectra of R99, R84, R127, and S61. The nebula around R127 is discussed in detail in Stahl (1987), Clampin *et al.* (1993), and Schulte-Ladbeck *et al.* (1993) and will not be discussed here. S119 is also discussed in detail in Nota *et al.* (1994). S61 was not included in our NTT observations.

2.1   R99

R99 has a systemic velocity of 272.3 km s^{-1}. The star spectrum also displays a number of nebular lines: in addition to nebular components superimposed on the stellar Balmer emission lines, we detect faint [O III] $\lambda\lambda$4959, 5007, [N II] λ6584, and [S II] $\lambda\lambda$6717, 6734. It is interesting to notice that Walborn (1982) discovered most nebular lines but mentioned the complete absence of [O III] emission in the spectrum of the star at the time of his observation, probably due to the low S/N of his spectra.

The nebular emissions appears also to be spatially extended within the slit width ($\sim$5''), approximately centered on the star). In Figure 1 we show the histogram of the measured radial velocity components, sampled every 0.345''. The histogram provides a good estimate of the significance of the measured components. In the case of R99, a constant velocity component is visible at $\sim$340 km s^{-1}, in addition to a radial velocity structure which is approximately centered on the star's systemic velocity. This could possibly indicate an association of the nebula with R99. The dynamics of the structure is such that the velocity dispersion seems to increase with distance from the star. Also, the radial velocity distribution appears symmetrical with respect to the star when we move from the East to the West region. Information on the density is provided by the [S II] λ6717/λ6731 intensity ratio. In the W region, we measure 10 – 100 cm^{-3}, quite low densities compared to a typical ejected nebula, while in the East region we find higher density values peaking at 800 cm^{-3} at a location of 2.5'' East from the star. Inspection of the corresponding plates of the deep Hα + [N II] atlas of Davies, Elliott, & Meaburn (1976) shows that R99 is also embedded in a region of extended Hα emission. On the same basis, Walborn (1982) concluded that the nebular lines in R99's spectrum are probably due to the associated HII region. In addition, Stahl (1987) attempted to resolve the R99 nebula but did not succeed. Coronographic imaging and a detailed abundance analysis will be required to establish if there is a connection.

2.2   R84

R84 shows strong nebular Hα, Hβ, and [N II] $\lambda\lambda$6548, 6583. The systemic velocity for the star is 247.9 km s^{-1}. We detect two different radial velocity components in the nebular Hα of R84. They are centered approximately on the star's systemic velocity, as we can see from Figure 1, where we show the histogram distribution of the measured radial velocity components. The nebular emissions appear to be spatially extended within the slit width ($\sim$5", approximately centered on the star). The distribution of radial velocity components is suggestive of a number of filaments moving with a maximum dispersion in velocity of $\sim$30 km s^{-1}. Given the similarity between stellar and nebular radial velocities, it would seem plausible to associate the nebular emission with the star itself. However, from inspection of the corresponding plates of the deep Hα + [N II] atlas of Davies, Elliott, & Meaburn (1976), it is clear that R84 is located in a region of extended Hα emission. In addition, Stahl (1987) failed to detect the nebula by direct imaging. We believe that on the basis of spectroscopic data alone no conclusion can be made about the physical association of the detected nebula with R84. Coronographic imaging will be necessary to investigate the nature of the nebular structure.

2.3   BE381

BE381 has strong nebular Hα, Hβ, [N II] $\lambda\lambda$6548, 6583, and also [O III] λ5007, and very weak [S II] $\lambda\lambda$6717, 6731. The systemic velocity for BE381 is 253.2 km s^{-1}. The nebular emissions appear to be spatially extended within the slit width. In Figure 1 we show a histogram of the measured radial velocity components. We find two different significant components in the Hα line profile. From the study of the overall radial velocity structure we can conclude that BE381 is surrounded by a shell, which expands at a velocity of $\sim$30 km s^{-1}. It is most likely associated with the star. Both [S II] lines appear very weak in the integrated spectrum, so it will not be possible to derive any density information as a function of position, but just an average value across the slit width. The density is of the order of 800 cm^{-3}, under the assumption of T$_e$ $\approx$ 10000 K.

2.4   HDE269927c

HDE269927c has strong nebular Hα, Hβ, [N II] $\lambda\lambda$6548, 6583, [O III] $\lambda\lambda$4959, 5007 and [S II] $\lambda\lambda$6717, 6731. The systemic velocity for this star is 276.8 km-s^{-1}. The nebular emissions appear to be spatially extended within the slit width. In this case the association of the detected nebular lines with HDE269927 is less obvious. In Figure 1 we show the histogram of the measured radial velocity components, derived by fitting the Hα line. We detect

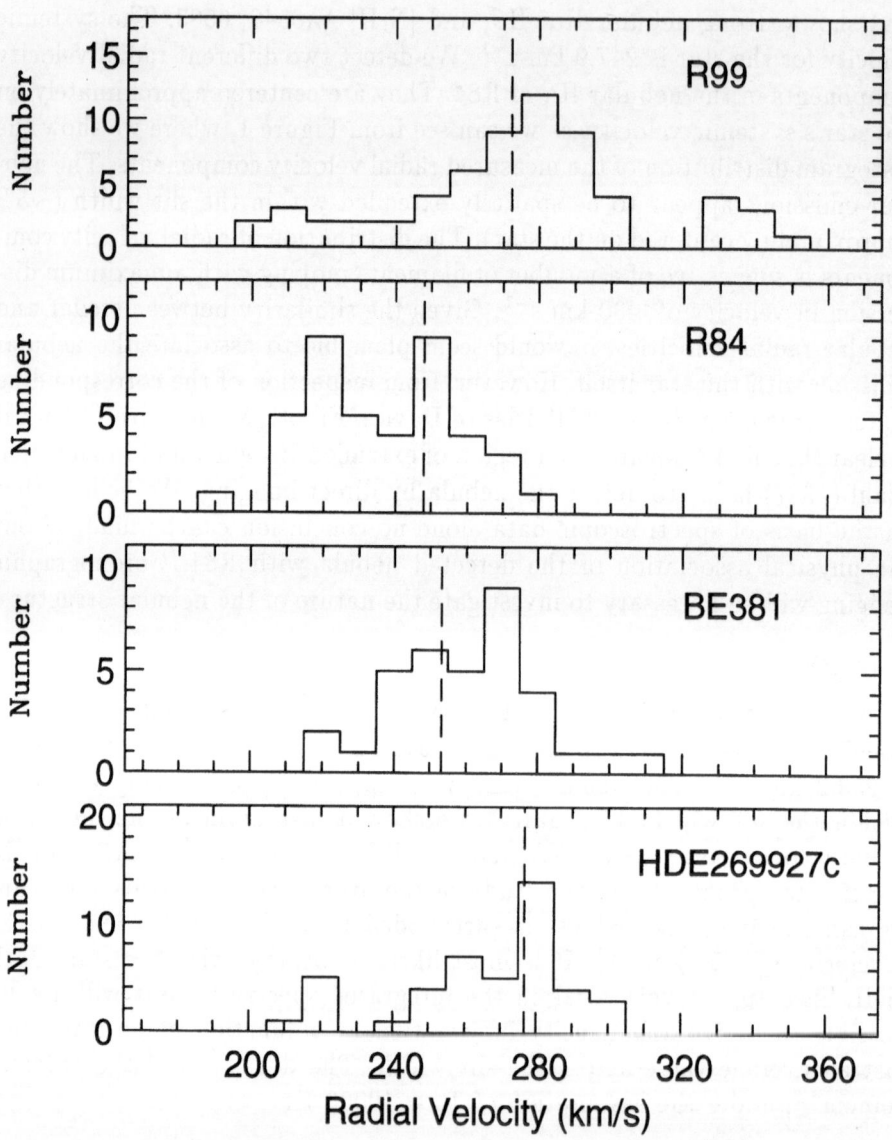

Fig. 1. Histograms of the radial velocity components for four Ofpe/WN9 stars. In each histogram, the vertical dashed line represents the star's systemic velocity.

a constant component in the Hα line (at $\sim$280 km s^{-1}), which could be due to the presence of an underlying H II region. It is interesting to notice that the average systemic velocity for HDE269927c is also 277 km s^{-1}. Probably this indicates a physical association with the H II region. Additional, weaker, Hα components are detected at $\sim$260 km s^{-1} and 220 km s^{-1}. Inspection of the corresponding plates of the deep Hα + [N II] atlas of Davies, Elliott & Meaburn (1976) shows that HDE269927c is located at the edge of a region of extended Hα emission. Again, coronographic imaging and a detailed abundance analysis will be required to establish if the nebular emission is generated within the underlying H II region and if it is associated with the nebula. The [S II] λ6717/λ6731 intensity ratio indicates similarly low densities as in R99. The resulting densities could be an additional hint that the nebular emission is indeed produced by the underlying H II region.

References

Bohannan, B. & Walborn, N.R.: 1989, Publ. of the ASP **101**, 520.

Clampin, M., Nota, A., Golimowski, D., Leitherer, C. & Durrance, S.T.: 1993, Astrophys. J. **410**, L35.

Davies, R.D., Elliott, K.H. & Meaburn, J.: 1976, Mon. Not. of the RAS **81**, 89.

Nota, A., Drissen, l., Clampin, M., Leitherer, C., Pasquali, A., Robert, C., Paresce, F. & Robberto, M.: 1994, in: *Circumstellar Media in the Late Stages of Stellar Evolution*, eds: R.E.S. Clegg, I.R. Stevens & W.P.S. Meikle, Cambridge: University Press, p89.

Schulte-Ladbeck, R.E., Leitherer, C., Clayton, G.C., Robert, C., Meade, M.R., Drissen, L., Nota, A. & Schmutz, W.: 1993, Astrophys. J. , **407**, 723.

Smith, L.J., Crowther, P.A. & Prinja, R.K.: 1994, Astron. & Astrophys. in press.

Stahl, O., Wolf, B., Klare, G., Cassatella, A., Krautter, J., Persi, P. & Ferrari-Toniolo, M.: 1983, Astron. & Astrophys. **127**, 49.

Stahl, O.: 1986, Astron. & Astrophys. **164**, 321.

Stahl, O.: 1987, Astron. & Astrophys. **182**, 229.

Walborn, N.R.: 1982, Astrophys. J. **256**, 45.

PLATE VII: The 1.2m Schmidt Telescope situated at Siding Spring Observatory in New South Wales, Australia. The telescope was first used in 1973 and was an outstation of the Royal Observatory until 1988 when it became part of the Anglo-Australian Observatory (AAO). All plates taken with the telescope are stored in the Plate Library at ROE.

WIND ANISOTROPY REVEALED BY DUST FORMATION
AROUND WOLF-RAYET STARS

P. M. WILLIAMS

Royal Observatories
Edinburgh UK

Abstract. The presence of heated circumstellar dust around WC type Wolf-Rayet stars requires the episodic or persistent condensation of carbon grains in their stellar winds. In order to survive in the stars' strong ultraviolet radiation fields, the grains must be located at least 100AU from the stellar surfaces. The densities in isotropic winds at such large distances are too low to allow grain growth and anisotropies such as clumps, disks or wind-collision wakes in colliding-wind binary systems are required to provide grain nurseries. Observational evidence for such features in grain-forming W-R stars is examined.

Key words: circumstellar grains – stars: Wolf-Rayet – stars: binary

1. Introduction — incidence and properties of dust

An early revelation of infrared astronomy was that dust grains condense around stars, particularly evolved late-type stars. It soon emerged that this phenomenon extended to hotter stars, of which the most extreme are the Wolf-Rayet (WR) stars (*e.g.* Allen, Harvey & Swings 1972; Cohen, Barlow & Kuhi 1975 and Williams, van der Hucht & Thé 1987 = WHT). These stars have dense, fast (1000–3000 km/s) stellar winds which carry the grains with them, causing the grains to cool rapidly and their emission to fade. The observation of *persistent* emission by most of the WC8–10 WR stars indicates that grains are continually condensing in their winds. Some, hotter, WC4–8 WR stars show infrared outbursts attributable to *episodic* dust formation and the insights they offer are discussed below.

Analysis of the infrared spectral energy distributions (SEDs) allows determination of the grain emissivity law. The early studies cited favoured graphitic grains but WHT preferred amorphous carbon (AC) from fitting the SED of WR 140. The difference in emissivity laws between graphite ($\kappa \propto \lambda^{-2}$) and AC ($\kappa \propto \lambda^{-1}$) is small and attempts to distinguish between them using observed SEDs is dependent on the temperature distribution of the grains. A less model-dependent diagnostic is the 11.52-μm graphite resonance predicted by Draine (1984). This was searched for but not seen in spectra of WR 104 (Glasse *et al.* 1986) or WR 140 (Williams 1994), supporting the AC identification. So far, the only spectroscopic feature known in WR dust emission is a broad one near 7.7μm observed in airborne and IRAS spectra and attributed to a carbonaceous carrier with small aromatic domains (Cohen, Tielens & Bregman 1989). Cherchneff & Tielens (1994) have discussed the chemical pathway to AC grain formation in WC type WR stars.

Astrophysics and Space Science **224**: 267–270, 1995.
© 1995 *Kluwer Academic Publishers.*

This differs from that operating in AGB stars because there is not enough hydrogen for C_2H_2 to be important and condensation is believed to occur via carbon chains, monocylic rings, polycyclic aromatic carbon (PAC) rings and curved molecules leading to the fullerene family. The presence of the $7.7\mu m$ feature supports this pathway but no-one has observed features from precursors such as C_2 or even C I. This puts severe limits on the abundances or, in the case of absorption lines, the location of these species.

The significance of dust emission by WR stars stems from the difficulty of the formation of dust grains near such hot stars. By considering the radiative equilibrium on a dust grain, we can determine the minimum distance from the star at which it can survive, which is about 100 AU for a WR star. At this distance, the WR stellar wind density ($\sim 10^{-17}$ g cm^{-3} for a uniform wind) is too low by several orders of magnitude to allow the processes discussed by Cherchneff & Tielens to operate, suggesting that grain formation is not isotropic or widespread in the wind but occurs in density enhancements of some sort. There is evidence the winds of some WR stars are clumpy rather than homogeneous (*e.g.* Moffat & Robert 1994, Antokhin 1994) and, from polarimetry, that some are non-spherical and have disk-like structures (*e.g.* Schulte-Ladbeck 1994).

The question is — are these effects necessary or sufficient to allow dust formation? Part of the answer lies in more detailed modelling of stellar winds and part in an comparison of the winds of stars which do and do not make dust. There is evidence that some dust clouds may be disk-like (*e.g.* Cohen & Kuhi 1977) but no evidence for systematic differences in clumpiness or wind-flattening between dusty and non-dusty stars. A more sensitive test may be provided by the episodic dust-makers. Conditions in these stars' winds are marginal for dust formation and discovery of the changes which trigger dust formation will provide valuable information on the broader picture.

2.   The episodic dust-maker WR 140 — freak or paradigm?

The prototype episodic dust-maker is WR 140 (HD 193793), which experienced dust formation episodes in 1977, 1985 and 1993. Models fitted to the 1985 infrared data by Williams *et al.* (1990 = W90) indicated that $2.8 \times 10^{-8} M_\odot$ of amorphous carbon condensed between 1985.21 and 1985.54, forming an expanding cloud. The rate of dust formation, $\sim 8 \times 10^{-8} M_\odot y^{-1}$, corresponds to 0.13% of the mass loss ($\sim 6 \times 10^{-5} M_\odot\ y^{-1}$) or $\sim 5\%$ of the total carbon being lost by the WC7 star. The repeatability of the light curve (Williams 1994) indicates that the same happened in 1977 and 1993, followed in each case by a slow fading of the emission as the newly formed dust was dispersed and cooled.

The significance of WR 140 is that the episodes are periodic and that the WC star is a member of a WC7+O4 binary system with the same period.

Both the WC7 and O4 stars have winds with velocities approaching 3000 km/s. Strong shocks occur where the winds collide, compressing the wind by a factor of 4, and further compression sufficient for dust formation can occur if the wind cools sufficiently, *e.g.* by X-ray emission (Usov 1991). This points to a third type of anisotropy, a wind-collision wake. The stellar winds must collide all the time but, because the orbit is eccentric ($e = 0.84$, W90), the separation of the stars and the pre-shock density vary drastically, especially around periastron passage. The occurrence of dust formation at this phase (W90) strongly supports the idea of dust formation in wind-collision wakes. There may be another effect: Becker & White (1994) have recently suggested from radio observations that the WC7 wind in WR 140 is flattened and that passage through this disk, which also occurs near periastron passage, triggers the dust formation — but this has yet to be studied in detail.

There are two other episodic dust making systems containing WC7 stars: WR 125 and WR 137 (HD 192641). Dust formation by WR 125 was observed during an infrared monitoring programme inspired by the manifest similarity of this system to WR 140 spectroscopically and in its X-ray and radio behaviour (Williams *et al.* 1992). So far, only one outburst has been observed and it is apparent from the monitoring that, if the outbursts are recurrent, the period must be greater than 15y. Photometry and radial velocity observations over many years will be needed to examine whether dust formation by this star is triggered by the same processes as in WR 140.

We may not have to wait so long for WR 137. Its last infrared maximum occurred in 1984.5 (Williams *et al.* 1985) and comparison of the subsequent fading with earlier observations suggests a recurrence time of $\sim 13y$, pointing to the next maximum in 1997–98.

3. Minor dust-formation episodes - dust formation in clumps?

Irregularities in the declining light curve of WR 137 since 1984 suggest further minor bursts of dust formation (Williams & van der Hucht 1994), possibly due to clumpiness in the wind. More intensive observations will be needed to characterize these irregularies and examine their origin.

A clearer example may be provided by the episodic dust-maker WR 48a, which has been fading slowly from a major dust formation episode in 1979. Observations at the SAAO in 1994 show evidence for a minor episode of dust formation this year. We don't yet know whether this is a harbinger of another major outburst or an isolated event, perhaps caused by condensation in a large clump in the wind. It is likely that, had this occurred when the system was brighter, say ten years ago, the infrared brightening caused by the minor dust formation episode would not have been observable against the stronger background flux.

4. Conclusion — application to the persistent dust-makers?

The episodic dust-maker WR 140 has revealed the importance of wind-compression wakes in colliding wind binaries as candidate locations for dust formation. To determine whether the persistent dust-makers form dust by a similar mechanism but continuously, perhaps because they have less eccentric orbits, it is first necessary to examine whether they are colliding wind binaries. There is some evidence for minor episodes of dust formation, perhaps attributable to wind clumpiness, but these may be difficult to observe against the strong infrared emission of the persistent dust-makers.

Acknowledgements

The infrared photometry underpinning this work was taken at UKIRT, ESO, SAAO, IAC and SPM for which I thank Tom Geballe, Karel van der Hucht, Patrice Bouchet, Patricia Whitelock, Brian Carter, Mark Kidger and Mauricio Tapia.

References

Allen, D.A., Harvey, P.M. & Swings, J.P.: 1972, Astron. & Astrophys. **20**, 333.
Antokhin, I.I.: 1994, in *Wolf-Rayet Stars: Binaries, Colliding Winds, Evolution, IAU Symposium 163*, eds: K.A. van der Hucht & P.M. Williams, Kluwer, p87.
Becker, R.H. & White, R.L., 1994, in: K.A. van der Hucht & P.M. Williams (eds) *Wolf-Rayet Stars: Binaries, Colliding Winds, Evolution, IAU Symp. 163*, (Kluwer), 438.
Cherchneff, I. & Tielens, A.G.G.M.: 1994, in *Wolf-Rayet Stars: Binaries, Colliding Winds, Evolution, IAU Symp. 163*, eds: K.A. van der Hucht & P.M. Williams, Kluwer, p346.
Cohen, M., Barlow, M.J. & Kuhi, L.V.: 1975, Astron. & Astrophys. **40**, 291.
Cohen, M & Kuhi, L.V.: 1977, Mon. Not. of the RAS **180**, 37.
Cohen, M., Tielens, A.G.G.M. & Bregman, J.D.: 1989, Astrophys. J. **344**, L13.
Draine, B.T.: 1984, Astrophys. J. **277**, L71.
Glasse, A.C.H., Towlson, W.A., Aitken, D.K. & Roche, P.F.: 1986, Mon. Not. of the RAS **220**, 185.
Moffat, A.F.J. & Robert, C.: 1994, Astrophys. J. **421**, 310.
Schulte-Ladbeck, R.E.: 1994, in *Wolf-Rayet Stars: Binaries, Colliding Winds, Evolution, IAU Symposium 163*, eds: K.A. van der Hucht & P.M. Williams, Kluwer, p176.
Usov, V.V.: 1991, Mon. Not. of the RAS **252**, 49.
Williams, P.M.: 1994, in *Wolf-Rayet Stars: Binaries, Colliding Winds, Evolution, IAU Symposium 163*, eds: K.A. van der Hucht & P.M. Williams, Kluwer, p335.
Williams, P.M., Longmore, A.J., van der Hucht, K.A., Talavera, A., Wamsteker, W.M., Abbott, D.C. & Telesco, C.M.: 1985, Mon. Not. of the RAS **215**, 23P.
Williams, P.M., van der Hucht, K.A. & Thé, P.S.: 1987, Astron. & Astrophys. **182**, 91 **(WHT)**.
Williams, P.M., van der Hucht, K.A., Pollock, A.M.T., Florkowski, D.R., van der Woerd, H. & Wamsteker, W.M.: 1990, Mon. Not. of the RAS **243**, 662 **(W90)**.
Williams, P.M., van der Hucht, K.A., Bouchet, P., Spoelstra, T.A.Th., Eenens, P.R.J., Geballe, T.R., Kidger, M.R. & Churchwell, E.: 1992, Mon. Not. of the RAS **258**, 461.
Williams, P.M. & van der Hucht, K.A.: 1994, in *The impact of Long-term Monitoring on Variable-Star Research*, eds: C. Sterken & M.J.H. de Groot, Dordrecht: Kluwer, p85.

DETECTION OF H_2 EMISSION TOWARDS THE
WOLF-RAYET NEBULA NGC 2359

N. ST-LOUIS, R. DOYON, F. CHAGNON AND D. NADEAU

Département de Physique, Université de Montréal
and Observatoire du Mont Mégantic
C.P. 6128, Succ. Centre Ville,
Montréal (Qc), H3C 3J7, Canada

Abstract. We report the first detection of molecular hydrogen emission in the vicinity of a Wolf-Rayet star and nebula. The spatial distribution of the excited molecular gas is filamentary and is not correlated with the distribution of the ionised gas as traced by optical emission lines. The typical H_2 surface brightness in the filaments is 5×10^{-5} ergs s^{-1} cm^{-2} str^{-1}. We demonstrate that the excitation mechanism can be shocks or fluorescence from the strong ultraviolet flux of the WR star.

Key words: WR Nebula – Molecular Hydrogen – Interstellar medium

1. Introduction

Wolf-Rayet (WR) stars, the descendants of the most massive O stars, have very dense stellar winds ($\dot{M} = 0.8 - 8.0 \times 10^{-5}$ $M_\odot$/yr; Abbott & Conti 1987) with high terminal velocities (1000 – 3000 km/s; Prinja, Barlow & Howarth 1990). Therefore, they have a significant impact on the surrounding interstellar medium. Not only do they supply enriched material to the interstellar gas, they also provide a large amount of energy and momentum.

Perhaps the most spectacular examples of the interaction of the winds of these hot stars with the interstellar medium are WR nebulae which have been classified in three different types: diffuse H II regions, ejecta-type nebulae and wind-blown bubbles. NGC 2359, which was one of the first three WR nebulae to be identified by Johnson & Hogg (1965), is the prototype wind-blown bubble. Recent abundance studies by Esteban *et al.* (1990) have shown that the nebula has, at most, only very little chemical enrichment.

NGC 2359 consists, in fact, of two distinct parts. The first, is a U-shaped diffuse H II region that was most likely shaped by the O-star ancestor of the WR star. Within this region which can be detected in the light of [SII] and [NII], lies the filamentary bubble blown by the wind of the WN4 star HD 56925 (=WR 7). This spherically shaped nebula is colliding with the diffuse H II region on its eastern side and can clearly be observed in the light of Hα or [OIII]. On the eastern side of this complex lie three molecular clouds identified by Schneps *et al.* (1981). The first borders the southern part of the U-shaped nebula and is observed to have the same velocity as the ambient interstellar medium around the nebula (V_{LSR}=55 km/s). The second bounds the bubble and the U-shaped nebula on their eastern side

Astrophysics and Space Science **224**: 271–274, 1995.
© 1995 *Kluwer Academic Publishers.*

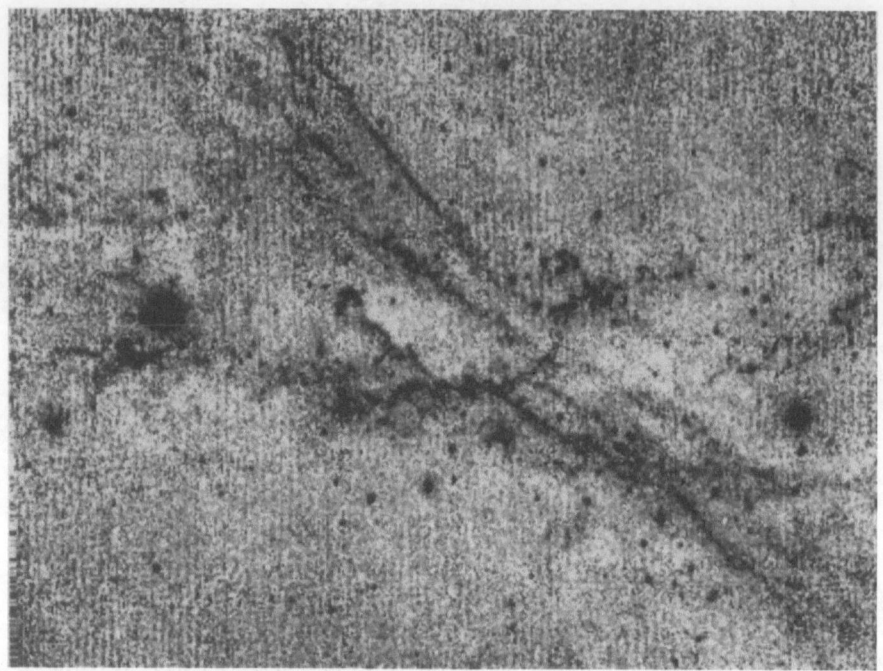

Fig. 1. Mosaic of H_2 emission in the vicinity of the WR nebula NGC2359. In this figure, North is to the top and East is to the left. This image is $4.5' \times 3.4'$.

and is moving at $V_{LSR}=37$ km/s. It is therefore thought to be compressed and accelerated by the wind-blown bubble. Finally the third cloud has a velocity of $V_{LSR}=67$ km/s and is probably foreground or background.

2.   Observations

We observed NGC 2359 in September/October 1993 with the Canada-France-Hawaii telescope on Mauna Kea, Hawaii, using the MONtreal Infrared CAmera (MONICA; Nadeau *et al.* 1994). With the setting used for the observations, the pixel scale of the detector was 0.25″ and therefore the 256×256 pixels of the camera corresponded to a field of view of approximately $1' \times 1'$. A preliminary scan of the entire $5' \times 7'$ nebula was made using a narrow band ($\Delta\lambda/\lambda = 1\%$; $\lambda_c=2.122\mu$m) H_2 filter. Emission was discovered in the southern part of the diffuse U-shaped H II region. Further observations were concentrated in this area. In total, five 60-second scans were made with a total area covered of $4.5' \times 3.4'$. Approximately 100 images were combined to form the final mosaic which is shown in Figure 1.

3. Results

Figure 1 shows that the excited H$_2$ gas in the vicinity of NGC 2359 lies in dense filaments. These do not overlap with Hα or [OIII] emission but rather are situated on the border of the ionised gas or in regions of very low emission. The typical surface brightness in the filaments is 5×10^{-5} ergs s^{-1} cm^{-2} str^{-1} with maximum values reaching 30×10^{-5} ergs s^{-1} cm^{-2} str^{-1}. This is comparable to what has been detected in planetary nebulae and supernova remnants (Graham et $al.$ 1987, 1993). The integrated flux in the brightest filament within a rectangular aperture of $35'' \times 9''$ is approximately 0.15 L$_\odot$ and we estimate the total H$_2$ luminosity for this region to be ~ 1 L$_\odot$.

The Hα/[NII] and Hα/[SII] line ratios clearly show that the ionised gas at various positions in the nebula, including the position where the H$_2$ was detected, is photoionised rather that being shock-excited ($cf.$ Goudis et $al.$ 1994). It is therefore natural to wonder whether fluorescence by UV photons from the WR star could excite the H$_2$ gas.

Black & van Dishoeck (1987) have calculated detailed models of fluorescent excitation of H$_2$ in interstellar clouds. In their Figure 5, they plot the total H$_2$ emission (I_{tot}) as a function of the ultraviolet scaling factor of the radiation field (I_{UV}) for clouds of various densities. From our observed flux in the 1–0 S(1) transition, we estimate the total flux emitted in the H$_2$ lines to be $I_{tot}=3 \times 10^{-3}$ ergs s^{-1} cm^{-2} str^{-1}, assuming model 14 of Black & van Dishoeck (1987). We have estimated the UV scaling factor of the WR radiation field as follows. From the number of ionising photons emitted by HD 56925, calculated by Esteban et $al.$ (1993), we have estimated the corresponding number of non-ionising photons capable of exciting H$_2$ (N_{UV}; between 912–1108 Å) to be roughly 8×10^{48} s^{-1}, from star properties derived from Kurucz (1979) model atmospheres ($cf.$ Puxley et $al.$ 1990). Adopting a distance of 5 kpc for the star we estimate the distance between the star and the H$_2$ gas (d) to be approximately 5 pc. We therefore estimate the UV scaling factor to be $I_{UV}=310(N_{UV}/10^{48}$ s$^{-1})(d/pc)^{-2} \sim 100$ assuming a mean background of 2.7×10^{11} photons s^{-1} m^{-2} measured in the solar neighborhood (Draine 1978). These values do not in fact correspond to any valid cloud model presented in Figure 5 of Black & van Dishoeck (1987). However, with a modest geometrical enhancement factor of ~ 5 for the H$_2$ emission, we can easily explain the observed flux with an interstellar cloud with a density of approximately 1000 cm^{-3}, which is close to the value estimated by Schneps et $al.$ (1981) from his CO observations. Therefore UV fluorescence is a valid candidate for the excitation mechanism of the observed molecular gas.

But if fluorescence is indeed the excitation mechanism, one can then wonder why the gas is distributed in filaments instead of having a more uniform distribution. In fact, the filamentary structure is reminiscent of what is observed for supernova remnants (i.e. Graham *et al.* 1991) in which shocks are identified as the excitation mechanism of the molecular gas. Following Graham *et al.* (1991) and using a typical density of 10^3 cm^{-3}, slow shock velocities of $\sim$ 15 km/s and a small geometrical enhancement factor of $\sim$ 2.5, we estimate that a radiative shock is indeed capable of reproducing the observed 1–0 S(1) H$_2$ flux. However, shocks should also produce considerable levels of [FeII]λ1.644 μm emission which we have not detected in CVF observations of one filament. This is a surprising result as the ratio of [FeII]/1-0 S(1) is usually much greater that unity (i.e. Graham *et al.* 1987) for shocks while, in our case, we estimate it to be smaller that 0.14 (3σ). One possible explanation for the lack of [FeII] emission is the presence of the strong UV flux from the WR star which could prevent the recombination of Fe behind the shocks, in which case we should expect significant emission from [FeIII].

4. Conclusions

We have presented IR line-emission observations of the WR nebula NGC 2359. We report the first detection of emission from H$_2$ gas the vicinity of a WR star and nebula. We have shown that either ultraviolet fluorescence from the strong radiation field of the star or shock excitation can explain the observed fluxes. Spectroscopic observations of several H$_2$ transitions are required to discriminate between both mechanisms and potentially model the shocks.

References

Abbott, D.C., & Conti, P.S.: 1987, Ann. Rev. of Astron. & Astrophys. **25**, 113.
Black, J.H., & van Dishoeck, E.F.: 1987, Astrophys. J. **322**, 412.
Draine, B.T.: 1978, Astrophys. J., Suppl. **36**, 595.
Esteban, C., Vílchez, J.M., Manchado, A. & Edmunds, M.G.: 1990, Astron. & Astrophys. **227**, 515.
Esteban, C., Smith, L.J., Vílchez, J.M., & Clegg, R.E.S.: 1993, Astron. & Astrophys. **272**, 299.
Goudis, C.D., Christopoulou, P.-E., Meaburn, J., & Dyson, J.E.: 1994, Astron. & Astrophys. **285**, 631.
Graham, J.R., Wright, G.S., & Longmore, A.J.: 1987, Astrophys. J. **313**, 847.
Graham, J.R., Wright, G.S., Hester, J.J., & Longmore, A.J.: 1991, Astron. J. **101**, 175.
Graham, J.R., Serabyn, E., Herbst, T.M., Matthews, K., Neugebauer, G., Soifer, B.T., Wilson, T.D., & Beckwith, S.: 1993, Astron. J. **105**, 250.
Johnson, H.M., & Hogg, D.E.: 1965, Astrophys. J. **142**, 1033.
Kurucz, R.L.: 1979, Astrophys. J., Suppl. **40**, 1.
Nadeau, D., Murphy, D.C., Doyon, R., & Rowlands, N.: 1994, Publ. of the ASP **106**, 909.
Prinja, R.K., Barlow, M.J., & Howarth, I.D.: 1990, Astrophys. J. **361**, 607.
Puxley, P.J., Hawarden, T.G., & Mountain, C.M.: 1990, Astrophys. J. **364**, 77.
Schneps, M.H., Haschick, A.D., Wright, E.L. & Barrett, A.H.: 1981, Astrophys. J. **243**, 184.

THE PRODUCTION OF CIRCUMSTELLAR ^{26}AL BY
MASSIVE STARS

N. LANGER, H. BRAUN and J. FLIEGNER

MPI für Astrophysik, D-85740 Garching, F.R.G.

Abstract. We investigate the production of ^{26}Al during hydrogen burning and its ejection by massive single and binary stars. Effects of convection and rotation are studied. We discuss the importance of RSGs, LBVs and WR stars to the total Galactic ^{26}Al production, and the detection probability of the ^{26}Al decay in individual objects as P Cygni, γ Velorum and η Carinae.

Key words: stellar evolution – massive stars – mass loss – γ-ray lines

1. Introduction

According to current models, Galactic very massive stars are believed to lose the major fraction of their initial mass in the form of stellar winds during core hydrogen and helium burning (*cf.* Schaller *et al.* 1992, Bressan *et al.* 1993, Woosley *et al.* 1993, Meynet *et al.* 1994, Langer *et al.* 1994). For example, a star with an initial mass of 60 $M_\odot$ is expected to end up with roughly $4-6\ M_\odot$ at the time of core collapse.

Since massive stars possess large convective cores, the loss of the major part of their initial mass implies that the ejected winds must be enriched by nuclear processed material. Again as an example (*cf.* Langer *et al.* 1994), a 60 $M_\odot$ star has initially a convective core of $\sim 43\ M_\odot$. Therefore, as soon as its actual mass decreases below 43 $M_\odot$, its wind will contain the nuclear products of core hydrogen burning. Due to the operation of the CNO cycle (*cf.* Arnould & Mowlavi 1993) the wind will thus be enriched in helium and nitrogen. However, at sufficiently high temperatures, which are actually obtained in massive star cores, also the NeNa- and the MgAl-cycles (*cf.* again Arnould & Mowlavi 1993) are partly activated, resulting in the production of Sodium, and — the key issue of the present paper — ^{26}Al . ^{26}Al is a radioactive isotope with a half life of $7.2\,10^5$ yr, and it β-decays to ^{26}Mg with the emission of a 1.809 MeV γ-photon.

This topic is of particular interest since the 1.809 MeV emission line of the ^{26}Al decay has been detected by the HEAO-C spacecraft (Mahoney *et al.* 1984), and currently detailed maps of the 1.8 MeV emission in the Galaxy are produced with the COMPTEL observatory on board the GRO satellite (Diehl *et al.* 1994). It is most intriguing that these maps show a very clumpy structure, pointing towards a young population of ^{26}Al producing objects, and moreover that even individual sources may be identified (Oberlack *et al.* 1994).

Astrophysics and Space Science **224**: 275–278, 1995.
© 1995 *Kluwer Academic Publishers.*

In the present paper, we report on modelling the ^{26}Al production during hydrogen burning and its ejection by massive stars, and investigate the dependence of the corresponding ^{26}Al yields from physical parameters as initial stellar mass and metallicity, stellar rotation, and the presence of a close binary companion. We want to point out that the detection of ^{26}Al emission from single stellar sources could provide an accurate tool to evaluate the enormous mass loss predicted by current evolutionary models for very massive stars. Finally, massive star ^{26}Al yields are required to arrive at an understanding of the total extended Galactic emission.

2.   Circumstellar ^{26}Al around massive stars

In Table 1 we summarize the results obtained from stellar evolution calculations for stars in the initial mass range $15-140\ M_\odot$, for various metallicities, and assuming either the Schwarzschild criterion for convection ("$\alpha_{sc} = \infty$") or semiconvection ("$\alpha_{sc} = 0.04$"; cf. Langer 1991). We used mass loss rates of Nieuwenhuijzen & de Jager (1990) and Langer (1989), OPAL opacities (Iglesias et al. 1992), and the ^{25}Mg(p,γ)^{26}Al-rate of Iliadis et al. (1990). Note that only the models at $Z = Z_\odot$ with $M_i \geq 40\ M_\odot$ evolve into Wolf-Rayet stars, while the others eject ^{26}Al either in a RSG stage (seq. # 2,3) or in the supernova explosion (seq. # 1, 9, 10, 11). Seq. #5 has a RSG stage preceding the final Wolf-Rayet phase.

From Table 1 we can see that the maximum circumstellar ^{26}Al mass obtained during the evolution increases steeply with increasing initial mass and metallicity. While semiconvection reduces the obtained yield by roughly a factor of 2 compared to plain convection according to the Schwarzschild criterion, convective core overshooting may reduce it even more (cf. also Prantzos 1991, Meynet & Arnould 1993).

Table 1 also contains two sequences where rapid rotation and thereby induced internal mixing processes are self-consistently taken into account (cf. Fliegner & Langer 1994). The initial equatorial rotation velocities were $420\ \mathrm{km\,s^{-1}}$ for seq. #14, and $430\ \mathrm{km\,s^{-1}}$ for seq. #15. Both sequences evolve into Wolf-Rayet stars during core hydrogen burning (cf. Maeder 1987, Langer 1992), and the total ^{26}Al yield as well as the maximum circumstellar amount is considerably increased.

Finally, two primary components of case B close binaries are considered in Table 1 (seq. #16 and 17). Note that M_{tot}, M_{wind}^{max}, and M_{wind}^{SN} are only upper limits in these cases, obtained by assuming that none of the ^{26}Al lost by the primary is accreted by the secondary star (i.e. "$\beta = 0$"; cf. Braun & Langer 1994). With this assumption, very massive primaries ($M_i \gtrsim 50\ M_\odot$) produce exactly the same amounts as single stars (cf. Langer 1994), while primaries with lower initial masses — which would evolve through a RSG phase as single stars — show a slightly increased ^{26}Al production.

TABLE I

^{26}Al yields of various stellar sequences. All numbers refer to ^{26}Al produced during hydrogen burning only, *i.e.* contributions from neon burning and to explosive processing during the SN outburst are ignored! M_{tot} represents the ^{26}Al mass corresponding to the total number of 1.8 MeV photons produced in the circumstellar medium during hydrostatic evolution and after the SN explosion. f_{IMF} is $M_{tot}/3.0\,10^{-6}\,M_\odot$ times $(15\,M_\odot/M_i)^{2.35}$. M_{wind}^{max} is the maximum circumstellar ^{26}Al mass occurring before SN explosion. M_*^{SN} is the amount of ^{26}Al present in the star at the time of core collapse, and M_{wind}^{SN} is the amount present in the circumstellar medium at that time. d_C and d_I give the maximum distances for which the considered stars would be detectable with COMPTEL (assuming detection threshold of $2.5\,10^{-5}\,s^{-1}\,cm^{-2}$) and INTEGRAL (assuming threshold of $5.0\,10^{-6}\,s^{-1}\,cm^{-2}$), respectively, adopting the circumstellar ^{26}Al mass as M_{wind}^{max}.

#	M_i $M_\odot$	Z %	α_{sc}	M_{tot} $M_\odot$	f_{IMF}	M_{wind}^{max} $M_\odot$	M_*^{SN} $M_\odot$	M_{wind}^{SN} $M_\odot$	d_C pc	d_I pc
1	15	2	0.04	3.0(-6)	1.0	0.0	3.0(-6)	0.0	-	-
2	25	2	0.04	1.1(-5)	1.1	5.1(-7)	1.1(-5)	5.1(-7)	16	35
3	30	2	0.04	1.7(-5)	1.1	3.0(-6)	1.3(-5)	3.0(-6)	38	87
4	40	2	0.04	6.0(-5)	2.0	5.0(-5)	0.0	4.9(-5)	160	360
5	50	2	0.04	1.4(-4)	2.8	1.2(-4)	0.0	1.0(-4)	240	540
6	140	2	0.04	1.0(-3)	1.7	8.0(-4)	0.0	8.0(-4)	620	1400
7	25	2	∞	2.1(-5)		2.0(-6)	1.9(-5)	2.0(-6)	31	90
8	30	2	∞	4.1(-5)		9.0(-6)	3.1(-5)	9.0(-6)	66	150
9	20	0.2	0.04	6.2(-8)		0.0	6.2(-8)	0.0		
10	25	0.2	0.04	1.3(-7)		0.0	1.3(-7)	0.0		
11	40	0.2	0.04	1.1(-6)		0.0	1.1(-6)	0.0		
12	25	4	0.04	1.7(-5)		3.1(-6)	1.3(-5)	3.1(-6)		
13	50	2	over.	4.3(-5)		3.4(-5)	0.0	3.2(-5)	130	280
14	20r	2	∞	2.2(-5)		1.2(-5)	0.0	1.2(-14)	77	170
15	40r	2	∞	1.6(-4)		1.4(-4)	0.0	4.6(-5)	260	590
16	40b	2	0.04	6.7(-5)		5.6(-5)	0.0	4.1(-5)	170	370
17	50b	2	0.04	1.4(-4)		1.2(-4)	0.0	1.0(-4)	240	540

3. Observational consequences

The last two columns of Table 1 display the maximum distance a considered star may have in order to be detectable in the light of the 1.8 MeV decay line of ^{26}Al. This means that the possible detection of γ Vel with COMPTEL (Oberlack *et al.* 1994) may be marginally consistent with our models, especially since its initial mass is potentially well above 50 $M_\odot$ (Braun & Langer

1994). Our $140\,M_\odot$ model which fits to $\eta\,$Car in many respects yields a maximum of $8\,10^{-4}\,M_\odot$ of circumstellar ^{26}Al, predicting $\eta\,$Car as individual object to be undetectable with COMPTEL.

The relative contributions of our models # 1–6 to the Galactic enrichment with ^{26}Al are indicated by the quantity f_{IMF} (see table caption). The insensitivity of f_{IMF} to the initial mass indicates that also stars with the lower considered masses, *i.e.* RSG winds and RSG envelopes ejected during the SN explosion, can not be neglected. The relative importance of the numbers shown in Table 1 depends on the amounts of ^{26}Al which are additionally produced in massive stars during neon burning and the SN explosion; these appear to be presently much more uncertain than the amounts produced from hydrogen burning (*cf.* Weaver & Woosley 1993, their Table 11).

Acknowledgements. We are grateful to R. Diehl and U. Oberlack for illuminating discussions and helpful comments. This work has been supported by the Deutsche Forschungsgemeinschaft through grant La 587/8-1.

References

Arnould M. & Mowlavi N.: 1993, in: *Inside The Stars*, eds: W.W. Weiss *et al.* , Proc. IAU-Colloq. No. 137, ASP Conf. Ser. **40**, p310.

Braun H. & Langer N.: 1994, in: *IAU-Symp. No. 163*, eds: K.A. van der Hucht & P.M. Williams, Kluwer, p305.

Bressan A., Fagotto F., Bertelli G.& Chiosi C.: 1993, Astr. & Astrophys. Suppl. **100**, 647.

Diehl R., *et al.* : 1994, Astron. & Astrophys. in press.

Fliegner J. & Langer N.: 1994, in: *IAU-Symp. No. 163*, eds: K.A. van der Hucht & P.M. Williams, Kluwer, p326.

Iglesias C.A., Rogers F.J. & Wilson B.G.: 1992, Astrophys. J. **397**, 717.

Iliadis Ch., *et al.* : 1990, Nucl. Phys. A **512**, 509.

Langer N.: 1989, Astron. & Astrophys. **220**, 135.

Langer N.: 1991, Astron. & Astrophys. **252**, 669.

Langer N.: 1992, Astron. & Astrophys. **265**, L17.

Langer N.: 1994, in: *IAU-Symp. No. 163*, eds: K.A. van der Hucht & P.M. Williams, Kluwer, p15.

Langer N., Hamann W.-R., Lennon M., Najarro F., Pauldrach A. & Puls J.: 1994, Astron. & Astrophys. in press.

Mahoney W.A., Ling J.C., Wheaton W.A. & Jacobsen A.S.: 1984, Ap. J. **286**, 578.

Maeder A.: 1987, Astron. & Astrophys. **178**, 159.

Meynet G. & Arnould M.: 1993, in: *Origin and Evolution of the Elements*, eds: N. Prantzos *et al.* , CUP, p540.

Meynet G., Maeder A., Schaller G., Schaerer D. & Charbonnel C.: 1994, Astron. & Astrophys. Suppl. **103**, 97.

Nieuwenhuijzen H. & de Jager C.: 1990, Astron. & Astrophys. **231**, 134.

Oberlack U., Diehl R., Montmerle T., Prantzos N. & von Ballmoos P.: 1994, Astrophys. J., Suppl. in press.

Prantzos N.: 1991, in: *Gamma Ray Line Astrophysics*, eds: N. Prantzos & P. Durouchoux, New York: AIP, p129.

Schaller G., Schaerer D., Meynet G. & Maeder A.: 1992, Astr. & Astrophys. Suppl. **96**, 269.

Weaver T.A. & Woosley S.E.: 1993, Phys. Rep. **227**, 65.

Woosley S.E., Langer N. & Weaver T.A.: 1993, Astrophys. J. **411**, 823.

SESSION G:

AGB STARS, PROTO-PLANETARY NEBULAE & PLANETARY NEBULAE

THE PHYSICAL AND CHEMICAL PROPERTIES OF

CIRCUMSTELLAR ENVELOPES

P. J. HUGGINS

Physics Department, New York University
4 Washington Place, New York, NY 10003, U.S.A.

Abstract. An overview is given of recent results on some basic properties of the circumstellar envelopes of evolved stars. The focus is on well studied examples which illustrate envelopes of different types and at different stages of evolution. The close connection between the physical and chemical properties of the envelopes is emphasized.

Key words: Mass-Loss – AGB Stars – IRC+10216 – α Ori – Planetary Nebulae

1. Introduction

Mass-loss by stars on the AGB is now established to be a key phase in the transition from red giants to white dwarfs. The mass-loss forms an extended ($R \gg R_*$) envelope of dust and gas around the star, with a typical expansion velocity $V_e \approx 5$–30 km s^{-1} and a typical mass loss rate $\dot{M} \approx 10^{-7}$–10^{-5} $M_\odot$ yr^{-1} (see, *e.g.*, Loup *et al.* 1993). Unless there is a chromosphere or a hot companion, the dense gas in the inner envelope is usually in the form of molecules or neutral atoms, but as the gas flows away from the star the molecules are eventually photo-dissociated and most atomic species are photo-ionized by the interstellar radiation field. Once formed, the envelopes undergo changes, especially as a result of the evolution of the central star, and during the post-AGB phase their character is completely altered by the onset of planetary nebula (PN) formation. A similar kind of mass-loss is also experienced by the coolest supergiants and contributes to the environment in which the star explodes when it becomes a supernova.

A major part of our understanding of these phases of mass-loss comes from observations of molecular and atomic line emission from the neutral circumstellar gas. In this paper we review some basic physical and chemical properties of the envelopes which are inferred from the observations, mainly with reference to well studied examples. We first describe recent observations and models for the archetypes IRC+10216 and α Ori, and comment on the limitations of our current picture. We also briefly discuss other AGB envelopes, and the effects of evolution in the post-AGB phase. More detailed treatment of several of the topics included here can be found in recent reviews by Omont (1993), Olofsson (1993, 1994), and Huggins (1993a,b).

Astrophysics and Space Science **224**: 281–292, 1995.

2. Archetypes

IRC+10216 (CW Leo) and α Ori (Betelgeuse) play important roles as archetypes in the study of mass loss since their circumstellar envelopes are among the best observed. IRC+10216 is probably the nearest carbon star (spectral type C 9,5), α Ori is the nearest supergiant (M2 Iab), and they are at approximately the same (uncertain) distance of 100–200 pc; at the latter distance $1'' = 3 \times 10^{15}$ cm. Their envelopes are very different in character and so illustrate a broad range of circumstellar processes.

2.1   IRC+10216

The circumstellar envelope of IRC+10216 exhibits a rich carbon chemistry with $\approx$ 44 molecular species now detected at infrared, millimeter, or radio wavelengths; for a list see Table 1 in Omont (1993) and add NaCN and MgNC. CO emission is seen out to $\approx$ 200" from the star (Huggins et al. 1988), but other species are not seen beyond $\approx$ 30" which is believed to be partly a consequence of their destruction by the interstellar radiation field. The large extent of CO is explained by its effective line self-shielding (Mamon et al. 1988).

The model by Kwan & Linke (1982) or variations of it has been widely used to describe the physical conditions in the circumstellar gas of IRC +10216. The model is based on a number of simplifying assumptions including spherical symmetry and a continuous, constant mass-loss rate, and it involves a self consistent calculation of the thermal structure of the gas. The kinetic temperature is governed primarily by adiabatic expansion, heating by dust-gas collisions, and line cooling by CO; recent discussions are given by Kwan & Webster (1993) and Groenewegen (1994). The emission in the millimeter CO lines depends on the CO abundance and the mass-loss rate, and is coupled to the kinetic temperature by collisions, so the comparison of observed and calculated CO lines constrains the model. The model found by Kwan & Linke has parameters $\dot{M} = 4 \times 10^{-5} M_\odot \, \text{yr}^{-1}$ and CO/H_2 $= 6 \times 10^{-4}$, with $T \approx 70 \, r_{16}^{-0.7}$ K and $n(H_2) = 4 \times 10^5 \, r_{16}^{-2}$ cm^{-3}, where r_{16} is in units of 10^{16} cm. The value for $\dot{M}$ depends on the adopted distance, and is actually close to the upper limit determined by Keady & Ridgway (1993) from their failure to detect absorption in the $v = 1 - 0$ S(1) line of H_2 at 2.12 μm. The model fails to account for the very extended CO emission seen out to 200" although the situation is somewhat improved by including heating effects of the interstellar radiation field (Huggins et al. 1988). However the model also fails in the inner envelope by underestimating the emission from high lying CO lines; this may be accommodated by appealing to changes in the mass loss rate (Sahai 1987, Truong-Bach et al. 1991) at the expense of introducing arbitrary parameters into the model. The uncertainties appear to be quite large: e.g. , at a few arc seconds from the star the temperature estimated by Truong-Bach et al. is ≈ 1000 K, a factor of 10 larger than that in the original model. In spite of these shortcomings, this model has been widely used to interpret observations of other envelopes, and as a basis for chemistry calculations.

There are currently a number of chemical models of IRC+10216 which are detailed developments of the layered carbon-rich chemistry outlined by Lafont *et al.* (1982) and Huggins & Glassgold (1982). Close to the star the high densities maintain abundances in thermodynamic equilibrium. In the inner wind, dust condensation takes place and gas phase abundances are frozen out in the flow. Farther out ($R \gtrsim 10^{16}$ cm), the shielding of external radiation decreases with the result that molecules are photo-ionized or photo-dissociated, forming a photochain of daughter species which terminates with the atoms and ions characteristic of the unshielded interstellar medium. In addition, secondary reactions within this region can lead to the synthesis of additional species through ion-molecule and fast neutral reactions (Cherchneff & Glassgold 1993, Howe & Millar 1990, and references therein).

The idea of a layered chemistry in IRC+10216 has been strikingly confirmed by high resolution imaging of millimeter emission from a number of molecular species (*e.g.* , Lucas 1993, Bieging & Tafalla 1993, and references therein). Maps of molecules present in the inner wind flow are centrally peaked, and are limited by photo-dissociation (*e.g.* , HCN) or depletion onto grains (*e.g.* , SiS); daughter species (*e.g.* , C_2H and CN) of abundant polyatomic molecules in the envelope (*e.g.* , C_2H_2 and HCN) form shells, at $R \approx 15''$, as do other species believed to be formed from secondary reactions in the photo-dominated region. The observed size of the photo-region is consistent with our understanding of the dust opacity in the envelope, and the abundances of quite a number of species can be matched by the models to within an order of magnitude, or better.

Several other developments add further insights into the envelope:

• HCO^+ and other molecular ions play a key role in the chemistry of interstellar clouds and their presence in IRC+10216 is of considerable interest. The early chemical models of Glassgold *et al.* (1987) and Nejad & Millar (1987) suggested that HCO^+ might be observable as a result of chemistry in the envelope initiated by cosmic ray ionization, and subsequently the $J=1-0$ line was tentatively detected by Lucas & Guelin (1990) with the IRAM 30-m telescope. More recently, Avery *et al.* (1994) found a line at the frequency of the $J=4-3$ line of HCO^+ but failed to find the $J=3-2$ line at the expected level. They conclude that neither of the tentatively assigned lines arise from HCO^+, but this needs to be checked. In any event, HCO^+ has an extremely low abundance in the envelope, and to account for this the chemical models of Cherchneff *et al.* (1993) require that the cosmic ray ionization rate in the envelope is lower than the rate usually used in models of interstellar clouds by at least a factor of 10. This may provide the first observational evidence for the exclusion of cosmic rays from circumstellar envelopes, which has been suggested by Tielens (1993).

• MgNC and NaCN are two recently detected molecules in IRC +10216 (Kawaguchi *et al.* 1993, Turner *et al.* 1994) which add to the list of metal bearing species in the envelope: NaCl, AlCl, KCl, and AlF (Cernicharo & Guelin 1987). NaCl (Lucas 1993) shows emission only within a few arc seconds of the star which suggests it is depleted onto grains farther out, and this is probably true of the other halides. MgNC, however, has recently been mapped by Guelin *et al.* (1993) and shows a shell structure of the same dimensions as other species in the photo-region. A shell structure for MgCN can reasonably be expected if MgCN forms via reactions of Mg^+ with HCN as suggested by Kawaguchi *et al.* (1993), since it will preferentially form in the narrow region where these species overlap in the HCN–CN and Mg–Mg^+ photochains. However, Guelin *et al.* (1993) have argued that the close coincidence of the shell sizes of several species points to a faster production mechanism like the release from grain surfaces.

• The 609 μm $^3P_1 \rightarrow ^3P_0$ fine structure line of neutral carbon has recently been detected in IRC+10216 by Keene *et al.* (1993) using the CSO, and affords important checks on the physical and chemical properties of the envelope. The on-source spectrum is shown in Figure 1. The double spiked profile shows that the C I is in a shell structure resolved by the telescope beam (15″), and this is confirmed by mapping data. The shell distribution is qualitatively in accord with the expectation that the C I is produced in the envelope from the photo-dissociation of carbon bearing molecules, but quantitative comparison with the chemical model of Cherchneff *et al.* (1993) showed that the model underestimated the line intensity by a factor of $\approx$50. Reasonable changes in the model parameters, however, can bring the C I abundance much closer into line with the observations (Glassgold & Huggins 1994) although a complete model fit for all species is problematic.

2.2 α Ori

The circumstellar envelope of α Ori is quite different from that of IRC+10216 in several respects. There is a deficiency of dust (*e.g.* , Skinner & Whitmore 1987) and of molecular gas; CO is the only molecular species detected so far in the envelope, and it appears to be under-associated (Huggins 1987). The mass loss rate is also about one order of magnitude lower than IRC+10216, and the O/C ratio is > 1. The lack of molecular gas and dust is presumably due to the presence of a warm 8000 K chromosphere (see, *e.g.* , Hartmann & Avrett 1984) which inhibits molecule formation in the immediate vicinity of the star, and is the source of intense UV radiation which has an influence over a large part of the envelope. Details of the interface between the chromosphere and the expanding envelope are still obscure, but progress is

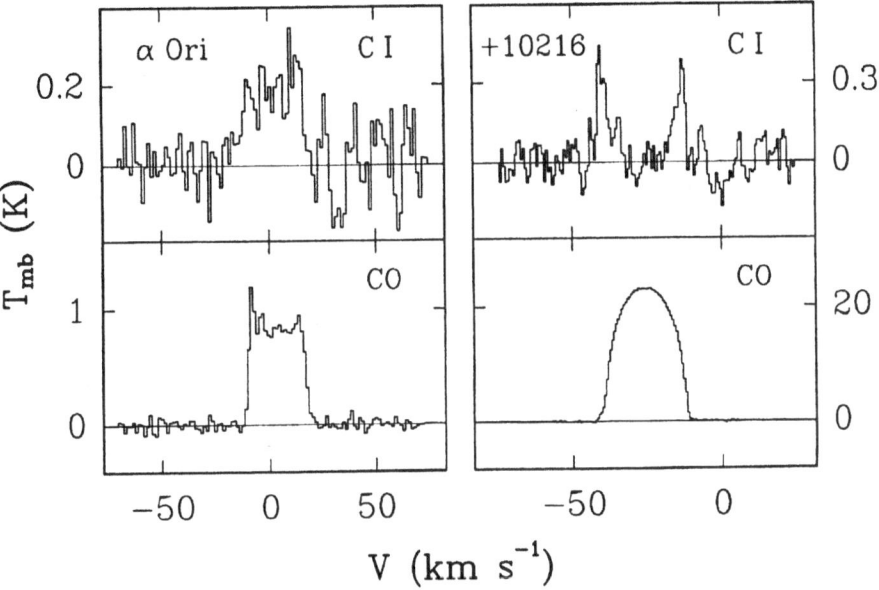

Fig. 1. Spectra of the C I ($^3P_1 \rightarrow {}^3P_0$) and CO ($J=2\rightarrow1$) lines in IRC+10216 and α Ori; the data are from Keene *et al.* (1993) and Huggins *et al.* (1994a), respectively.

being made in modeling this region (*e.g.* , Beck *et al.* 1992).

The detection of H I 21 cm emission in α Ori (Bowers & Knapp 1987) confirms the importance of atomic gas in the envelope, and provides an estimate for the mass-loss rate of $2\times 10^{-6} M_\odot \, \mathrm{yr}^{-1}$, for d=200 pc. The gas temperature has been modelled by Rodgers & Glassgold (1991) along the lines described earlier for IRC+10216, and they find that the temperature drops rapidly from chromospheric temperatures down to $\approx$ 100 K within 2×10^{15} cm of the star. The O I 63 μm fine structure line is an important contributor to the cooling in these inner regions, and a detection of this line at the predicted level has recently been reported (Haas & Glassgold 1993; however, see also Danchi *et al.* 1994). In the extended envelope, the ionization structure has been modelled by Glassgold & Huggins (1986), and in combination with the temperature model and mass-loss rate given here, provides a consistent account of the K I resonance line scattering which is seen out to $\approx$ 60″ (Mauron 1990). The excitation temperatures seen in CO in absorption in the infrared (Bernat *et al.* 1979) and UV (Wahlgren *et al.* 1992) range from 70–500 K, but the observed levels are probably radiatively excited.

The 609 μm C I line has recently been detected in α Ori (Huggins *et al.* 1984a) and provides useful information on the properties of the envelope as in the case of IRC+10216. The single component spectrum shown in Fig. 1

indicates that the C I emission is found closer to the star than in IRC+10216 and it is much stronger relative to CO. In fact the C I abundance in α Ori is larger than that of CO, and it appears to be the primary C-bearing species formed in the inner envelope by recombination of the chromosphere. A model of the emission using the photospheric C/H ratio provides an estimate of the mass loss rate of $2 \times 10^{-6} M_\odot \, yr^{-1}$, in good agreement with the value obtained from the H I observations. The large amount of C I in the envelope also provides a ready explanation of the anomalous behavior of the C I UV 2 multiplet near 1657 Å seen in GHRS spectra obtained with the Space Telescope (Carpenter *et al.* 1994). This resonance multiplet is seen in absorption in α Ori whereas other lines arising from the same upper state are seen in emission from the chromosphere. This can be accounted for by absorption of the resonance multiplet by ground state C I in the extended envelope. In fact the large amount of C I observed suggests that there will also be a continuum discontinuity in the spectrum at the C I ionization limit of $\lambda = 1100$ Å due to absorption in the envelope.

2.3   COMMENTS ON THE CURRENT PICTURE

All the model interpretations of the observations mentioned so far have been made on the assumptions that the envelopes are homogeneous and spherically symmetric, but it is important to examine to what extent this is a valid description. The envelope of α Ori has mild asymmetries on the large scale seen in K I (Mauron *et al.* 1984), CO shows two velocity components in absorption (Bernat *et al.* 1979), and the dust emission is time variable close to the star (*e.g.*, Danchi *et al.* 1994). IRC+10216 is better documented. It has long been known to have an elongated infrared core on arcsecond scales (*e.g.*, Kastner & Weintraub 1994, and references therein) and the deep V-band image by Crabtree *et al.* (1987) shows that the extended envelope is slightly flattened along the core axis. This image also dramatically reveals the presence of multiple, fragmented shell structures on scales of 5″–30″. It is perhaps, therefore, not too surprising that the most recent, high resolution molecular images are beginning to show similar effects. Maps in MgNC and C_4H by Guelin *et al.* (1993) reveal that the shells have openings along the core axis, and are fragmented on the size scale of the observations ($\lesssim 5''$); the mass scale of these structures is $\approx 5 \times 10^{-4} M_\odot$, and if typical there will be $\gtrsim 1000$ in the whole envelope. It is of interest to point out that the molecular "shell" approximately coincides with one of the dust shells seen in reflection in Crabtree's V–band image, so that a physical shell structure may in fact contribute to the appearance of the molecular images.

The presence of such structure has important consequences for physical and chemical models of the envelopes. Density inhomogeneities imply temperature inhomogeneities through changes in the heating and cooling, as well as changes in the radiative transport through the envelope. Relative to homogeneous models for IRC+10216, one can imagine that such a picture can more easily accommodate the main observations, such as the extended envelope seen in CO and the similarity of the shell at 15″ seen in many species. Such a picture is quite complicated, of course, but appears to be required by the observations. The importance of structure has been more readily recognized in some other AGB stars and PNe.

3.   AGB Envelopes

The properties of ordinary AGB envelopes are known in less detail than the examples given above, and for most observations they are spatially unresolved. There are a few extreme cases where the structures appear dominated by an equatorial density enhancement (e.g. , V Hya, Kahane et al. 1993), but CO maps of the nearby sample observed at resolutions $\gtrsim 12″$ by Bujarrabal & Alcolea (1991) mostly show roughly circular contours. However, as in IRC+10216, unresolved structure on small scales may be common. Recent observations in the scattered KI resonance line at arcsecond resolution by Plez & Lambert (1994) suggests that asymmetries may in fact be the norm. On the smallest size scales, the first, full, high-resolution (1 mas) images of the SiO masers in O-rich envelopes show that the masers form at a few stellar radii in small spots $\approx 2 \times 10^{13}$ cm in size (Diamond et al. 1994); for a density of 10^{10} cm^{-3} the mass scale of these regions in 10^{-7} $M_\odot$.

Several multi-line millimeter surveys have recently been carried out to systematically study chemical abundances in AGB envelopes for the small number of species which can be widely detected (e.g. , Bujarrabal et al. 1994, Olofsson et al. 1993, Bieging & Latter 1994). In general the line ratios behave roughly as expected with respect to the spectral type or photospheric abundances of the central star. For example, Olofsson et al. (1993) find the abundance of HCN in the envelopes of bright carbon stars correlates with the photospheric abundance (but with large systematic differences) and the average CN/HCN ratio is ≈ 0.6; this is consistent with the freeze out of HCN in the inner envelope and photo-production of CN farther out as in IRC+10216. In these cases, however, a detailed interpretation is not possible as the structure of the envelopes is not spatially resolved, but important progress is being made in this direction (see Lindqvist et al. , this volume).

One especially interesting class of envelopes are those with detached shells, and in some cases these can be resolved, even with single dish observations. The best studied is the carbon star S Sct which has been mapped in CO (see Olofsson et al. 1992, and references therein), and modelled by Eriksson & Stenholm (1993), Yamamura et al. (1993), and Bergman et al. (1993), with a reasonable consensus on the conditions in the shell. The main shell has radius $R=5\times 10^{17}$cm (68″) with $\Delta R \lesssim 1 \times 10^{17}$cm, and $V_e = 17$ km s^{-1}, which corresponds to an expansion time of 10^4 yr. The density

$n(H_2)$ is in the range 10^2–10^3 cm^{-3} with $T \approx 10$–100 K, although the lower temperatures are preferred if one tries to model the heating and cooling. If the shell is homogeneous, the observations force the shell to be geometrically very thin, but Bergman *et al.* (1993) have modelled the intensity variations across the shell, and their preferred model is one with a large number ($\approx$ 1000) of small clumps of mass $\approx 10^{-5}$ $M_\odot$ (although they are not individually resolved) which may have originated in the mass-loss process ($\dot{M} \gtrsim 10^{-5}$ $M_\odot$ yr^{-1}) which formed the shell. The presence of CO in the shell so far from the star is accounted for by line self-shielding, although other more fragile molecular species are expected to be dissociated much closer in. It is interesting to note that the physical and chemical conditions in the shell are not very different from those in the model by Huggins *et al.* (1988) for IRC+10216 at the same distance from the star (170″ at 200 pc). The clump model for S Sct is certainly in line with the clumpy picture of IRC+10216 sketched in §2.3.

4.   Post-AGB Envelopes

During the post-AGB phase circumstellar envelopes undergo fundamental changes in structure and chemistry. The cessation of heavy mass-loss means that homogeneous envelopes rapidly become transparent ($A_V \approx 800/t$ radially through the envelope for an object like IRC+10216, where t is the time in years after the mass-loss stops); the gas may be shocked and entrained by fast winds which appear most prominent in the proto-PN phase (although the statistics on this are still very incomplete); and the rising temperature of the central star eventually leads to photo-dissociation and photo-ionization of the envelope.

The two best studied examples in transition to the PN phase are CRL 2688 and CRL 618, which have central stars of spectral type F and B, respectively. Both have optical bipolar reflection nebulae, fast winds, and fairly intact AGB envelopes with a rich complement of molecular gas; CRL 618 also has a very compact ionized core. There is an extensive recent literature on these complex objects including diagnostic observations of the shocks induced by the fast winds (Cox *et al.* 1994a, Hora & Latter 1994, Shibata *et al.* 1993, Lucas 1993, Martin-Pintado *et al.* 1993, Latter *et al.* 1993a, Trammell *et al.* 1993, Neri *et al.* 1992), and space does not allow a detailed discussion here. Judging by the amount of molecular gas in these objects they are extreme cases likely to evolve into the class of PNe in which molecular gas survives for a significant part of the PN phase. The molecular gas seen in CO in more typical post-AGB envelopes is essentially destroyed by the time a compact nebula forms (Huggins *et al.* 1994b).

The structure in envelopes is accentuated during PN formation by the fast winds and by the ionizing radiation which will etch the structure in the neutral gas. The presence of massive toroids of neutral gas around the waists of bow-tie PNe and along the bright rims of elliptical PNe (Huggins 1993b, and references therein) is a natural consequence of the trapping of the ionization fronts in the densest neutral gas. A similar situation occurs on much smaller size scales, the best example being the well known globules in the Helix, which is one of the nearest PNe, and is probably typical of PNe of similar type. The globules have recently been shown to consist of dense clumps of molecular gas (Huggins et $al.$ 1992; Meaburn et $al.$ 1992). These clumps have sizes of $\approx 2 \times 10^{15}$ cm and masses of $\approx 10^{-5}$ $M_{\odot}$, and are in approximate pressure equilibrium with the ionized gas. Dyson et $al.$ (1989) have argued that these clumps originated in inhomogeneities in the AGB wind close to the star which give rise to the SiO maser spots, and survived the subsequent changes in environment. The mass scale in the maser spots may be too low (see §3), but there is also evidence for larger mass scales ($e.g.$, §2.3) so this general scenario is still very plausible.

The physical conditions in the neutral gas which survives into the PN phase have important differences from those in AGB envelopes. The intense UV radiation field from the star at $\lambda > 912$ Å penetrates beyond the ionization fronts and forms warm Photo-Dissociation Regions (PDRs) at the surface of the neutral gas. Models of PDRs have been extensively discussed in connection with interstellar clouds, and their relevance to PNe has been reviewed by Tielens (1993), although no detailed models have yet been made for C-rich gas. The best studied example is NGC 7027 where recent high resolution imaging (Cox et al. 1994b; Graham et $al.$ 1993) shows the detailed structure of the PDR gas in the near infra-red lines of H_2. These lines are thermalized at ≈ 1000 K in the warm PDR gas and reveal a well defined torus with extended loops of emission along the periphery of the ionized nebula. Other signatures of PDR gas such as C I and C II fine structure lines are also seen in NGC 7027 ($e.g.$, Young et $al.$ 1994) and indicate in this case that a relatively small fraction ($\approx 10\%$) of the neutral circumstellar gas is within the PDR; most is still in the cooler, more extended, molecular envelope. The fraction of neutral gas in the PDRs is expected to increase with time as the molecular gas is destroyed, and evidence for this is seen in the Ring nebula. The Ring has only a thin, fragmented neutral shell and Bachiller et $al.$ (1994a) have recently shown that most (≈ 90 %) of the neutral carbon in the shell is in C I rather than CO.

An unusual chemistry accompanies the unusual physical conditions in the neutral gas during PN formation. Besides CO and H_2, NGC 7027 is detected in HCN, HNC, CN, C_3H_2, HCO^+, N_2H^+, and CO^+ (see refs. in Omont 1993, Cox et $al.$ 1993, and Latter et $al.$ 1993b). This array of species is quite different from that typically found in C-rich AGB envelopes. CO^+ and N_2H^+ are not seen in IRC+10216, and HCO^+ is marginal, as discussed in §2.1. It was shown by Cox et $al.$ (1992) that CN, HNC, and HCO^+ are enhanced in the PNe IC 4406 and NGC 6072, as well as in NGC 7027, and similar results are found in M1–16 (Sahai et al. 1994). The presence of these species in PNe has recently been studied systematically by Bachiller et al. (1994b), and they are now seen, for example, in the molecular fragments in the Helix nebula $380''$ (8×10^{17}cm) from the central star, much farther out than molecules (other than CO) are seen in IRC+10216. These observations also confirm the large densities in the shell fragments ($\gtrsim 10^4$ cm^{-3}) which are necessary to excite the lines and to provide sufficient column density for the shielding of the molecules.

The origin of the unusual chemistry in PNe shells is probably related, at least in part, to the enhanced radiation field in the dense gas. This is supported by the molecular ions observed, and by the high resolution imaging of NGC 7027 by Cox et al. (1994a) who find the HCO$^+$(1–0) line emission lies in a fragmented shell lying around the PDR gas seen in H$_2$. An outline of the chemistry has been given by Cox et al. (1992) and Howe et al. (1992), but detailed models based on a realistic treatment of the physical conditions in the PDR gas have not yet been developed.

5. Concluding Remarks

The physical and chemical properties of circumstellar envelopes are closely linked. The most extreme physical conditions and an unusual chemistry are found in the dense, clumped, neutral structures in PNe. These structures are likely to be seeded in the AGB phase, and there is growing evidence that they are significant features of this earlier phase. In particular, it appears that physical and chemical models of the archetype IRC+10216 need to take into account the detailed structure in the envelope in order to fully account for the observations now becoming available.

Acknowledgements

It is a pleasure to acknowledge collaborative programs with R. Bachiller, P. Cox, and T. Forveille which form the basis of part of this paper. Thanks also to D.R. Crabtree and J. Keene for useful information and A.E. Glassgold for many helpful discussions. This work was supported in part by a grant from the NSF.

References

Avery, L.W., et al. : 1994, Astrophys. J. **426**, 737.
Bachiller, R., Huggins, P.J., Cox, P. & Forveille, T.: 1994a, Astron. & Astrophys. **281**, L93.
Bachiller, R., et al. : 1994b, in preparation.

Beck, H.K.B., Gail, H.P., Henkel, R. & Sedlmayr, E.: 1992, Astron. & Astrophys. **265**, 626.

Bergman, P., Carlstrom, U. & Olofsson, H.: 1993, Astron. & Astrophys. **268**, 685.

Bernat, A.P., Hall, D.N.B., Hinkle, K.H. & Ridgway, S.T.: 1979, Astrophys. J. **233**, L135.

Bieging, J. & Latter, W.B.: 1994, Astrophys. J. **422**, 765.

Bieging, J. & Tafalla, M.: 1993, Astron. J. **105**, 576.

Bowers, P.F. & Knapp, G.R.: 1987, Astrophys. J. **315**, 305.

Bujarrabal, V. & Alcolea, J.: 1991 Astron. & Astrophys. **251**, 536.

Bujarrabal, V., Fuente, A. & Omont, A.: 1994, Astron. & Astrophys. **285**, 247.

Carpenter, K., Robinson, R.D., Wahlgren, G.M., Linsky, J.L. & Brown, A.: 1994, Astrophys. J. **428**, 329.

Cernicharo, J. & Guelin, M.: 1987, Astron. & Astrophys. **183**, L10.

Cherchneff, I. & Glassgold, A.E., 1993, Astrophys. J. **419**, L41.

Cherchneff, I., Glassgold, A.E. & Mamon, G.A.: 1993, Astrophys. J. **410**, 188.

Cox, P., Omont, A., Huggins, P.J., Bachiller, R. & Forveille, T.: 1992, Astron. & Astrophys. **266**, 420.

Cox, P., Bachiller, R., Huggins, P.J., Omont, A. & Guilloteau, S.: 1993, in *Planetary Nebulae*, eds: R. Weinberger & A. Acker, Kluwer, Dordrecht, p227.

Cox, P., *et al.* : 1994a, in preparation.

Cox, P., *et al.* : 1994b, in preparation.

Crabtree, D.R., McLaren, R.A. & Christian, C.A.: 1987, in *Late Stages of Stellar Evolution*, eds: S. Kwok & S.R. Pottasch, Reidel, Dordrecht, p145.

Danchi *et al.* : 1994, Astron. J. **107**, 1469.

Diamond, P.J., *et al.* : 1994, Astrophys. J. **430**, L61.

Dyson, J.E., Hartquist, T.W., Pettini, M. & Smith, L.J.: 1989, Mon. Not. of the RAS **241**, 625.

Eriksson, K. & Stenholm, L.: 1993, Astron. & Astrophys. **271**, 508.

Glassgold, A.E. & Huggins, P.J.: 1986, Astrophys. J. **306**, 605.

Glassgold, A.E. & Huggins, P.J.: 1994, in preparation.

Glassgold, A.E., Mamon, G.A., Omont, A. & Lucas, R.: 1987, Astr. Astrophys. **180**, 183.

Graham, J.R., *et al.* : 1993, Astron. J. **105**, 250.

Groenewegen, M.: 1994, Astron. & Astrophys. in press.

Guelin, M., Lucas, R. & Cernicharo, J.: 1993, Astron. & Astrophys. **280**, L19.

Haas, M.R. & Glassgold, A.E.: 1993, Astrophys. J. **410**, L111.

Hartmann, L. & Avrett, E.H.: 1984, Astrophys. J. **284**, 238.

Hora, J.L. & Latter, W.B.: 1994, Astrophys. J. in press.

Howe, D.A. & Millar, T.J.: 1990, Mon. Not. of the RAS **244**, 444.

Howe, D.A., Millar, T.J. & Williams, D.A.: 1992, Mon. Not. of the RAS **255**, 217.

Huggins, P.J.: 1987, Astrophys. J. **313**, 400.

Huggins, P.J.: 1993a, in *Planetary Nebulae*, eds: R. Weinberger & A. Acker, Kluwer, Dordrecht, p147.

Huggins, P.J.: 1993b, in *Mass Loss on the AGB and Beyond*, ed: H.E. Schwarz, ESO, Garching, p365.

Huggins, P.J. & Glassgold, A.E.: 1982, Astrophys. J. **252**, 201.

Huggins, P.J., Olofsson, H. & Johansson, L.E.B.: 1988, Astrophys. J. **332**, 1009.

Huggins, P.J., Bachiller, R., Cox, P. & Forveille, T.: 1992, Astrophys. J. **401**, L43.

Huggins, P.J., Bachiller, R., Cox, P. & Forveille, T.: 1994a, Astrophys. J. **424**, L127.

Huggins, P.J., Bachiller, R., Cox, P. & Forveille, T.: 1994b, in *Circumstellar Media in the Late Stages of Stellar Evolution*, eds: R. Clegg, P. Meikle, I. Stevens, Cambridge University Press, in press.

Kahane, C., Audinos, P., Barnbaum, C. & Morris, M.: 1993, in *Mass Loss on the AGB and Beyond*, ed: H.E. Schwarz, ESO, Garching, p437.

Kastner, J.H. & Weintraub, D.A.: 1994, Astrophys. J. in press.

Kawaguchi, K., Kagi, E., Hirano, T., Takano, S. & Saito, S.: 1993, Astrophys. J. **406**, L39.

Keady, J.J. & Ridgway, S.T.: 1993, Astrophys. J. **406**, 199.

Keene, J., Young, K., Phillips, T.G., Buttgenbach, T.H. & Carlstrom, J.E.: 1993, Astrophys. J. **415**, L131.

Kwan, J.Y. & Linke, R.A.: 1982, Astrophys. J. **254**, 587.

Kwan, J.Y. & Webster, Z.: 1993, Astrophys. J. **419**, 674.

Lafont, S., Lucas, R. & Omont, A.: 1982, Astron. & Astrophys. **106**, 201.

Latter, W.B., Hora, J.L., Kelly, D.M., Deutsch, L.K. & Maloney, P.R.: 1993a, Astron. J. **106**, 260.

Latter, W.B., Walker, C.K. & Maloney, P.R.: 1993b, Astrophys. J. **419**, L97.

Loup, C., Forveille, T., Omont, A. & Paul, J.F.: 1993, Astron. & Astrophys. Suppl. **99**, 291.

Lucas, R.: 1993, in *Astronomy with Millimeter and Sub-millimeter Interferometry*, eds: M. Ishiguro & J. Welch, in press.

Lucas, R., Guelin, M.: 1990, in *Submillimeter Astronomy*, eds: G.D. Watt & A.S. Webster, Kluwer, Dordrecht, p97.

Mamon, G.A., Glassgold, A.E. & Huggins, P.J.: 1988, Astrophys. J. **328**, 797.

Martin-Pintado, J., Gaume, R., Bachiller, R. & Johnson, K.: 1993, Astrophys. J. **419**, 725.

Mauron, N.: 1990, Astron. & Astrophys. **227**, 141.

Mauron, N. *et al.* : 1984, Astron. & Astrophys. **130**, 341.

Meaburn, J., *et al.* : 1992, Mon. Not. of the RAS **255**, 177.

Nejad, L.A.M. & Millar, T.J.: 1987, Astron. & Astrophys. **183**, 279.

Neri, R., *et al.* : 1992, Astron. & Astrophys. **262**, 544.

Olofsson, H.: 1993, in *Molecular Opacities in the Stellar Environment*, ed: U. Jorgensen, in press.

Olofsson, H.: 1994, in *Circumstellar Media in the Late Stages of Stellar Evolution*, eds: R. Clegg, P. Meikle & I. Stevens, Cambridge University Press, in press.

Olofsson, H., Carlstrom. U., Eriksson, K. & Gustafsson B.: 1992, Astron. & Astrophys. **253**, L17.

Olofsson, H., Eriksson, K., Gustafsson, B. & Carlstrom, U.: 1993, Astrophys. J., Suppl. **87**, 305.

Omont, A.: 1993, J. Chem. Soc. Faraday Trans. **89**, in press.

Plez, B., Lambert, D.L.: 1994, Astrophys. J. **425**, L101.

Rodgers, B. & Glassgold, A.E.: 1991, Astrophys. J. **382**, 606.

Sahai, R.: 1987, Astrophys. J. **318**, 809.

Sahai, R., Wootten, A., Schwarz, H.E. & Wild, W.: 1994, Astrophys. J. **428**, 237.

Shibata, K.M., Deguchi, S., Hirano, N., Kameya, O. & Tamura, S.: 1993, Astrophys. J. **415**, 708.

Skinner, C.J. & Whitmore, B.: 1987, Mon. Not. of the RAS **224**, 335.

Tielens, A.G.G.M.: 1993, in *Planetary Nebulae*, eds: R. Weinberger & A. Acker, Kluwer, Dordrecht, p155.

Trammell, S.R., Dinerstein, H.L. & Goodrich, R.W.: 1993, Astrophys. J. **402**, 249.

Truong-Bach, Morris, D. & Nguyen-Q-Rieu: 1991, Astron. & Astrophys. **249**, 435.

Turner, B.E., Steimle, T.C. & Meerts. L.: 1994, Astrophys. J. **426**, L97.

Wahlgren, G.M., Robinson. R.D. & Carpenter, K.: 1992, in *Cool Stars, Stellar Systems, and the Sun*, eds: M. Giampapa & J. Bookbinder, ASP Conf. Series, **26**, p37.

Yamamura, I., Onaka, T., Kamijo, F., Izumiura, H. & Deguchi, S.: 1993, Publ. of the ASJ **45**, 573.

Young, K., *et al.* : 1994, in preparation.

PLATEAU DE BURE OBSERVATIONS OF IRC+10216:

HIGH SENSITIVITY MAPS OF SIC$_2$, SIS, AND CS

R. LUCAS and M. GUÉLIN

IRAM, Grenoble, France

C. KAHANE and P. AUDINOS

Observatoire de Grenoble, France

and

J. CERNICHARO

Centro Astronómico de Yebes (IGN), Guadalajara, Spain

Abstract. We have obtained high angular resolution ($\sim 3''$), and high sensitivity maps of IRC+10216. SiC$_2$ is found both in a spherical shell and in the very central region, indicating it is formed both in the inner envelope close to the star, and in the outer shell. The molecules SiS and CS are mostly found in the inner parts of the envelope, but are still detectable in the outer region ($r \sim 15''$) where the products of photochemistry are found. The maps show that IRC+10216 has a very clumpy envelope, with strong departures from spherical symmetry; an axis oriented NS-SW (P.A. 20°) can be seen in all maps. The radial brightness distribution of CS has secondary maxima, at the radius where the SiC$_2$ shell has its peak emission. A preliminary map shows CN in the same shell, but also in a still larger outer shell. Time variations in the mass loss rate, could be invoked to explain the multiple shell structure of this envelope.

Due to considerable mass loss, red giants at the end of their evolution are surrounded by circumstellar envelopes, ending in the form of pre-planetary nebulae (PPN) then planetary nebulae. Chemically, one may separate two envelope regions: the inner layers where chemistry is dominated by neutral reactions, grain processes, and shocks, and outer layers where photon processes (due to external UV radiation) lead to photodissociation of CO, ionization, ion-molecule reactions and radical reactions, yielding a very rich chemistry with complex molecules. For an introduction, see the review by Huggins in these proceedings. We present here detailed, sensitive observations of silicon and sulfur containing molecules in the archetypical carbon-rich envelopes, IRC+10216.

The IRAM 30-m telescope at Pico Veleta (Spain) was used to obtain single-dish maps of several transitions of the following molecules: SiS J=5–4, J=8–7, J=9–8, and J=12–11 at 90.8, 145.2, 163.4 and 217.8 GHz), SiC$_2$ 4_{23}-3_{22}, 6_{06}-5_{05}, 6_{24}-5_{23}, 9_{27}-8_{26} and 10_{29}-9_{28} at 94.2, 137.2, 145.3, 222.0, and 232.5 GHz; C^{34}S J=2–1, J=3–2, and J=5–4, at 96.4, 144.6, and 241.0 GHz; and CS J=2–1 at 98.0 GHz.

We used the IRAM Plateau de Bure interferometer, on three separate observing periods, to map the $4_{23} - 2_{22}$ transition of SiC$_2$, the J=5–4 transition of SiS, and the J=2–1 transition of CS. For all observations, a single

Astrophysics and Space Science **224**: 293–296, 1995.

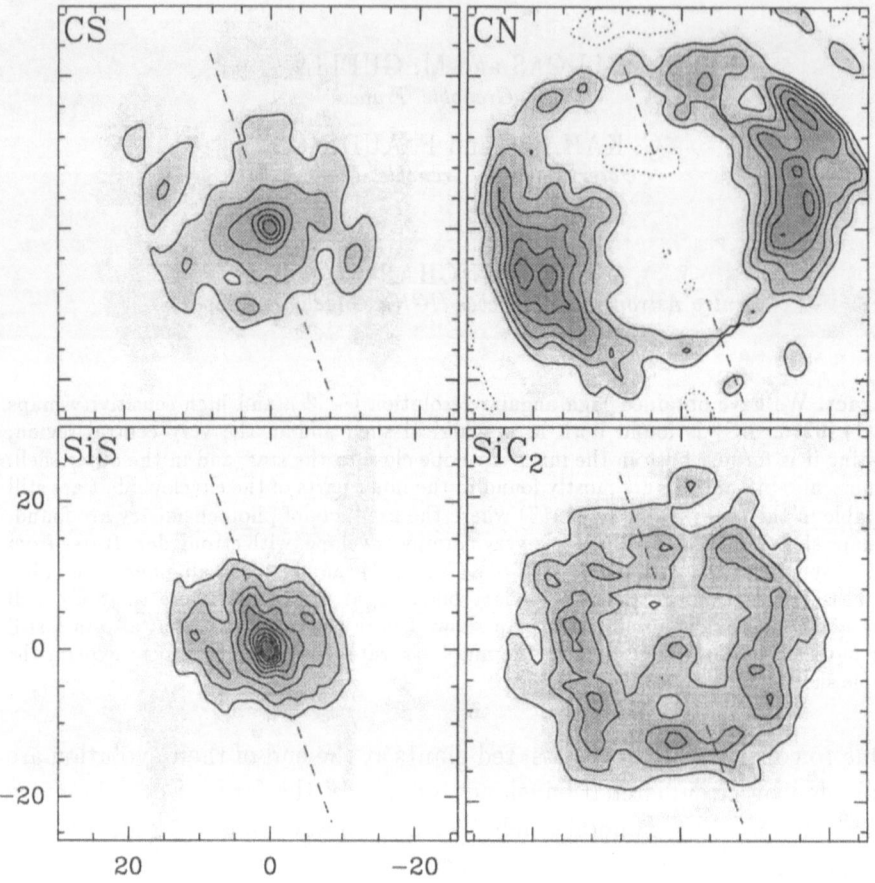

Fig. 1. Observed brightness distributions of SiS, CS, and SiC$_2$, averaged over the velocity range −28.5 to −21.5 km/s. The coordinates are in arc seconds, offset from the stellar position. A preliminary map of CN (N=1-0), obtained by observing a mosaic of seven fields, is included for comparison (Lucas *et al.* 1995, in preparation). The nebula axis (Position angle 20°), is shown. The angular resolutions are 3.1″, 3.6″, 3.9″, and 3.6″ for the SiS, CS, SiC$_2$ and CN maps.

field centered on the central continuum source was observed. For all three maps, the total flux and short spacing visibilities were extracted from the 30-m maps and included in the visibility tables to produce maps containing all the source flux (Fig. 1).

SiS shows a centrally peaked distribution, confirming results by Bieging & Tafalla 1993. Due to our higher angular resolution and sensitivity we can see finer details not present in the previous maps: the main source is elongated along an axis with P.A. $\sim$ 20° and is surrounded by a second envelope peaking $\sim$ 7″ from the star, in the perpendicular (P.A.=70°). The PA=20° axis and elongation are similar to that found by Kastner and Weintraub (preprint) in the near infrared. The CS maps also show a centrally peaked distribution, but more extended than that of SiS, and surrounded by an extended envelope more intense along (P.A.=70°); a ring of secondary maxima is seen at a radius of $\sim 15''$. The SiC$_2$ map presents a clumpy ring at a radius $\sim 15''$, displaced towards East, just like the rings of MgNC and C$_3$H, found by Guélin, Lucas and Cernicharo (1993); the emission peaks in this ring are well correlated with the emission maxima in the CS outer ring. In addition, emission is found at the stellar position, with about the same brightness as the emission in the ring.

In order to compare our results with the predictions of chemical models, we have tried to estimate the radial dependance of the molecular abundances. Audinos et al. (1994) have shown, in the case of HC$_3$N, that the strong variation of the infrared excitation with distance influences the brightness distribution. To disentangle abundance from excitation effects, we observed several lines of different excitation for each molecule. We then constructed a spherically symmetrical model of the excitation of each species, and adjusted the radial abundance curves to reproduce the central spectra and radial brightness distributions of the lines observed.

The model included infrared excitation up to the first vibrational levels by radiation from the star, at $\sim 10\mu$m, followed by spontaneous emission (Morris 1975). For SiC$_2$ the ν_2 mode at 840.6 cm^{-1} was used (Si–CC vibration, with $\mu_{IR} = 0.2$ debye). Collisional excitation by H$_2$ molecules was also included. The distance to IRC+10216 was estimated to 200 pc, and its mass loss rate to 3 x 10^{-5} M$_\odot$ yr^{-1}. The temperature in the envelope ranged from 200 K close to the star (Boyle et al. 1994), to 22 K at 10^{17} cm from the star (Truong-Bach et al. 1991). The radiative transfer was treated locally using the Sobolev approximation, but the emerging profiles are computed using a finite local line width, as in Morris, Lucas & Omont (1984). The best fit abundance curves are given in Figure 2.

The SiS abundance was predicted by Lafont, Lucas & Omont (1982) to be large ($4x10^{-5}$) on the basis of equilibrium chemistry calculations; SiS was indeed recently observed in the infrared by Boyle et al. (1994). The abundance decreases from $5x10^{-5}$ to $4x10^{-6}$ between 1 and 12 R_* (7 x 10^{13} to 8.4 x 10^{14} cm). Glassgold, Lucas & Omont (1986) proposed that SiC$_2$ is formed by reactions of Si$^+$ with C$_2$H$_2$ or C$_2$H in the outer envelope, while Howe & Millar (1990) assumed that SiC$_2$ is a parent molecule, present in the inner layers. The silicon chemistry in IRC+10216 has been reinvestigated by Glassgold & Mamon (1991); they proposed that SiS is mostly depleted onto grains, but some SiS molecules react with acetylenic ions C$_2$H$_2^+$ and C$_2$H$_3^+$ to form SiC$_2$H$^+$, giving SiC$_2$ by dissociative recombination. SiC$_2$ is then photodissociated to form SiC.

Our observations show that SiS is, as previously known, a parent molecule; SiC$_2$ is also a parent molecule, with a much lower initial abundance ($\sim 10^{-7}$), but its abundance is considerably enhanced in the outer envelope (at $\sim 15''$),

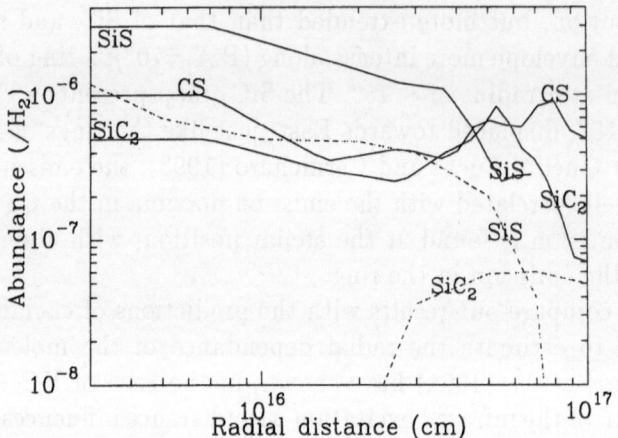

Fig. 2. The best fit abundance curves of SiS, CS, and SiC$_2$. The theoretical abundances of Glassgold & Mamon (1992) are shown as dashed curves.

possibly by ion-molecule reactions with carbon chains. We also find that CS is a parent molecule, as proposed by Howe & Millar (1990), who showed that CS cannot be produced from SiS in the observed amounts. The CS abundance decreases from 10^{-6} to a few 10^{-7} in the outer envelope.

However, the detailed structure of the emission (Fig. 1) shows clumps in the CS and SiC$_2$ shells that are probably density peaks rather than molecular abundance peaks. The CN brightness distribution (Fig. 1), shows two concentric shells, with a typical spacing of $\sim 5''$, or 1.5×10^{16} cm. The inner CN shell coincides more or less with the C$_4$H, C$_3$H, MgNC, and the outer SiC$_2$ shell; the outer shell lies further out and is mostly visible in the equatorial plane. The morphology of the envelope, as seen through these various molecular tracers, strongly suggests variations in the mass loss rate, with a time scale of order ~ 300 years.

References

Audinos, P, Kahane, C. & Lucas, R.: 1994, Astron. & Astrophys. **287**, L5.
Bieging, J.H. & Tafalla, M.: 1993, Astron. J. **105**, 576.
Boyle, R.J., Keady, J.J., Jennings, D.E., Hirsch, K.L. & Wiedemann, G.R.: 1994, Astrophys. J. **420**, 863.
Glassgold A.E., Lucas, R. & Omont, A.: 1986, Astron. & Astrophys. **157**, 35.
Glassgold A.E. & Mamon, G.A.: 1991, in *Chemistry and spectroscopy of interstellar molecules*, ed N. Kaifu, Tokyo, p.261.
Guélin, M., Lucas R. & Cernicharo, J.: 1993, Astron. & Astrophys. **280**, L19.
Howe, D.A. & Millar, T.J.: 1990, Mon. Not. of the RAS **244**, 444.
Lafont, S., Lucas, R. & Omont, A.: 1982, Astron. & Astrophys. **106**, 201.
Morris, M.: 1975, Astrophys. J. **197**, 603.
Morris, M., Lucas, R. & Omont, A.: 1984, Astron. & Astrophys. **142**, 107.
Truong-Bach, Morris, D. & Nguyen.-Q. Rieu: 1991, Astron. & Astrophys. **249**, 435.

GASEOUS REFRACTORY-ELEMENT MOLECULES IN IRC10216

B. E. TURNER

National Radio Astronomy Observatory *

Abstract. The list of detected refractory-element (RE) species in IRC10216 is now large enough to try to assess their chemistry, and the fraction of each that escapes in gas phase to the ISM. The former may tell us how grains are formed, the latter whether mass-loss from evolved stars is important in determining interstellar elemental depletions as distinct from accretion processes in the ISM. We expect that much of the Si chemistry is now understood and about 25% of Si escapes as a gas. Other REs are less well understood but most should be more volatile than Si. For many of the REs, O-rich CSEs should behave similarly to C-rich ones.

1. Introduction

The study of gaseous refractory-element (RE) molecules in mass-losing circumstellar envelopes (CSEs) is of interest to the astrochemistry of the REs, the depletion of the elements, the composition of interstellar grains and the corresponding gas/dust mass ratio. Observed gas/dust ratios (*e.g.* Knapp 1985) require some O and C in the solid form as well as the REs, so a detailed relation between the depletions of the elements and the gas/dust ratio must involve the gas-phase chemistry of the heavy elements. Heavy elements expelled from CSEs in atomic or molecular form (the latter quickly photodissociating) are important in determining the resulting depletion factors, at least in the warm low-velocity diffuse ISM where accretion may be minimal. K and Na have recently been found not to condense significantly in CSEs (Mauron & Caux 1992; Mauron & Guilain 1994). The rich Si-chemistry recently observed in IRC10216 suggests that even this "highly-refractory" element may remain largely in the gas phase in CSEs. To understand RE chemistry, CSEs (IRC10216) are proving much more useful than the ISM, because of the relative wealth of RE species now being observed in IRC10216.

2. Current Observations of RE Molecules In IRC10216

Table 1 displays observed RE species according to increasing refractivity of the RE downward, vs. chemical species horizontally. All species have been searched to sensitive limits and represent those currently having well-determined molecular constants.

* Operated by Associated Universities, Inc., under cooperative agreement with the National Science Foundation

Underlined species are detected (AlC and KCN are tentative). Species in square brackets are predicted to dominate in regions of thermochemical equilibrium (TE) occurring in the dense, hot innermost envelope or stellar atmosphere; a superscript 'o' refers to an O-rich CSE only, otherwise the brackets refer to both C- and O-rich cases. If the dominant TE species is atomic, then no molecular species is assigned brackets. Note the dominance of Si species among those detected and Ca species among those undetected.

We have made an LTE analysis on all detected species for source diameters of 12″ and 36″, utilizing all lines currently observed. The 12″ models describe inner-core species believed to have formed under TE. The 36″ models refer either to species which occur at all radii out to 18″ (e.g. HC_3N, SiC_2) or to species concentrated in a shell with peak abundance typically at radius 14″ ± 5″ (e.g. HC_7N, MgNC). In a few cases (e.g. SiC) a source diameter of 50″ is used. Derived column densities N are those centered on the star. Classic TE species (NaCl, AlCl, KCl) have $T_{rot} \geq 30K$ in the IE, while well-defined OE species (SiC, SiC_2, SiC_4, MgNC) have $T_{rot} \sim 15 \pm 5K$ in the OE. T_{rot} results suggest that SiN, NaCN, PC are OE rather than IE species. Interferometry shows that SiC_2 is both IE and OE and a detailed analysis (Avery et al. 1992) reveals two distinct T_{rot} regimes.

TABLE I
Observed Refractory-Element Species

NaH			NaO		[NaCl]	NaF	NaOH	[NaCN]
					[KCl]			KCN?
PH$_3$	PC	[PN]	PO	[PS]	PCl			
MgH			MgO	MgS			[MgOH]°	MgNC
	SiC							
SiH$_2$	SiC$_2$	SiN	[SiO]	[SiS]	SiCl			
SiH$_4$	SiC$_4$							
			[FeO]°					
	AlC?		AlO	AlS	[AlCl]	[AlF]		
		TiN	[TiO]					
CaH			CaO	CaS	CaCl		[CaOH]°	CaNC

Fractional abundances X are determined by dividing N by the column density $N(H_2)$ found by integrating the distribution $n(r) = 2(37)r^{-2}cm^{-3}$, corresponding to a mass loss rate of $2(-5)M_\odot/yr$, over suitable ranges of r, corresponding to the angular distributions and a distance of 200 pc. The inner cutoff radius $r = 1.3(14)cm$ (Lafont et al. 1982) is where n(r) falls below $1(13)cm^{-3}$, hence where TE ceases. It produces uncertainty in values of X for the core model.

3. TE, Condensation and Adsorption

For C-rich objects, Tsuji (1987, 1991 priv.comm.) has calculated the TE abundances for Si, P, Na, Mg and Fe. SiS and SiO dominate for $500 \leq T \leq 1100K$, while SiH_4 becomes important for $T \leq 1200K$. Phosphorus forms PN with $\sim 100\%$ efficiency for $T > 600K$ and PS similarly for $T \leq 550K$. PN and PS together utilize all phosphorus at all T. NaCN dominates NaCl at all T although NaCN includes only 4% of cosmic Na at most. MgH is the dominant Mg species but has negligible abundance at all T ($MgH/Mg \leq 1(-7)$). Fe forms no compounds at even this level. Thus virtually all Na, Mg and Fe remains in atomic form under TE. Tsuji (1973) finds TiO comprises $\sim 60\%$ of all Ti for $T < 1200K$, even in C-rich objects. No models are available for Al or Ca.

For C-rich systems the elements are expected to condense into solids in the order graphite (2000), TiC (?), SiC (1660), CaC_2 (1600), TiN (?), AlN (1300), Fe_2C (1260), and MgC_2(?) where the condensation temperature in K is given. Thus C, Ti, Si, Ca, Al, Fe, Mg are (partly) removed in order, those with lower values of T_{cond} being less likely to go to completion because of the decrease in n(r) with increasing r. The order of elemental removal differs for O-rich envelopes (Turner 1991), being Al, Ti, Ca, Fe, (Mg,Si), P, (Na,K), S.

Elements which do not condense may be adsorbed onto grains. We have expanded the list of adsorption fractions calculated by Jura & Morris (1985) from the binding energies of pure substances. In order of decreasing volatility, we find the fractions remaining in the gas phase for sticking probabilities 0.1 and 1.0 respectively are N_2 (1.0,1.0), CO (1.0,1.0), SiH_4 (1.0,0.97), P (1.0,0.97), PH_3 (1.0,0.96), HCN (0.99,0.93), NH_3 (0.98,0.85), S (0.94,0.55), K (0.84,0.17), SiO (0.06,< 1(−10)) and Al (0.05,< 1(−10)). For modest sticking factors, the results for K (and Na) agree with the findings of Mauron and coworkers. For similar sticking factors the calculations for SiO agree with the results of Bieging & Rieu (1988) who find that $\sim 2\%$ of SiS (similar to SiO?) remains gaseous at r = 3(16) cm where photodissociation begins to dominate.

4. Information From Abundance Ratios

Several conclusions can be made by comparing the relative abundances of the REs within each chemical group (*e.g.* sulfides) with cosmic abundances of the REs themselves.

1) Chlorides (Al, Na, K): The observed ratios are consistent with TE ratios (O-rich objects) for $T \sim 1300K$. $X(NaCl) = 1(-8)$ for C-rich TE models. The observed X(NaCl) = 4.8(-11), implying severe adsorption of NaCl for $r < 1.3(15)cm$.

2) Phosphides (C,N): The observed PC/PN ratio (> 11) suggests that

CP is not formed under TE. Also, the TE value for X(PN) is much greater than the observed value (PN is not detected). If PN were formed under TE and strongly adsorbed, there would be insufficient remaining P to form CP. It appears that PN may not form efficiently under TE, contrary to the models.

3) Carbides (Si,Al,P): If SiC, SiC_2 and SiC_4 are coextensive and the carbide chemistry is equally efficient, Si is adsorbed more than P, as expected.

4) Sulfides (Si, Mg, Al, Ca, P): Interferometry shows that 31% of SiS survives adsorption at $1''$ radius, 4% at $12''$, and $\sim 2\%$ at $24''$. Upper limits on MgS and PS are consistent with C-rich TE models at $T \leq 600K$. Limits on AlS and CaS are consistent with O-rich TE models at $T < 1000K$.

5) Cyanides (Mg, Na, Ca, K): Na and K should have similar chemistry and adsorption should be low. These expectations are consistent with the observed ratios. Mg and Ca are less efficient chemically, or are adsorbed more severely. The former is more likely for Mg.

5. Inner Vs. Outer Envelope Distributions

Silicon: The abundance of SiS+SiO escaping to the OE is X = 3.7(-7) (Bieging & Rieu 1988). For the OE, $X(SiC_2) = 1.0(-6)$. Thus the IE provides marginally insufficient Si atoms even if all SiS+SiO is converted to SiC_2. However, SiH_4 is observed under TE conditions (Goldhaber & Betz 1984) with $N(SiH_4) \geq 2(15)cm^{-2}$; under TE, $X(SiH_4) \approx 1(-5)$, the same as for SiS and SiO (Tsuji 1973). SiH_4 is not expected to be adsorbed in the outflow. It photodissociates at 75% efficiency to SiH_2, a species we apparently detect with $X = 1.3(-6)$ in the OE. The sum of $SiS + SiO + SiH_4$ leaving the IE is enough to produce $SiC_2 + SiH_2$ in the OE.

Sodium: For NaCl, X(IE) = 4.8(-11) and X(TE) = 1(-9) - 1(-8) so NaCl is consistent with TE formation and heavy adsorption. For NaCN, X(IE) = 3(-10) and X(TE) = 6(-8), implying the same conclusion if NaCN is an IE species. However, the low values of T_{rot} for NaCN and the presence of MgNC in the OE suggest that NaCN is an OE species. If so, either NaCN is not formed by TE, or since (NaCl + NaCN)/Na(TE) = 4% under TE, NaCN could be adsorbed in the IE and the remaining atomic Na could form NaCN in the OE.

Magnesium: In the OE, $X(MgNC) = 6.7(-9)$, $\gg X(MgH)$, the dominant molecule under TE. Essentially all Mg remains atomic under TE and the ratio X(MgNC)/X(cosmic Mg) = 2(-4) in the OE represents inefficient formation of MgNC and/or adsorption of Mg, with the former more likely since Mg is expected to be less refractory than Si.

Phosphorus: All P is in the form PN + PS under TE in C-rich objects. TE could still explain CP in the IE, since CP comprises only 0.1% of total cosmic P. Or, at least 7.4% of PS + PN must convert to CP in the OE (more, if any PS + PN is adsorbed). An OE formation of CP appears most likely.

Aluminum: For an IE distribution, X(AlCl+AlF) = 9.2(-10) while X(TE) = 1(-7) for O-rich objects. C-rich models are needed to confirm a TE origin, which is expected. AlC is stable and strongly bound, so it might form efficiently in C-rich cases under TE. If so, 1.6% of Al must form AlC and survive as such into the OE if the tentatively observed AlC has an OE location. AlC is probably strongly adsorbed.

Potassium: There are no C-rich TE models. If KCN is confirmed, its IE abundance exceeds that of KCl by a factor of 2.5 while an OE distribution has X(KCN) = 110X(KCl). These arguments and the presence of MgNC and NaCN in the OE suggest KCN is formed in the OE also.

6.   Outer Envelope Chemistry

Under TE in C-rich objects, CH_4, C_2H_2, CO and HCN have large abundances and are highly volatile, so they will dominate the chemistry in the OE. REs may react with them via radiative association reactions (e.g. Mg^+ + $HCN \rightarrow MgNCH^+ + h\nu$, leading to MgNC), by ion-molecule reactions (e.g. $Mg^+ + HCN \rightarrow MgNC^+ + H$, leading to MgNC), or by radical-radical reactions which involve photodissociation products of the abundant species (e.g. $Mg+CN \rightarrow MgNC$). In similar fashion carbides and acetylides form from CH_4 and C_2H_2, and nitrides from the less abundant NH_3. Thus PC and AlC form from CH_4; SiC_2, MgC_2, NaCCH and perhaps AlC from C_2H_2; MgNC and NaCN from HCN; and SiN from NH_3. Despite expected differences in reaction efficiency among the REs, the observed relative abundances decrease in the order of products of CH_4, C_2H_2, HCN and NH_3, respectively, i.e. in the same order as those of the abundant reactants, lending credence to this picture. The species MgC_2 and NaCCH are unstudied spectroscopically, but both are known to be highly stable and may be expected to dominate the Mg and Na chemistry in the OE. P is the only RE whose ion is known to react efficiently with CH_4. Except for Si, it is likely that the reactants CH_4, C_2H_2 and HCN account for all RE species of significant abundance in the OE. Si has the additional contribution from SiH_4/SiH_2, which originates under TE.

7.   Ejection of Refractory Elements To The ISM

Table 2 compares gas-phase RE abundances observed in the OE of IRC10216 with depleted values observed in the warm diffuse ISM (Turner 1991). IRC10216 values of X for each element are relative to H and are assumed to represent the gaseous abundances ejected into the ISM. The percentage of the depleted warm diffuse ISM abundance accounted for by the IRC10216 ejecta is given in the last column.

TABLE II

Interstellar and Ejected Abundances of Refractory Elements

Order of Volatiltity	X/X_{cosmic}		% of Gaseous ISM Component Ejected
	$IRC10216$	ISM	
P	7.4(-2)	8.0(-1)	9.2
Na	4.5(-3)	3.1(-1)	1.5
Mg	1.2(-4)	6.3(-1)	0.02
Si	5.9(-2)	2.5(-1)	23.6
Al	8.0(-3)	1.0(-1)	8.0

If IRC10216 were typical of all mass-losing CSEs, a value of 100% in column (3) would mean that both grain accretion and disruption in the ISM are unimportant, a possibility for quiescent regions of the warm diffuse ISM. Values much less than 100% refute this possibility and/or reveal incomplete accounting of the gas-phase RE component ejected from IRC10216. Values in column (3) would be expected to correlate with the relative volatility of the RE only if all gas-phase species were accounted for. The largest value occurs for Si whose molecular component we believe is essentially accounted for, while the smallest values occur for Na and Mg whose major molecular forms we believe have not been assessed (carbides, acetylides). Atomic components in the OE are likely unobservable in C-rich objects owing to the obscuration of the central star and can only be assessed through a complete understanding of the chemistry. For Si, the escaping fraction can range from 5.9% $(SiC_2 + SiH_2)$ to 33% $(SiH_4 + SiS + SiO)$, assuming all Si originating from SiH_4 escapes. The latter exceeds the observed depleted ISM abundance. Even more P, Na, Mg would be expected to escape in the gas-phase.

Would the ejecta of O-rich CSEs differ markedly from those of C-rich objects? Likely not for Si. In either case, SiH_4, formed under TE, escapes almost completely from the inner core (forming SiC_2 and SiO in the OEs of C- and O-rich cases). The abundance of gaseous Si in the OE is essentially that of SiH_4, which is the same in either case. Similarly, Fe, Mg, and Na form molecules very inefficiently under TE in either O-rich or C-rich environs and since at least Mg and Na are highly volatile, differences between silicate grains (O-rich) and graphite grains (C-rich) may not matter much. Conversely, differences in adsorption by graphite and silicate grains could be decisive in the case of P (i.e. PN and PS) and Al (AlOH and AlC for O-rich and C-rich, respectively).

For Si, Mg, Na, K we conclude that O-rich and C-rich CSEs should behave similarly and that a lower limit of 25% of Si (higher for Mg, Na, K) is ejected into the ISM in gas phase. For these elements the ejecta of evolved stars are important, but perhaps not decisive, in determining their depletions in the warm diffuse ISM. Other REs are more difficult to assess.

References

Avery, L., *et al.* : 1992, Astrophys. J., Suppl. **83**, 363.
Bieging, J.H. & Rieu, N-Q.: 1988, Astrophys. J. **343**, L25.
Goldhaber, D.M. & Betz, A.L.: 1984, Astrophys. J. **279**, L55.
Herbst,E., *et al.* : 1989, Astron. & Astrophys. **222**, 205.
Jura, M. & Morris, M.: 1985, Astrophys. J. **292**, 487.
Knapp, G.R.: 1985, Astrophys. J. **293**, 273.
Lafont, S., Lucas, R. & Omont, A.: 1982, Astron. & Astrophys. **106**, 201.
Mauron, N. & Caux, E.: 1992, Astron. & Astrophys. **265**, 711.
Mauron, N. & Guilain, C.: 1994, Astron. & Astrophys. in press.
Tsuji, T.: 1973, Astron. & Astrophys. **23**, 411.
Tsuji, T.: 1987, in: *Astrochemistry*, eds: M.S. Vardya & S.P. Tarafdar, Reidel, p409.
Turner, B.E.: 1991, Astrophys. J. **376**, 573.

PLATE VIII: The 3.8m UK Infrared Telescope (UKIRT) located atop the 4200 metre high dormant volcano Mauna Kea in Hawaii. Because of its high altitude, dry atmosphere and stable weather, Mauna Kea is the best site in the world for infrared and submillimetre astronomy and UKIRT shares the mountain with a number of other major telescopes. The UKIRT was set up by the ROE in 1979. It is now operated by the Joint Astronomy Centre on behalf of the UK Particle Physics and Astronomy Research Council.

OBSERVATIONS OF HIGH CO ROTATIONAL LINES IN POST-AGBS AND PN

K. JUSTTANONT, A.G.G.M. TIELENS

NASA Ames Research Center, MS 245-3, Moffett Field, CA 94035, USA

C.J. SKINNER

IGPP, L-413, Lawrence Livermore National Laboratory, PO Box 808, Livermore, CA 94551, USA

and

M.R. HAAS

NASA Ames Research Center, MS 245-6, Moffett Field, CA 94035, USA

Abstract. We present a preliminary interpretation of high CO rotational line data obtained from KAO. The possibility of either a PDR or a shock model is considered in order to explain the observations.

Key words: Molecular processes – shock waves – stars : AGB and post-AGB

1. Introduction

A significant fraction of the initial stellar mass is lost when a star is on the Asymptotic Giant Branch (AGB). Mass loss rates range from 10^{-7} $M_\odot$/yr for early AGB stars to a few 10^{-4} $M_\odot$/yr for stars at the tip of the AGB. Dust grains condense from the outflows as the gas expands and form dust shells around the central stars, which can heavily obscure the star itself. A superwind ($\sim 10^{-4} - 10^{-3}$ $M_\odot$/yr) is thought to terminate the AGB phase (Renzini 1981). At that point, the star evolves to a higher effective temperature, the mass loss decreases ($\sim 10^{-8}$ $M_\odot$/yr), but the wind velocity increases (~ 1000 km/s). The interaction of this wind with the dense AGB wind shapes the nebula. During this evolution, dust and gas are exposed to harsher radiation and when T_{eff} reaches about 30,000K, the nebula is ionised and becomes a planetary nebula (PN).

The photons from the central star can create a photodissociation region (PDR) in the expanding superwind. Gas can be heated through the photo-electric effect working on small grains and polycyclic aromatic hydrocarbons (PAHs) (*e.g.* , Tielens & Hollenbach 1985; Bakes & Tielens 1994). The gas can cool via the atomic fine structure lines of OI 63μm, and CII 158μm as well as the rotational lines of CO. Alternatively, gas can be heated by shocks driven by the interacting winds.

2. Observations

We observed GL2688, GL618 and NGC7027 in November 1993 using the Ames Cryogenic Grating Spectrometer (CGS) on the KAO (Erickson *et al.* 1985). Three rotational lines of CO were observed : J=13-12, 17-16 and 22-21. The relative detector response was removed by dividing each spectrum by a calibration spectrum of Saturn at the same wavelength. The absolute flux calibration was done by multiplying the ratio spectra by the flux of Saturn. The line flux is then calculated from fitting the data points with a gaussian profile and a flat continuum.

The three objects represent different evolutionary stages. The central star of GL2688 has a spectral type of F5Iae ($T_{eff} \sim 6,000$ K), while GL618 has a B0 star ($T_{eff} \sim 30,000$ K) and NGC7027 is a young PN ($T_{eff} \sim 100,000$ K).

3. PDR model

A PDR is defined as a region where UV radiation dominates the heating and chemistry. Heating is done by absorption of a UV photon by a grain. This results in ejection of an electron which transfers its energy by collision with the gas. The heating efficiency depends on the UV field, G_0, the gas temperature, T, and density, n. The cooling is via gas-grain collisions and via atomic and molecular lines. Main assumptions in the model are constant hydrogen nuclei density; the pressure is dominated by turbulence so that the PDR is in pressure equilibrium. The chemistry includes that of C, O, S, Si, Fe, and Mg and the molecular species of H_2, CH, OH, H_2O, and CO.

We assume a distance of 1 kpc for all three objects. The terminal velocity of each star is taken from a catalogue of CO J=1-0 and 2-1 observations by Loup *et al.* (1993). Various models with different UV fields and densities were calculated. Higher T_{eff} leads to intensities of high CO rotational lines become more significant. However, if T_{eff} is too high, CO molecules will be dissociated by FUV photons from the star. Higher density results in higher CO line intensities for J>20.

4. Shock model

When a star is on the AGB, it loses mass which forms a dust shell, moving at typically 10-20 km/s. Shocks can develop in the circumstellar envelope if there is a fast wind ($v \geq 50$ km/s) sweeping through it. This can photo-dissociate molecules which form in the ambient red giant wind. The main infrared signature of a shock is the OI 63μm line.

There is evidence for the presence of fast winds in PPN. Burton & Geballe (1986) report a line width for the v=1-0 S(1) molecular hydrogen line of 250

TABLE I

Line intensities of CO and other species as calculated from PDR and shock models. The line intensity is in the unit of $erg/cm^2/s/sr$. The values in the square brackets are the observed values. Input parameters for each model is also listed.

	PDR			Shock
	GL2688	GL618	NGC7027	
CO J=13-12	3.7E-3 [3.5E-2]	4.4E-3 [2.5E-2]	1.8E-3	2.9E-3
CO J=17-16	7.0E-3 [3.6E-2]	8.1E-3 [2.8E-2]	1.5E-3 [1.5E-3]	6.4E-3
CO J=22-21	9.5E-3 [2.4E-2]	1.0E-2 [1.4E-2]	4.1E-4	1.1E-2
OI 63μm	5.2E-2	2.2E-1	1.8E-1 [1.2E-1]	8.1E-2
OI 145μm	2.1E-3	6.9E-3	6.0E-3 [5.0E-3]	4.5E-3
CII 158μm	1.0E-4	2.8E-3	2.8E-3 [1.3E-2]	2.2E-4
SI 25μm	3.1E-2	2.8E-4	−3.2E-4	1.9E-1
SiII 35μm	1.3E-2	2.2E-2	1.5E-2	–
H$_2$ 1-0	–	thermal	1.0E-3	5.3E-4
H$_2$ 2-1	–	thermal	3.0E-5	2.6E-4
n (cm^{-3})	1.0E+7	1.0E+7	1.0E+6	1.0E+7
Ω (sr)	2.0E-10	2.0E-10	8.0E-10	1.5E-9
G (Habing)	1.0E+8	3.0E+7	2.0E+6	
T (K)	6000	30000	30000	
FIR (erg/cm^2/s)	2.5E-7	5.0E-7	2.0E-7	
v_s (km/s)				100

km/s for GL618, while Young *et al.* (1992) observe CO and HCN profiles from GL2688 to have high velocity component of 100 km/s.

We compared the results of CO rotational lines from the dissociative J shock models calculated by Hollenbach & McKee (1989) which takes the chemistry of molecular and atomic species into account. Their Figure 6 shows predicted line intensities of CO as a function of the rotational levels for different shock velocities.

For a comparison, we calculated line intensities for a dissociative shock model with density of 10^7 cm^{-3} and a shock velocity of 100 km/s. The adopted source sizes for GL2688 and GL618 are in good agreement with the observed H$_2$ emission regions (presumably due to shocks). Note that the CO line intensities are very similar to those from PDR models.

Assuming the source size and an AGB wind of 20 km/s, the mass loss rate can be calculated from the calculated density. For both GL618 and GL2688, the mass loss rate is 3×10^{-3} M$_\odot$/yr.

5. Conclusion

Intensities of high rotational line CO emission can be used to estimate the density and the stellar effective temperature (for PDRs) or the shock velocity (shock models). It cannot be used to differentiate either model. One way to distinguish between PDR and shock models is to measure the atomic line intensities of OI, CII, SI, SiII (see table 1). Signature of the CII line is a good indication of PDR at work.

Acknowledgements

We would like to thank Dave Hollenbach in assiting us in calculating the effect of the dissociative shocks.

References

Bakes, E.L.O. & Tielens, A.G.G.M.: 1994, Astrophys. J. **427**, 822.
Burton, M.G. & Geballe, T.R.: 1986, Mon. Not. of the RAS **223**, 13P.
Erickson, E.F., Houck, J.R., Harwit, M.O., Rank, D.M., Haas, M.R., Hollenbach, D.J.,
    Simpson, J.P. & Augason, G.C.: 1985, Infrared Physics **25**, 513.
Graham, J.R., Serabyn, E., Herbst, T.M., Matthews, K., Neugebauer, G., Soifer, B.T.,
    Wilson, T.D. & Beckwith, S.: 1993, Astron. J. **105**, 250.
Hollenbach, D. & McKee, C.F.: 1989, Astrophys. J. **342**, 306.
Loup, C., Forveille, T., Omont, A. & Paul, J.F.: 1993, Astron. & Astrophys. Suppl. **99**,
    291.
Tielens, A.G.G.M. & Hollenbach, D.J.: 1985, Astrophys. J. **291**, 722.
Young, K., Serabyn, G., Phillips, T.G., Knapp, G.R., Gusten, R. & Schulz, A.: 1992,
    Astrophys. J. **385**, 265.

STELLAR EVOLUTION AND MASS LOSS ON THE ASYMPTOTIC GIANT BRANCH

ALBERT A. ZIJLSTRA

European Southern Observatory, Karl Schwarzschild Strasse 2, D-85748 Garching bei München, Germany

Abstract. Mass loss dominates the stellar evolution on the Asymptotic Giant Branch. The phase of highest mass-loss occurs during the last 1–10% of the AGB and includes the so-called Miras and OH/IR stars. In this review I will discuss the characteristics and evolution of especially Miras, and discuss how they are linked to the mass loss. There are indications that high mass-loss rates are only reached for relatively young stars with massive progenitors. The mass loss rates vary both on long and short time scales: the short-term variations are likely linked to luminosity variations associated with the thermal-pulse cycle. The influence of mass loss in the post-AGB phase is also discussed.

1. The Asymptotic Giant Branch

The Asymptotic Giant Branch (AGB) marks the end point of the evolution of stars with initial masses $\lesssim 8 M_\odot$. Stars on the AGB are characterized by a degenerate C/O core, with nuclear burning taking place in a shell; the total AGB life time is of order 10^6 yr. A recommended review of AGB evolution can be found in Iben & Renzini (1983), although many of the numerical results have been updated since (*e.g.* Vassiliadis & Wood 1993).

During the first 90% of the total AGB life time the energy is provided by helium shell burning. This phase is called the early AGB, and is marked by a steady increase in luminosity. The early AGB ends when the helium, itself the remnant of earlier hydrogen burning, is finally exhausted and hydrogen is re-ignited in a thin shell. This starts the thermal-pulsing AGB: after sufficient helium has formed, the helium ignites with a flash (the thermal pulse; TP) followed by a new phase of helium burning. After the helium is again exhausted the cycle reruns. Typical time scales between thermal pulses are from 10^4 yr for the most massive stars to $> 10^5$ yr at low masses.

During the AGB the mass of the envelope surrounding the nuclear-burning shell is steadily reduced, by nuclear burning on the inside (increasing the mass of the C/O core) and mass loss through a stellar wind on the outside. When the remaining envelope mass has dropped below $0.05 M_\odot$, the stellar temperature begins to increase and the so-called post-AGB phase starts. The temperature can reach values of 10^5 K (depending on mass) before nuclear burning ceases and the star enters the white-dwarf cooling track.

The best-known AGB stars are the Miras and OH/IR stars. Miras have typical luminosities slightly below $10^4 L_\odot$. They are large-amplitude pulsators with periods of 200–600 days. The periods are generally stable, but

Astrophysics and Space Science **224**: 309–320, 1995.

TABLE I
Mira bolometric magnitudes

population	M_{bol}	ref.
Galactic Centre	-4.5 → -6	Jones *et al.* 1994
Bulge	-3.5 → -5	Whitelock *et al.* 1991
Galactic Cap	-3.5 → -5	Whitelock *et al.* 1994
Metal-rich globular clusters	-3.0 → -4.5	Menzies & Whitelock 1985
LMC	$\geq$ -7.2	Wood *et al.* 1992

the amplitudes may vary in time. The observed luminosities of Miras, shown in Table I, indicate that they are located at the tip of the AGB. They are consistent with expected values for the TP-AGB even for the faint globular cluster Miras (Vassiliadis & Wood 1993, their table 2). The OH/IR stars are optically obscured stars, with very high mass-loss rates and long periods of 1000–3500 days. It has been a long-standing problem whether they are more massive than the Miras or form a later evolutionary phase (*e.g.* Habing 1989).

In this review I will first discuss the luminosity variations on the AGB and the pulsations. The mass loss is discussed in Section 4, followed by some comments on the initial–final mass relation and the post-AGB evolution.

2. Luminosity variations

The luminosity during shell burning is in principle a monotonous function of core mass. Core mass–luminosity relations have been derived by many authors: a review which is still useful can be found in Boothroyd & Sackman (1988b). From this one derives the classical AGB limit of $M_{bol} = -7.2$ for the highest possible core mass of $1.4 M_\odot$. This relation is calculated for quiescent hydrogen burning. Thus, it cannot be used for phases of helium burning, and during the first thermal pulses the quiescent luminosity is in fact not reached. The core mass here refers to both the C/O core and the helium shell.

Strong deviations from the $M_c L$ relation occur at times. First, there is a strong effect on L during the TP, as shown in Figure 1. A substantial luminosity dip occurs during quiescent helium burning which lasts up to 30% of the cycle (Boothroyd & Sackmann 1988a, Vassiliadis & Wood 1993). For stars with low envelope mass (*i.e.* all low-mass stars) the TP causes a brief but large luminosity increase lasting $\lesssim 10^3$ yr. For stars with high envelope mass this increase is much less pronounced. Second, for massive stars with high-mass envelopes, the base of the convective envelope may penetrate

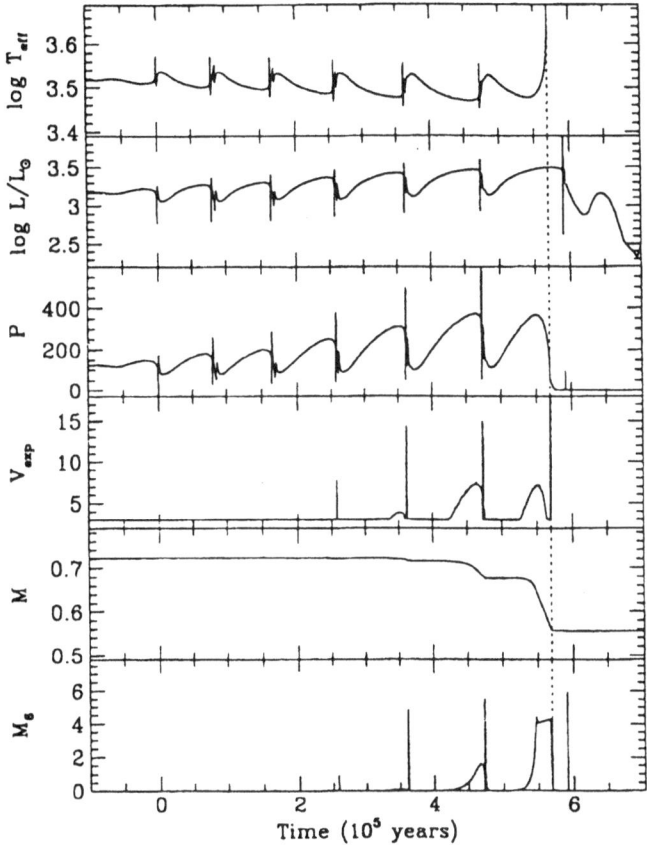

Fig. 1. (Taken from Vassiliadis & Wood 1993) Variations of stellar parameters during the thermal-pulse cycle. The period is calculated assuming fundamental mode; V_{exp} is derived from the period. $\dot{M}_6$ is the mass-loss rate in units of $10^{-6} M_\odot yr^{-1}$

the hydrogen-burning layer. This will temporarily increase the luminosity of the star (Blöcker & Schönberner 1991), although by how much is not clear (Vassiliadis and Wood 1993). When the envelope mass decreases, deep convection ceases and the star will appear to evolve *down* the AGB. Because of these deviations it is dangerous to derive a core mass from observed AGB luminosities for individual stars.

Wood *et al.* (1992) argue that there is no evidence for overluminous AGB stars in the LMC which might be expected from envelope burning. This is based on the lack of stars with $M_{bol} < -7.2$. However, AGB core masses possibly do not exceed $1.1 M_\odot$, as indicated by initial–final mass relations (*e.g.* Weideman 1987). If this is the case, the limiting magnitude derived from the luminosity–core mass relation is closer to -6.9 and there would be some overluminous stars in the LMC.

TABLE II

(From Feast 1989) Mira velocity dispersions

Period range (days)	asymmetric drift (km/s)	σ_T (km/s)	no. of stars
< 140	-33 ± 13	81	22
145–200	-111 ± 22	180	46
200–250	-61 ± 22	101	71
250–300	-33 ± 10	88	77
300–350	-32 ± 6	69	83
350–410	-23 ± 8	58	54
>410	-15 ± 8	50	35

3. Pulsations and the PL relation

Optical Miras show a strong period–luminosity relation, in the sense that higher luminosity coincides with longer periods. The relation is quite narrow in the K-band; it shows more scatter when the bolometric luminosity is used. The relation has been interpreted as an evolutionary sequence, but this is unlikely for several reasons: (1) Mira life times ($\sim 10^5$ yr) are too short to give an appreciable evolution in L. (2) Miras in individual globular clusters show a very small range in luminosity. (3) For Galactic Miras the velocity dispersion is a function of period (Table II, from Feast 1989) indicating older progenitors for shorter-period Miras. (The shortest-period Miras ($P <$ 145 days) have lower velocity dispersion than expected and may be a different group of objects, possibly higher-overtone pulsators.) Thus, it appears that the relation traces stars with different progenitor mass, with longer periods implying younger, more massive stars. Metallicity may also play a role.

Whitelock (1986) has shown that in globular clusters, semiregular variables and Miras define a sequence in the PL diagram which is much shallower than the Mira PL relation (Figure 2). The sequence has the same slope as found in recent theoretical calculations (Vassiliadis & Wood 1993, their figure 20) although it is shifted towards shorter periods. It seems likely that the Whitelock relation is the true evolutionary sequence and that the semi-regular variables form a pre-Mira phase.

During the thermal-pulse cycle, the changing characteristics of the star will affect the periods. This is shown in Figure 1: the star will move up and down (but not away from) the PL relation. This behaviour would of course not be observed if the Mira phase is confined to a single phase of the thermal-pulse cycle. A large shift up the PL could occur during the luminosity spike following a TP. Even though this spike is very brief, Mira

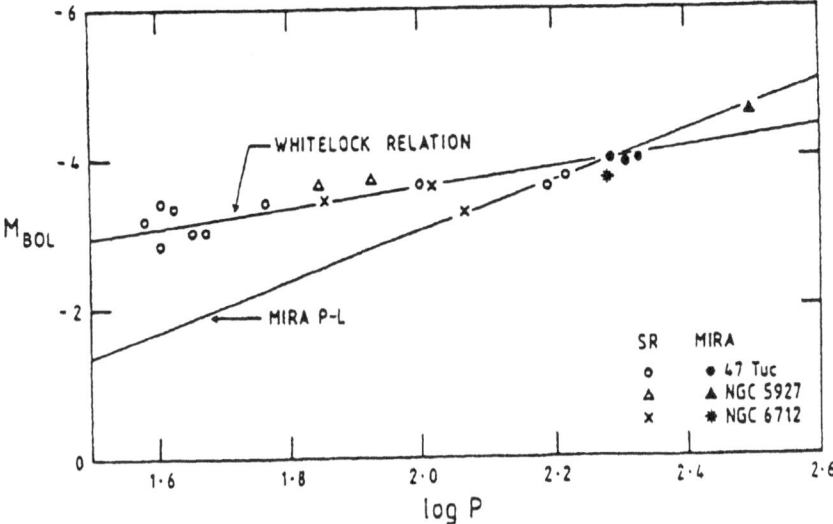

Fig. 2. (Taken from Feast 1989) The Miras and SR variables in globular clusters, together with the pre-Mira evolutionary track suggested by Whitelock (1986). The Mira period–luminosity relation (Feast *et al.* 1989) is shown for comparison.

pulsations could be triggered by a resonance because the nuclear time scale of the pulse is of the same order as the pulsation. In conclusion, although the PL relation does not appear to be an evolutionary sequence, movement up and down the relation could still occur.

The *PL* relation holds for $P < 400$ days. For longer periods the scatter in P becomes very large and in fact it is not obvious whether any relation still exists (*e.g.* Wood *et al.* 1992, their figure 8). The relation could be broadened by envelope removal through mass loss which would increase the period; this could be especially important for more massive stars. A mode change to a lower pulsation mode would also have this effect. In contrast, contraction of the star, which occurs at the start of the post-AGB evolution will shorten its period.

Pulsation mode. Whether Miras pulsate in the fundamental mode or in an overtone has been a long-standing problem. The existence of a clear P-L relation implies that all Miras with $P < 400$ days have the same pulsation mode. (For the long-period OH/IR stars the absence of a clear relation does not allow one to extrapolate this conclusion.) In recent years the fundamental mode has generally been favoured: the evidence in favour is reviewed by Wood (1989) and is mainly based on observed shock velocities. However, the observed T_{eff} of the stars is lower than predicted from the fundamental mode. This was recently confirmed by angular diameter measurements of R Leo (Tuthill *et al.* 1994) which can exclude the fundamental mode for this Mira. There are also a few Miras known with multiple periods. In the case of IZ Peg, the periods are 488 and 345 days (Whitelock et al. 1994): if these are both modes of the star, the fundamental mode would have a period close to 1000 days. Finally, from a Fourier analyses of two Miras, Barthés & Tuchman (1994) favour first-overtone pulsations. Thus, although the situation is by no means clear, the balance is shifting to Miras being overtone pulsators.

Whitelock (1986) has suggested that metal-poor stars pulsate in the fundamental mode, while metal-rich stars pulsate in the first overtone. Since only metal-rich clusters contain Miras, this would imply that all Miras pulsate in the overtone.

4.   Mass loss

Mass loss occurs along both the first giant branch (RGB) and the AGB, but only reaches catastrophic values at the tip of the AGB. Here values of $10^{-5} M_\odot yr^{-1}$ or higher are reached, which means that the envelope is removed much faster by mass loss than by nuclear burning. Thus, catastrophic mass loss will terminate the AGB, and the time at which it occurs determines the final core mass.

The IRAS colours of AGB and RGB stars show a gap between stars with essentially stellar colours at 12, 25 and 60μm, and 'dusty' stars with significant excess (van der Veen & Habing 1988). Fitting the infrared spectra with a dust model, it appears that the gap coincides with mass-loss rates $\lesssim 10^{-7} M_\odot yr^{-1}$. In contrast, optical and UV observations show that essentially all red-giant stars show mass loss, often with mass-loss rates as low as $10^{-10} M_\odot yr^{-1}$ (e.g. Judge & Stencel 1991). These low mass-loss rates are not seen in the infrared and appear to be 'dustless'.

All Miras have 'dusty' IRAS colours and thus exhibit substantial mass-loss rates. Semi-regular variables, on the other hand, are found in both categories. If we identify these with pre-Mira stars, the conclusion is that 'dusty' mass loss begins during the semi-regular phase and that the mass-loss rate correlates with the amplitude of the pulsation. Very high mass-loss rates are only found for long-period Miras with $P > 400$ days, as shown by Jura (1986) for carbon stars and Whitelock et al. (1994) for oxygen stars. Combining this with the results from the previous section, implies that high mass-loss rates are in general not reached for the lowest-mass stars; for instance, it is doubtful whether our Sun will ever become such a long-period Mira. However, low-mass stars could mimic this behaviour during the brief high-luminosity phase following a thermal pulse.

At present there is no theoretical model which can actually predict the mass loss. However, it is generally supposed that mass loss occurs in two stages. One mechanism such as pulsations or other long-term variability heats the atmosphere and causes it to extend to several stellar radii. At that distance dust forms and the radiation pressure on the dust will accelerate the shell and amplify the mass-loss rate. This model naturally leads to the expectation that there are two evolutionary phases: (1) At low mass-loss rates, the densities are so low that even if dust still forms, it will not be dynamically coupled to the gas. In this case mass loss is driven solely by the mechanism which heats the envelope; (2) At higher mass-loss rates the gas couples to the gas and mass loss becomes more efficient and larger. The first phase would then be associated with the 'dustless' RGB and early AGB, the second phase is when the mass loss suddenly becomes catastrophic. Calculations on the dust-to-gas coupling have been done by MacGregor & Stencel (1992) and Netzer & Elitzur (1993). In either case, in principal the mass-loss rate would be determined by the fundamental stellar parameters, and could be quantified.

Such parametrizations are indeed available. The oldest is the well-known Reimers law (1975):

$$\dot{M} = -4 \times 10^{-13} \frac{LR}{M} \quad M_\odot yr^{-1}$$

with L,M,R in solar units. This relation was derived from data on K stars and is known to predict too low rates from late AGB stars. Fitting model results from Bowen (1988), Blöcker (1993) has proposed a stronger dependence on L and M:

$$\dot{M} = -4.8 \times 10^{-9} \frac{L^{2.7}}{M^{2.1}} \frac{LR}{M} \quad M_\odot yr^{-1}$$

This, however, predicts mass-loss rates close to $10^{-3} M_\odot yr^{-1}$ which are not observed. Although it is possible that the phase with such high mass-loss rates lasts too short to be observable, from stellar population studies in the LMC, Groenewegen and de Jong (1994b) and Groenewegen et al. (1994) conclude that this formula overestimates mass-loss rates by factors of 3–10.

The best observationally-established relation is given by Judge & Stencel (1991) for a sample of RGB and AGB stars. They find a good correlation between $\dot{M}$ and $\log g$, which can be converted to a similar relation as above:

$$\dot{M} = (2.3 \pm 1.3) \times 10^{-14} \left(\frac{R^2}{M} \right)^{1.43 \pm 0.23} \quad M_\odot yr^{-1}$$

This is close to Reimers' law with L replaced by R, a reasonable substitution since the two appear to be correlated during this phase of the evolution. Although the observational spread in the relation is significant, and especially the luminous carbon stars at the tip of the AGB fall above the relation, it is probably the best one to use in model calculations.

Vassiliadis & Wood (1993) have suggested that momentum in the wind should not exceed that in the radiation field, which for AGB stars implies $\dot{M} \lesssim 3 \times 10^{-5} M_\odot yr^{-1}$. They calculate mass-loss rates from a PL relation which is very steep, and Groenewegen & de Jong (1994b) find that their results do not fit the observed stellar population in the LMC. However, it is possible that the momentum limit should be included in the above relations.

Mass-loss variations. The above formulae predict that (1) $\dot{M}$ increases with time on the AGB, (2) significant variations should occur following a thermal pulse due to the change in surface parameters. Thus, both long-term and short-term variations are expected. The short-term variations occur on time scales comparable to or shorter than the expansion of the nebula and should be observable.

The luminous OH/IR stars, which show very high mass-loss rates, often show very faint CO emission (Heske *et al.* 1990). Schutte & Tielens (1989) interpret this as evidence that the outer shell, where the CO emission originates, traces a lower mass-loss phase. The increase in $\dot{M}$ would date back to $\sim 10^4$yr ago. These time scales are consistent with those of the thermal pulse cycle, although it is not proven that the increase is related to this.

Figure 3 shows the IRAS colours for all M stars from the GCVS, using only data with good quality at all plotted bands ($q = 3$). The vertical lines connect points with the same mass-loss rates; horizontal lines connect points with the same outer radius, converted to a time using an expansion velocity of 10 km/s. There are at most very few sources for which the mass loss has started less than 10^3 yr ago. At 100μm almost all sources show a significant excess: this is either explained by a very shallow emissivity index (Rowan-Robinson *et al.* 1986) or an outer shell caused by earlier mass loss (*e.g.* van der Veen *et al.* 1994). Thus, mass loss generally continues for 10^3yr or longer.

Interruptions of mass loss appear to be a common feature of optical carbon stars, which all show 60μm excess indicating a detached, cool shell. Willems & de Jong (1986) have proposed that this occurs when due to a thermal pulse the C/O ratio becomes unity, with the resulting chemistry leading to a decrease in dust formation. This scenario predicts that the detached shells are oxygen rich. However, Bujarrabal & Cernicharo (1994) and Groenewegen & de Jong (1994a) find evidence that at least for some carbon stars the shells are carbon rich. Zijlstra *et al.* (1992) and Hashimoto (1994) show that some oxygen stars also show detached shells. This suggests that mass-loss interruptions may be a common phenomenon on the AGB and are not limited to carbon stars. It would seem logical to associate these interruptions with the phase of quiescent helium burning during the TP cycle, when the luminosity is lowest (*e.g.* Olofsson *et al.* 1990). Zijlstra *et al.* conclude that mass loss on the AGB contains a *periodic* component.

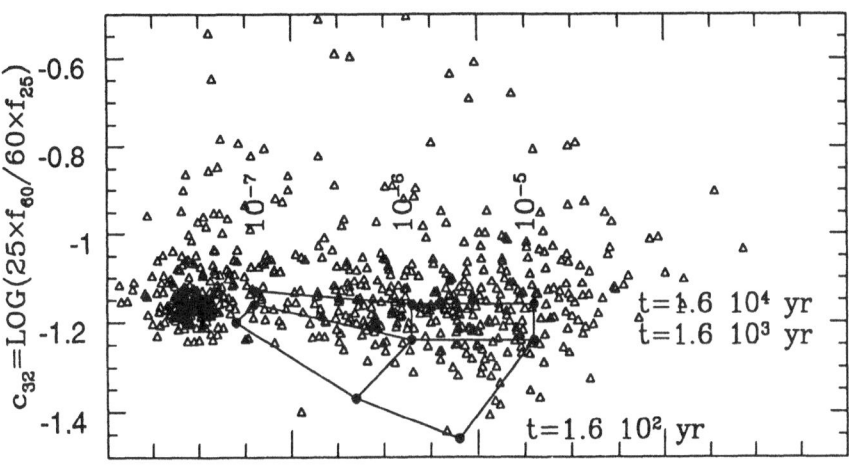

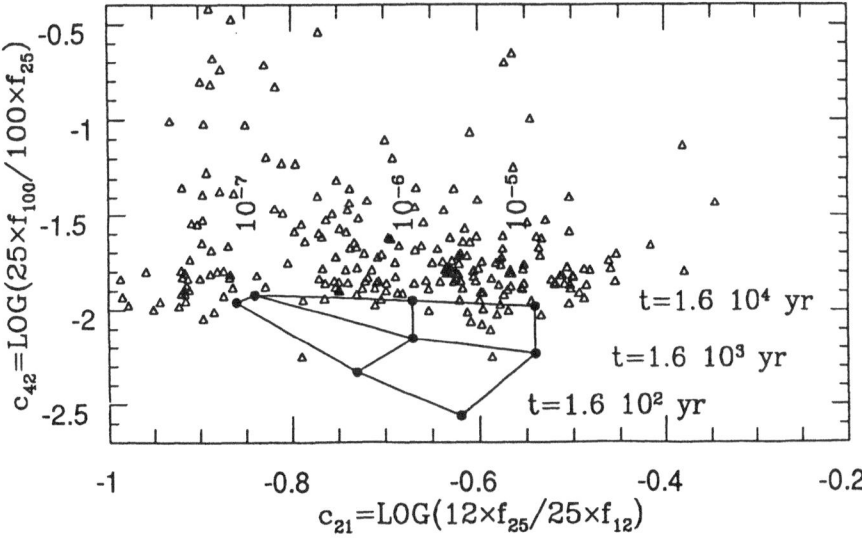

Fig. 3.   IRAS colour-colour plots for M-stars taken from the GCVS. These will be mainly (but not exclusively) AGB and RGB stars. Model tracks at constant $\dot{M}$ (vertical lines) and constant time scale (horizontal lines) are shown. (The absolute values for $\dot{M}$ can change with assumed gas-to-dust ratio.)

Implications. In conclusion, high mass-loss rates coincide with high luminosity. This indicates that the primary factor determining the mass-loss rate is stellar mass, although low-mass stars could briefly obtain these luminosities during a thermal pulse. If mass loss varies through the TP cycle, as is likely, stars may pass through several mass-loss episodes, each lasting of order 10^4yr. Because of the dust/gas decoupling, there may be a limit below which $\dot{M}$ may drop by one or more orders of magnitude. This could amplify the effect of the TP-cycle.

5. Initial–final mass relation

The mass loss determines the mass of the core of the final white dwarf, and thus the initial–final mass relation. This relation is poorly constrained by observations: the only limits are given by a few white dwarfs in young clusters (Weideman 1987, Vassiliadis & Wood 1993). Applying the mass-loss relations above to theoretical models of AGB stars (Groenewegen& de Jong 1994b) leads to results consistent with these limits. Specifically, the relation is flat for 1–$3M_\odot$ stars which all end up as white dwarfs with masses around $0.6M_\odot$. Higher-mass stars end up as heavier white dwarfs, but the Chandrasekhar limit is never reached: the most massive AGB stars leave 1–$1.2M_\odot$ remnants. This has important consequences for the white-dwarf mass distribution. Han *et al.* (1994), following earlier suggestions, have shown that the observational constraints are well matched if one assumes that the envelope is lost shortly after the binding energy of the envelope becomes positive. Their relation could possibly be taken as a lower limit to the initial–final mass relation.

6. Post-AGB

The AGB ends when the stellar temperature begins to rise: the following phase is normally called the post-AGB. The mass loss continues into the early post-AGB phase as indicated by observational time scales: if the decrease in envelope mass (which determines how fast the stellar temperature increases) were only due to nuclear burning, it would take so long to reach the planetary-nebula phase that the nebula would long have disappeared. Schönberner (1983) has assumed that the high mass loss ends when the star first reaches $T_{eff} = 5500$ K. This is consistent with the fact that there are many optically visible post-AGB stars of type F and later, but almost none of K (Waters *et al.* 1989). However, van der Veen *et al.* (1994) present a few cases where a significantly cooler star has a detached shell. The details of how the mass loss ends are still very unclear.

In a close binary system where one of the components passed through the AGB, it is possible that part of the shell may be retained in a circum-binary disk. This mechanism has been invoked to explain the extreme metal deficiency of some post-AGB stars (Van Winckel et al. 1992, Waters et al. 1992). It could also explain why highly obscured post-AGB stars invariably have bipolar shells: only objects where part of the shell remains close to the star will show high obscuration during the post-AGB phase (e.g. Sieben-morgen et al. 1994), and the binary nature of these objects could cause the non-sphericity of the shells.

Because the helium-burning phase in the TP cycle is relatively short, it is commonly assumed that almost all post-AGB stars have left the AGB as hydrogen burners. However, Vassiliadis (1993) argues that 25–50% of all post-AGB stars are helium burners, based on his evolutionary models. The reason is that for high-mass stars, the time the post-AGB star is visible is much longer for helium burners than for hydrogen burners, thus biasing the statistics. In addition, for low-mass stars the mass loss may strongly peak immediately after a pulse thus increasing the likelihood the envelope will be lost at this time. Therefore, both the highest- and lowest-mass stars may often be helium burners. The expected relative numbers of helium burners is strongly depended on poorly understood details of the mass-loss process.

For hydrogen-burning stars, there is a chance that a final thermal pulse will occur during the post-AGB evolution or even in the white-dwarf phase. It is thought that this may lead to a final phase of high mass loss, in which the entire remaining hydrogen layer is removed. The mass loss in this phase will be hydrogen-poor leading to strong infrared (dust) emission. Although a rare phenomenon, there are in fact stars which appear to have very recently entered this phase, in particular FG SGe (e.g. van Genderen 1994) and N66 in the LMC (Peña et al. 1994). Especially the latter is a good candidate for studying this final episode of mass loss.

References

Barthès, D. & Tuchman, Y.: 1994, Astron. & Astrophys. **289**, 429.
Blöcker, T. & Schönberner, D.: 1991, Astron. & Astrophys. **244**, L43.
Blöcker, T.: 1993, Acta Astronomica **43**, 305.
Boothroyd, A.I. & Sackmann, I.-J.: 1988a, Astrophys. J. **328**, 632.
Boothroyd, A.I. & Sackmann, I.-J.: 1988b, Astrophys. J. **328**, 641.
Bowen, G.H.: 1988, Astrophys. J. **329**, 299.
Bujarrabal, V. & Cernicharo, J.: 1994, Astron. & Astrophys. in press.
Feast, M.W.: 1989, in: *The Use of Pulsating Stars in Fundamental Problems of Astronomy*, ed: E.G. Schmidt, Cambridge University Press, p.205.
Feast, M.W., Glass, I.S., Whitelock, P.A. & Catchpole, R.M.: 1989, Mon. Not. of the RAS **241**, 375.
Groenewegen, M.A.T. & de Jong, T.: 1994a, Astron. & Astrophys. **282**, 115.
Groenewegen, M.A.T. & de Jong, T.: 1994b, Astron. & Astrophys. **283**, 463.
Groenewegen, M.A.T., van den Hoek, L.B. & de Jong, T.: 1994, Astron. & Astrophys. in press.

Habing, H.J.: 1989, in: *From Miras to Planetary Nebulae: which Path for Stellar Evolution?*, (Editions Frontières, Gif sur Yvette), p.16.

Han, Z., Podsiadlowski, Ph. & Eggleton, P.P.: 1994, Mon. Not. of the RAS **270**, 121.

Hashimoto, O.: 1994, Astron. & Astrophys. Suppl. , in press.

Heske, A., Forveille, T., Omont, A., van der Veen, W.E.C.J. & Habing, H.J.: 1990, Astron. & Astrophys. **239**, 173.

Iben Jr., I. & Renzini, A.: 1983, Ann. Rev. of Astron. & Astrophys. **21**, 271.

Jones, T.J., Gehrz, R.D., Lawrence, G.F. & McGregor, P.J.: 1994, University of Minnesota preprint.

Judge, P.G. & Stencel, R.E.: 1991, Astrophys. J. **371**, 357.

Jura, M.: 1986, Astrophys. J. **303**, 327.

MacGregor, K.B. & Stencel, R.E.: 1992, Astrophys. J. **397**, 644.

Menzies, J.W. & Whitelock, P.A.: 1985, Mon. Not. of the RAS **212**, 783.

Netzer, N. & Elitzur, M.: 1993, Astrophys. J. **410**, 701.

Olofsson, H., Carlström, U., Eriksson, K., Gustafsson, B. & Willson, L.A.: 1990, Astron. & Astrophys. **230**, L13.

Peña, M., Torres-Peimbert, S. & Peimbert, M.: 1994, Astrophys. J. **428**, L9.

Reimers, D.: 1975, in: *Problems in Stellar Atmospheres and Envelopes*, eds: Basheck *et al.* , Springer, Berlin, p.229.

Rowan-Robinson, M., Lock, T.D., Walker, D.W. & Harris, S.: 1986, Mon. Not. of the RAS **222**, 273.

Schönberner, D.: 1983, Astrophys. J. **272**, 708.

Schutte, W.A. & Tielens, A.G.G.M.: 1989, Astrophys. J. **343**, 369.

Siebenmorgen, R., Zijlstra, A.A. & Krügel, E.: 1994, Mon. Not. of the RAS in press.

Tuthill, P.G., Haniff, C.A., Baldwin, J.E. & Feast, M.W.: 1994, Mon. Not. of the RAS **266**, 745.

van der Veen, W.E.C.J. & Habing, H.J.: 1988, Astron. & Astrophys. **194**, 125.

van der Veen, W.E.C.J., Waters, L.B.F.M., Trams, N.R. & Matthews, H.E.: 1994, Astron. & Astrophys. **285**, 551.

van Genderen, A.M.: 1994, Astron. & Astrophys. **284**, 465.

Van Winckel, H., Mathis, J.S. & Waelkens, C.: 1992, Nature **356**, 500.

Vassiliadis, E.: 1993, Acta Astronomica **43**, 315.

Vassiliadis, E. & Wood, P.R.: 1993, Astrophys. J. **413**, 641.

Waters, L.B.F.M., Trams, N.R. & Waelkens, C.: 1992, Astron. & Astrophys. **262**, L37.

Waters, L.B.F.M., Waelkens, C. & Trams, N.R.: 1989, in: *From Miras to Planetary Nebulae: which Path for Stellar Evolution?*, Editions Frontières, Gif sur Yvette, p.449.

Weidemann, V.: 1987, Astron. & Astrophys. **188**, 74.

Whitelock, P.A.: 1986, Mon. Not. of the RAS **219**, 525.

Whitelock, P.A., Feast, M.W. & Catchpole, R.M.: 1991, Mon. Not. of the RAS **248**, 276.

Whitelock, P.A. *et al.* : 1994, Mon. Not. of the RAS **267**, 711.

Willems, F.J. & de Jong, T.: 1986, Astrophys. J. **303**, L39.

Wood, P.R.: 1989, in: *From Miras to Planetary Nebulae: which Path for Stellar Evolution?*, Editions Frontières, Gif sur Yvette, p67.

Wood, P.R., Whiteoak, J.B., Hughes, S.M.G., Bessell, M.S., Gardner, F.F. & Hyland, A.R.: 1992, Astrophys. J. **397**, 552.

Zijlstra , A.A., Loup, C., Waters, L.B.F.M. & de Jong, T.: 1992, Astron. & Astrophys. **265**, L5.

THE CIRCUMSTELLAR SHELLS OF CARBON MIRAS

MARTIN GROENEWEGEN
Institut d'Astrophysique de Paris, 98 bis Boulevard Arago, F-75014 Paris, France

Abstract. The relation between mass loss rate and pulsation period in carbon Miras is discussed. The dust mass loss rate is very low (about 2 x 10^{-10} $M_\odot$/yr) up to about P = 380 days, where there is a sudden increase. For $P > 400$ days there is a linear relation between log $\dot{M}$ and P. The change in the mass loss rate near 380 days may be related to radiation pressure on dust becoming effective in driving the outflow.

Key words: AGB stars – mass loss – long period variables

1. Introduction

The mass loss rate in AGB stars determines in most cases the evolution of these stars, since the time scale on which mass loss takes place is usually much shorter than the time scale of the nuclear processes.

From a theoretical point of view, the mass loss rate is often parameterized in terms of poorly known stellar quantities (*e.g.* Reimers law $\dot{M} \sim LR/M$). It is more convenient to describe the mass loss rate as a function of pulsation period, a quantity which can be determined by monitoring the optical and/or infrared light curves. The $\dot{M} - P$ relation has been studied in quite some detail for oxygen-rich Miras (Schild 1989, Wood 1990, Whitelock 1990, Whitelock *et al.* 1994) but never for carbon-rich Miras. In this contribution such a relation is presented.

2. Results

Thirty-two carbon Miras were selected from the General Catalog of Variable Stars, Jones *et al.* (1990) and Le Bertre (1992). The spectral energy distributions (SEDs) and IRAS LRS spectra were fitted using the dust radiative transfer model of Groenewegen (1993). The model simultaneously solves the radiative transfer and radiative equilibrium equation for the dust in spherical geometry. Ten of the 32 stars have been fitted previously by Groenewegen (1994a) as part of a study into the dust properties of infrared carbon stars in general.

The parameters of the model are: (1) T_c = the temperature of the dust at the inner radius, (2) SiC/AMC = the ratio by mass of silicon carbide to amorphous carbon dust, and (3) $\dot{M}$ = the (total) mass loss rate. The luminosities have been determined from a newly derived period-luminosity relation for carbon Miras in the LMC (based on data in Feast *et al.* 1989 and Hughes & Wood 1990).

Astrophysics and Space Science **224**: 321–324, 1995.
© 1995 *Kluwer Academic Publishers.*

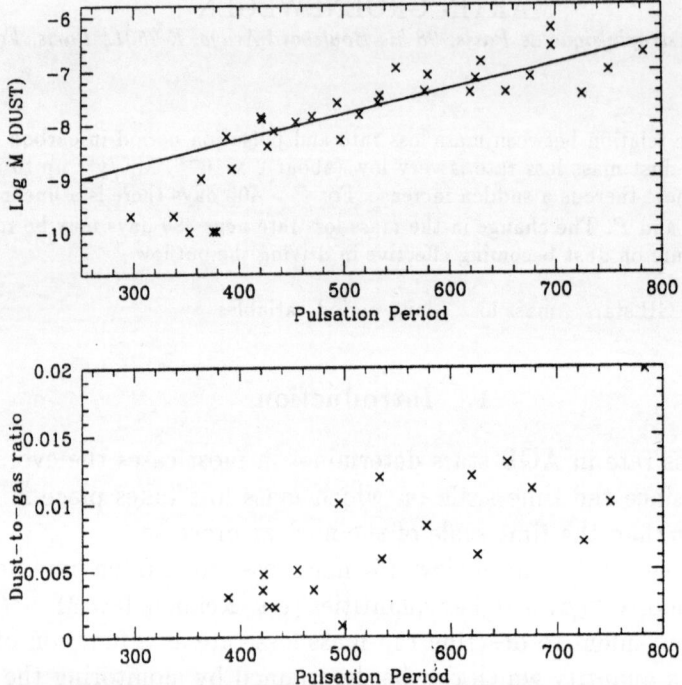

Fig. 1. The dust mass loss rate (top panel) and dust-to-gas ratio (bottom panel) in the sample of carbon Miras. The solid line is a least-squares fit for $P > 400$ days.

The relation between the dust mass loss rate and pulsation period is shown in Fig. 1 (top panel). Two features stand out: there is a tight correlation above 400 days, and there appears to be a significant increase in the dust mass loss rate near $P = 380$ days. Unfortunately there are no infrared data and/or LRS spectra available for carbon Miras with shorter periods.

Recently, Netzer & Elitzur (1993) and Habing et al. (1994) predicted that a lower limit of about 10^{-7} $M_\odot$/yr to the (total) mass loss rate is needed to sustain an outflow driven by radiation pressure on dust. For lower mass loss rates the coupling between the gas and the dust is too weak. Using a dust-to-gas (DTG) ratio of 0.003 near $P = 380$ days (see below), a total mass loss rate of 8×10^{-8} $M_\odot$/yr is derived, close to this lower limit. Thus, the increase in the dust mass loss rate near 380 days may be related to radiation pressure on dust becoming effective and hence provide observational evidence for the theoretical prediction of Netzer & Elitzur and Habing et al.

A different tracer of the dust formation process is the DTG ratio. In Fig. 1 (bottom panel) the DTG ratio is plotted versus P. The gas mass loss rates as derived from CO measurements were taken from Loup et al. (1993) and scaled to the adopted distances. The DTG ratio increases with period for $P \gtrsim 400$ days, indicating a more efficient dust production; there is no data

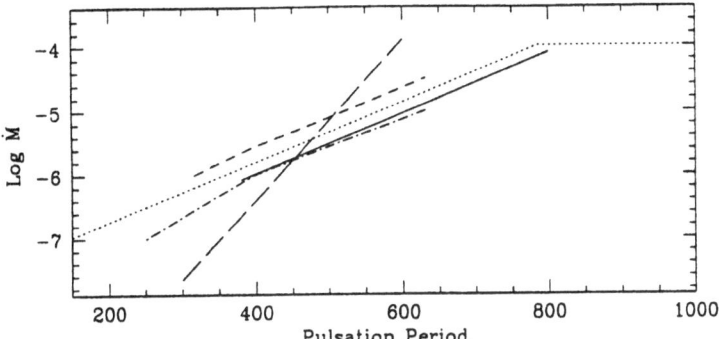

Fig. 2. The mean $\dot{M} - P$ relation for carbon Miras (solid line), and oxygen-rich Miras: in the solar neighbourhood (Schild 1989, dotted line; Wood 1990, long-dashed line), Galactic Bulge (Whitelock 1990, short-dashed line), South Galactic cap (Whitelock *et al.* 1994, dot-dashed line).

available for shorter periods. The scatter in this relation is larger than in the $\dot{M}(\mathrm{dust})-P$ relation. This is probably due to the fact that the gas mass loss rates were derived with a simple formula from the CO measurements. The most accurate method to determine the total mass loss rate and the DTG ratio is by modelling both the SED and CO line profiles (See my poster contribution in this volume, and Groenewegen 1994b).

In Fig. 2 the mean $\dot{M} - P$ relation for carbon Miras with $P > 400$ days is compared to relations derived for oxygen-rich Miras in several locations of the Galaxy. For the carbon Miras the transformation of the dust mass loss rate to a total mass loss rate is based a gas-to-dust ratio of 200. Based on Fig. 1 this is accurate to about a factor of 3. The scatter in each relation is about 1 dex. for a given P. Except for the relation of Wood (1990), the four other relations agree very well with each other over the period range they have in common, probably (and not surprisingly) indicating that it is the same physical mechanism which drives the outflow in both carbon-rich and oxygen-rich Miras when $P > 400$ days. The relation presented by Wood (1990) has a much steeper slope and is not in agreement with the other relations. It was this relation (+ an extension for longer periods) which was used in the evolutionary models of Vassiliadis & Wood (1993). One of the conclusions of that paper was that mass loss predominantly occurs during the last thermal pulse. The present study suggests that if one of the other $\dot{M} - P$ relations is used that mass loss is probably found to be important during the last few thermal pulses.

As a by-product of the modelling the dust temperature at the inner radius and the ratio of SiC to amorphous carbon are derived. In Groenewegen

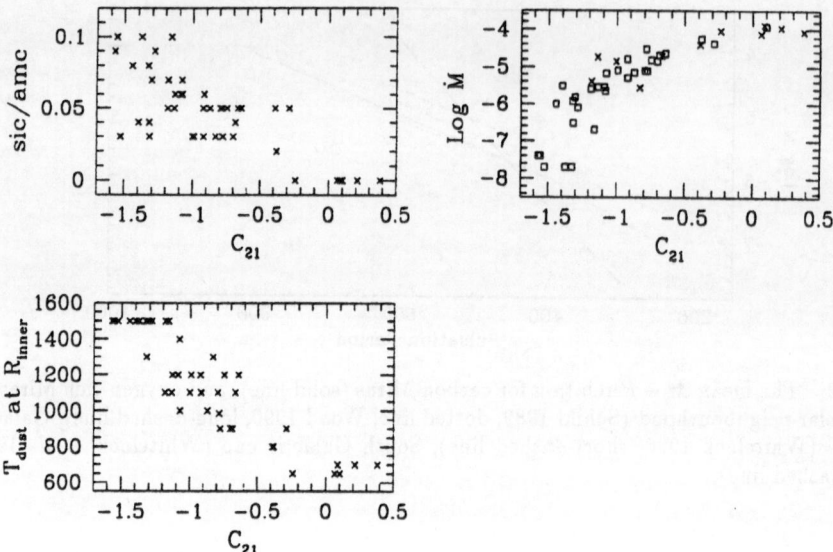

Fig. 3.   The SiC/AMC ratio (top left), the mass loss rate (top right) and the dust temperature at the inner radius (bottom left) plotted against $C_{21} = 2.5 \log(S_{25}/S_{12})$ color. In the top-right panel the squares indicate stars for which the luminosity is based on a period-luminosity relation; for stars represented by the crosses a luminosity of 7050 $L_\odot$ was assumed (see Groenewegen 1994a). The derived mass loss rates scale like $L^{0.5}$.

(1994a) these quantities (and the mass loss rate) were studied explicitly for infrared carbon stars. With the present sample, which is twice as large and extents over the entire range of IRAS C_{21} color ($= 2.5 \log(S_{25}/S_{12})$) occupied by carbon stars these results can be updated (see Fig. 3). The trends in T_c and the SiC/AMC ratio with color are confirmed and extended to bluer IRAS colors. For a discussion I refer to Groenewegen (1994a).

References

Feast M.W., *et al.* : 1989, Mon. Not. of the RAS **241**, 375.
Groenewegen M.A.T.: 1993, Ph.D. thesis, University of Amsterdam.
Groenewegen M.A.T.: 1994a, Astron. & Astrophys. in press.
Groenewegen M.A.T.: 1994b, Astron. & Astrophys. in press.
Habing H.J., Tignon J. & Tielens A.G.G.M.: 1994, Astron. & Astrophys. **286**, 523.
Hughes S.M.G. & Wood P.R.: 1990, Astron. J. **99**, 784.
Jones T.J., *et al.* : 1990, Astrophys. J., Suppl. **74**, 785.
Le Bertre: 1992, Astron. & Astrophys. Suppl. **94**, 377.
Loup C., Forveille T., Omont A. & Paul J.F.: 1993, Astron. & Astrophys. Suppl. **99**, 231.
Netzer N. & Elitzur M.: 1993, Astrophys. J. **410**, 701.
Schild H.: 1989, Mon. Not. of the RAS **240**, 63.
Vassiliadis E. & Wood P.R.: 1993, Astrophys. J. **413**, 641.
Whitelock P.: 1990, Publ. of the ASP **11**, 365.
Whitelock P., *et al.* : 1994, Mon. Not. of the RAS **267**, 711.
Wood P.R.: 1990, in *From Miras to Planetary Nebulae*, ed: Frontieres: Gif-sur-Yvette, p67.

THE CIRCUMSTELLAR ENVIRONMENT OF THE
PECULAR F HYPERGIANT IRC +10420

RENÉ D. OUDMAIJER
Kapteyn Astronomical Institute
P.O. Box 800
9700 AV Groningen, The Netherlands

Abstract.
IRC +10420 is to date the only object that has been proposed to be in the transition from the Red Supergiant Phase to the Wolf-Rayet phase. In this contribution we report on new high resolution optical spectra of IRC +10420.

Key words: spectroscopy - IRC +10420 - mass loss - supergiants

1. Introduction

IRC +10420 is a peculiar F hypergiant with a large far-infrared excess attributed to circumstellar dust, and is one of the warmest central stars associated with an OH maser. While extensively studied at radio frequencies, not many spectroscopic studies at optical or near-infrared wavelengths have yet been made.

Based on its large inferred distance of 5 kpc, which implies luminosities in excess of 500,000 times solar, and its large infrared excess due to a previous mass losing phase, Jones *et al.* (1993) proposed that IRC +10420 is in the evolutionary phase between the Red SuperGiant phase (RSG) and Wolf-Rayet phase.

The object appears to be variable on timescales less than decades; Lewis *et al.* (1986) found the OH strength linearly variable with time, the onset of a rapid ionization of the stellar wind was found by the discovery of Hα and near-infrared lines that went into emission less than a decade ago (Irvine, 1986; Oudmaijer *et al.* 1994), and finally the optical part of the spectral energy distribution has seen considerable changes over the last 20 years (Jones *et al.* 1993).

Since our discovery of the near-infrared emission lines in the spectrum of IRC +10420 we have collected a large set of data on the object. Here we report on some of the results.

2. The distance of IRC +10420.

It is hard to determine the distance to an object to which no luminosity can be attributed on first principles. So far, the distance indicators are rather indirect, a situation that will probably not change in the future. These

Astrophysics and Space Science **224**: 325–328, 1995.

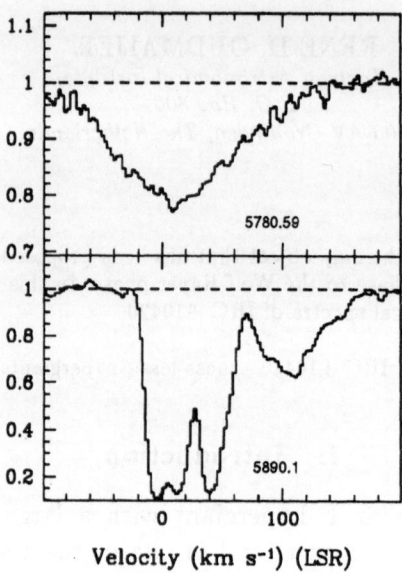

Fig. 1.   Examples of interstellar absorption lines in the spectrum of IRC +10420. In the
lower panel the Na D2 line is plotted, and in the upper panel the well-known λ578.0 DIB.

include the radial velocity and the large interstellar reddening of at least 4
magnitudes that Jones *et al.* inferred from their value of the interstellar
polarization towards the object. More indicators are listed by Jones *et al.*
(1993).

From our high resolution optical spectra, obtained with the Utrecht
Echelle Spectrograph mounted on the 4.2m William Herschel Telescope, La
Palma, Spain, we find that many interstellar absorption lines are present in
the data. We mention specifically the first high resolution observations of
the Na D lines, and a large set of Diffuse Interstellar Bands (DIB). In fact,
basically all absorption lines longward of 600 nm are interstellar!
In Fig. 1 we show the Na D2, and the well-known DIB λ578.0 absorption
line.

The photospheric component of the Na D line is well separated from the
interstellar line, that appears to be saturated. The DIB is displaced from
the stellar velocity, and is likely to be interstellar. We find from these lines
that the clouds responsible for the absorption have $A_V \sim 5$ mag. This value
is consistent with the results of Jones *et al.* (1993) who found $A_V > 4$.
The observed A_V as determined from fitting the photometry with Kurucz
(1979) atmospheric models amounts to 6-7 magnitudes (Oudmaijer *et al.*
1994). This implies that the bulk of the reddening towards IRC +10420 is
interstellar in origin. The large extinction indeed suggests a large distance
to the object, yielding luminosities consistent with the star being massive.

Paschen lines

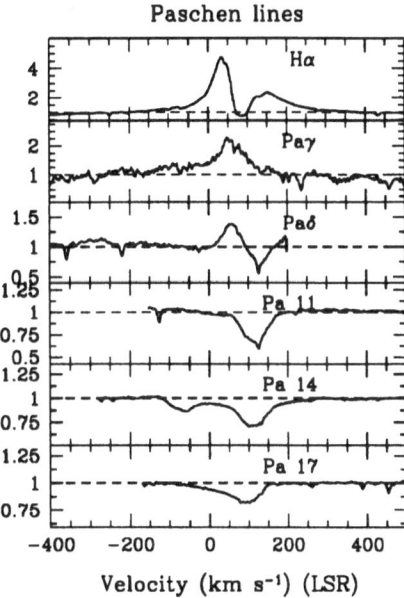

Fig. 2. A selection of Paschen lines plotted together with Hα. Note the trend, where the strongest line shows two peaks, the weaker lines show only blue emission, to blue emission with a red absorption component, to pure absorption.

3. The circumstellar environment

In Fig. 2 we show a selection of hydrogen recombination lines that we measured in the optical spectra of IRC +10420. What catches the eye are the peculiar shapes of the lines. The faintest line shows only absorption. The medium strong lines show blue shifted emission lines with respect to the stellar velocity, and the strong Hα is double peaked, with the red and blue peaks symmetrically placed around the stellar rest velocity as traced by CO.

The lines are identical to our spectra of the near-infrared Brα, Brγ and Pfγ lines as reported on in Oudmaijer *et al.* (1994). So far, these lines are typical for the rest of the emission lines we found in the spectra; there seems to be a trend in that strong lines, as for example the CaII triplet at 850 nm are double peaked, and that moderately strong lines only show blue-shifted emission.

These lines originate in an entirely different region, which is located much closer to the star than the better studied OH maser and (submm) molecular lines (*e.g.* Bachiller *et al.* 1988; Nercessian *et al.* 1989; Oloffson *et al.* 1982; Omont *et al.* 1993). A geometry for these regions was derived by Nedoluha & Bowers (1992) who found that an oblate spheroid, with an axial ratio tilted into the line of sight was the best representation of their high quality OH mapping.

It is hard however to interpret the emission lines as in Fig. 2. Although the blue shifted lines could indicate an inverse P Cygni type of behaviour, it proves very difficult to interpret the red emission that is present in the Hα line.

A possible geometry of the ionized region is a bipolar flow with a rather small opening angle. The reasoning is as follows. A bipolar outflow, beamed roughly into the line of sight, will show two peaks, and a minimum in the emission at the stellar velocity. This minimum is not an absorption dip, but arises purely from the fact that there is no or hardly any gas at the projected velocity of the star.

The higher level lines, that are formed in a region close to the star, will only show the approaching part of the flow since the star eclipses nearly all redshifted emission. In the case of Hα not all redshifted emission is eclipsed, because it is formed in a region larger than the star, consistent with the observed V/R ratio, where the red peak is still obscured by the star.

The bipolar scenario can explain the observed line profiles. It also explains the observed V/R ratio in Hα. The high velocities (in excess of 100 km s^{-1}) seen in Hα, when compared to the lower expansion velocities derived from CO and OH measurements ($\sim$ 40 km s^{-1}, Nedoluha and Bowers, 1992; Bachiller et al. 1988; Lewis et al. 1986), imply that a new high velocity outflow started recently, possibly less than 10 years ago, at the same time as the hydrogen lines went into emission.

Acknowledgements

It is a pleasure to thank my collaborators Rens Waters, Tom Geballe and Kailash Sahu, for their continuing enthusiasm and support in our study of this enigmatic object.

References

Bachiller R., Gómea-González J., Bujarrabal V. & Martin-Pintado J.: 1988, Astron. & Astrophys. **196**, L5.
Irvine C.E.: 1986, IAU Circ. 4286.
Jones T.J. et al. : 1993, Astrophys. J. **411**, 323 (**J93**).
Kurucz R.L.: 1979, Astrophys. J., Suppl. **40**, 1.
Lewis B.M., Terzian Y. & Eder J: 1986, Astrophys. J. **302**, L23.
Nedoluha G.E. & Bowers P.F.: 1992, Astrophys. J. **392**, 249.
Nercessian E., Guilloteau S., Omont A. & Benayoun J.J.: 1989, Astron. & Astrophys. **210**, 225.
Olofsson H., Johansson L.E.B., Hjalmarson A. & Rieu N.Q.: 1982, Astron. & Astrophys. **107**, 128.
Omont A., Lucas R., Morris M. & Guilloteau S.: 1993, Astron. & Astrophys. **267**, 490.
Oudmaijer R.D., Geballe T.R., Waters L.B.F.M. & Sahu K.C.: 1994, Astron. & Astrophys. **281**, L33.

INFRARED APPEARANCE OF DYNAMICAL MODELS FOR CIRCUMSTELLAR DUST SHELLS AROUND LONG–PERIOD VARIABLES

J. M. WINTERS, A. J. FLEISCHER and E. SEDLMAYR

Institut für Astronomie und Astrophysik, Technische Universität Berlin, Sekr. PN 8-1, Hardenbergstraße 36, D-10623 Berlin, FRG

and

A. GAUGER

Los Alamos National Laboratory, Group T-4, MS B268, Los Alamos, NM 87545, USA

Abstract. Synthetic brightness profiles resulting from consistent dynamical models for circumstellar dust shells around long–period variables are presented and discussed with respect to a corresponding observation of IRC +10216.

Key words: hydrodynamics – radiative transfer – circumstellar shells – carbon stars – long–period variables – dust – stars: IRC +10216

1. Introduction

Long–period variables and Miras are cool pulsating stars located on the Asymptotic Giant Branch in the HR diagram. These objects generally show high mass loss rates (up to $\approx 10^{-4} M_\odot \mathrm{yr}^{-1}$) caused by effective dust formation. In fact, they are the main sources of the galactic dust production and also of a variety of complex chemical species.

2. The Physical Problem

The consistent description of a dust forming envelope around a Mira variable requires the simultaneous solution of the coupled physical problem consisting of time dependent hydrodynamics, thermodynamics, chemistry, and of the dust formation complex (*cf.* Winters *et al.* 1994a). The occurrence of dust introduces a strong nonlinear coupling among these different subsystems. By a consistent treatment of this problem, the resulting models are *completely* specified by the four fundamental stellar parameters *stellar mass* $M_\star$, *stellar luminosity* $L_\star$, *photospheric temperature* $T_\star$ and chemical abundances ϵ_i. To describe the pulsation of the star, two additional parameters, the pulsation period P and the velocity amplitude Δu_P at the inner boundary (located inside the photosphere), have to be prescribed (*cf.* Fleischer *et al.* 1992).

Astrophysics and Space Science **224**: 329–332, 1995.
© 1995 *Kluwer Academic Publishers.*

3.   Modelling Method

The system of partial differential equations describing the dust forming stellar atmosphere is written in Lagrangian coordinates, assuming spherical symmetry, and is solved by an explicit integration in time with an internal implicit solution of the energy equation. The numerical method and the behaviour of the dynamical models are discussed in detail in Fleischer *et al.* (1992). As a result of this calculation the complete radial structure of the dust shell model is obtained at each instant of time. In order to determine the equilibrium temperature, these dynamical models only include an approximate treatment of the radiative transfer problem, which yields no information about the detailed radiation field. Therefore, at selected instants of time the angle and frequency dependent stationary radiative transfer equation is solved in spherical geometry. The method and the results for stationary dust shell models are discussed in detail in Winters *et al.* (1994a), the application to the dynamical models and the resulting lightcurves are presented in Winters *et al.* (1994b).

4.   Results

4.1   MODEL PARAMETERS

The model presented here is specified by the following parameters, which are typical for a carbon–rich long–period variable star: $M_\star = 1M_\odot$, $L_\star = 10^4 L_\odot$, $T_\star = 2600$ K, and a carbon to oxygen ratio of $\epsilon_C/\epsilon_O = 1.80$, while all other elements have solar chemical abundances ϵ_i. To simulate the pulsation of the star a sinusodial velocity variation of the inner boundary is assumed with period $P = 650$ days and velocity amplitude $\Delta u_P = 2\,\mathrm{km\,s^{-1}}$.

4.2   RADIAL STRUCTURE

The radial structure of this model is shown in Fig. 1 at a given instant of time. As can be seen from the stratification of the flux–mean extinction coefficient (upper diagram), the dust is not distributed homogeneously across the shell but is concentrated in rather thin discrete layers. This dust distribution is caused by the fact, that the physical conditions (i.e. low temperatures and simultaneously sufficiently high densities), neccessary for effective dust formation and growth are only met for a limited time interval during the pulsation cycle and within a narrow spatial region of the shell. For details we refer to Fleischer *et al.* (1992). This striking dust distribution shows up to have a significant influence on the hydrodynamical and thermodynamical structure of the shell. Caused by the onion–like dust distribution, the radiative acceleration of the dust grains is also concentrated in discrete layers, as can be seen from the long–dashed curve labelled α (middle diagram, r.h.s. ordinate) which is the outward directed radiative acceleration in units of the local gravitational deceleration. The radiative acceleration acting on the dust grains is closely connected to the velocity structure of the shell (solid line). The strong shocks present in the atmosphere are solely produced by radiative acceleration of the discrete dust layers. Since α exceeds unity close to the star, the material is accelerated to velocities above the escape velocity v_{esc} already at radii below $5\,R_0$. Due to it's opacity, the dust also leads to a pronounced backwarming of the material inside the dust layers as is evident from the steps present in the temperature stratification around $2\,R_0$ and $7\,R_0$ (lower diagram). In summary we find, that the dust completely determines the internal structure of the circumstellar shell, a result that is only revealed by the consistent treatment of the involved physical processes.

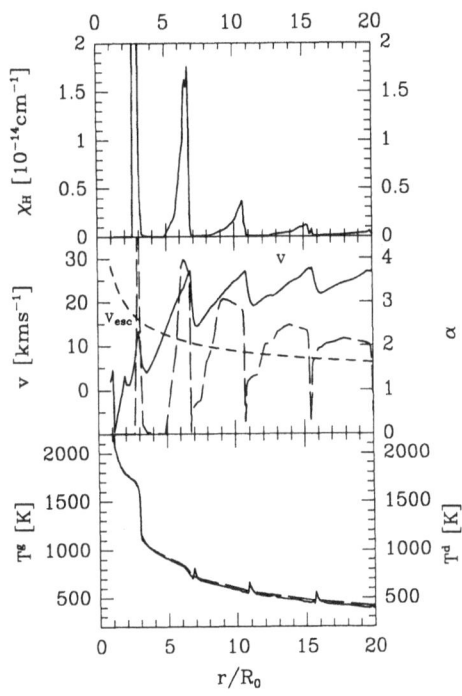

Fig. 1.   Radial structure of the model.

4.3   INTENSITY PROFILE AND BRIGHTNESS DISTRIBUTION

The discrete spatial dust distribution also leads to directly observable implications, in producing *e.g.* a step–like distribution of the synthetic *intensity profiles* of the dust shell. In Fig. 2a the intensity profiles are plotted at three different wavelengths as a function of the impact parameter p (in units of the stellar radius, lower abscissa) and on an angular scale (upper abscissa). Thereby a distance of 200 pc is assumed. For this distance, the step–like structures subtend an angle $\phi < 0.1\,\mathrm{arcsec}$, which is hardly resolved by a single ground–based infrared telescope. A common solution of this problem is of course to use interferometric techniques, which measure the Fourier transform of the intensity distribution, i.e. the visibility function of the shell. A more direct way to get information about the intensity profile is to reconstruct the *brightness distribution* of the shell (i.e. the 2–D intensity profile integrated over one spatial coordinate) from a lunar occultation lightcurve. The resulting synthetic brightness distributions of the model are shown in Fig. 2b.

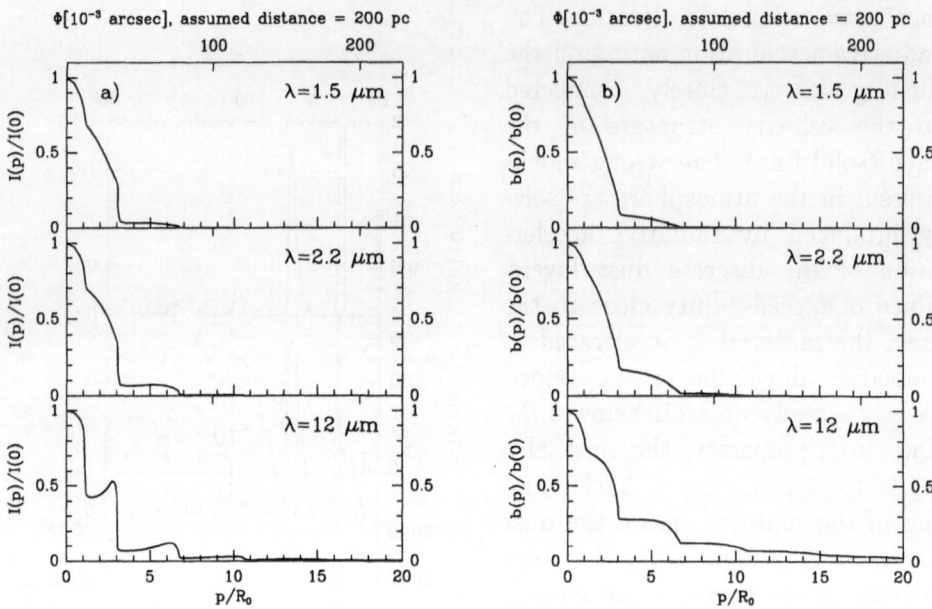

Fig. 2.   a): Intensity profiles and b): brightness distribution of the dust shell model at different wavelengths.

5.   Discussion

Fortunately, the best observed dust enshrouded C-star, IRC +10216, is located in a region of the sky, that is sporadically crossed by the lunar disk. Very recently, the brightness profile of IRC +10216 was reconstructed from a lunar occultation lightcurve with an angular resolution of $2 \cdot 10^{-3}$ arcsec by Dr. Andrea Richichi (1993, private communication). In this observed brightness profile several quite regular steps are present, which look very similar to the structures shown in the lower diagram of Fig. 2b. Although our calculation was not performed to model any individual object, synthetic and observed brightness profile show rather good agreement. In particular, the angular separation between successive steps occurs on the same scale of a few stellar radii both, in the model and in the observation as well. This fact puts strong support to the reliability of the consistent model calculations and leads to the conclusion, that these models possibly can *physically* explain the structures present in the observed brightness profiles of dust shells around long–period variable stars whithout introducing arbitrary *ad–hoc* assumptions.

References

Fleischer A. J., Gauger A., Sedlmayr E.: 1992, Astron. & Astrophys. **266**, 321.
Winters J. M., Dominik C., Sedlmayr E.: 1994a, Astron. & Astrophys. **288**, 255.
Winters J. M., Fleischer A. J., Gauger A., Sedlmayr, E.: 1994b, Astron. & Astrophys. in press.

THE STRUCTURE OF CIRCUMSTELLAR ENVELOPES

G.R. KNAPP

Princeton University, Princeton, NJ 08544, USA

Abstract. Copious mass loss on the Asymptotic Giant Branch dominates the late stages of stellar evolution. Maps of extended circumstellar envelopes provide a history of mass loss and trace out anisotropic mass loss. This review concentrates on observations of millimeter wavelength molecular line emission, on high resolution maps of maser emission and on observations of submillimeter, millimeter and radio wavelength continuum emission. Radio continuum observations show that AGB stars are larger at radio than at optical wavelengths. The extended chromospheres indicated by these observations extend to distances from the star large enough for dust to form, thereby initiating mass loss. Molecular line maps have found time-variable mass loss for some stars, including detached shells indicating interrupted mass loss and evidence for a rapid increase in the mass loss rate at the end of the AGB phase. Maps of circumstellar envelopes show evidence of flattening, bipolar outflow and angular variations in both the mass loss rate and the outflow velocity. As stars evolve away from the AGB and planetary nebula formation begins, these structures become more pronounced, and fast bipolar molecular winds are observed. The time scales derived from the dynamical times of these winds and from the expansion rates of the central planetary nebulae are very rapid in some cases, about 100 years, in agreement with the predictions of stellar evolution theory.

Astrophysics and Space Science **224**: 333, 1995.
© 1995 *Kluwer Academic Publishers.*

CIRCUMSTELLAR MOLECULAR LINE ABSORPTION AND EMISSION

IN THE OPTICAL SPECTRA OF POST-AGB STARS *

ERIC J. BAKKER and HENNY J.G.L.M. LAMERS
SRON Laboratory for Space Research Netherlands, SRON-Utrecht
Astronomical Institute, University of Utrecht

L.B.F.M. WATERS
Astronomical Institute 'Anton Pannekoek', University of Amsterdam
SRON Laboratory for Space Research Netherlands, SRON-Groningen

and

TON SCHOENMAKER
Kapteyn Sterrenwacht Roden

Abstract. We present a list of post-AGB stars showing molecular line absorption and emission in the optical spectrum. Two objects show CH^+, one in emission and one in absorption, and 10 stars show C_2 and CN in absorption. The Doppler velocities of the C_2 lines and the rotational temperatures indicate that the line forming region is the AGB remnant. An analysis of the post-AGB stars of which CO millimeter data is available suggests that the C_2 expansion velocity is of the same order as the CO expansion velocity. HD 56126 has been studied in detail and we find a mass-loss rate of $\dot{M}=2.8\times10^{-4}$ $M_\odot yr^{-1}$, $f_{C_2} = 2.4 \times 10^{-8}$ and $f_{CN} = 1.3 \times 10^{-8}$. The mass-loss derived from C_2 is significantly larger than $\dot{M}=1.2 \times 10^{-5}$ $M_\odot yr^{-1}$ derived from CO. We find that all objects with the $21\mu m$ feature in emission show C_2 and CN absorption, but not all objects with C_2 and CN detections show a $21\mu m$ feature.

Key words: molecules – physical conditions in AGB remnant – mass-loss history on AGB

1. Introduction

The study of molecular lines in the optical spectra of post-AGB stars started with the paper by Waelkens *et al.* (1992) in which they presented a band of narrow emission lines in the spectrum of the Red Rectangle (HD 44179). These emission lines were identified as the $(0,0)$ $A^1\Pi_u - X^1\Sigma_g^+$ band of CH^+ by Balm and Jura (1992). The presence of the Swan and Phillips bands of C_2 (Fig. 1) and red system bands of CN (Fig. 2) in the spectrum of HD 56126 studied by Bakker (1994), showed that the outflow velocity and the excitation conditions of C_2 suggest that these lines are formed in the AGB remnant. This opens new possibilities to study the physical conditions in AGB remnants. Hrivnak (1995) presents a list of post-AGB stars showing C_2 and C_3 in absorption. Here we present the observations of post-AGB stars that show molecular lines in the optical spectrum. For several stars

* Based on observations with the WHT/UES

Astrophysics and Space Science **224**: 335–338, 1995.
© 1995 *Kluwer Academic Publishers.*

TABLE I

Expansion velocities of the AGB remnant

| Object | Id[*] | $\delta V_{optical}$ | δV^+ | T_{rot} | $N10^{14}$ | Remark |
		[km s^{-1}]		[K]	[cm^{-2}]	
IRAS 04296+3429	C_2	-7.5	-12	138	14	
IRAS 05113+1347	C_2	-5.2	no	198	12	not observed in CO
HD 56126	C_2	-8.8	-10.0	240	20	
IRAS 08005-2356	C_2	-38.1	-50.0	150	31	OH maser
IRC +10216	C_2	-17.9	-15.7			AGB star
AFGL 2688	C_2	-15.0	-22.8	56	26	
HD 44179	CH^+	-4.5 em.	-3	267	2	
HD 213985	CH^+	-5.4 abs.	nd			not detected in CO

[*] all stars with C_2 do also show CN absorption.
[+] from CO millimeter emission. For IRAS 08005-2356 there is no CO data available and
we used the velocities from the OH maser line.

the excitation temperature and outflow velocity have been determined. The
C_2 absorption in HD 56126 is studied in detail by modeling the population
density over the rotational quantum number (Section 3).

2. Detection of Molecules

In Table 1 we list the first results of our work. For each star the 3th and
4th column give the expansion velocity derived from respectively the optical (C_2,CN,CH$^+$) and the CO millimeter or OH maser line. The 5th and
6th column give the rotational temperature and column density derived
from C_2 (3,0) or CH$^+$(0,0) band. Negative velocities are due to outflow
of material. The C_2 line absorption is formed closer to the star than the
CO line emission, the two different molecules trace different material and
hence different stages of the AGB evolution. Our data may indicate that
the expansion velocity decreases as the star evolves along the AGB (Table
1). However a larger sample is needed to quantify this result. Stars showing
the unidentified 21μm emission feature all exhibit C_2 and CN absorption,
strongly suggesting that the 21μm feature is from carbon-rich material, but
not all stars with C_2 absorption show the 21μm feature.

3. Modeling the C_2 absorption Bands of HD 56126

From the rotational diagram of the C_2 bands (Fig. 3) we find T_{rot}=240K
and for the CN (1,0) band we find T_{rot}=24K with a column density of
$N_{C_2} = 20 \times 10^{14}$ cm^{-2} and $N_{CN} = 11 \times 10^{14}$ cm^{-2}. Unlike CN, C_2 is a
homo nuclear molecule having no permanent dipole moment and the rotation

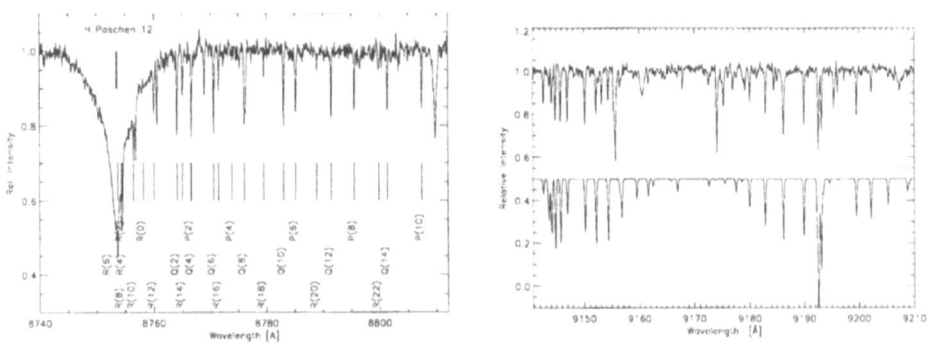

Fig. 1. Left: The Phillips (2,0) band in the spectrum of HD 56126. The C_2 absorption lines are not resolved

Fig. 2. Right: The CN $A^2\Pi - X^2\Sigma^+$ (1,0) band in the spectrum of HD 56126. The computed synthetic spectrum is shifted down by 0.5

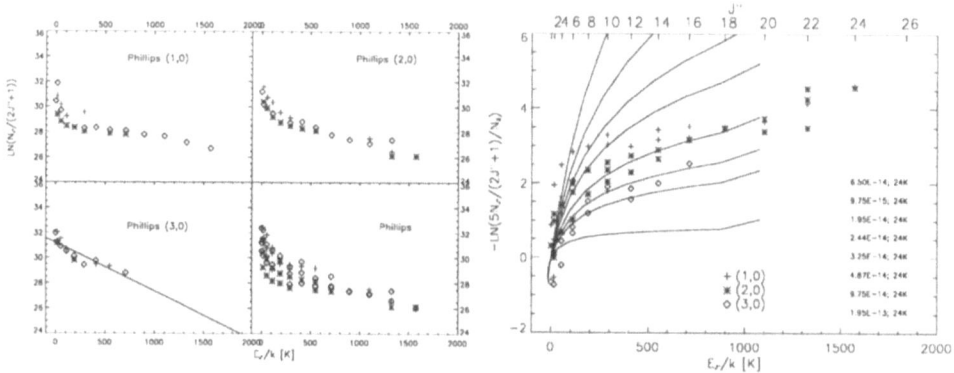

Fig. 3. Left: Rotational Diagram of the C_2 $A^1\Pi_u - X^1\Sigma_g^+$ (1,0), (2,0), and (3,0) absorption band. The P, Q, and R branches are denoted by respectively a plus, asterisk, and a square. Only the weakest rotational band (3,0) is optically thin and gives an almost linear relation in the rotational diagram with T_{rot}=240K and N= 20 × 10^{14} cm^{-2}

Fig. 4. Right: Relative rotational diagram of the C_2 $A^1\Pi_u - X^1\Sigma_g^+$ (1,0), (2,0), and (3,0) absorption band of HD 56126 with the model fit for T_{kin}= 24K superimposed

temperature is not a good indicator of the kinetic temperature because of pumping by the stellar radiation field. We use the CN rotational temperature as a measure of the kinetic temperature of the gas.

We have tried to fit the relative rotational diagram (Fig. 4) with the population density distribution derived from modeling the C_2 excitation by taking into account radiative pumping and collisional de-excitation (Van Dishoeck and Black 1982). The best fit is reached for $n_c\sigma/I = 3.25\ 10^{-14}$. Taking the latest H_2-C_2 cross section of $\sigma = 7.8 \times 10^{-16}$ cm^{-2} (Phillips 1994) the mass-loss rate can be calculated using Eq. 1. This gives a mass-loss rate

TABLE II
Overview of derived mass-loss rates for HD 56126

Modeling C_2 excitation	2.8×10^{-4} $M_\odot yr^{-1}$	this study
Infrared excess from dust	1.9×10^{-4} $M_\odot yr^{-1}$	this study; gas/dust=100
CO millimeter emission	1.2×10^{-5} $M_\odot yr^{-1}$	$^{12}CO(2-1)$ Omont et $al.$ 1993

of $\dot{M}_{C_2} = 2.8 \times 10^{-4}$ $M_\odot yr^{-1}$.

$$\dot{M} = 2.8 \times 10^{-4} \cdot \left(\frac{v_{exp}}{8.8} \cdot \frac{n_c}{1.7 \times 10^7} \cdot \frac{\sigma}{7.8 \times 10^{-16}} \cdot \frac{4.1 \times 10^5}{I} \right) M_\odot \ yr^{-1} \quad (1)$$

Fitting an optically thin dust model (Waters et $al.$ 1988) to the the spectral energy distribution of HD 56126 yields a dust inner radius of $2 \times 10^3 R_*$ and M=1.9×10^{-4} $M_\odot yr^{-1}$. Assuming that the molecular line absorption originates at this dust inner radius we find a particle abundance relative to H_2 of $f_{C_2} = 2.4 \times 10^{-8}$ and $f_{CN} = 1.3 \times 10^{-8}$. Table 2 summarizes the mass-loss rates derived for HD 56126, where we have scaled the different mass-loss rates to R_*=50$R_\odot$, T_{eff}=6500K (logL=3.6) and D=2.7kpc. The mass-loss rate derived from C_2 and from the IR excess are significantly higher than that derived from the CO millimeter emission, which might indicate that the mass loss rate increased dramatically towards the end of the AGB.

Acknowledgements

The authors want to thank Ewine van Dishoeck and Christoffel Waelkens for the stimulating and constructive discussions on this work. EJB was supported by grant no. 782-371-040 by ASTRON, which receives funds from NWO. LBFMW acknowledges financial support from the Royal Dutch Academy of Arts and Sciences. This research has made use of the Simbad database, operated at CDS, Strasbourg, France.

References

Bakker, E.J.: 1994 in: $Circumstellar$ $Media$ in the $Late$ $Stages$ of $Stellar$ $Evolution$, eds: R.E.S. Clegg, I. Stevens & W.P.S. Meikle, CUP, pxxx.

Balm, S.P. & Jura, M.: 1992, Astron. & Astrophys. **261**, L25.

Van Dishoeck E.F. & Black J.H.: 1982, Astrophys. J. **258**, 533.

Hrivnak, B.J.: 1995, Astrophys. J. **438**, in press.

Omont, A., Loup, C., Forveille, T., te Lintel Hekkert, P., Habing, H. & Sivagnanam, P.: 1993, Astron. & Astrophys. **267**, 515.

Phillips, T.R.: 1994, Mon. Not. of the RAS accepted.

Waelkens, C., van Winckel, H., Trams, N.R. & Waters, L.B.F.M.: 1992, Astron. & Astrophys. **256**, L15.

Waters, L.B.F.M., Coté, J. & Geballe, T.R.: 1988, Astron. & Astrophys. **203**, 348.

LONG BASELINE INTERFEROMETRIC OBSERVATIONS
OF LONG PERIOD VARIABLE STARS

W.C. DANCHI and M. BESTER

Space Sciences Laboratory, University of California, Berkeley, CA 94720-7450 USA

Abstract. Recent observations of long-period variable stars at spatial resolutions from approximately 1 arcsec to several milli-arcsecs have provided new insights into pulsation, dust formation, and mass-loss of AGB stars. These insights have come from long baseline interferometric observations obtained across a wide range of wavelengths, from the optical, through the infrared, to wavelengths as long as several millimeters. The present status and recent results from long baseline interferometry, particularly at optical and infrared wavelengths, are discussed. Such results include diameters and limb-darkening, surface features, mode of pulsation, location of SiO masers, inner radii of dust shells, physical conditions in the dust formation zone and of the inner regions of the dust shells. The results are interpreted in terms of present models of dust formation and mass-loss.

Key words: interferometry – mass loss – stellar pulsation – stellar diameters – masers – dust shells

1. Introduction

It is well known from millimeter wavelength observations of emission lines of carbon monoxide and other molecules that long period variable stars lose considerable mass to the interstellar medium via slow winds (*e.g.* Knapp & Morris, 1985). Previous reviews in this conference by Knapp (1994) and Huggins (1994) have focussed on the physical and chemical properties of the molecular envelopes created by these cool winds, which can have angular sizes larger than one arcminute. The time scales corresponding to these angular sizes are on the order of hundreds or thousands of years (see also the review by Ziljstra 1994). This review is concerned with the properties of the envelopes of mass losing stars for angular sizes below one arcsecond, corresponding to phenomena with time scales shorter than about 50-100 years. This includes processes occuring in the material very close to the stellar photospheres or that involve the photospheres themselves such as pulsation, convection, and dust formation. Results from long baseline interferometry at infrared and optical wavelengths are discussed from the standpoint of physical models of mass loss.

Examples of mass loss mechanisms and motivation for work at very high angular resolution are discussed in Section 2. Since the field of high resolution astronomy at infrared and optical wavelengths is rapidly developing, the present status of instrumentation is also discussed (Section 3). Section 4 surveys recent results including stellar diameters and limb darkening, asymmetries and surface features, modes of pulsation, inner radii of dust shells, and the relationship of masers (SiO, H_2O, & OH) to the dust shells.

Astrophysics and Space Science **224**: 339–352, 1995.

2. Mass Loss Mechanisms

A variety of theoretical models have been developed to explain the physical
origin of cool winds from asymptotic giant branch stars. Perhaps best known
is the pulsation driven dusty wind model. In this model stellar pulsations
create shock waves. Energy in the shocks is dissipated in the atmospheres
of the stars and causes them to be greatly extended compared to those of
non-variable stars. If the velocity amplitude of pulsation (piston amplitude)
is large enough, it is then possible for the gas in the extended atmosphere
to have a high enough density and low enough temperature to nucleate
dust grains. Once formed the dust grains are accelerated by radiation pressure (and gravity). Momentum is transferred from the dust grains to gas
molecules by collisions, thus creating the observed molecular wind. This
model is most critically dependent on where the dust condenses, which is
expected to be within a few stellar radii of the photosphere, perhaps even
within the extended atmosphere of the star (*cf.* Sedlmayr 1990).

Although this picture is conceptually attractive there are other possibilities worth discussing. One is the radiatively driven wind model. In this
model, dust can be formed in the atmosphere of the star even in the absence
of stellar pulsations and once formed radiation pressure on the dust drives
the wind as before. A problem with this model is that the scale height for
hydrostatic atmospheres is $\sim 0.02\ R_*$, which implies that the density is
extremely low at distances where the temperature is low enough to nucleate
the dust ($\sim 3.5\ R_*$) (*cf.* Tielens 1983).

Another possibility is the pulsation (only) driven wind, *i.e.* without dust
formation (*cf.* Hill & Willson 1979, Wood 1979). In this case the atmosphere
is greatly extended but very little mass is lost even though a shock wave
accelerates individual mass elements upward because these elements have
sufficient time to return to their original locations before a new shock arrives
at the next cycle. Thus little net momentum is actually transferred to these
mass elements so they do not reach escape velocity (Bowen 1988).

A third possibility is Alfvén wave driven winds. This mechanism depends
most strongly on the dissipation length for the waves and the density scale
height in the wind. Long dissipation lengths imply winds that are very fast,
however if the dissipation length is about one stellar radius, which is roughly
the density scale height assumed, then the velocity of the wind can be closer
to that observed. Problems with this model include extreme sensitivity to the
dissipation length and larger than expected magnetic fields at the surfaces
of the stars.

Finally sound wave driven winds are another possibility. This mechanism
can work if the sound waves driving the winds have wavelengths much smaller than the density scale height, and it assumes the dissipation length is a
constant input parameter. Under these conditions winds can be produced

with the expected final velocities. These winds accelerate slowly over long distances and do not require dust formation. However dust increases the final velocity of the wind if present.

To decide between the mechanisms Hearn (1990) concludes that two crucial questions must be answered from observations:
- How close to the star does the dust form?
- Is mass loss spherically symmetric?

Clearly high spatial resolution observations can play an important role in answering these questions. The necessary resolution is easily estimated. From observations it is known that the largest stars have angular diameters of the order of 50 milli-arcsec (mas). Therefore if the dust forms within a few stellar radii of the photosphere, then the inner diameter of the dust shell should be of the order of 100-200 mas. This is smaller than the diffraction limit of even a 10 m telescope when observing at 10 μm, *i.e.* near the peak of the silicate emission feature. A natural choice of instrumentation for this problem is long-baseline interferometry at infrared or optical wavelengths. The present status of instrumentation in this field is reviewed below.

3. Instrumentation

The status of long baseline interferometry at infrared and optical wavelengths can be viewed with the perspective gained from many years of experience with radio interferometers. Clearly, aperture-synthesis techniques have played a central role in the development of radio astronomy in the past thirty years. Presently there are approximately ten major radio interferometers in use (Figure 1). They cover the spectral range from centimeter to submillimeter wavelengths, with maximum resolution typically $\sim 0''.1$; the highest resolution of course comes from the Very Long Baseline Array (VLBA) with maximum resolution $\sim 0''.0001$. These instruments are the result of more than 40 years of research and development and a capital investment of several hundred million dollars.

Long-baseline interferometry at optical and infrared wavelengths is at a much earlier stage of development than radio interferometry. One important reason is that the relative variation in path lengths due to refractivity fluctuations in the atmosphere is approximately 5-10 μm for a baseline of 10 m, which is 20 wavelengths at 500 nm. This can be compared to fluctuations at radio wavelengths, such as have been measured at the VLA. For baselines approximately 20 km in length the fluctuations vary between about 2-10 mm, which is typically a fraction of a wavelength. Thus interferometers operating at optical and infrared wavelengths have to cope with an atmosphere that fluctuates much more wildly than it does in the radio region. Atmospheric fluctuations and their implications for interferometry are discussed in the review by Masson (1994).

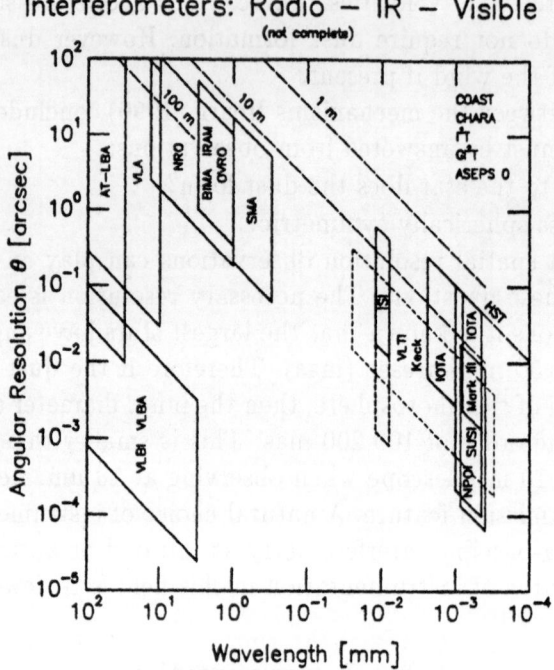

Single Aperture: $\theta = 1.2\,\lambda/D$
Multiple Apertures: $\theta = 0.5\,\lambda/B$

Fig. 1.   Comparison of angular resolution and wavelength coverage of radio, infrared, and visible wavelength long baseline interferometers.

Despite the difficulties caused by a strongly fluctuating atmosphere, considerable progress has been made in the past decade in long-baseline interferometry in the optical and infrared. A number of two-element single baseline interferometers have been built and quite a number of significant observational results have been obtained. The right hand part of Figure 1 displays the resolution and wavelength coverage of some of these instruments. Instruments that have operated or are currently operating at optical wavelengths include the Mark III Interferometer (USA, Shao *et al.* 1988), GI2T (France, Labeyrie *et al.* 1986), Sydney University Stellar Interferometer (SUSI) (Australia, Davis *et al.* 1994). The Infrared/Optical Telescope Array (IOTA) interferometer operates at 2.2 μm (USA, Carleton *et al.* 1994) and was preceded by the Infrared Michelson Array (IRMA) array (USA, Dyck *et al.* 1993). The I2T interferometer was also successfully used at 2.2 μm (France, Di Benedetto *et al.* 1992). The Infrared Spatial Interferometer (ISI) (USA, Bester *et al.* 1994) operates in the spectral window in the mid-infrared (9-12 μm).

A number of instruments are currently under construction or in the planning stages. The University of Cambridge (UK) is building the Cambridge Optical Aperture Synthesis Telescope (COAST) instrument, a 4 element interferometer for the optical and near-infrared (Baldwin *et al.* 1994). Three of four telescopes have been constructed and phase closure has been demonstrated with it (Baldwin *et al.* 1994). Four elements will allow for amplitude closure as well as phase closure (Pearson & Readhead 1984). The Navy Prototype Optical Interferometer (NPOI) consists of two separate interferometers sharing common subsystems. The NPOI includes the Imaging Array (IA), a 6 element optical interferometer of the Naval Research Laboratory (Armstrong *et al.* 1994), and the Astrometric Interferometer (AI), a 4 element optical interferometer of the Naval Observatory (Hutter *et al.* 1994). As part of its Mission to Planet Earth NASA (JPL) is constructing a special two-element interferometer for the near-infrared, called ASEP0, which is designed for very high precision narrow-angle astrometry to address the problem of planet detection (Colavita *et al.* 1994).

Currently in the planning or proposal stages are the Very Large Telescope Interferometer or VLTI (EU) for the infrared (von der Lühe *et al.* 1994), and the Center for High Angular Resolution Astronomy (CHARA) array proposed by Georgia State University (USA) for the optical and near infrared (McAlister *et al.* 1994). These systems both involve a large number of telescopes (≥ 4) but their designs are not yet finalized.

4. Observational Results

4.1 STELLAR DIAMETERS & LIMB DARKENING

The present generation of two-element inteferometers has greatly added to the number of stars for which stellar diameters have been measured and also the precision of the measurements. The Mark III interferometer has been particularly productive in this area. So far more than 70 stellar diameters have been obtained by the Mark III team mostly in the continuum band at 700 nm, the TiO band at 710 nm, and the continuum band at 800 nm. The precision is typically about 1% (*cf.* Armstrong 1994). Diameters of stars of different spectral types have been measured including cool giants in and out of TiO bands (Quirrenbach *et al.* 1993a), carbon stars with optically thin dust emission (Quirrenbach *et al.* 1994a), Be stars (Quirrenbach *et al.* 1993b), Nova Cygni 1992 (Quirrenbach *et al.* 1993c), and Mira (Quirrenbach *et al.* 1992). Visual orbits of more than 25 spectroscopic binaries have been obtained with major axes between 4 and 200 mas. Mass determinations and orbits for five double-lined systems were made with major axes as small as 5 mas (*cf.* Armstrong 1994 and references therein).

The I2T system observed from 1983 to 1987 in the near-infrared K to M bands (Di Bennedetto & Foy 1992). The diameters of 17 cool stars were measured, of which 6 were supergiants and 11 were giants. The precision of these measurements is about 5%. More recently the GI2T system (Mourard *et al.* 1994) has been operating at optical wavelengths. Their emphasis has been on Be stars such as γ Cas, β Lyr, and o And, luminous blue variables such as P Cyg, and eclipsing binaries like β Aur. The observations of the Be stars emphasize the Hα emission line. Results for γ Cas support the idea that the emission lines are created in an equatorial disk, while electron scattering in the disk causes the high linear polarization of the star (Quirrenbach *et al.* 1993b, Vakili *et al.* 1994).

The IRMA system, a prototype of the IOTA interferometer, operated from 1990 to 1992. Diameters at 2.2 μm were obtained for o Cet (Ridgway *et al.* 1992), α Ori (Dyck *et al.* 1992), and α Her (Benson *et al.* 1991).

Recently the first diameters at 11 μm were obtained with the ISI for α Ori and α Sco (Danchi *et al.* 1994b). It is interesting to compare results for stars that have been observed at more than one wavelength. At 11 μm the diameter of α Ori is 53$\pm$4 mas (95% confidence limit, CL). This can be compared to a non-redundant aperture masking result of Wilson *et al.* (1992) at 700 nm, in which a uniform disk of 49$\pm$3 mas and an unresolved spot with 10% of the total flux was the best fitting model. Using static model atmosphere calculations of Manduca (1979), limb darkening corrections are +9.7% at 700 nm and only +1.4% at 11 μm. This corrects the result of Wilson *et al.* (1992) to 53.8 mas while the ISI result becomes 53.7 mas in excellent agreement. This can be contrasted with the result at 2.2 μm of Dyck *et al.* (1992), who found a uniform disk diameter of 44.0$\pm$0.2 mas. At 2.2 μm the limb darkening correction is +4.6%, giving a corrected diameter of 46.2 mas. Thus the ratio of the uniform disk diameters at optical and mid-infrared wavelengths to the near-infrared is 1.17, which is larger than expected from static model atmospheres. A similar result is obtained for α Her, in which the uniform disk diameter at 2.2 μm is 32$\pm$1 mas, while at 700 nm it is 36 mas, and at 710 nm (TiO) it is 43 mas. Alpha Sco has been measured at 700 nm and at 11 μm with uniform disk diameters of 33$\pm$3 mas and 42$\pm$6 mas respectively. The ratio of the 11 μm diameter to that at 700 nm is 1.27, and is also much larger than would be expected on the basis of static model atmosphere calculations. A direct measurement of limb darkening for α Boo was made with the Mark III interferometer. It was found that the visibility data past the first zero could be fit with a 85% limb-darkened disk at 550 nm (Quirrenbach 1994b).

Some generalizations can be made concerning diameters in specific spectral and luminosity classes. For example for stars from K5III to M0III the diameters are the same in the TiO absorption band at 712 nm and in the continuum at 754 nm. Cooler stars, *i.e.* from M3III to M5III are about 10% larger on average at 712 nm than at 754 nm (Quirrenbach *et al.* 1993a). For the supergiants this ratio is even larger still as noted above.

4.2 ASYMMETRIES & SURFACE FEATURES

Aspects of the two dimensional brightness distribution of these stars, such as surface features and asymmetries, are just beginning to be explored primarily at optical wavelengths. One method has been to observe with Earth rotation synthesis using a single baseline and with that baseline changed in discrete steps from 4 m to 32 m (Quirrenbach *et al.* 1992). Aperture masking techniques have also been used, whereby a long baseline interferometer is created from a filled aperture telescope by masking off most of the telescope aperture and interfering only discrete sub-apertures either with redundant or non-redundant baseline spacings (Haniff *et al.* 1987). Using the latter technique Buscher *et al.* (1990), Wilson *et al.* (1992), and Tuthill *et al.* (1994) have reported bright spots on stars such as α Ori and α Her, and asymmetries in the shapes of stars like o Cet and R Cas.

For α Ori, Wilson *et al.* (1992) reported the evolution of unresolved surface features. In 1989 February there was a single bright feature on the surface containing about 10% of the total flux, at an offset of 8 mas from the center, and with position angle (PA) 270°. About 2 years later in 1991 January, there were two unresolved spots. One was at PA 105°, offset by 9 mas, containing 12% of the flux, with the second spot at PA 305°, offset by 9 mas, containing 11% of the flux. A similar bright spot was reported for α Her by Tuthill *et al.* (1994) for observations from 1992 July. A single bright spot was observed at a position angle of 90°, 9 mas offset from the center, containing about 7% of the flux.

For mira stars, including Mira itself, surface features have not been observed, but there are several observations of non-spherical shapes. Wilson *et al.* (1992) fit an elliptical shape to their aperture masking data for o Cet. The best fitting model was an elliptical gaussian shape with a major axis 30 mas in length (FWHM), axial ratio of 0.85, and position angle 120°. Similar asymmetries for o Cet were reported by Quirrenbach *et al.* (1992) using the Mark III interferometer and also by Karovska *et al.* (1991) using speckle techniques. Recent non-redundant aperture masking observations by Tuthill *et al.* (1994) show that R Cas is elongated in the North-South direction with a significantly larger axial ratio than o Cet. R Cas does not have a close companion so the asymmetry cannot be ascribed to an interaction with the companion, tidal or otherwise.

A possible source of asymmetries in the observations at optical wavelengths is scattering of the stellar light by dust in an asymmetrical dust shell surrounding the star. Reasonable agreement between visibility data at 700 nm (Wilson *et al.* 1992, Haniff *et al.* 1992) and visibility curves from a dust shell model was reported by Danchi *et al.* (1994a). This dust model was the best fitting model for the 11 μm ISI observations but was recomputed for the 700 nm wavelength with all other parameters unchanged. Variations in visibility data at 11 μm with position angle were fit to a model with the dust in an inclined disk by Lopez *et al.* (1994). The shape of the image at optical wavelengths computed from this model agrees well with the shape observed by Karovska *et al.* (1991).

On the other hand asymmetries in the atmosphere of *o* Cet are apparent in the difference image shown in Figure 3(c) of Haniff *et al.* (1992). This image was obtained by subtracting the 700 nm continuum map from the 710 nm TiO map. Clearly TiO emission is much stronger in the direction facing the companion. Note also that the radial location of the TiO emission coincides with the inner radius of the dust shell. Both of these issues are discussed in some detail in Danchi *et al.* (1994a).

Clearly these stars can no longer be regarded as smooth featureless spheres. Note that another suggestion for the cause of the asymmetries in the shapes of stars like *o* Cet has been that of non-radial pulsations (*e.g.* Karovska *et al.* 1991, Haniff *et al.* 1992). The mode of pulsation of mira variables remains controversial and is discussed in the next section.

4.3 MODE OF PULSATION

The modes of pulsation of mira stars is still an unresolved issue for which competing observational techniques provide different answers. Spectral line observations tend to favor fundamental mode pulsation because of the discrepancy between observed and predicted pulsation velocities (Wood 1990). Recent aperture masking results for R Leo are inconsistent with fundamental mode pulsation because the true linear radius of the star, $495 \pm 83 \, R_\odot$, is too large to be consistent with expected sizes based on fundamental model pulsation for a reasonable range of masses (*i.e.* $\sim 226 \, R_\odot$). This linear size came from an accurate diameter measurement plus a recent parallax measurement of Gatewood (1992). The measured size is approximately consistent with expectations for a star of similar luminosity pulsating in the first overtone mode (Fox & Wood 1985). Further observations are required to determine if other miras are as large as R Leo. Perhaps most important is a set of accurate parallaxes for a large enough sample of these stars to cover a range of periods and luminosities.

4.4 INNER RADII OF DUST SHELLS

Measurements of the inner radii of dust shells surrounding 13 late-type stars were reported by Danchi *et al.* (1994a) using the Infrared Spatial Inteferometer at 11 μm. Inner radii of two additional stars were reported by Danchi *et al.* (1994b,c). Detailed discussions of analyses and radiative transfer modeling are in Danchi *et al.* (1994a). Figure 2 displays a graphical summary of these results. The location of inner radii divide into groups in a natural way that depends on the spectral and luminosity classes of the stars.

One of the most important results of this work is that the mira variables

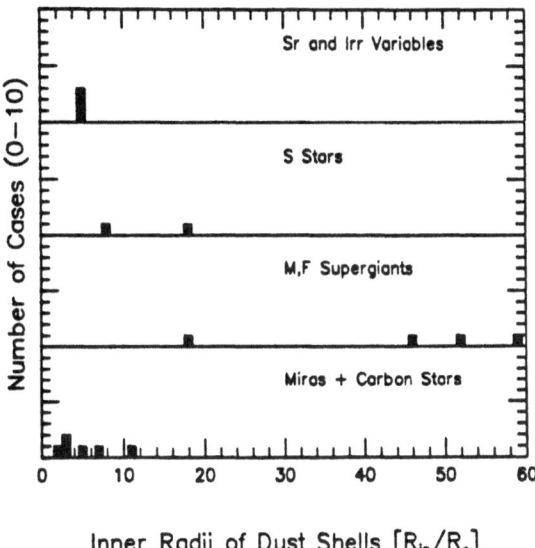

Inner Radii of Dust Shells [R$_{in}$/R$_*$]

Fig. 2. Summary of inner radii of dust shells determined from visibility data and modeling. Note the wide variation of inner radii which is dependent on spectral type.

and carbon stars (*e.g.* *o* Cet, R Leo, IK Tau, and IRC +10216) have inner radii close to the star. The mean inner radius for these stars is 3.4±1.4 R_* from the center of the star. In these cases the dust is located about as close to the star as it can condense fully, and it is assumed material is emitted sometime during each cycle of the luminosity variation (see also Danchi *et al.* 1990, Bester *et al.* 1991). The supergiants VY CMa, VX Sgr, and NML Cyg also have dust within about 6 R_*, and hence also produce dust during approximately every cycle of luminosity variation.

The S stars χ Cyg and W Aql have dust shells with inner radii much farther from the photosphere at 19 R_* and 8 R_*, respectively. This indicates a time scale for dust emission episodes of the order of 20 years.

Early M supergiants α Ori, α Sco, and α Her, and the F supergiant IRC +10420 all have inner radii relatively far from their photospheres. Estimates for most recent dust emission episodes are 48 years for α Ori, 52 years for α Sco, and $\sim$ 9 years for α Her. The average time between episodes is of the order of 30 years.

An important conclusion from these observations is that for the mira variables and the late M supergiants (semi and irregular variables) all have dust close to their photospheres. This gives strong support to the pulsation driven dusty wind model, since the location of the inner radius of the dust is the central issue, as discussed by Hearn (1990). What is most surprising however is that the inner radii of the dust shells varies so much with spectral type and that a variety of time scales are indicated that are long in comparison to the pulsation period but relatively short in comparison to time scales for thermal pulses. Dust is also believed to condense close to the photospheres of these stars, but special conditions are required to trigger it, such as unusually large radial velocity or photometric variations (Danchi *et al.* 1994a). It is interesting to note that dust is not close to the photospheres of χ Cyg and W Aql, despite strong pulsations, indicating that the chemistry of the atmospheres must also be conducive to dust formation for it to occur regularly.

4.5 MASS LOSS RATES

The inner radii of the dust shells discussed in the previous section can be combined with the 11 μm optical depth, and with the velocity of the molecular material at large distances from the stars, to calculate mass loss rates and the total mass of the dust and gas within a given radius from the star. Danchi *et al.* (1994a) found good agreement between mass loss rates obtained from measurements of the dust and those obtained from CO or other molecules (*cf.* Figs. 28 and 29). Their conclusions are that the dust and gas must move at close to the same velocity, consistent with the expectation that radiation pressure on dust grains drives the molecular outflows, and that for most of their sample of stars the average mass loss rates have not changed appreciably in the past 100 years or so.

4.6 MASERS

Several conclusions have been made concerning the relative location of the inner radii of the dust shells and the OH, H_2O, and SiO masers (Danchi *et al.* 1994a). The masers farthest from the photospheres of the stars, the OH masers, for the most part are located well outside the inner radius of the dust shells for stars in their sample. An example is IK Tau, for which the OH maser shell is at an angular radius of $\sim 1''\!.65 \pm 0''\!.14$ or 33 R_0 (Bowers *et al.* 1989). The symbol R_0 denotes the angular distance of the inner radius of the dust shell from the center of the star. The OH masers for the supergiant VY CMa are at a distance greater than 42 R_0. On the other hand the OH maser shell surrounding U Ori is at only 2.6 R_0, which is rather closer to the inner radius than the others. In general the OH shells are ≥ 30 R_0 away from the star, or equivalently ≥ 100 R_*, however there is certainly considerable variability.

Compared to the OH masers, the H_2O masers lie much closer to the inner radii of the dust shells, but are generally outside of it. For example IK Tau has two H_2O maser shells, one with radius 93 mas, the other at 218 mas (Lane *et al.* 1987). The inner radius of the dust shell is 50 mas, hence the closest H_2O masers are at 1.9 R_0 from the inner radius. Examination of other stars shows that generally the H_2O lie within a few R_0 away from the inner radius of the dust shell, *i.e.* ≤ 4 R_0. This is in the region where dust is accelerated by radiation pressure. It has for some time been recognized that intense H_2O masers require a substantial non-equilibrium input of power to operate. Near the inner radii of dust shells either shock waves or intense radiation from newly formed dust could serve this purpose. Although it is not clear that shock waves can be generated by radiative acceleration of dust in this region (Tielens 1983), shock wave excitation of H_2O masers has been modeled by Hollenbach et al. (1993) in densities of $10^6 - 10^8$ cm^{-3}, which is the range of densities found for the H_2O maser regions based on the dust models (cf. Danchi et al. 1994a).

The SiO masers are less well understood than the OH and H_2O masers. Elitzur (1992) concludes that for a collisional pump mechanism, the hydrogen density is probably in the range of $10^9 - 10^{10}$ cm^{-3}. Given a critical density for nucleation of dust of $\sim 10^{11}$ cm^{-3} from homogeneous nucleation theory (Draine 1981), it would seem that SiO masers should lie just outside the location where dust is formed. This is contrary to most expectations, since once the dust is formed the gas phase SiO should be depleted since it should be incorporated into the grains. Recent observations of the dust and SiO masers help to clarify and resolve this apparent conflict. For VX Sgr, Greenhill et al. (1994a,b,1995) found that the SiO masers were located in a ring at roughly 1.3 R_* from the center of the star, which is well inside the inner radius of the dust shell, 4.6 R_*. The hydrogen density at the inner radius is 4×10^7 cm^{-3}. If the densities at the maser region and at the inner radius of the dust shell are connected by an exponential density law, the corresponding scale height is 1 R_*, with an uncertainty of about 0.3 R_*, because of intrinsic uncertainties in the contributing densities (cf. Greenhill et al. 1995). This is approximately the scale height expected from the pulsation-driven shock wave models of the atmospheres of mira variables. For piston amplitudes of 2-4 km s^{-1} these models give atmospheric scale heights in the range of 0.6 to 0.7 R_* (Bowen 1988) which is close to the observational result. If homogeneous nucleation theory is not applicable to the formation of dust (cf. Cherchneff & Tielens 1994), then the conflict can be resolved.

Recently Diamond et al. (1994) measured the locations of SiO masers around TX Cam and U Her and found they were at $\sim$ 2-4 R_*. However, the inner radii of the dust shells for these stars are not known, nor are there any infrared or optical diameters available for these stars. Diamond et al. (1994) suggest the diameters used could be at least a factor of 2 too small. This implies that the SiO masers could be $\leq$ 2 R_*, which is approximately consistent with the results of Greenhill et al. (1994a,b). Improvements to the interpretation of the results for TX Cam and U Her will be possible when direct measurements of the dust shells and diameters become available.

5. Conclusions

In the past few years long baseline interferometry at infrared and optical wavelengths has become a powerful tool for studies of mass loss phenomena occuring close to the photospheres of AGB stars, including pulsation, surface features, and dust formation. Probably the most important next step is for infrared and optical interferometry to produce high-quality maps, and for this step, the construction of instruments with three or more telescopes is critical. Beyond that it is important to bring the sensitivity to the point where all objects of interest can be observed. It seems possible that given sufficient funding levels, long baseline interferometry can play as important a role in infrared and optical astronomy as radio interferometry has played in radio astronomy.

Acknowledgements

Long baseline interferometry at the University of California, Berkeley is supported in part by the Office of the Naval Research (N00014-89-J-1583, N00014-93-0775), and by the National Science Foundation (AST-9315485, AST-9321289, & AST-9221105). One of the authors (WCD) thanks the Royal Observatory Edinburgh for partial financial support which enabled him to attend this conference.

References

Armstrong, J.T.: 1994, Proc. SPIE **2200**, 62.

Baldwin, J.E. *et al.* : 1994, Proc. SPIE **2200**, 231.

Benson, J.A., Dyck, H.M., Ridgway, S.T., Mason, W., & Howell, R.R.: 1991, Astron. J. **102**, 2091.

Bester,M., Danchi, W.C., Degiacomi, C.G., Townes, C.H., & Geballe, T.R.: 1991, Astrophys. J. **367**, L27.

Bester, M., Danchi, W.C., Degiacomi, C.G. & Bratt, P.R.: 1994 Proc. SPIE **2200**, 274.

Bowen, G.H.: 1988, Astrophys. J. **329**, 299.

Bowers, P.F., Johnston, K.J., & de Vegt: 1989, Astrophys. J. **330**, 339.

Buscher, D.F., Haniff, C.A., Baldwin, J.E., & Warner, P.J.: 1990, Mon. Not. of the RAS **245**, 7.

Carleton, N.P. *et al.* : 1994, Proc. SPIE **2200**, 152.

Cherchneff, I., & Tielens, A.G.G.M.: 1994, in: *Circumstellar Media in Late Stages of Evolution*, eds: R.E.S. Clegg & I. Stevens, Cambridge: University Press, in press.

Colavita, M.M. *et al.* : 1994, Proc. SPIE **2200**, 89.

Danchi, W.C., Bester, M., Degiacomi, C.G., McCullough, P.R. & Townes, C.H.: 1990, Astrophys. J. **359**, L59.

Danchi, W.C., Bester, M., Degiacomi, C.G., Greenhill, L.J. & Townes, C.H.: 1994a, Astron. J. **107**, 1469.

Danchi, W.C., Bester, M., Degiacomi, C.G., Greenhill, L.J. & Townes, C.H.: 1994b, Proc. SPIE **2200**, 286.

Danchi, W.C., Bester, M., Greenhill, L.J., Degiacomi, Geis, N., Hale, D., Lopez, B. & Townes, C.H.: 1994c, these proceedings.

Davis, J., Tango, W.J., Booth, A.J., Minard, R.A., Owens, S.M., & Shobrook, R.R.: 1994, Proc. SPIE **2200**, 231.

Di Benedetto, G.P. & Foy, R.: 1992, in: *High Resolution Imaging by Interferometry II*, ESO Conference and Workshop Proceedings **29**, 691.

Diamond, P.J., Kemball, A.J., Junor, W., Zensus, A., Benson, J., & Dhawan, A.: 1994, Astrophys. J. **430**, L61.

Draine, B.T.: 1981, in: *in Physical Processes in Red Giants*, eds: I. Iben & A. Renzini, Reidel:Dordrecht, p317.

Dyck, H.M., Benson, J.A., Ridgway, S.T., & Dixon, D.T.: 1992, Astron. J. **104**, 1982.

Dyck, H.M., Benson, J.A., & Ridgway, S.T.: 1993, Publ. of the ASP **105**, 610.

Elitzur, M.: 1992, in: *Astrophysical Masers*, Kluwer: New York.

Fox, M.W. & Wood, P.R.: 1985, Astrophys. J. **297**, 455.

Gatewood, G.: 1992, Publ. of the ASP **104**, 23.

Greenhill, L.J., Moran, J.M., Backer, D.C., Bester, M., & Danchi, W.C.: 1994a, Proc. SPIE **2200**, 304.

Greenhill, L.J., Moran, J.M., Backer, D.C., Danchi, W.C., & Bester, M.: 1994b, these proceedings.

Greenhill, L.J., Moran, J.M., Backer, D.C., Danchi, W.C., & Bester, M.: 1995, Astrophys. J. in press.

Haniff, C.A., Mackay, C.D., Titterington, D.J., Sivia, D., Baldwin, J.E., & Warner, P.J.: 1987, Nature **328**, 694.

Haniff, C., Ghez, A.M., Gorham, P.W., Kulkarni, S.R., Mathews, K., & Neugebauer, G.: 1992, Astron. J. **103**, 1662.

Hearn, A.: 1990, in: *From Miras to PN: Which Path for Stellar Evolution?*, Editions Frontieres: Gif sur Yvette, p121.

Hill, S.J. & Willson, L.A.: 1979, Astrophys. J. **229**, 1029.

Hollenbach, D.H., Elitzur, M., & McKee, C.F.: 1993, in: *Astrophysical Masers*, eds: A.W. Clegg & G.E. Nedoluha, Springer: New York, p159.

Huggins, P.J.: 1994, these proceedings.

Hutter, D.J.: 1994, Proc. SPIE **2200**, 81.

Karovska, M., Nisenson, P., Papaliolios, C. & Boyle, R.P.: 1991, Astrophys. J. **374**, L51.

Knapp, G.R. & Morris, M.: 1985, Astrophys. J. **292**, 640.

Knapp, G.R.: 1994, these proceedings.

Labeyrie, A., Schumacher, G., Dugue, M. *et al.* : 1986, Astron. & Astrophys. **162**, 359.

Lane, A.P., Johnston, K.J, Bowers, P.F., Spencer, J.H., & Diamond, P.J.: 1987, Astrophys. J. **323**, 756.

Lopez, B., Danchi, W.C., Bester, M., Townes, C.H., & Lefevre,: 1994, in: *Proceedings of the ESO Workshop on Science with the Very Large Telescope Interferometer*, Garching, Germany.

von der Lühe, O., Ferrand, D., Koehler, B., Neng-hong, Z. & Reinheimer, T.: 1994, Proc. SPIE **2200**, 168.

Manduca, S.: 1979, Astron. & Astrophys. Suppl. **36**, 411.

Masson, C.R.: 1994, in: *Very High Angular Resolution Imaging*, eds: J.G. Robertson & Tango, IAU Symposium 158, Kluwer: Dordrecht, p1.

McAlister, H.A. *et al.* : 1994, Proc. SPIE **2200**, 129.

Mourard, D. *et al.* : 1994, Astron. & Astrophys. **283**, 705.

Pearson, T.J. & Readhead, A.C.S.: 1984, Ann. Rev. of Astron. & Astrophys. **22**, 97.

Quirrenbach, A., Mozurkewich, D., Armstrong, J.T., Johnston, K.J., Colavita, M.M. & Shao, M.: 1992, Astron. & Astrophys. **259**, L19.

Quirrenbach, A., Mozurkewich, D., Armstrong, J.T., Buscher, D.F., & Hummel, C.A.: 1993a, Astrophys. J. **406**, 215.

Quirrenbach, A., Hummel, C.A., Buscher, D.F., Armstrong, J.T., Mozurkewich, D., & Elias, N.M.: 1993b, Astrophys. J. **416**, L25.

Quirrenbach, A., Elias, N.M., Mozurkewich, D., Armstrong, J.T., Buscher, D.F., & Hummel, C.A.: 1993c, Astron. J. **106**, 118.

Quirrenbach, A., Mozurkewich, D., Hummel, C.A., Buscher, D.F., & Armstrong, J.T.:
    1994a, Astron. & Astrophys. in press.
Quirrenbach, A.: 1994b, in: *Very High Angular Resolution Imaging*, eds: J.G. Robertson
    & W.J. Tango, Kluwer: Dordrecht, p407.
Sedlmayr, E.: 1990, in: *From Miras to PN: Which Path for Stellar Evolution?*, Editions
    Frontieres: Gif sur Yvette, p179.
Shao, M. *et al.* : 1988, Astron. & Astrophys. **193**, 357.
Tielens, A.G.G.M.: 1983, Astrophys. J. **271**, 702.
Tuthill, P.G., Haniff, C.A. & Baldwin, J.E.: 1994, in: *Very High Angular Resolution Imag-
    ing*, eds: Robertson & Tango, Kluwer: Dordrecht, p395.
Vakili, F., Mourard, D., & Stee, P.: 1994, in: *Pulsation, Rotation and Mass Loss in Early-
    Type Stars*, eds: Balona *et al.* , Kluwer: Dordrecht, p435.
Wilson, R.W., Baldwin, J.E., & Buscher, D.F & Warner, P.J.: 1992, Mon. Not. of the
    RAS **257**. 369.
Wood, P.R.: 1979, Astrophys. J. **227**, 220.
Ziljstra, A.: 1994, these proceedings.

ANALYSIS OF THE IR AND SUB–MM EMISSION OF FOUR POST–AGB STARS

CARSTEN KÖMPE and JOACHIM GÜRTLER
Astrophysikalisches Institut und Universitäts-Sternwarte,
Schillergäßchen 2, D–07745 Jena, Germany

and

THOMAS HENNING
MPG–Arbeitsgruppe "Staub in Sternentstehungsgebieten",
Schillergäßchen 3, D–07745 Jena, Germany

Abstract. The goal of this work is to derive the physical properties of dust envelopes around post–AGB stars by means of radiative transfer calculations. The model spectral energy distributions (SEDs) have been compared with observational data of the post–AGB stars IRAS 10215–5916, 16342–3814, 17150–3224, and 19500–1709 in the wavelength range from 0.4 to 1300 μm. The match between our model SEDs and the observational data is very satisfactory. As a result, we have obtained estimates of the inner and outer radii, the density, the temperature, and the mass of the envelopes of the four objects.

Key words: Post–AGB Stars – circumstellar envelopes – radiative transfer

1. Introduction

The evolution of low– and intermediate–mass stars (≈ 1 to ≈ 8 $M_\odot$) from the asymptotic giant branch (AGB) to the planetary nebulae stage is characterized by intense stellar wind. AGB stars evolve on a very short time scale (several thousand years) via a phase where the star is hidden from view by a dense dust shell (*e.g.* OH/IR star) towards the post–AGB phase (an optically visible star surrounded by a cold dust/gas shell). Eventually, the effective temperature of the star reaches values that are sufficiently high so that the stellar radiation ionizes the inner parts of the envelope, and a planetary nebula appears.

Knowledge of the physical and chemical conditions in the dust/gas envelopes around post–AGB stars is not only interesting for its own sake, but it also constrains evolutionary models of AGB and post–AGB stars. Most of these objects are, however, too small and too distant to be resolved by mm or IR telescopes. Another way to explore the physical conditions in post–AGB envelopes is to carry out model calculations in order to obtain theoretical spectral energy distributions (SEDs) that can then be compared with observed SEDs. We have applied this technique to four IRAS point sources previously suggested to be post–AGB objects. Our results are presented in the following sections.

Astrophysics and Space Science **224**: 353–356, 1995.
© 1995 *Kluwer Academic Publishers.*

2. The Data

For this work, we have selected four objects: IRAS 10215−5916, 16342−3814, 17150−3224, and 19500−1709. The first three objects have been proposed to be post–AGB objects on the basis of their IRAS colours; the last object has been proposed based on being a high–latitude super–giant with IR excess (Kwok 1993, van der Veen *et al.* 1989). The observational SEDs of these objects have been constructed using 1.3 mm data obtained with the Swedish–ESO Sub–mm Telescope (SEST) on La Silla, Chile, low–resolution (LRS) IRAS spectra, IRAS point source fluxes, and IR fluxes in the range from 1 to 8 μm from the literature (Gezari *et al.* 1993).

We show (Fig. 1) the SED of IRAS 17150−3224 as a representative example. It shows the double peaked structure typical of post–AGB stars: The peak at near IR wavelengths corresponds to the radiation of the central star; the second peak at wavelengths longer than $\approx$ 8 μm is dominated by the emission of the dust particles in the envelope. The 10 μm silicate feature is clearly seen in absorption indicating a substantial optical depth in the envelope.

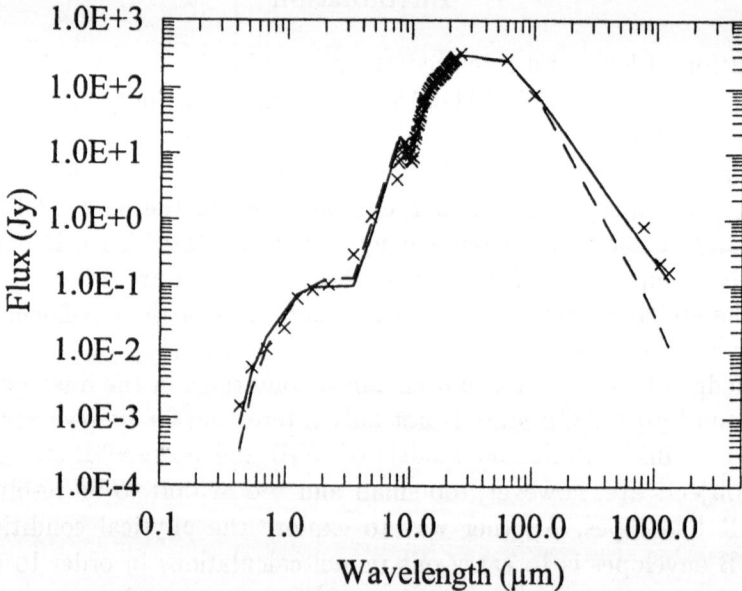

Fig. 1.   Spectral energy distribution of IRAS 17150−3224. The crosses show the observational data points, the continuuous line the model curve, and the dashed line corresponds to a model curve that has been calculated with optical constants taken from Draine (1985).

3. Description of the Model

To analyze our data, we have used a radiative transfer code that has been adapted from a program originally developed by E. Krügel (MPIfR, Bonn) to model the environment of compact HII regions (Chini *et al.* 1986). In the model, we assume that an evolved star is located in the centre of a spherically symmetric envelope of dust grains. The main input parameters are those of the central star (effective temperature, luminosity), those of the dust shell (inner and outer radius, density law $n_d \propto r^{-x}$), and the parameters of the dust grains (spherical shape, minimum and maximum radius, size distribution $n_d \propto a^{-3.7}$, optical constants n, k). According to the spectral type and/or the presence of the 10 μm silicate feature, we assumed the dust to consist either of silicate or carbon. The optical constants (n,k) have been taken from Draine (1985) and for amorphous carbon from Blanco *et al.* (1991) (BE–curve in Blanco *et al.*) and Rouleau & Martin (1991). The absorption and scattering properties (extinction efficiency Q_{ext}) of the dust grains have been calculated using Mie's theory. Finally, the program produces as output the spectral energy distribution F_λ (Jy) that can be compared to the observed data.

4. Results and Discussion

We have run the radiative transfer code using appropriate input parameters for our post–AGB objects. The model SEDs produced by the program match the observed data very satisfactorily. This way, we have derived a consistent set of physical parameters (see Table I) for each of our sources. Within the limits of the model, these parameters are expected to represent the real conditions of the envelopes keeping in mind, however, that the distances of the objects are not known. In this work, we have adopted a standard distance of 1 kpc.

It should be noted that the exponent of the density law has consistently a value of x = −1 for the oxygen–rich objects. This result is not compatible with a stationary model in which outflow velocity and mass–loss rate are constant that implies an exponent of x = −2. Still assuming a constant outflow velocity, an exponent of x = −1 indicates a decrease of the mass–loss rate with time.

Another important point to discuss is the FIR wavelength dependence of the extinction efficiency Q_{ext} of the dust particles. The optical constants from Draine (1985) that we have used for the oxygen–rich objets imply an extinction efficiency $Q_{ext} \propto \lambda^{-2}$. However, only in the case of IRAS 10215−5916 do we obtain a satisfactory match between the model and the data. For the other two objects, the model predicts a flux at 1.3 mm that is lower than the observed flux by an order of magnitude (see Fig. 1). In order to predict the observed long wavelength fluxes correctly, we have used an extinction efficiency law of $Q_{ext} \propto \lambda^{-0.7}$ for wavelengths longer than 230 μm. This law had also to be used for the carbon–rich object IRAS 19500–1709. We thus find an exponent somewhat smaller than predicted for amorphous, layered dust materials or very small amorphous grains (Tielens & Allamandola 1986).

TABLE I

Physical parameters derived from model calculations.

Source:	10215–5916	16342–3814	17150–3224	19500–1709
Star:				
T_{eff} (K)	4750	8200	15000	7500
L ($L_\odot$)	$1.0 \ 10^4$	640	$1.5 \ 10^3$	$2.0 \ 10^3$
Envelope:				
r_{in} (AU)	550	230	120	1600
r_{out} (AU)	7650	7860	4700	3530
n (cm^{-3})	$1.3 \ 10^{22} \ r^{-1}$	$1.5 \ 10^{24} \ r^{-1.1}$	$3.0 \ 10^{22} \ r^{-1}$	$2.0 \ 10^{38} \ r^{-2}$
M ($M_\odot$)	0.9	2.3	0.75	0.013
T_{max} (K)	200	209	401	159
T_{min} (K)	60	35	44	82
Dust:	Silicate	Silicate	Silicate	Carbon
Interstellar medium:				
A_V	2.0 mag	2.0 mag	1.25 mag	0 mag

Acknowledgements

This work was financially supported by a grant from the German Bundesministerium für Forschung und Technologie (Förderungsnummer 05 2JN-13A).

References

Blanco, A., Fonti, S. & Rizzo, F.: 1991, Infrared Physics **31**, 167.
Chini, R., Krügel, E. & Kreysa, E.: 1986, Astron. & Astrophys. **167**, 315.
Draine, B.T.: 1985, Astrophys. J., Suppl. **57**, 587.
Gezari, D.Y., Schmitz, M., Pitts, P.S. & Mead, J.M.: 1993, *Catalog of Infrared Observations*, NASA RP-1294.
Kwok, S.: 1993, in *Planetary Nebulae (IAU Symp. No. 155)*, eds: R. Weinberg & A. Acker, Kluwer, Dordrecht, p263.
Rouleau, F. & Martin, P.G.: 1991, Astrophys. J. **377**, 526.
Tielens A.G.G.M. & Allamandola L.J.: 1986, in *Interstellar Processes*, eds: D.J. Hollenbach & H.A. Thronson, Reidel, Dordrecht, p397.
van der Veen, W.E.C.J., Habing, H.J. & Geballe, T.R.: 1989, Astron. & Astrophys. **226**, 108.

CIRCUMSTELLAR ABSORPTION AND EMISSION
IN THE POST-AGB STARS HR4049 AND HD213985

C. WAELKENS
Instituut voor Sterrenkunde, Leuven, Belgium

L.B.F.M. WATERS
Universiteit van Amsterdam, The Netherlands
Space Research Laboratory, Groningen, The Netherlands

and

H. VAN WINCKEL and K. DAEMS
Instituut voor Sterrenkunde, Leuven, Belgium

Abstract. The evolved C-rich low-mass stars HR 4049 and HD 213985 present variable circumstellar extinction and emission. Observations of HR 4049 strengthen the link between the far-UV extinction rise and the presence of small C-rich particles. Variability of the UV absorption bump in HD 213985 is reported here for the first time and is discussed.

Key words: Post-AGB stars – circumstellar matter – HR 4049 – HD 213985

1. Introduction

The stars HR4049 and HD213985 are A-type high-galactic-latitude supergiants with an IR excess. Current interpretation is that both objects are low-mass stars in a late (post-AGB?) evolutionary stage. The dust around both stars appears to be carbon-rich: HR4049 displays the UIR features; the high photospheric carbon abundance of HD213985 suggests that the rather featureless IR emission of this star is due to amorphous carbon grains.
The circumstellar grains of both stars cause variable extinction in the optical and the UV. These stars then offer us rather exceptional opportunities to probe cosmic dust at various wavelengths. In this paper we explore what the correlation between UV, optical and IR dust features of these stars teaches us about dust properties and composition.

2. HR 4049

HR 4049 is a 429-day binary with an unseen low-mass companion (Van Winckel and Waelkens, these proceedings). The optical variability of this star can be explained by periodic obscuration by a dust ring that surrounds the system (Waelkens *et al.* 1991) and that is inclined with respect to the observer: at inferior conjunction obscuration is maximal, while the extinction is low, or sometimes vanishing, when the star is at its most distant position and is seen above or below the part of the ring toward the observer.

Astrophysics and Space Science **224**: 357–360, 1995.
© 1995 *Kluwer Academic Publishers.*

This model also accounts for the linear component of the UV extinction.

We have obtained low-resolution IUE data for HR4049 at five epochs. These data can be expressed as circumstellar extinction curves if the interstellar reddening and the stellar UV flux are known. The interstellar reddening can be accurately estimated at E(B-V) = 0.15: this is the value found for some main-sequence stars near the line-of-sight toward HR 4049 and is also consistent with the observed colors at maximum brightness and the typical B-V for a star with the atmospheric parameters of HR 4049. We have estimated the intrinsic UV flux from an extrapolation of a Kurucz model with parameters $T_{eff} = 7500\ K$, $log\ g = 1.0$ and $Z = -2.5$ normalized on the dereddened optical fluxes at maximum brightness.

As already pointed out by Waters *et al.* (1988) and Buss *et al.* (1989) the circumstellar UV extinction consists of a linear component and a far-UV rise. The uncertainty on the UV fluxes of the star may induce a systematic quantitative error on the determination of the UV extinction, but the two-component model must be basically correct. A possible interpretation is that the far-UV rise is linked to the small particles that cause the IR PAH-like emission bands. An interesting new result is that the far-UV rise of HR 4049 appears to be variable: it was much higher than average on November 30, 1993. Remarkably, there have been some unpublished claims (Le Bertre, Private Communication; Geballe, Private communication) that also the strength of the UIR bands is variable.

3.   HD 213985

The optical photometric behavior of HD213985 is similar to that of HR 4049 (Waelkens *et al.* 1987, Whitelock *et al.* 1989). Combining all our Geneva photometric data obtained with the Swiss Telescope at La Silla, Chile, with the radial-velocity measurements we have obtained with the CES Spectrograph attached at the CAT Telescope at ESO, Chile, we have found that the same inclined-binary model applies to HD 213985. Maximum brightness occurs again at superior conjunction, and minimum brightness when the star is nearest to us and circumsystem obscuration is maximal. Unlike HR 4049, however, HD 213985 describes a circular orbit. Also, the mass function implies that the mass of the secondary is of the order of two solar masses.

Due to the high galactic latitude of HD 213985, the interstellar reddening is low. A fortuitous circumstance is the presence near the line of sight of the early-B supergiant HD 214080, for which the E(B-V) is 0.15. This reddening is again consistent with the spectral type of HD 213985 (A2 Ia) and the colors at maximum brightness. The description of the intrinsic UV flux of HD 213985 with the Kurucz model derived from the optical atmosphere analysis is probably fairly secure, since the chemical composition of this star differs much less from the solar composition as in the case of HR 4049. Since the optical variations can be explained by extrinsic effects, there is no reason to assume that the intrinsic UV flux of the star is variable.

The resultant UV absorption curves (Figure 1) show three components: a fairly flat 'linear' component, indicating that the grains responsible for the optical absorption are rather large, a (probably constant) far-UV rise, and - most interestingly - an absorption bump of variable strength. The central wavelength of the absorption bump falls in the range between 226 and 240 nm, *i.e.* it is redder that the interstellar 220 nm feature but bluer than the circumstellar bump observed for R Coronae Borealis (Hecht 1986). The width of the bump is intermediate between that of the interstellar feature and that of R CrB. We have not found a clear correlation between the strength of the bump and orbital phase. Nevertheless, no definite conclusion on this point can be made from our data, since unfortunately no simultaneous UV and optical data are available, and since the intrinsic scatter in the optical light curve is large.

We finally also report the surprising detection of CH^+ absorption lines in the optical spectrum of HD 213985. These lines are stationary, and show an expansion velocity of 5 km/s with respect to the systemic velocity. It is not clear whether these absorption lines are caused by the torus of matter near the system or by the larger and cooler dust cloud surrounding the system.

4. Discussion

The configurations of HR 4049 and HD 213985 as binary stars are very similar. In addition, the dust around both stars is carbon rich: HR 4049 displays the IR features, and HD 213985 displays CH^+ absorption. Yet, the extinction and emission properties of both stars are qualitatively different. The most striking difference is the absence of any circumstellar absorption bump for HR 4049 and the prominent presence of a variable bump for HD 213985.
    Laboratory results (Blanco *et al.* 1993) suggest that a 220 nm absorption bump may be caused by the dehydrogenation of amorphous carbon grains by UV irradiation. The featureless and broad near-IR excess of HD 213985 may indeed be due to amorphous carbon grains. We point out that HD 213985 is significantly hotter than HR 4049.

Could it be that the presence of CH^+ absorption lines is the signature of this dehydrogenation process? For HR 4049, where carbon-rich circumstellar material is also present, no CH^+ absorption is observed. On the other hand, CH^+ *emission* is observed for the Red-Rectangle star HD 44179 (Balm & Jura 1992), which has a temperature similar to that of HR 4049. The cir-

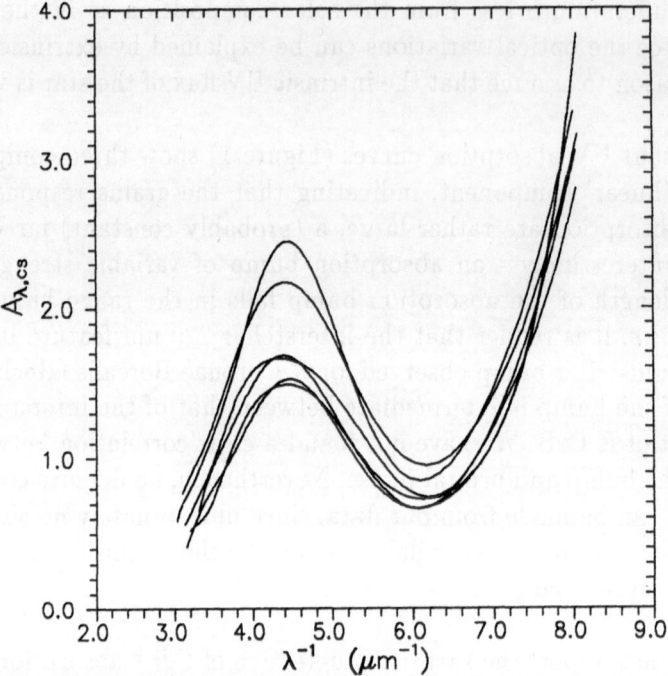

Fig. 1.   The UV circumstellar extinction curves of HD 213985 on 19/10/86; 22/05/87;
11/11/88; 19/05/92; 4/12/92 and 12/05/94.

cumstellar bump, if any, for HD 44179 is extremely weak (Webster 1993),
but we remark that it also doesn't show variable circumstellar extinction in
the optical.

The variability of the circumstellar absorption and emission of HR 4049
and HD 213985 clearly offers remarkable opportunities for further probing
the dust around both stars. The orbits of both objects, which are the natural
time scales on which variability may be expected to occur, are shorter than
the expected lifetime of ISO. Simultaneous ISO and IUE observations of
HR 4049 and HD 213985 may yield several clues to the solution of several
problems concerning the carriers of prominent dust features.

References

Balm, S.P. & Jura, M.: 1992, Astron. & Astrophys. **261**, L25.
Blanco, A., *et al.* : 1993, Astrophys. J. **406**, 739.
Buss, R.H., *et al.* : 1989, Astrophys. J. **347**, 977.
Hecht, J.: 1986, Astrophys. J. **305**, 817.
Waelkens, C., *et al.* : 1987, Astron. & Astrophys. **406**, 739.
Waelkens, C., *et al.* : 1991, Astron. & Astrophys. **242**, 433.
Waters, L.B.F.M., *et al.* : 1989, Astron. & Astrophys. **211**, 208.
Webster, A., *et al.* : 1993, Mon. Not. of the RAS **262**, L59.
Whitelock, P., *et al.* : 1989, Mon. Not. of the RAS **241**, 393.

THE MID-INFRARED STRUCTURE OF THE BIPOLAR

NEBULAE AFGL 915, AFGL 618, AND AFGL 2688

JOSEPH L. HORA

Institute for Astronomy, 2680 Woodlawn Drive, Honolulu, HI 96822

LYNNE K. DEUTSCH

FCAD/University of Massachusetts, Amherst, MA 01003

WILLIAM F. HOFFMANN

Steward Observatory, University of Arizona, Tucson, AZ 85721

and

GIOVANNI G. FAZIO

Smithsonian Astrophysical Observatory, 60 Garden St., Cambridge, MA 02138

Abstract. We have obtained near diffraction-limited images of three bipolar PPN at UKIRT in October, 1993: AFGL 915 (the "Red Rectangle"), AFGL 618, and AFGL 2688 (the "Egg Nebula"). Images were taken at unidentified infrared (UIR) emission feature wavelengths and at several continuum wavelengths in the 10 and 20 μm atmospheric windows. In all three PPN the emission is dominated by a central point source with fainter emission extending for several arcsec. In AFGL 2688, the mid-IR emission is extended in the same direction as the main optical lobes. In AFGL 915, the UIR feature emission is spatially separated from the central source. The "spikes" that have been observed at 2 μm and give the nebula its rectangular appearance are also visible at 10 μm.

Key words: planetary nebulae: individual (AFGL 915, AFGL 618, AFGL 2688) – planetary nebulae: General – planetary nebulae: Infrared

1. Introduction

We are investigating the distribution of the 10 and 20 μm emission from proto-planetary nebulae (PPN) to gain an understanding of their formation and structure. The primary goals of the study are to image the continuum, fine-structure line emission, and unidentified infrared (UIR) emission features (usually attributed to PAHs or other hydrocarbon molecules) to determine the distribution of the IR-emitting material in the objects. Previous observations have shown that these PPN are strong sources of mid-IR emission, yet they lacked the spatial resolution to determine the structure of the PPN.

2. Observations and Reduction

The three nebulae were observed at the United Kingdom Infrared Telescope (UKIRT) on 23 and 28 October 1993 using the UA/SAO/NRL MIRAC mid-infrared camera (Hoffmann *et al.* 1993). The pixel scale used was 0''.5

Astrophysics and Space Science **224**: 361–364, 1995.

pixel^{-1}. The 8–13 μm images were taken using the 2% CVF, and a 8% filter was used at 20.2 μm. Short (0.1 sec) chopped integrations were taken in order to "freeze" the seeing. Several hundred images were obtained at each wavelength. The short integration times resulted in nearly diffraction-limited images (0''6–0''8 FWHM). The 1σ noise in the final 8–13 μm images is on the order of 0.15 Jy arcsec^{-2} or better. The individual chopped frames were shifted and coadded to obtain the final images. Standard stars (β And, α Tau, α CMa) were observed in the same way and were used as flux and point source standards.

3.   AFGL 618

The bipolar nebula AFGL 618 has two lobes of roughly 2''– 3'' diameter separated in the E-W direction by 7''. The bright mid-IR source is located between the lobes. This object is likely a young PN as it has a region of ionized emission in the central 0''4 (Kwok & Bignell 1984). The mid-IR emission is relatively featureless, with a color temperature of 275 K (Westbrook *et al.* 1975).

Observations of AFGL 618 were made at 10.0, 11.2, 13.2, and 20.2 μm. The source was extremely compact at all wavelengths. Comparison with β And showed evidence for extension in the E-W direction on the order of 0''1 – 0''2. The object is not as extended in the N-S direction, with the source size <0''05 at 10.0, 11.2, and 13.2 μm, and <0''1 at 20.2 μm.

4.   AFGL 2688

The PPN AFGL 2688 has two optically bright lobes separated by about 8'', along a bipolar axis oriented ∼15°E of N. The oval lobes have horns or spikes that extend from the extreme ends of the lobes (*e.g.* Latter *et al.* 1993). The centrally located compact mid-IR source has a relatively featureless spectrum that can be fitted by 200 K and 120 K blackbody curves in the 8–13 μm and 16–24 μm ranges, respectively (Forrest *et al.* 1975).

Mid-infrared images were obtained with MIRAC at 10.0 and 11.2 μm. The source is 1''0×1''6 FWHM, consistent with earlier measurements (Ney *et al.* 1975). Figure 1 shows the 11.2 μm image. Also shown is the color temperature image calculated from the 10.0 and 11.2 μm images. The temperature is centrally peaked, with a maximum of about 200 K. The region outside the central 2'' is reasonably flat, with an average temperature of about 150±15 K.

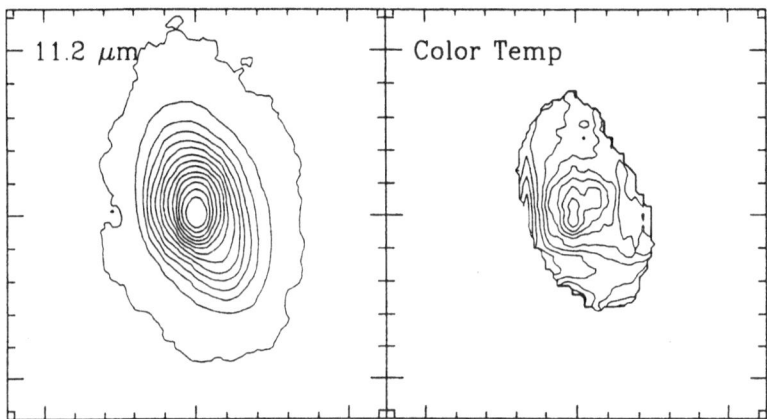

Fig. 1. Contour plots of AFGL 2688. The left figure shows the 11.2 μm image. Ticks along edge are arcsec, N is up and E to the left. Contours are logarithmically spaced with a minimum and maximum value of 2 and 200 Jy arcsec^{-2}. The 1σ noise in this image is 0.13 Jy arcsec^{-2}. The plot on the right is the color temperature image, with linearly spaced contours, with minimum, maximum, and spacing of 120, 260, and 8 K, respectively.

5. AFGL 915

The object AFGL 915 consists of a B9-A0 star surrounded by a rectangular-shaped nebulosity extending roughly 40"N-S (Cohen *et al.* 1975). There are spikes extending from the N and S ends at the corners of the rectangle. No radio continuum has been detected from this source. The mid-IR spectra is dominated by strong UIR feature emission superposed on a continuum that can be fit by a 280 K blackbody (Kleinmann *et al.* 1978).

MIRAC images of AFGL 915 were obtained at 8.0, 8.6, 10.0, 11.2, 12.7, 13.2, and 20.2 μm. The source consists of an unresolved bright core and a rectangular extended emission similar to that seen in the near-IR (Leinert & Haas 1989). The source size at 10.0 μm is $\sim 0\overset{''}{.}8 \times 1\overset{''}{.}25$ FWHM (not deconvolved). The distribution of emission at the UIR feature wavelengths is different than the continuum flux. The FWHM source size at 11.2 μm is $0\overset{''}{.}8 \times 1\overset{''}{.}5$. The images show that the UIR emission is more prominent along the spikes than the continuum emission. This is apparent in the continuum-subtracted 11.2 μm UIR feature image in Figure 2. In this image, the peak of UIR emission is offset from the center of the object. Also, the spikes are prominent and can be seen to extend several arcsecs from the center. The color temperature image calculated from the 10.0 and 13.2 μm images shows a flat temperature distribution over most of the source of 210 ± 16 K, with a bright peak near the center of 450 K.

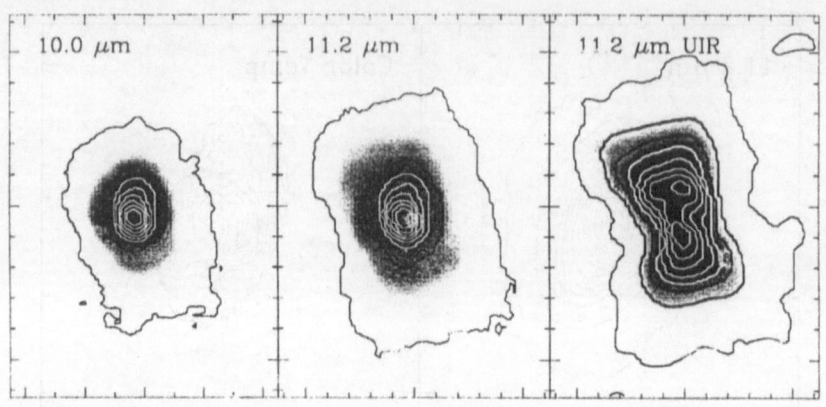

Fig. 2.   Contour/grayscale plots of AFGL 915. The plots left and center are the images obtained at 10.0 and 11.2 μm. Ticks along edge are arcsec, N is up and E to the left. Contours are logarithmically spaced between a minimum, maximum of 2, 166 and 2, 186 Jy arcsec^{-2} for the 10.0 μm and 11.2 μm images, respectively. The 1σ noise in these images is 0.1 Jy arcsec^{-2}. The right-most plot is the UIR feature image where a scaled continuum image was subtracted from the 11.2 μm image. Contours in the UIR feature plot are linearly spaced with a minimum, maximum, and spacing of 0.4, 20, 2 Jy arcsec^{-2}, respectively.

6.   Conclusions

In all three PPN, the mid-IR emission is seen to be very compact and central-ly located. Where extended emission is detected, it is most extended along the major lobes of the nebulae. The mid-IR emission therefore does not directly resolve any torus or disk responsible for creating the bipolar mor-phology. The extended emission in AFGL 2688 and AFGL 915 has a fairly flat color temperature distribution, with peaks near the central source.

References

Cohen, M. *et al.* : 1975, Astrophys. J. **196**, 175.

Forrest, W. J., Merrill, K. M., Russell, R. W., & Soifer, B. T.: 1975, Astrophys. J. **199**, L181.

Hoffmann, W. F., Fazio, G. G., Shivanandan, K., Hora, J. L., & Deutsch, L. K.: 1993, in *Infrared Detectors and Instrumentation*, ed: A. M. Fowler, Proc. SPIE 1946, p449.

Kleinmann, S. G. *et al.* : 1978, Astron. & Astrophys. **65**, 139.

Kwok, S., & Bignell, R. C.: 1984, Astrophys. J. **276**, 544.

Latter, W. B., Hora, J. L., Kelly, D. M., Deutsch, L. K., & Maloney, P. R.: 1993, Astron. J. **106**, 260.

Leinert, C., & Haas, M.: 1989, Astron. & Astrophys. **221**, 110.

Ney, E. P., Merrill, K. M., Becklin, E. E., Neugebauer, G., & Wynn-Williams, C. G.: 1975, Astrophys. J. **198**, L129.

Westbrook, W. E. *et al.* : 1975, Astrophys. J. **202**, 407.

SESSION H:

SHOCKS
IN
CIRCUMSTELLAR WINDS

SESSION II.

SHOCKS
IN
CIRCUMSTELLAR WINDS

CIRCUMSTELLAR SHOCKS:
COLLIDING STELLAR WINDS

IAN R. STEVENS

School of Physics and Space Research, The University of Birmingham, Edgbaston, Birmingham, B15 2TT, UK.

Abstract. In this paper I will review some recent developments in the field of circumstellar shocks, particularly as they relate to colliding stellar winds. I shall review the basic physics of colliding winds and shocks, and discuss recent developments in hydrodynamic modelling of colliding winds. I shall also report on recent X-ray observations of shock emission in Wolf-Rayet binary systems where high resolution X-ray spectra of colliding wind shock emission is being seen. I will discuss the occurrence of colliding winds to such diverse systems as Wolf-Rayet binaries, pre-main sequence binaries, symbiotic stars as well as the Galactic center object IRS 7, where recent results on interacting winds are yielded insight into the structure of winds in general.

Key words: stars:Wolf-Rayet – stars:X-rays – stars:colliding winds – stars:symbiotic

1. Introduction

Important examples of circumstellar shocks occur in binary systems where both stars have substantial stellar winds; *i.e.* colliding stellar winds. Colliding stellar winds are now being recognised as a common phenomenon in a diverse class of objects, ranging from the highest mass systems in late stages of stellar evolution to lower mass systems as well as, potentially, binaries in early stages of stellar evolution.

In this paper I discuss some of the basic physics of colliding winds, before reviewing recent developments in the hydrodynamic modelling of colliding winds. Recent X-ray results colliding wind systems will be discussed, as will recent results on colliding winds in symbiotic stars, PMS binaries and IRS 7 (a Galactic Center supergiant). I will not discuss at all the role of colliding winds in single stars, which are important in the context of planetary nebulae and supernovae (see Clegg, Stevens & Meikle 1994).

2. Basic Physics

The basic physics of colliding winds will be discussed here primarily in the context of Wolf-Rayet + O-star binary systems. A more comprehensive review of colliding wind physics can be found in Stevens *et al.* (1992). In these systems, both stars have massive supersonic winds. These wind collide, forming a double shock structure with a region of hot gas sandwiched between the shocks, subdivided by a contact discontinuity (see Fig. 1 for a schematic view). For material passing through a strong shock, the post-shock

Astrophysics and Space Science **224**: 367–376, 1995.
© 1995 *Kluwer Academic Publishers.*

IAN R. STEVENS

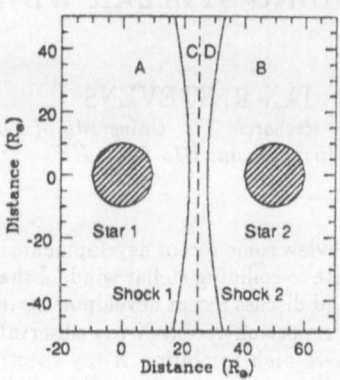

Fig. 1. Sketch of a colliding wind binary with identical stars with isotropic winds. The unshocked parts of the winds of stars 1 and 2, labelled A and B respectively, are bounded by two opposing shock fronts represented by the solid lines. In the hot region in between, regions C and D of shocked material from stars 1 and 2, respectively, are separated by a contact discontinuity represented by the dashed line.

gas temperature is $kT_s = \frac{3}{16}\bar{m}v^2$ with $\bar{m}$ the mean particle mass ($\sim 10^{-24}$ g for solar abundances). This expression can be restated as $kT_s(\text{keV}) \sim 1.2\, v_8^2$ with v_8 being the wind velocity at the shock in units of 10^8 cm s^{-1}.

In WR+O systems, where wind velocities are $\sim 1000 - 2000$ km s^{-1}, the characteristic post-shock temperature is of order a few keV. Consequently the shocked gas will radiate mainly at X-ray energies. Thus, while colliding winds will have signatures at different energies (γ-rays, Eichler & Usov 1993; UV, St-Louis et al. 1993) colliding winds will be directly visible by their thermal X-ray emission. However, because of potential confusion with X-rays produced by instabilities in all radiatively driven stellar winds, it has proved difficult to unambiguously relate X-ray emission from binary systems with colliding wind emission. As we shall see later, recent improvements in X-ray technology has now made this possible.

WIND MOMENTUM RATIO $\mathcal{R}$

There are two basic parameters that can be used to give an overall description of colliding winds. The momentum ratio $\mathcal{R}$ of the winds in a colliding wind binary is defined as:

$$\mathcal{R} = \left[\frac{\dot{M}(WR)v_\infty(WR)}{\dot{M}(O)v_\infty(O)}\right]^{1/2} \tag{1}$$

In WR+O-star systems the WR wind tends to have more momentum ($\mathcal{R} \sim 3-6$), and the O-star wind is contained within a cone of opening half-angle θ. Eichler & Usov (1993) give a useful analytic formula for θ

$$\theta\ (\text{degrees}) \sim 120 \left(1 - \frac{\mathcal{R}^{-2/5}}{4}\right)\mathcal{R}^{-2/3} \tag{2}$$

COOLING PARAMETER χ

Another important factor in determining the structure of colliding winds is the degree to which the post-shock wind can cool. If the post-shock flow-time is much greater than a dynamical timescale then the post-shock flow is unable to cool substantially (*i.e.* is essentially adiabatic) the post-shock region will consist of a pressure supported region of hot diffuse gas. On the other hand, if the cooling timescale is short compared to a dynamical timescale then the post-shock regions collapses into a dense shell of material, primarily constrained by ram pressure. We define a cooling parameter χ as

$$\chi = \frac{t_{cool}}{t_{dyn}} \approx \frac{v_8^4 d_{12}}{\dot{M}_{-7}} \tag{3}$$

where d_{12} is the distance to the contact discontinuity in units of 10^{12} cm, and $\dot{M}_{-7}$ is the mass loss rate in units of $10^{-7} M_\odot$ yr^{-1}. For dense post-shock material (*i.e.* close binaries) the wind material will cool, while for wide binaries the post-shock material will be adiabatic.

3.   Hydrodynamic Models

There have substantial recent progress concerning the ongoing development of hydrodynamic models and simulations of colliding stellar winds.

Stevens *et al.* (1992) presented the first fully self consistent models of colliding winds, including radiative cooling. The hydrodynamic calculations were also used as a basis for calculating spectral models of colliding wind binaries, in particular V444 Cyg and HD 193793. These calculations were 2-D and did not include binary rotation. Stevens *et al.* (1992) found a total of three instabilities at the interface between the colliding winds, depending on the degree to which the post-shock gas was able to cool. In the adiabatic-adiabatic case the Kelvin-Helmholtz instability will occur when there is a velocity shear across the contact discontinuity. When one or both sides of the post-shock material can cool two further ram-pressure instabilities were found to occur.

Walder (1994) has gone even further than this, performing 3-D wind collision calculations, and including binary rotation effects. In particular, he finds that the inclusion of binary rotation effects, does not significantly affect the level of X-ray production at the shock surface, but given the alterations that it makes in the system geometry it will substantially change the observed X-ray flux. In Figure 2 we show a simulation by Walder (1994) of the WR+O-star system V444 Cyg where binary rotation has been included.

One long-standing problem is that the predicted X-ray temperature of colliding wind systems is higher than that actually observed. Recently, Stevens & Pollock (1994) have published 1-D radiation hydrodynamic calculations of

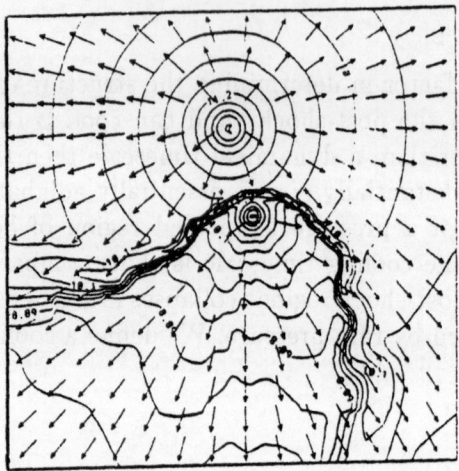

Fig. 2.   A hydrodynamic simulation of V444 Cyg (Walder 1994), where binary rotation
has been included. The WR star (near the top) has a much stronger wind than the O-star,
and the O-star wind is compressed into a small 'cone-like' region. In this diagrams density
contours and velocity vectors are shown. Note how binary rotation effects have already
begun to distort the shape of the shocks.

colliding winds, including the effects of two-radiation fields on the dynamics. They found that the inclusion of the second radiation field inhibited
the wind acceleration and substantially lowered the wind velocity at the
shock surface and hence lowered the X-ray temperature. This mechanism
only works in comparatively close binary star. We discuss this in the light
of the new X-ray observations reported later.

4.   ASCA Observations of Colliding Winds

The Japanese ASCA satellite has excellent spectral resolution, approximately an order of magnitude better than that of the ROSAT PSPC. In additional
the wider energy range of $\sim 0.5 - 10$ keV provides an excellent opportunity
to study colliding wind phenomena in early-type binary system.

Two important colliding wind systems HD193793 (Koyama *et al.* 1994)
and Gamma Velorum (Stevens *et al.* 1994, this volume) have been observed
with ASCA. The results on Gamma Velorum are presented in Stevens *et al.*
(this volume).

4.1   HD 193793

HD 193793 (WR 140) is a most puzzling object, displaying a wide range of
interesting phenomena from radio to X-ray energies (Williams *et al.* 1990).
It consists of WC7 Wolf-Rayet star and an O4-5 companion. It has an orbital
period of 7.9 years and a very eccentric orbit ($e \sim 0.9$). It is observed to have
roughly constant thermal radio emission consistent with a massive stellar

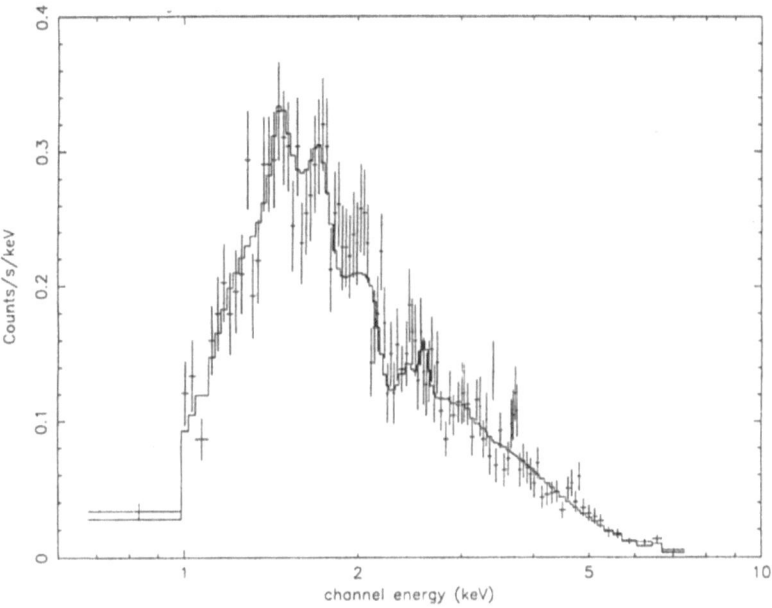

Fig. 3.   The ASCA spectrum of HD 193793. The best fit model is a single temperature Raymond-Smith plasma with $kT = 3.3$ keV with $N_H = 3.3 \times 10^{22}$ cm^{-2}. In addition, enhanced C and Ne abundances are also found.

wind from the WR star and highly variable non-thermal radio emission, largely absorbed by the WR wind, but visible at epochs when we observe the system through the more diffuse WR wind.

Also seen are large infrared outbursts around periastron passage ($\times 10$ increase in flux on a timescale of weeks). This has been interpreted as being due to dust formation coinciding with periastron passage (though how this occurs is the subject of much debate, though it may be related to radiative cooling in the post-shock region giving rise to dense clumps which can provide the sites for dust formation). WR 140 has also been observed as a strong X-ray source by EXOSAT (Williams *et al.* 1990).

ASCA observations of HD 193793 have been made by Koyama *et al.* (1994) and Pollock *et al.* (1994, in preparation), and the X-ray spectrum at one epoch is shown in Figure 3. The spectrum is well fitted by a single temperature Raymond-Smith plasma model with $kT \sim 3$ keV and $N_H \sim 3 \times 10^{22}$ cm^{-2}. Better fits can be obtained by varying element abundances, with N(C) = $30\times$ solar, N(Ne) = $100\times$ solar, and N(Fe) = $0.8\times$ solar. These abundances are in reasonable agreement with the expectations of stellar evolution models and results from line-profile modelling of WC stars. It does however show the potential that of resolution X-ray spectroscopy in determining abundances in WR stars. While ASCA is not particularly sensitive below 0.5 keV, which limits the extent to which ASCA can determine abundances, the next generation of X-ray instruments will be, and will still have excellent spectral resolution.

In general terms, the observed spectrum of WR 140 is in reasonable agreement with the spectral models of Stevens et al. (1992), though modelling work is still in progress. There is, however, little evidence of temporal variability between the ASCA observations and earlier GINGA/EXOSAT observations, which is in contradiction to the hydrodynamical models of HD 193793.

5.   Pre-Main Sequence Binary Systems

Pre-main sequence binaries, such as T-Tauri and Herbig Ae/Be stars, are seen to be X-ray emitting objects with characteristic $kT \sim 1$ keV and $L_x \sim 5 \times 10^{28} - 10^{32}$ ergs s^{-1} . The X-ray emission from PMS stars is believed to be basically due to enhanced solar-type activity driven by a dynamo mechanism. While this is plausible for the low-mass T-Tauri objects, it is less so for the higher mass systems, and other mechanisms have been proposed (see Preibisch & Zinnecker 1994).

Recently, Zhekov et al. (1994) have suggested that colliding stellar winds may play an important role in PMS binaries. Young stars are often found in binary systems, and often exhibit strong outflows in the form of winds and jets, and hence colliding winds should be expected.

To test this hypothesis, Zhekov et al. (1994) have calculated the expected X-ray emission from a PMS binary for a range of reasonable parameters. They find that for close binaries, with $D_{sep} < 50$ AU, that colliding wind X-ray emission is comparable with observed values, with $kT < 1$ keV. Further to this, Zhekov et al. (1994) suggest that even when there is a substantial stellar contribution to the X-ray flux, it should be possible to distinguish between this and colliding wind emission by the X-ray spectra, with the colliding wind emission giving rise to a soft X-ray excess, which should act as a diagnostic for colliding winds.

In summary, colliding winds may well play a role in producing the observed X-ray emission from some PMS binaries, particularly the Herbig Ae/Be systems. A direct test of the colliding wind hypothesis would be time-resolved spectral monitoring of these systems throughout an orbit, as colliding winds will show a characteristic variation in the absorbing column throughout the orbit.

6.   Colliding Winds in the Galactic Center

The Galactic Center (GC) is a complex and confusing region. The M2 supergiant IRS 7, which is only ~ 6 arcsec from Sgr A*, seems to be a colliding wind system. Observations by Yusef-Zadeh & Morris (1991) and Serabyn et

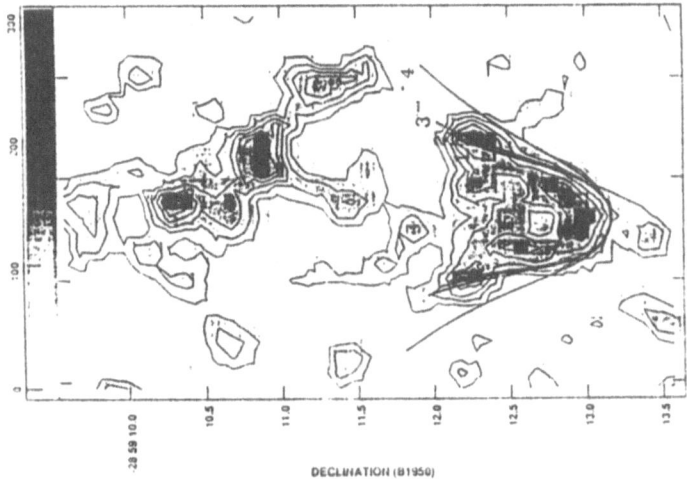

Fig. 4.   Colliding winds in the Galactic Center around the M2 supergiant IRS 7, from Yusef-Zadeh & Melia (1992). The bow-shock like structure and tail are clearly seen. The lines show the bow-shock models calculated by YM92, which fit the bow-shock region well, but produce very short stubby tails and cannot account for the long tail observed.

al. (1992) found evidence for a bow-shock structure around IRS 7 and a long ($\sim$ 3 arcsec) cometary tail (see Figure 5). The explanation, discussed by Yusef-Zadeh & Melia (1992, hereafter YM92) was that these features are a consequence of colliding winds, with the interaction between a slow massive wind from IRS 7 and a Galactic wind, probably emanating from IRS 16 (a cluster of peculiar early-type stars). The estimated values of the wind parameters for these systems; for IRS 7, $\dot{M} \sim 10^{-5} M_\odot$ yr^{-1} and $v_\infty \sim 20 - 30$ km s^{-1} , and for the IRS 16 cluster $\dot{M} \sim 3 \times 10^{-3} M_\odot$ yr^{-1} and $v_\infty \sim 500 - 700$ km s^{-1} .

While colliding winds may easily account for the bow-shock structure seen around IRS 7 it is more difficult to see how a simple colliding wind model could explain the long tail. The reason for this is that the energy deposited per unit area in the wind collision falls off very strongly away from the system line-of-centers (YM92). If θ is the angle from the line-of-centers of IRS 7 and IRS 16, then for values of $\theta > 150°$ the post-shock gas temperature will be less than 10^4K and will likely not contribute a substantial free-free flux, and a colliding wind model would predict a short, stubby tail.

As a consequence of this, Dyson & Hartquist (1994) have proposed a different form of the colliding wind scenario to explain the tail. If instead of being homogeneous, the wind of IRS 7 is very clumpy, with little inter-clump material, then the Galactic wind would sweep through the RSG wind, interacting directly with, and forming little bow-shocks around each clump. The interaction between the Galactic wind and the clumps would result in substantial amounts of material being ablated from the clumps and mass-loading the wind. While the issue is not closed yet, and more detailed calculations are required, it is interesting to note that the ability to spatially resolve colliding winds has led to potentially new insights into the dynamics and structure of stellar winds. In most of the other binary systems we have discussed this will not be possible, though in some of the wider Wolf-Rayet binaries there may be some potential for spatially resolving the system with long-baseline interferometry (*cf.* WR 147, Moran *et al.* 1989).

7.   Symbiotic Stars

For a long time symbiotic stars were an enigma. It is now clear that symbiotics are well separated, but interacting binary systems, containing a red-giant star and a white dwarf (or sub-dwarf) luminous enough to ionize the red giant wind. In addition, a sub-class of symbiotics undergo novae-like events (symbiotic novae, *i.e.* HM Sge, PU Vul). However, in spite of progress in our understanding them, symbiotic stars display a diverse range of puzzling phenomena and exceptions to the rules are common.

One common thread that has occurred in recent years is the notion that colliding winds are important in symbiotics. Wallerstein *et al.* (1984) and Girard & Willson (1987) suggested that binary wind collision was an important feature of symbiotic novae. More recently, there is now clear evidence of a wind from the hot star in several symbiotic systems, with wind velocities $\sim 500\,\mathrm{km\,s^{-1}}$ (for example, AG Peg, Vogel & Nussbaumer 1994; EG And, Vogel 1993), so that now there is strong evidence that both stars have substantial winds, and consequently there is no doubt that colliding wind phenomenon play an important role in the structure and evolution of these systems. In particular, two important features of symbiotics stars that colliding winds might explain, emission line profiles and X-ray emission are discussed below.

7.1   EMISSION LINE PROFILES

Recently, Walder and co-workers have undertaken 2-D and 3-D simulations of a particular symbiotic system, EG And, and using these calculations they have calculated emission line profiles and the expected X-ray spectra.

Symbiotic stars are observed to have varied and complex emission line profiles in lines such as Hβ, He II λ1640 and C IV λ1548. Simulations by Nussbaumer & Walder (1993) have shown clearly that a colliding wind model can reproduce this wide range of features, including double-peaked structures and asymmetric profiles. While these simulations are rather 'generic', future detailed modelling of individual systems offers great potential for a detailed understanding of these systems.

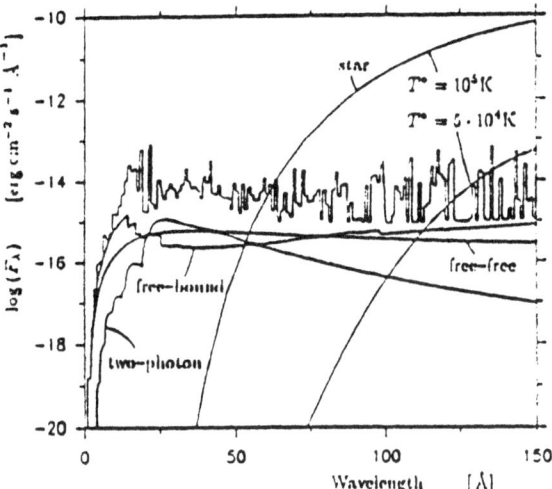

Fig. 5. The expected X-ray spectrum of a symbiotic star with colliding winds taken from Walder & Vogel (1993). At low energies the stellar contribution dominates, but at wavelength s $\lambda \leq 50$Åthe colliding wind contribution from free-free and line emission from the hot post-shock gas dominates.

7.2   X-RAY EMISSION

X-ray emission has been observed from symbiotic stars (Garcia 1986). In ROSAT observations of RR Tel (Jordan *et al.* 1994), most of the flux comes from the photospheric emission from the hot star, but in addition, there is a hard component with $L_x \sim 2 - 3 \times 10^{32}$ ergs s^{-1} which probably is associated with colliding winds, and is due to emission from shocked material from the hot star wind. Simulations by Walder & Vogel (1993) show that colliding wind X-ray emission should be visible from symbiotic systems as an excess at $E \geq 0.2$ keV (see Figure 6). Below this energy the emission will be completely dominated by the photospheric component. X-ray spectral observations in conjunction with hydrodynamic modelling should enable a determination of hot star wind parameters.

8.   Summary

In this paper we have discussed the phenomenon of colliding stellar winds in a diverse range of objects, ranging from high-mass Wolf-Rayet binary systems, through pre-main sequence objects, through to comparatively low-mass systems such as symbiotic stars, taking in such a peculiar object as IRS 7. Colliding winds are increasingly being recognised as an important feature in a wide variety of circumstances. Also, they are now being observed throughout the spectrum, from radio observations through to X-rays and γ-rays.

In particular, I have highlighted recent developments in X-ray spectro-

scopic observations of colliding winds in WR binaries and in the increasingly
sophisticated hydrodynamical modelling of such systems. Further developments in both these fields will enable significant progress in understanding colliding winds. Colliding winds are intrinsically interesting objects, the
wide range of phenomena associated with them (hydrodynamics, instabilities, radiative transfer, UV variability, X-ray properties) make them fascinating in their own right. The relation of colliding stellar winds to dust formation in Wolf-Rayet binaries is a particularly interesting question, where
dynamical simulations may yield insight into dust formation mechanisms
(and vice versa). In addition, they have potentially a great deal to tell us
about other much broader topics. For instance, they are a particularly good
'test-bed' for studying astrophysical fluid dynamics, as well as offering us
a good chance to study astrophysical shocks (X-ray emission from shocks,
shock acceleration). They might also provide a diagnostic for binarity in
PMS systems. Detailed study of colliding wind systems also gives us the
potential for understanding the dynamics of stellar winds in general.

References

Clegg, R.E.S., Stevens, I.R. & Meikle, W.P.S.: 1994, *Circumstellar Media in the Late
    Stages of Stellar Evolution*, CUP.
Dyson, J.E. & Hartquist, T.W.: 1994, Mon. Not. of the RAS **269**, 447.
Eichler, D. & Usov, V.: 1993, Astrophys. J. **402**, 271.
Garcia, M.: 1986, Astron. J. **91**, 1400.
Girard, T. & Willson, L.A.: 1987, Astron. & Astrophys. **183**, 247.
Jordan, S., Murset, U. & Werner, K.: 1994, Astron. & Astrophys. in press.
Koyama. K. *et al.* : 1994, Publ. of the ASJ **46**, L87.
Moran, J.P. *et al.* : 1989, Nature **340**, 449.
Nussbaumer, H. & Walder, R.: 1993, Astron. & Astrophys. **278**, 209.
Preibisch, T. & Zinnecker, H.: 1994, Astron. & Astrophys. submitted.
Prilutskii, O. & Usov, V.: 1976, Soviet Astronomy **20**, 2.
St-Louis, N., Willis, A.J. & Stevens, I.R.: 1993, Astrophys. J. **415**, 298.
Serabyn, E., Lacy, J.H. & Achtermann, J.M.: 1991, Astrophys. J. **378**, 557.
Stevens, I.R., Blondin, J.M. & Pollock, A.M.T.: 1992, Astrophys. J. **386**, 265.
Stevens, I.R. & Pollock, A.M.T.: 1994, Mon. Not. of the RAS **269**, 226.
Vogel, M. & Nussbaumer, H.: 1994, Astron. & Astrophys. **284**, 145.
Vogel, M.: 1993, Astron. & Astrophys. **274**, L21.
Walder, R.: 1994, in *Wolf-Rayet Stars: Binaries, Colliding Winds, Evolution (IAU Sym-
    posium 163)*, eds: K. van der Hucht & P.M. Williams, Kluwer.
Walder, R, & Vogel, M.: 1993, in *Cataclysmic Variables and Related Physics*, eds: O. Regev
    & G. Shaviv, 2nd Technion Haifa Conference.
Wallerstein, G. *et al.* : 1984, Astron. & Astrophys. **133**, 137.
Willis, A.J., Schild, H. & Stevens, I.R.: 1994, Mon. Not. of the RAS submitted.
Williams, P.M. *et al.* : 1990, Mon. Not. of the RAS **243**, 662.
Yusef-Zadeh, F. & Morris, M.: 1991, Astrophys. J. **371**, L59.
Yusef-Zadeh, F. & Melia, F.: 1992, Astrophys. J. **385**, L41 (**YM92**).
Zhekov, S., Palla, F. & Myasnikov, A.V.: 1994, Mon. Not. of the RAS in press.

CIRCUMSTELLAR DYNAMICS AND TRANSFER

JEAN-PIERRE J. LAFON

Observatoire de Paris-Meudon, DASGAL, URA D0335, F-92195 Meudon Cedex,
FRANCE

and

NICOLE BERRUYER

Observatoire de la Côte d'Azur, URA 1362, B.P. 229, F-06304 Nice Cedex 4, FRANCE

Abstract. This paper is both a review and a presentation of new models.

Observation and modelization of circumstellar envelopes of early type or late type stars are now quickly evolving because of new techniques and facilities for observations, and increased power of computers.

More and more complex physical phenomena involved in mass driving can now be modelized, at many different size scales. While most of models were previously based on informations derived from spectrophotometric data only or on measurements concerning objects observed with no spatial resolution, observations at much increased angular resolution can provide constraints on models of these phenomena.

Theory and modelization must take this new situation into account. Two approaches are possible and effectively used. On the one hand, dynamical/physical self consistent models can be built; on the other hand, elaborate semi-empirical models including complicated distributions of matter with asymmetries (3D models) can be built and fitted for direct comparison with results of High Angular Resolution Measurements.

Adding such constraints to classical constraints leads to a new insight in the physics of circumstellar matter and, through it, of stellar and interstellar evolution.

Two examples have been chosen, in which new models are presented and assuming or not spherical symmetry is carefully discussed:

- Circumstellar matter around evolved stars
  - Shock waves propagating in the circumstellar matter around evolved stars.

References

Lafon J.-P. J.: 1991 in: *The Infrared Spectral Region of Stars*, Proc. Int. Colloquium held in Montpellier (France), eds: C. Jaschek & Y. Andrillat, Cambridge University Press, p. 193-209.

Lafon J.-P. J. & Berruyer N., 1991, Astron. Astrophys. Reviews, **2**, 249.

Lafon J.-P. J.: in: *Target for Space-based Interferometry*, Beaulieu (France), ESA SP-354, p. 265.

Astrophysics and Space Science **224**: 377, 1995.
© 1995 *Kluwer Academic Publishers.*

PLATE IX: The James Clerk Maxwell Telescope (JCMT) is a 15m telescope designed for observation of millimetre and submillimetre wavelength range of the spectrum. The facility was set up by ROE, opened in 1986 and is now operated by the Joint Astronomy Centre, Hilo on behalf of the UK Particle Physics and Astronomy Research Council, the National Research Council of Canada, and the Nederlandse Organisatie voor Wetenschappelijk Onderzoek.

FORMING DUST PRECURSORS IN THE INNER
ENVELOPES OF CARBON-RICH AGB STARS

ISABELLE CHERCHNEFF
Mathematics Department, UMIST, PO Box 88, Manchester M60 1QD - England
& Royal Observatory, Blackford Hill, Edinburgh EH9 3HJ - Scotland

Abstract. The formation of polycyclic aromatic hydrocarbon (PAH) molecules is studied in the inner envelope of a typical carbon-rich AGB star. The deep envelope is formed of layers of gas that experience the passage of strong periodic shocks forming close to the stellar photosphere. The parcels of gas then follow quasi-ballistic trajectories which are characterized by high gas densities. A chemical scheme based on combustion chemistry is applied to shocked layers of gas, and a PAH formation yield is calculated. PAHs up to coronene ($C_{24}H_{12}$) survive shocks with strengths of ~ 10 km s^{-1}, and they accumulate in the gas parcel over several stellar pulsations. This result illustrates that any C-rich AGB star can nucleate dust precursors in its envelope.

1. Introduction

Dust grains represent a very important component of AGB circumstellar envelopes, as they participate both to the dynamics and the chemistry of stellar winds. Because of their crucial role, several studies have tried to shed light on their formation mechanism, and have concentrated essentially on dust in carbon-rich objects. Gail and Sedlmayr (1987, 1988) have investigated mainly the condensation stage of dust formation, assuming heterogeneous nucleation theory. A chemical kinetic approach was first adopted by Frenklach and Feigelson (1989) to describe the nucleation stage of dust formation via the formation of polycyclic aromatic hydrocarbons (PAHs) as dust precursors. A more thorough study of the nucleation processes was done by Cherchneff *et al.* (1991, 1993), who concluded that it was unlikely that PAHs could form in large amounts in the inner stellar outflow of a C-rich AGB star. However, a recent study by Cadwell *et al.* (1994) shows that the condensation of dust grains from precursor "seeds" is very fast and efficient, if small amounts of precursors are initially present in the inner flow. All these studies have ignored the true dynamics of the inner envelopes of AGB stars, which experience the passage of strong, periodic shocks close to the photosphere. The aim of this contribution is to present preliminary results on the formation of PAH molecules in the deep layers of a typical AGB envelope, subjected to shocks. In Section 2, the model for the inner envelope is described; conditions on PAH survival are given in Section 3; and results and conclusions are presented in Section 4.

Astrophysics and Space Science **224**: 379–382, 1995.
© 1995 *Kluwer Academic Publishers.*

2.   A Model for the Inner Envelope

The deep layers of the envelope experience the passage of pulsational, driven shocks that form at the surface of the photosphere. The energy deposition due to these periodic shocks is lost via cooling of the gas by expansion, because the gas densities in these regions are too low to provide cooling by radiative processes. As a result, the parcels of gas are accelerated upwards, decelerate under the influence of stellar gravity, and eventually fall back towards their initial position. This cycle will repeat itself with the next pulsation, resulting in oscillatory motions of the gas layers. These trajectories are illustrated in Figure 2 of Bowen's dynamical modelling of AGB stellar atmospheres (Bowen, 1988).

Two steps are then necessary to study the formation of PAHs in the shocked inner envelope. Firstly, PAH formation is investigated in the shocks, and the chemistry describing PAH nucleation is independently applied to pre-existing models of shock structures, as PAH molecules have no effect on postshock processes. Shock structure models for various shock strengths from Fox & Wood (1985) were used in our calculations. The shocks encountered in AGB stars are weakly subsonic (Mach number M $\sim$ 4 − 6), and there is no significant ionization or excitation by electron and atomic hydrogen collisions (and consequent Lyman continuum and Balmer line emissions) in the postshock gas. The only destruction process for molecular hydrogen is dissociation by collisions from atomic hydrogen, but this process takes place for the fastest shocks.

Secondly, the gas parcel excursions are modelled semi-analytically using the study of Bertschinger & Chevalier (1985) on periodically-shocked Mira atmospheres. Lagrangian profiles of the density, temperature, velocity and radius are obtained for a typical parcel of gas, and the chemistry of PAH formation is applied to such parameter profiles. The results of the shock calculation are entered as the molecular concentration inputs for the excursion calculations. The inverse greenhouse effect (IGE) on small PAHs is taken into account (see for details, Cherchneff *et al.* 1991), and PAH growth up to coronene is modelled for a gas parcel excursion.

3.   Conditions on PAH Survival in the Shocks

The growth of PAHs molecules described in this study consists of applying recurrently a set of chemical reactions that characterizes the closure of a newly formed aromatic ring (Cherchneff *et al.* 1993, Cherchneff & Tielens 1994). This set consists of an acetylene addition reaction, followed by a hydrogen abstraction reaction, to form a radical species, and a new acetylene addition reaction, to form and close a new benzene ring. The only destruction mechanism in our PAH formation chemical scheme is represented by the reverse of the last acetylene addition reaction, *i.e.* the unimolecular decomposition of a PAH made of $n + 1$ aromatic rings, that gives back an acetylene molecule and a PAH species containing n benzene rings. Such a decomposition reaction is temperature dependent, and, in order to investigate PAH survival in a shock, the net chemical rate for this reaction needs to be compared with the inverse of the dynamical time $\tau = v/r$ of the postshock region (v is the flow velocity and r is the position after the shock front). If the net chemical rate of the PAH decomposition is smaller than τ^{-1} in the postshock region, then PAH decomposition will not prevail over the closure of a new aromatic ring, and PAH desctruction will be hindered. This survival condition is true for gas temperatures ranging from 4000 K to 8000 K, depending on PAH sizes. Therefore, PAH molecules will survive shocks characterized by such temperature jumps, which correspond to shock velocity strengths of 10 to 13 km s^{-1}. Such values are smaller than the shock strengths derived from observations ($\Delta v \sim 20$ km s^{-1}). However, the shock velocity strength decreases with increasing radius and scales as the escape velocity (Willson & Bowen 1986, Cherchneff *et al.* 1993), therefore such low strength values can be reached at $\sim 2 - 3$ R$_*$. This implies the existence of a region not too far from the stellar photosphere where dust precursors, *i.e.* PAH molecules, can survive weak periodic shocks.

4. Results and Conclusions

To assess PAH growth, a PAH formation yield Y is defined as the ratio of the final number of carbon atoms locked up in PAHs to the number of carbon atoms initially in the form of hydrocarbon molecules. Figure 1 displays PAH formation yield values for a parcel of gas first shocked (shock 1) and expanding (parcel 1), then shocked again (shock 2), and going through a second excursion (parcel 2). The shock strength for this special case is 10 km s^{-1}, which corresponds to a Mach number M of 4.5, and a shock temperature jump of ~ 4000 K. No initial PAH concentrations were assumed for the first shock, and almost no PAHs are formed in the postshock region of shock 1. During the first parcel excursion, the IGE applies, and the PAH formation yield rises steeply once the temperature window for which PAH nucleation is possible (see Cherchneff *et al.* 1993) is reached. During the passage of the second shock, the PAHs formed previously are not destroyed (*i.e.* the PAH survival condition is satisfied for a shock with M = 4.5). Finally, during the second excursion, PAH growth continues, and the formation yield Y nearly doubles.

The present study shows that there exists a region in the inner envelope of AGB stars, located around $2 - 3$ R$_*$, where dust precursors (*i.e.* PAH species) can form and survive the passage of the periodic shocks induced by stellar pulsations. PAHs then accumulate over several stellar pulsations. The

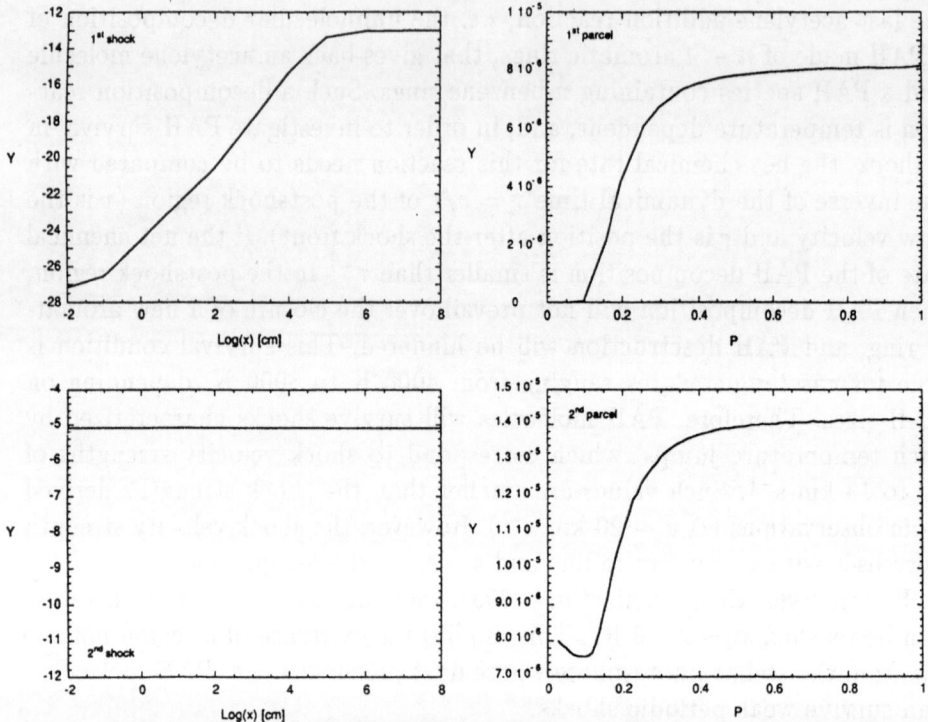

Fig. 1.   PAH formation yield Y for a shock and parcel excursion over two stellar pulsation periods. The abscissa x represents the distance after the shock front; the abscissa P is the phase (*i.e.* P=t/pulsation period).

PAH concentrations obtained in this study are in good agreement with the concentrations of nucleation "seeds" derived by Cadwell *et al.* (1994) which are necessary to induce dust condensation and the resulting large stellar outflows. These results therefore illustrate that any C-rich AGB star can naturally nucleate dust precursors in its inner envelope.

References

Bertschinger E. & Chevalier R.A.: 1985, Astrophys. J. **299**, 167.
Bowen G.H.: 1988, Astrophys. J. **329**, 299.
Cadwell, B.J., Wang, H., Feigelson, E.D., & Frenklach, M.: 1994, Astrophys. J. **429**, 285.
Cherchneff, I., Barker, J.R., & Tielens, A.G.G.M.: 1991, Astrophys. J. **377**, 541.
Cherchneff, I., Barker, J.R., & Tielens, A.G.G.M.: 1993, Astrophys. J. **413**, 445.
Cherchneff, I. & Tielens, A.G.G.M.: 1994 in: *Circumstellar Media in the Late Stage of Stellar Evolution*, eds. R. Clegg, I. Stevens & W.P.S. Meikle, CUP, p232.
Frenklach, M. & Feigelson, E.D.: 1989, Astrophys. J. **341**, 372.
Fox M.W. & Wood P.R.: 1985, Astrophys. J. **297**, 455.
Gail, H. & Sedlmayr, E.: 1987 in: *Physical Processes in Interstellar Clouds*, Dordrecht: Reidel, p275.
Gail, H. & Sedlmayr, E.: 1988, Astron. & Astrophys. **206**, 153.
Willson, L.A. & Bowen, G.H.: 1986 in: *Cool Stars, Stellar Systems, and the Sun*, Vol 254, Berlin: Springer, p385.

THE ONSET OF AXIAL SYMMETRY IN
PROTOPLANETARY NEBULAE

C. J. SKINNER

IGPP, L-413, Lawrence Livermore National Laboratory

M. MEIXNER

Astronomy Dept, University of Illinois

M. J. BARLOW, K. JUSTTANONT*

Dept of Physics & Astronomy, University College London

and

J. F. ARENS, J. G. JERNIGAN

Space Sciences Laboratory, University of California, Berkeley

Abstract. We have obtained resolved IR images, at wavelengths from 1.2μm to 12.5μm, of a small sample of protoplanetary nebulae. The results suggest that all PPNe are axially, not spherically, symmetric, that spherical symmetry ends at the tip of the AGB, and that AGB evolution terminates with a burst of remarkably rapid, equatorially concentrated mass-loss.

Key words: PPN – mass-loss – bipolar nebulae

1. Introduction

Shklovskii suggested in 1956 that PN descend from AGB stars. This is now accepted, but problems remain in our understanding of the transition from the AGB to PN phases. In particular, whilst most AGB winds are spherical (*e.g.* Habing & Blommaert 1992; Herman *et al.* 1985), more than 80% of PN display axial, and not spherical, symmetry (Zuckerman & Aller 1986). When does such axial symmetry arise – on the AGB, or during post-AGB evolution, perhaps by some interaction between AGB and post-AGB winds? In this paper we present some IR images of protoplanetary nebulae (PPNe) which shed some light on this issue.

2. Observations

Images were obtained in 1991 and 1992 at the NASA IRTF and UKIRT. Mid-IR images were taken with the Berkeley-Livermore mid-IR camera, using a Hughes 10×64 pixel, photoconducting Ga:As array, through a 10% bandwidth circular variable filter, and show thermal dust emission. Plate scales at both telescopes were 0.4″, and the spatial resolution of order 1.0″. Sources observed include an evolutionary sequence of typical C-rich stars,

---

* Present address: NASA-Ames Research Center

Astrophysics and Space Science **224**: 383–386, 1995.
© 1995 *Kluwer Academic Publishers.*

with effective temperatures from 4,500 K to 30,000 K. We also include one low-mass O-rich PPN, and two high-mass C-rich sources. For the latter we also obtained near-IR images at UKIRT with the facility camera IRCAM2, and we obtained high spatial resolution 6 cm images of GL618 using the MERLIN interferometer.

2.1 AN EVOLUTIONARY SEQUENCE FOR TYPICAL C-RICH PPNE

The 21μm objects are PPN discovered by Kwok *et al.* (1989) with IR spectra dominated by the Unidentified InfraRed (UIR) bands, and a very strong, broad feature around 21μm. We took mid-IR images of three of these, and two further sources belonging to the same evolutionary sequence (Justtanont *et al.* 1994) – the PPN SAO163075 and the young PN IRAS21282+5050. All five have extended, elongated mid-IR nebulae, and two of them, IRAS21282 and IRAS07134 (Fig. 1), show fully resolved tori. The other three may be elliptical or have unresolved tori. The drop in surface brightness at the outer edge of each is abrupt and unresolved. The dynamical age of the nebula can be estimated from the inner dimension of the torus and wind velocity, if the sources are assumed to evolve at constant, and equal, luminosity. Table 1 summarises the results.

2.2 HD161796

We include for comparison one example of a low-mass O-rich PPN, HD161796, published by Skinner *et al.* (1994). This source also has a torus with an abrupt outer edge. Full radiative transfer models by Skinner *et al.* (1994) show that it left the AGB with a burst of very rapid mass-loss ($\sim 3 \times 10^{-4}$M$_\odot$/yr). This is probably the first direct evidence of a superwind terminating the AGB, as suggested by Iben & Renzini (1985).

TABLE I

Results for low-mass C-rich & O-rich post-AGB sources

Source	Spectral type	Dynamical age [yrs]	Morphology	Dimensions [arcsec]
IRAS22272+5435	G5Ia	<190	Elliptical	1.6 × 1.3
IRAS04296+3429	G0Ia	<260	Elliptical	~ 0.5 × 0.4
IRAS07134+1005	F5I	190	Torus	3.6 × 3.3
SAO163075	F2Ia	<200	Elliptical	1.8 × 1.4
IRAS21282+5050	O(f)[WC11]	700	Torus	6 × 3
HD161796	F2Ia	240	Torus	2.2 × 2.0

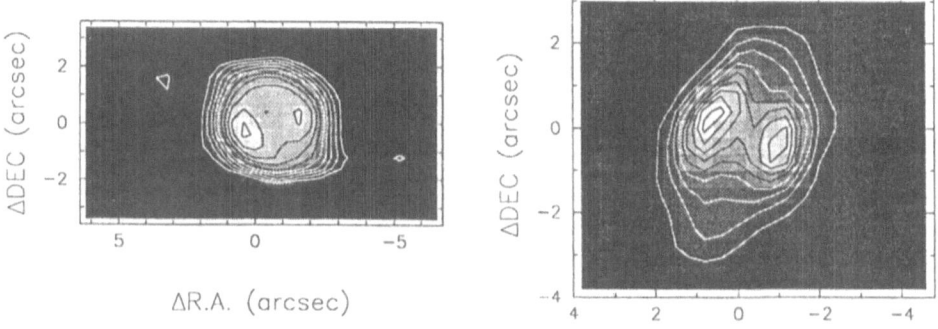

Fig. 1.   Images of (a) IRAS07134 at 11.8μm and (b) IRAS21282 at 12.5μm.

2.3   GL2688 & GL618

GL2688 has long served as the prototype PPN. Unfortunately, our data show it to be highly complex, and the simple bipolar flow model usually invoked for it should no longer be regarded as valid. We present images of it at 1.2μm, and 2.122μm in the shock-excited 1-0 S(1) line of H_2. The 1.2μm image reveals that the reflection nebula does not have a single axis of symmetry - each lobe has its own symmetry axis, the two being offset by 2″. Images at wavelengths from 3.5 to 12.0μm show a box-like structure, rotated somewhat from the apparent optical bipolar axis. The H_2 image reveals six main blobs of emission. Blobs to the N and E are connected by faint structure, probably a limb-brightened hoop, and the S and W blobs present a similar structure. These hoops may be the interface between a fast biconical wind and a slower AGB remnant wind. The fast flow is terminated by two fainter blobs to the ENE and WSW, which may be shocked ansae. Full details of this model will be presented by Skinner et al. (in preparation).

GL2688 is bizarre, but not unique. We have images of the similar source GL618 from 1.2 to 12.0μm, and at 6 cm. The 6 cm image shows a bipolar nebula tilted by about 20° with respect to the 1.6μm lobes. Careful inspection shows that it is no single axis in any of the images. Perhaps GL618 has either bent or helical precessing outflows, but it is certainly not the simple biconical nebula usually depicted: like GL2688, it has a very complex structure.

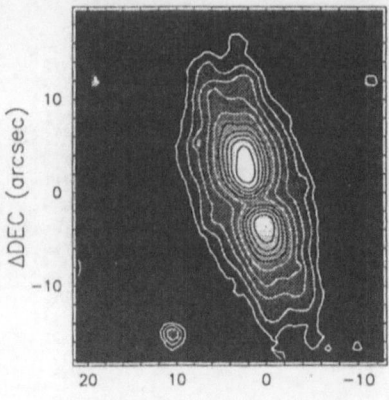 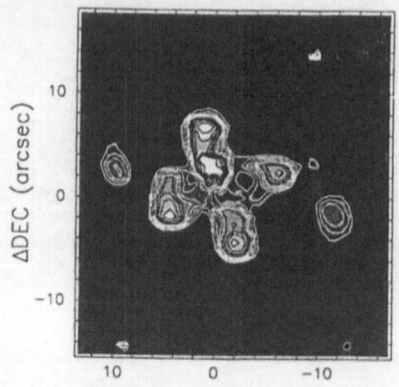

Fig. 2.   GL2688 seen (a) in reflection at 1.2μm and (b) in shock excited H_2.

3.   Discussion

Every PPN we have imaged is bipolar, toroidal or elliptical. It appears that
at the tip of the AGB mass-loss undergoes a major change of mode, from
moderate, isotropic mass-loss to very rapid, equatorially concentrated mass-
loss. From our images and models we estimate that this episode lasts no
more than a thousand years. Spectacular bipolar nebulae such as GL2688
are very rare and probably represent only the most massive AGB stars: our
observations indicate that they are also far more complex than would be
suspected from previous work. We suggest further that the 21μm objects,
originally regarded as oddities, may represent a common stage in evolution
of C-rich sources, and the evolutionary sequence we have described may be
typical for C-stars.

References

Habing, H. J. & Blommaert, J. A. D. L.:1992, Carbon- and Oxygen-Rich Progenitors of
    Planetary Nebulae, in *Planetary Nebulae (IAU Symposium 155)*, eds: R. Weinberger
    & A. Acker, Kluwer:Dordrecht, p243.
Herman, J., Baud, B., Habing, H. J. & Winnberg, A.: 1985, Astron. & Astrophys. **143**,
    122.
Iben, I. & Renzini, A.: 1983, Ann. Rev. of Astron. & Astrophys. **21**, 271.
Justtanont, K., Barlow, M. J., Skinner, C. J., Roche, P. F., Aitken, D. K. & Smith, C. H.:
    1994, Astron. & Astrophys. submitted.
Kwok, S., Volk, K. & Hrivnak, B. J.: 1989, Astrophys. J. **345**, L51.
Shklovskii, I. S.: 1956, Astron. Zh. **53**, 315.
Skinner, C. J., Meixner, M., Hawkins, G., Keto, E., Jernigan, J. G. & Arens, J. F.: 1994,
    Astrophys. J. **423**, L135.
Zuckerman, B. & Aller, L. H.: 1986, Astrophys. J. **303**, 772.

SESSION I:

MAIN
SEQUENCE
DISKS

FOLLOW-UP OBSERVATIONS OF β-PIC-LIKE STARS

H. J. WALKER

DRAL, Rutherford Appleton Laboratory, Chilton, Didcot, Oxon, OX11 0QX, UK

and

H. M. BUTNER

Department of Terrestrial Magnetism, Carnegie Institution of Washington, 5241 Broad Branch Road N.W., Washington, D.C. 20015, USA

Abstract. Following the discovery, by IRAS, of the dust disc around Vega and three other main sequence stars, searches have been made for other candidates. The β-Pic-like candidates have 12μm excesses and 100μm fluxes (unlike the Vega-like candidates), so they can be further investigated using ground-based techniques. Data are presented here, comprising 10μm spectroscopy and sub-mm observations, for several candidates from the Walker & Wolstencroft list, showing that the stars have silicate dust, and optically thick dust discs even at 1300μm.

Key words: Dust discs – main sequence stars – Vega-like stars – β-Pic-like stars

1. The Prototypes

Vega (α Lyrae) was observed by IRAS and excess emission (over that expected for a stellar black body) at 60μm was found (Aumann *et al.* 1984). The excess was attributed to dust left over from the formation of the star, which had formed a thin disc around the star. The reasons for this conclusion were that the emission was slightly extended (when compared to a point source), Vega showed no evidence of mass-loss, and the model required dust grains which were larger than normal interstellar dust grains. Harper *et al.* (1984) observed Vega at 193μm using the KAO and found that the flux fell below the simple, single temperature, black-body model used for the dust.

Three other normal, main-sequence stars were found, during the IRAS mission, to have this same type of excess (β Pic, α PsA, ϵ Eri), and Gillett (1986) reviewed their properties. The IRAS emission arises from cool dust, typically around 100K. The discs have low luminosity and low mass. The IRAS data do not set an upper limit on the grain size, so masses are difficult to define. If large grains are permitted (*i.e.* rocks or asteroids) the mass in the disc of β Pic, which is the most massive disc of the four, can rise from $3 \times 10^{-8} M_\odot$ (or $0.01 M_\oplus$) to $0.02 M_\odot$ ($7000 M_\oplus$). Smith & Terrile (1984) showed that the β Pic disc is viewed edge-on; Vega is viewed face-on. All four prototypes have slightly extended dust emission when observed by IRAS, a few 10's of arcsec across, implying dust discs a few hundred AU across. For three of the prototypes, there is less 12μm excess than might be expected, implying an inner dust free region in the dust discs. Wolstencroft addresses the issue of the β Pic disc in more detail. There has been a long series of observations of the gas in the β Pic system (see, for example, Lagrange-Henri et al. 1988), showing that matter (perhaps in the form of comets) falls on the star as often as two of three times a week. Neff & Cheng show work on other stars, looking for the same phenomenon.

2. The Candidates

Following the discovery of the four prototypes, various searches were made for other similar systems. Aumann (1985) found 8 stars within 25 parsecs with an excess of the same type as Vega. The candidates proposed by Walker & Wolstencroft (1988) were further away, and had stronger fluxes at the IRAS wavelengths. They resembled β Pic, with its 12μm excess, rather than Vega, which has no 12μm excess and a small 25μm excess, hence the term used in the title. Since the emission at 12μm is dominated by dust emission, the dust in these systems can be further studied from the ground at 10μm. The systems also had significant flux at 100μm, unlike those in Aumann's list, which meant they could be observed at longer wavelengths. However, the dust emission is most prominent in the range between around 20μm and 100μm which is inaccessible from the ground, leaving follow-up mainly to satellites like ISO. An exception to this statement is the KAO, used by Harvey et al. to investigate the spatial extent of Vega-like/β-Pic-like stars at 50μm and 100μm.

3. The Spectra

Before modelling the dust discs, it is essential to know something about the composition of the dust. Knacke et al. (1993) recently observed β Pic and compared the silicate feature to that found in comets by Hanner et al. (1990). They suggested that β Pic had a silicate feature similar to that found in some comets, for example Comet Halley, in that there is a feature around 11.2μm attributed to crystalline silicates. This is the only non-solar-system object where it has been detected; Hanner (1994) failed to observe it in selected young stellar objects. The silicate emission feature has been reported in SAO179815 (HD98800) by Skinner et al. (1992), and in 51 Oph by Fajardo-Acosta et al. (1993).

We recently (June 1994) observed 51 Oph, SAO193956, and SAO184124 at IRTF using HIFOGS to obtain high resolution 10μm spectra. The figure shows the spectrum obtained, the continuum emission has not been subtracted. Our spectrum of 51 Oph does not show the extra 11.2μm crystalline silicate feature, just a normal weak circumstellar silicate feature (Walker et al. 1994). By comparison, the other two sources have an extended red wing to the silicate feature when compared to 51 Oph, which may be indicative of the crystalline silicate feature. Sylvester discusses the spectra of this class

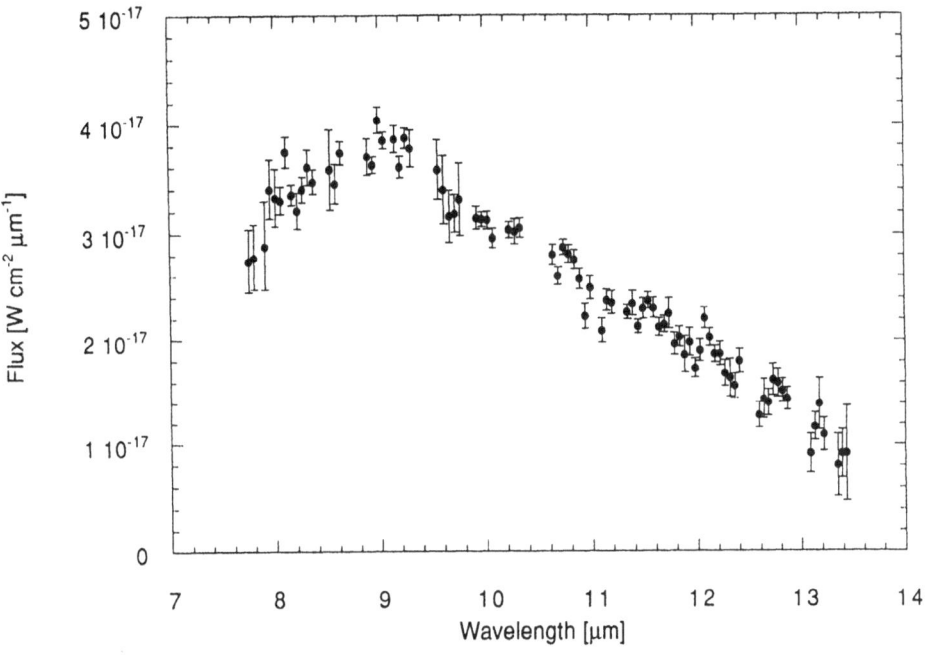

Fig. 1. The 10μm silicate feature in 51 Oph

of source in more detail.

4. The Long Wavelength Energy Distributions

Zuckerman & Becklin (1993) and Chini *et al.* (1991) have observed the prototypes at sub-mm wavelengths. They find (as did Harper *et al.* (1984) for Vega) that that the emission falls below the single-temperature black-body line. There is no evidence for a substantial population of cold dust around the prototypes. The tables give the list of sources we have observed at sub-mm wavelengths; the sources have $S/N > 2.5$ for at least one sub-mm measurement (Butner *et al.* 1994). We also obtained upper limits for Vega, SAO5496, SAO22268, SAO26804, SAO35498, and SAO99809. In addition to the data at 800μm and 1300μm, we also observed all the accessible sources at MWO for CO. Where CO was detected, this was interpreted as evidence of mass-loss which would mean the system was less likely to be a β-Pic-like one. Some sources have a revised spectral type, which makes the dust emission more likely from mass-loss than β-Pic-like. The 1300μm point from IRAM is generally lower than that from CSO, but the cause of this needs investigation. These results are similar to those for the prototypes, in that they suggest the models will need large grains of silicates and ices.

TABLE I

Data on β Pic candidates with S/N > 2.5

Source SAO	CSO data				IRAM data		Remarks
	800μm		1300μm		1300μm		
	flux (mJy)	rms (mJy)	flux (mJy)	rms (mJy)	flux (mJy)	rms (mJy)	
ϵ Eri					12.7	3.90	
77144					35.4	2.18	? CO
87856	19.4	42.3	16.0	6.2	11.2	1.91	? M*
112630					18.1	1.90	
140789	366.6	104.9	12.4	10.7	9.39	3.72	
147886					13.9	2.48	
179815	282.0	10.0	30.7	8.2	54.4	3.61	
183956	524.0	238.5	144.3	14.2	138.1	9.90	
183986	199.7	93.6	68.0	8.3	51.9	5.33	
184124	-21.1	71.5	44.6	10.7	36.7	2.52	
186777	630.3	89.8	237.6	11.6	161.8	9.21	
206462	653.0	459.9	157.0	31.2			
226057	97.7	116.0	182.7	29.5			
226389	4160.5	466.4	1190.3	33.3			?F6III

5.   Discussion

The data are sufficiently recently acquired that the modelling work is still under way. Sylvester *et al.* (1994) show that SAO179815 is quite close (18pc) and has a more massive dust disc than β Pic. There are also suggestions that the star may not yet have quite reached the main sequence, from the Li abundance. Generally, the Walker & Wolstencroft sources have more flux than the other candidates, so they can be followed-up from the ground. However, this implies the dust discs are more massive and these systems may be a transition phase between the massive T Tau star discs and the prototype Vega-like dust disc.

Acknowledgements

We would like to thank Diane Wooden for her speedy initial data reduction of the spectra for this meeting. We also acknowledge, with pleasure the assistance of our other collaborators; Ray Wolstencroft, Elizabeth Lada,

Steve Beckwith, Fred Witteborn.

References

Aumann, H.H.: 1985, Publ. of the ASP **97**, 885.

Aumann, H.H. *et al.* : 1984, Astrophys. J. **278**, L23.

Butner, H.M, Walker, H.J., Lada, E.A. & Beckwith, S.V.W.: 1994, in preparation.

Chini, R., Krügel, E., Shustov, B., Tutukov, A. & Kreysa, E.: 1991, Astron. & Astrophys. **252**, 220.

Fajardo-Acosta, S.B., Telesco, C.M. & Knacke, R.F.: 1993, Astrophys. J. **417**, L33.

Gillett, F.C.: 1986, in *Light on Dark Matter*, ed: F.P. Israel, Reidel: Dordrecht, p61.

Hanner, M.S., Newburn, R.L., Gehrz, R.D., Harrison, T., Ney, E.P. & Hayward, T.L.: 1990, Astrophys. J. **348**, 312.

Hanner, M.S.: 1994, private communication.

Harper, D.A., Loewenstein, R.F. & Davidson, J.A.: 1984, Astrophys. J. **285**, 808.

Harvey, P.M., Smith, B. & DiFrancesco, J.: 1994, in this volume.

Knacke, R.F., Fajardo-Acosta, S.B., Telesco, C.M., Hackwell, J.A., Lynch, D.K. & Russell, R.W.: 1993, Astrophys. J. **418**, 440.

Lagrange-Henri, A.M., Vidal-Madjar, A. & Ferlet, R.: 1988, Astron. & Astrophys. **190**, 275.

Neff, J.E. & Cheng, K-P: 1994, in this volume.

Skinner, C.J., Barlow, M.J. & Justtanont, K.: 1992, Mon. Not. of the RAS **255**, 31P.

Smith, B.A. & Terrile, R.J.: 1984, Science **226**, 1421.

Sylvester, R.J., Barlow, M.J. & Skinner, C.J.: 1994, Astrophys. & Space Sci. **212**, 261.

Sylvester, R.J.: 1994, in this volume.

Walker, H.J., Butner, H.M., Wooden, D.H. & Witteborn, F.C.: 1994, in preparation.

Walker, H.J. & Wolstencroft, R.D.: 1988, Publ. of the ASP **100**, 1509.

Wolstencroft, R.D.: 1994, in this volume.

Zuckerman, B. & Becklin, E.E.: 1993, Astrophys. J. **414**, 793.

PROPERTIES OF THE β PICTORIS DISC DEDUCED FROM OPTICAL IMAGING POLARIMETRY

RAMON D. WOLSTENCROFT
Royal Observatory, Edinburgh EH9 3HJ, UK

and

S.M. SCARROTT and T.M.GLEDHILL *
Department of Physics, University of Durham, Durham, DH1 3LE, UK.

Abstract. The asymmetries in the brightness structure and levels of polarization between the two sides of the β Pic disc are discussed. The possibility of these differences being caused by a planet, or planets, within the circumstellar disc is investigated.

Key words: Stars: Circumstellar matter – Stars:Individual (β Pictoris)

1. Introduction

The FIR and MIR emission radiated by some nearby MS stars is believed to originate in thermal re-radiation by circumstellar dust. In the case of β Pic starlight scattered by dust shows that the circumstellar material is in the form of an edge-on disc with a radius >1000AU (Smith & Terrile 1984). Spectroscopy in the 8 - 13μm region of the central part of the disc reveals a silicate feature similar to the one found in Comet Halley (Knacke *et al.* 1993). Two-component grain models in which the dust properties and spatial distribution differ within and beyond about 80AU provide a viable fit to IR and optical data (Backman, Gillett & Witteborn 1992). Further support for this model comes from recent optical imaging studies by Golimowski, Durrance & Clampin (1993) which show a clear change in the radial slope of surface brightness(NE side of disc) from $r^{-3.5}$ at r>100AU to $r^{-2.4}$ at r<100AU. It seems that the change may be due to the sublimation of the ice mantles from silicate grains at r~100AU and that a planet, or planets, within ~40AU may be involved.

2. Imaging Polarimetry of the Disc

Our R-band observations for the mid-plane of the disc show that the polarization varies relatively slowly between 15 and 30″(240 to 480AU) from the star in the range 14 to 21% (Gledhill, Scarrott & Wolstencroft 1991). These levels are comparable to those seen in the zodiacal light.

A model based on Mie scattering by silicate grains with a size distribution of the form a^{-n} and radial number density fall-off r^{-m} was found to give a

---

* Now at the University of Hertfordshire

Astrophysics and Space Science **224**: 395–398, 1995.

satisfactory fit to these data (Scarrott, Draper & Gledhill 1992). Polarization is the more sensitive observable in limiting n to 4.0 ± 0.3, whereas the surface brightness gradient is more important in constraining m to 2.75 ± 0.25. The effective grain size deduced from this model is a$\sim$ 0.4μm. While this result is compatible with the value a$\sim$ 0.2 $-$ 0.3μm deduced by Telesco *et al.* (1988) from 10 and 20μm imaging of the disc, the two component models of Backman, Gillett & Witteborn (1992) and Knacke *et al.* (1993) suggest a grain size for the outer zone (r>80AU) of a$\sim$ 1 $-$ 2μm. It is a feature of Mie scattering with power law size distributions that grains with sizes $\sim \lambda$ are most influential in determining the polarization properties of radiation at a given wavelength λ so that observations in various regions of the optical and IR spectrum may be sampling different parts of the grain population.

Since the above analysis we have made polarization measurements of the β Pic disc in the BVRI wavebands and the results are shown in fig. 1. The polarization on the NE side of the disc is essentially constant (negative offset in fig. 1), yet on the SW side the polarization changes slightly, but significantly, with both offset and wavelength. The *grey* polarization observed for the NE side of the disc is totally consistent with the power law size distribution discussed above whereas the variation in polarization to the SW suggests that on this side of the disc we are encountering slightly different grain properties and/or scattering geometry. In the optical images of the disc (*e.g.* fig. 1 of Lagrange 1994 and Gledhill, Scarrott & Wolstencroft 1991) there are differences in the brightness structure and, perhaps more importantly, the two sides are not colinear.

High resolution imaging of the inner parts of the disc (r<70AU) at 10μm shows that the SW side of the disc is considerably brighter (Lagage & Pantin 1994). This image shows a fall-off in the surface brightness within about 40AU of the star which these authors believe is due to the presence of a planet 20AU from the star. Simulations by Roques *et al.* (1994) of the resonant trapping of dust by a planet orbiting within the disc lead Lagage & Pantin (1994) to suggest that a planet in an eccentric orbit could give rise to arc-like density enhancements that travel with the planet and hence to a NE/SW asymmetry. Resonant trapping of dust by the Earth may also be the cause of asymmetries in the zodiacal light at IRAS wavelengths (Dermott *et al.* 1994). The idea that a planet somewhere in the disc could be responsible for the asymmetries is attractive but it seems difficult to believe that a planet at 20AU could have such a large influence on the disc well away from the planet, *i.e.* at r>100AU. However a planet further out in the disc could explain the observed asymmetry in polarization and brightness.

The efficiency of the resonant trapping mechanism depends on grain size since smaller grains can move through the trapping zone more rapidly than large ones because of the stronger Poynting-Robertson drag on the former, thus the size distribution would be preferentially depleted of the larger grains

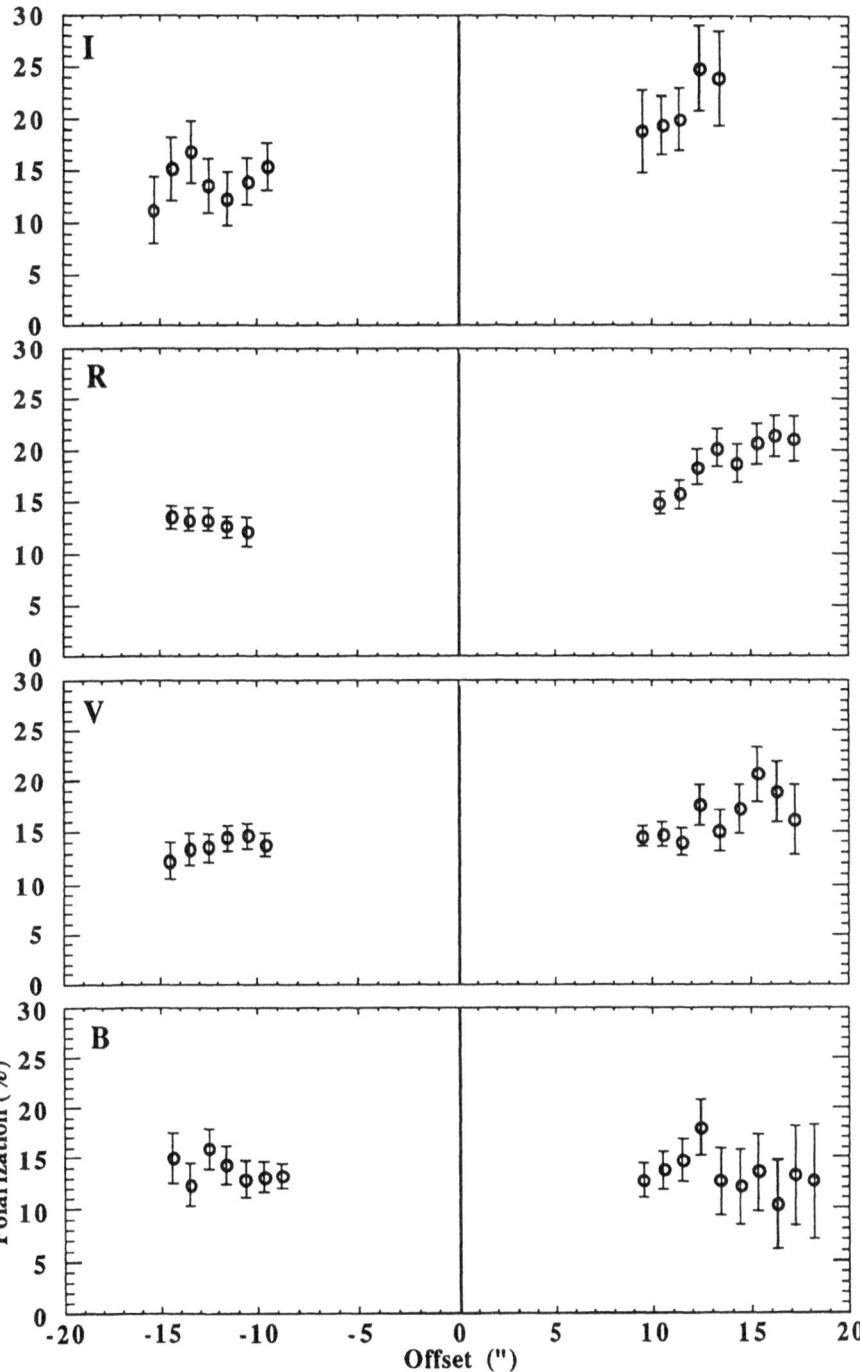

Fig. 1. Multi-waveband measurements of the polarization along the disc of β Pic.

on the side of the disc containing the planet. This depletion could lead to different grain size distributions on either side of the disc and, since the light scattering properties of grains depend on their sizes, this would give rise to brightness and polarization differences between opposite sides of the disc. The lack of colinearity between the two sides of the disc may mean that it is warped and this could be produced by a planet whose orbit is inclined to the disc symmetry plane. A similar explanation for the asymmetry was made by Whitmire, Tomley & Matese (1987) who considered a brown dwarf rather than a planet as the perturbing influence.

References

Backman,D.E., Gillett,F.C. & Witteborn,F.C.: 1992, Astrophys. J. **385**, 670.

Dermott, S.F., Jayaraman, S.,Xu, Y.L.,Gustafson, B.A.S. & Liou,J.C.: 1994, Nature **369**, 719.

Gledhill,T.M., Scarrott, S.M. & Wolstencroft, R.D.: 1991, Mon. Not. of the RAS **252**, 50P.

Golimowski,D.A., Durrance, S.T. & Clampin,M.: 1994, Astrophys. J. **411**, L41.

Knacke,R.F., Fajardo-Acosta,S.B., Telesco,C.M., Hackwell,J.A., Lynch,D.K. & Russell, R.W.: 1993, Astrophys. J. **418**, 440.

Lagage,P.O. & Pantin,E.: 1994, Nature **369**, 628.

Lagrange,A.-M.: 1994, ESO Messenger **76**, 23.

Roques,F., Scholl,H., Sicardy,B. & Smith,B.A.: 1994, Icarus **108**, 37.

Scarrott,S.M., Draper,P.W. & Gledhill,T.M.: 1992, in: *Dusty Discs*, ed: P.M.Gondhalekar, (RAL-92-084), p8.

Smith,B.A. & Terrile,R.J.: 1984, Science **226**, 1421.

Telesco,C.M., Becklin,E.E., Wolstencroft,R.D. & Decher,R.: 1988, Nature **335**, 51.

Whitmire,D.P., Tomley,L., & Matese,J.J.: 1987, Bull. Amer. Astron. Soc. **19**, 830.

MULTI-COLOR CORONOGRAPHIC IMAGING OF THE
BETA PICTORIS DISK

M. CLAMPIN & F. PARESCE

Space Telescope Science Institute, 3700 San Martin Drive, Baltimore, MD 21218, USA

M. ROBBERTO

Oss. Torino, ITALY

and

A. MACCIONI

Oss. Firenze, ITALY

Abstract. In this paper we present multicolor images of the Beta Pictoris disk, obtained with a new Coronograph, built at STScI, for ESO's New Technology Telescope. These B, V, and R observations probe the Beta Pictoris circumstellar disk in to a distance of 35 AU (2") from the central star. We derive surface brightness profiles for the disk and discuss the results in the context of two-component disk models (*e.g.* Backman *et al.* 1992), which are based on fits to available optical and infrared data for Beta Pictoris's disk. These models generally imply that the disk is composed of 1) an outer disk extending from 100-1000 AU, with a midplane particle density decreasing as $n(r) \propto r^{-3}$, 2) a transition region extending from 10-100 AU with lower density and either smaller grains or a less steep spatial gradient than the outer component, 3) an innermost "clear" zone with even lower density.

THE INFRARED COLORS OF MAIN SEQUENCE STARS:
HOW MUCH CIRCUMSTELLAR DEBRIS IS NORMAL?

ROBERT E. STENCEL

Department of Physics & Astronomy, University of Denver, Denver CO 80208 U.S.A

and

DANA E. BACKMAN

Department of Physics & Astronomy, Franklin & Marshall College, Lancaster PA 17604 U.S.A.

Abstract. We have been awarded NASA Key Project observing time on ISO, in order to establish the true frequency of far-infrared excesses in a volume-limited sample of main sequence and related stars, and address the relative success or failure of single stars in processes related to the forming of planetary systems. For a volume-limited subset of main sequence and related stars, PHT03 measurements at 3.6, 11.5, 20 and 60 micron will be obtained, using a 120 arsec aperture in all cases to eliminate possible companion confusion with differing apertures, to ascertain spectral energy distributions. For the M dwarfs, 100 micron observations will also be obtained. For some, brighter sources, more extensive wavelength coverage and improved spatial resolution will be attempted, using CAM and SWS. Spatially over-sampled PHOT observations will be made at 60 micron of the brightest and nearest Vega-like sources to measure the characteristic sizes of the emitting regions and obtain some information regarding their shapes and orientations. The goal is not a map, but scan profiles along 3 position angles which can be deconvolved to find the intrinsic size and shape of the half-maximum contour of the emitting region. Photometry of selected lines of sight through the zodiacal dust will also be carried out to look for outer solar system (Kuiper Belt) material. Observation at a range of wavelengths, ecliptic latitudes and at 2 epochs is designed to help untangle foreground Zodiacal from background Kuiper flux, not necessarily to look for individual macroscopic objects.

Key words: circumstellar material, protoplanets, infrared astronomy, space astronomy.

1. Introduction

The Infrared Space Observatory [ISO] satellite is an ESA project which will place a supercooled 64 cm aperture, f/15 Richey-Cretien telescope, diffraction limited at 5 microns, into an elliptical orbit (Kessler *et al.* 1992). As discussed by Stencel & Backman (1994), some of the questions that we believe ISO observations of Vega-like stars can address, are these: (1) Does β Pic represent success or failure in forming planets? (2) How frequent are less dense, less extensive versions of the β Pic disks among normal main-sequence stars? (3) Does disk density anti-correlate with stellar age? (4) Do Kuiper belts and Oort clouds play a role in post-main sequence evolution, as stars begin their ascent of the red giant branch, in terms of brief but significant augmentation of observed infrared excesses?

Astrophysics and Space Science **224**: 401–404, 1995.
© 1995 *Kluwer Academic Publishers.*

2.   Observing Plans with ISO

The IRAS discovery of far-infrared excesses among seemingly normal main-sequence stars have been interpreted in terms of disks of cold material. The aim of our effort is to establish the true frequency for far-infrared excesses in a volume-limited sample of main sequence and related stars using PHT-P, PHT-C, CAM and SWS measurements, in order to address the success or failure of single stars in processes related to the forming of planetary systems. For brighter sources, more extensive wavelength coverage and spatial resolution will be attempted, with PHT-P, PHT-C, PHT-S and SWS. Finally, observations of Kuiper Belt objects will be attempted.

2.1 SPECTRAL ENERGY DISTRIBUTIONS

For a volume-limited sample of main sequence and related stars, PHT03 measurements at 3.6, 11.5, 20 and 60 microns will be obtained, using a 120 arcsec aperture in all cases to eliminate possible companion confusion with differing apertures. For the M dwarfs, 100 micron observations will also be obtained. Rectangular chopping will be used for sources above and below 10 degrees galactic or ecliptic latitude; triangular chopping otherwise, with a throw of 120 arcsec. For selected bright stars, CAM01 observations using microscans with 5 second exposures and 6 arcsec pfov and 6 arcsec steps, for 10 linear steps using the LW3 filter are planned. The signal to noise ratios are predicted, from the PHOT Cookbook to be in excess of 1000 for 32 second observations of sources brighter than 0.7 Jy at 3.6 microns, decreasing to 300 at 11.5 and 20 microns and declining to 2.5 at 60 microns, even at the ecliptic equator. Therefore the minimum 32 sec for each PHT-P filter will be sufficient for the continuum measurements at 3.6, 11.5 and 20 microns, and 64 seconds for the 60 micron observations. The 100 micron observations for the M stars will require chopping and a minimum of 128 sec to obtain S/N of >2. Total time per star is therefore $(3 * 32 \text{ s} + 64 \text{ s} (+128 \text{ s})) * 2$ for chopping, + 180 s acquisition + 155 s other overheads = 655 s (921 s). PHT-S, CAM and SWS observations are more exploratory, with S/N in excess of 3 are based on comparisons with Central Programme estimates.

2.2 SPATIALLY-RESOLVED SOURCES

We request micro-scanning observations with the best equivalent of AOT P12, either PHT17-19 or PHT32 (C60) at a wavelength of 60 microns of the brightest and nearest Vega-like sources to measure the characteristic sizes of the emitting regions and glean some information regarding their shapes and orientations. The sample is limited to main sequence stars found to have IR excesses in IRAS data, with total 60-micron flux greater than 0.5 Jansky and parallax distances less than 20 parsecs, plus a few very nearby stars of special interest, plus a few stellar point source standards. The observing plan in each case will be to perform photometry with the P3-60 or C60 detector and filter and the diffraction limited aperture, diameter 52 arcseconds. The goal is not a "map" 'per se' but rather scan profiles along 3 position angles which can be deconvolved to find the intrinsic size and shape of the FWHM contour of the emitting region.

There will be a first integration at the target position, then a series of micro-scan steps by the focal plane chopper extending radially away from the target position in equi-spaced 3 legs of a "Y", allowing reduction of ambiguity regarding source shape. On EACH leg there will be 9 positions spaced 5 to 10 arcseconds apart. The minimum spacing for P12 or equivalent P17-19 or P32) micro-scan steps with the 52 arcsec aperture is 5 arcsec (Table 7, ISOPHOT Observer's Manual). The spacing between micro-scan positions is approximately the expected (deconvolved) angular scale of the emission. We expect 9 diffraction-sized aperture positions on each leg extending from the target to radii of approx. 1 to 2 Airy disk diameters should allow deconvolution of the source sizes along those position angles (using point source observations as references). IN ADDITION, to properly measure the background, there will be 3 more measurements with the aperture positioned 200 arcseconds from the target along each of the scan legs. The largest of the target sources may have some measurable 60 micron flux as far as 50 arcseconds from the star, so the true background must be measured at greater distances than this. Some of the sources are near the galactic plane so 3 background measurements rather than 1 seems prudent. An integration time as short as 16 sec seems possible given that successive measurements along the scan legs will be viewing almost exactly the same source flux and background, thus "settling" time of the detector after each measurement should be small. The Signal/Noise ratio in 16 sec on a 0.5 Jy source at 60 microns with P3-60 in the 52 arcsec aperture with pessimistic background assumptions (ecliptic latitude = 0) should be 45.

2.3   KUIPER BELT OBSERVATION

We request photometry of lines of sight through the zodiacal dust into possible outer solar system (Kuiper Belt) material. To increase chances of distinguishing KB emission from hotter foreground zodiacal emission, we plan to observe: 1) at wavelengths from 3.6 to 200 μm; 2) at 2 epochs, such that the same VOLUME ELEMENT in the KB is observed through different pieces of the zodiacal cloud; 3) at two ecliptic latitudes, on the assumption that nearby and distant material will have different latitude distributions.

Observation at 2 epochs is designed to help untangle foreground from background flux, not to look for moving objects. The volume sampled by ISO will not be large enough for practical hopes of detecting macroscopic KB objects. Our schematic target definition includes: 2 volume elements with the following HELIOCENTRIC ecliptic coordinates: 1) longitude L = 0.0

deg, latitude B = 3.0 deg, distance R = 50.0 AU and 2) L = 0.0, B = 13.0 and R = 50 AU. Ecliptic longitude of 0.0 degrees represents a region of the ecliptic in the neighborhood of the S. galactic pole, thus reducing galactic background. Symmetric to these locations at ecliptic longitude of 180 deg, are equivalently galactic-free locations which would also be acceptable for this experiment. The tolerance on the chosen longitude is large, ~15 degrees. This tolerance is meant to provide flexibility in scheduling the initial observations. The second epoch observations must return as exactly as possible to whatever longitude was observed at the first epoch. The latitudes are chosen to lie on one side of the likely plane of symmetry of the outer solar system. Neptune's orbit, which is a reasonable first guess as to the location of that plane, is tilted 1.7 degrees relative to the earth's orbit. The range of longitudes is meant to allow comparison of the vertical gradients of warm and cold material. The GEOCENTRIC ecliptic coordinates of the chosen volume elements will depend on the earth's position in its orbit. These can in turn be converted to geocentric RA and Dec.

Our trial scenario involves: (A) observe both locations when their solar phase angle first falls within ISO's pointing tolerance, *i.e.* phi $\leq$ 120.0°. This occurs when the earth's heliocentric longitude is 59.00°, approximately Nov. 21. The GEOCENTRIC RA and Dec of the 3 KB volume elements at this time will be: 1) RA = +2.11° Dec = +2.39°; 2) RA = +6.07° Dec = +11.34°; (B) observe the same two locations in the KB when the line of sight through the foreground zodiacal dust differs as much as possible from the first epoch, *i.e.* heliocentric longitude of the earth = 90.00 degrees, solar phase angle = 88.85 degrees, 1 MONTH AFTER THE FIRST OBSERVATIONS. The GEOCENTRIC RA and Dec of the same two KB locations this time will be: 1) RA = +2.25° Dec = +2.30°; 2) RA = +6.16°, Dec = +11.16°; Thus, total time at one latitude (incl. chop to 2nd location), one epoch: 1391 seconds. Total experiment = 2 epochs x 2 latitudes x 1391 seconds = 5564 sec = 1.55 hr. The integration times are designed to yield S/N of 5-10 at 100-200 microns in detection of emission from KB dust with an optical depth of $\sim 10^{-7}$ and a temperature ~40 K.

We are pleased to acknowledge partial support for these efforts under NASA grant NAGW-3680 to the University of Denver. Useful discussions with Carol Grady, Joseph Nuth and Francesco Paresce are gratefully acknowledged, as are the heroic efforts of NASA personnel in creating the possibility for US observers to use ISO.

References

Backman, D. & Gillett, F.: 1986, in *Fifth Cambridge Workshop on Cool Stars*, eds: J. Linsky & R. Stencel, Springer-Verlag, Heidelberg, p.340.

Kessler, M., Metcalfe, L. & Salama, A.: 1992, Space Science Reviews **61**, 45.

Stencel, R. & Backman, D.: 1991, Astrophys. J., Suppl. **75**, 905.

Stencel, R. & Backman, D.: 1994, Astrophys. & Space Sci. **212**, 417.

OBSERVATIONS OF DUST EMISSION
FROM VEGA-EXCESS STARS

R. J. SYLVESTER and M. J. BARLOW

Department of Physics and Astronomy
University College London
Gower Street, London WC1E 6BT

and

C. J. SKINNER*

Institute of Geophysics and Planetary Physics, L-413
Lawrence Livermore National Laboratory
P.O Box 808, Livermore, CA94551-9900 U.S.A.

Abstract. Observations have been made of a large sample of Vega-excess stars. Mid-infrared spectra show silicate dust features and also the UIR bands, implying the presence of hydrocarbon material. Millimetre-wave photometry indicates the presence of large amounts of cool material, while near-IR photometry reveals an excess in several stars, which we ascribe to transiently-heated very small grains.

Key words: Stars, main-sequence — spectra, infra-red — dust — UIR bands

1. Introduction

Vega-excess stars are main sequence stars discovered by *IRAS* to show excess infrared emission due to dust distributed in circumstellar discs. We have observed a large sample (over 20 objects) of Vega-excess candidates, taken mainly from the lists of Walker & Wolstencroft (1988) and Stencel & Backman (1991), at wavelengths spanning over three orders of magnitude from the optical to the millimetre-wave region. Each type of observation is complementary in that it can provide information about the sources that could not be gleaned from the other observations. Table 1 summarises the number of stars observed with each technique and the type of information which such observations could reveal. In addition to their individual contributions, the data when assembled together with the *IRAS* data define the optical–mm spectral energy distribution (SED) of the system, against which the results of modelling can be compared.

2. Observations

Millimetre-wave observations with the JCMT show that many of the stars exhibit strong excess emission at long wavelengths, with spectral slopes shallower than those expected for Rayleigh-Jeans type emission, implying the

---

* Present address: Space Telescope Science Institute, 3700 San Martin Drive, Baltimore, MD 21218, U.S.A.

Astrophysics and Space Science **224**: 405–408, 1995.
© 1995 *Kluwer Academic Publishers.*

TABLE I

Summary of observations

Observation	No. of sources	Science
Optical Photometry (CCD)	8	Photospheric flux, reddening
1–5 μm Photometry (UKT9)	18	Photosphere, 'hot' emission
Near-IR Spectroscopy (CGS4)	1	Nature of near-IR excess
10 & 20 μm Spectroscopy (CGS3)	13	Grain compositions/sizes
10 μm Imaging (BerkCam)	1	Spatial distribution of dust
mm/sub-mm Photometry (UKT14)	21	Large, cool grains

presence of large amounts of cool material, and probably large grains (radius ≥ 0.1 mm).

Mid-infrared spectra obtained with CGS3 provide information about grain composition and, via modelling, grain size distribution. Some sources, for example SAO 179815 and SAO 184124 (Figure 1) show broad emission features at 9.7 and 18 μm, indicating the presence of silicate dust. Others, such as SAO 186777, show the UIR bands at 7.7, 8.7, and 11.3 μm, characteristic of carbonaceous material, such as polycyclic aromatic hydrocarbons (PAHs), or hydrogenated amorphous carbon.

Near-IR photometry shows that 9 of the 23 stars for which we have data show a strong excess in the 1–5 μm region; for the others the near-IR photometry agrees with the expected photospheric flux. The near-IR excess energy distribution is consistent with emission from grains at temperatures of around 1500 K, which is too hot for dust in thermal equilibrium to survive, but similar to the temperatures found by Sellgren et al. (1983) for some reflection nebulae. The near-IR excess may therefore be due to very small grains (~ 5 Å) transiently heated to high temperatures by single-photon absorption. In order to understand better the nature of the near-IR excess, we obtained a 3.2–3.4 μm spectrum of SAO 186777, which shows the 3.3 μm UIR feature superposed on a strong continuum. This implies that the broad-band near-IR photometry does indeed show an excess continuum, rather than line emission. The presence of the UIR bands supports the small-grain hypothesis, since PAHs are small and excited by single photons.

A 10 μm BerkCam image of SAO 26804 (Skinner et al. 1995) shows extended emission which is resolved compared to the point-spread function along one axis, but unresolved along the other, consistent with emission from an inclined dust disc. This is the first Vega-excess star other than β Pic to be shown to have a disc geometry.

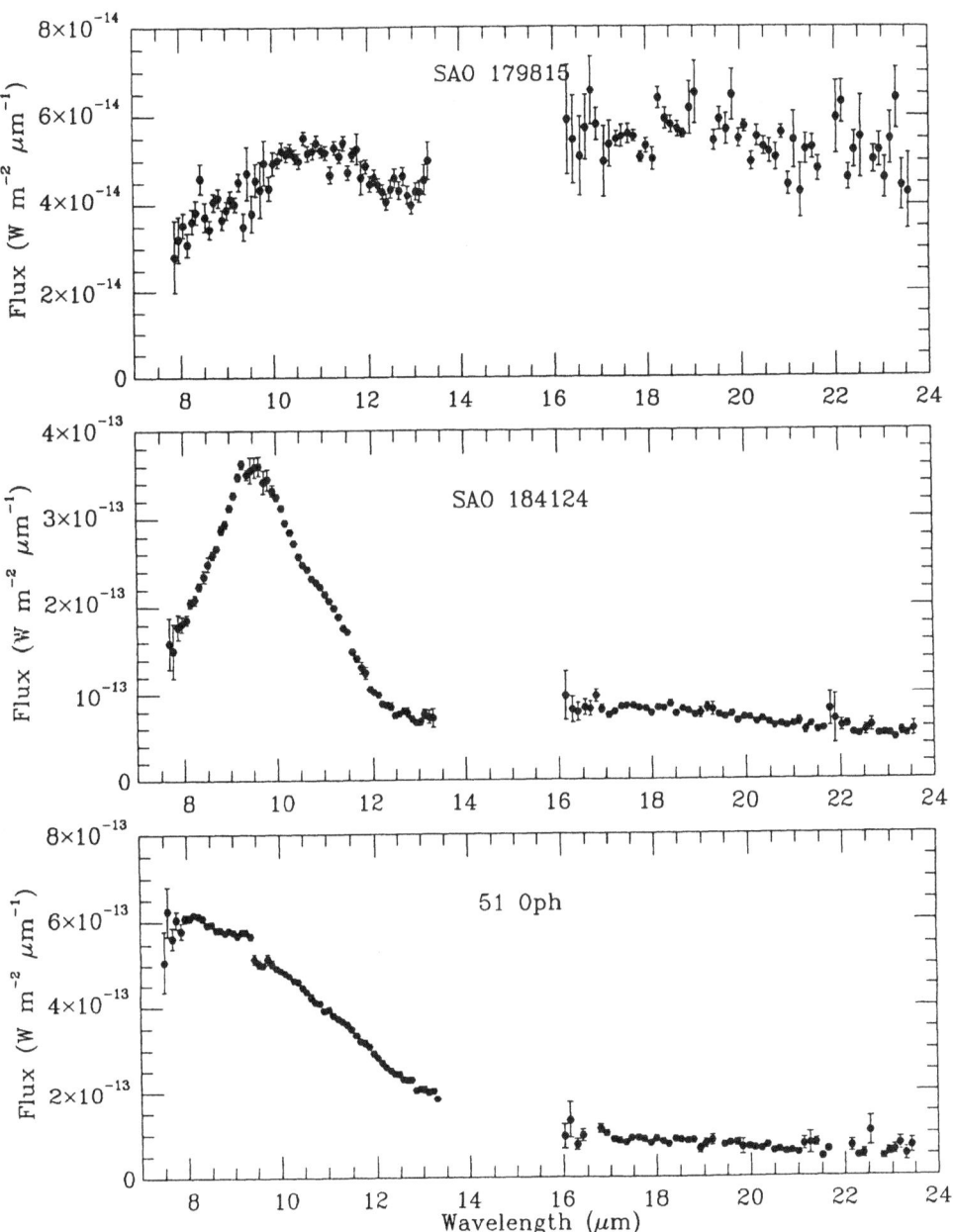

Fig. 1.   CGS3 spectra showing silicate features

Acknowledgements

UKIRT and the JCMT are operated by the Joint Astronomy Centre, Hilo. We are grateful to the PATT for observing time, and to the UKIRT and JCMT staff for their help, and for making Service observations for us.

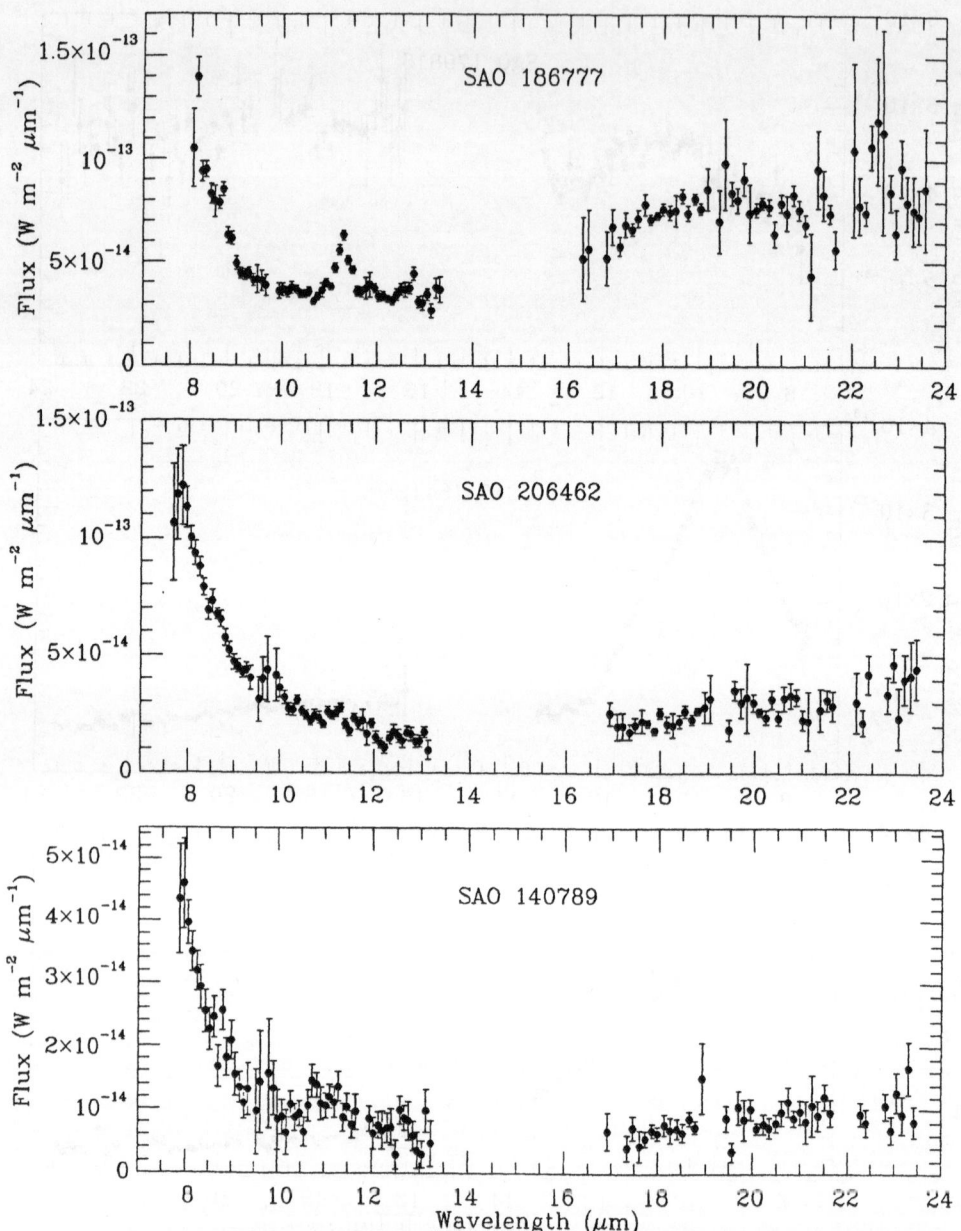

Fig. 2.   CGS3 spectra showing UIR bands

References

Sellgren, K., Werner, M.W. & Dinerstein, H.L.: 1983, Astrophys. J. **271**, L13.

Skinner, C.J., Sylvester, R.J., Graham, J.R., Barlow, M.J., Meixner, M., Keto, E., Arens, J.F. & Jernigan, J.G.: 1995, Astrophys. J. in press.

Stencel, R.E. & Backman, D.E.: 1991, Astrophys. J., Suppl. **75**, 905.

Walker, H.E. & Wolstencroft, R.D.: 1988, Publ. of the ASP **100**, 1509.

POSTER

CONTRIBUTIONS

SHARP IMAGES OF YOUNG STELLAR OBJECTS

N. AGEORGES, A. ECKART, R. HOFMANN

Max Planck Institut für Extraterrestrische Physik

Abstract. We present high resolution infrared images of the pre-main sequence star V536 Aql and polarization maps of the reflection nebula IRN Cha observed with the SHARP camera on the ESO NTT. Extended structures are observed near V536 Aql. The polarization pattern of IRN Cha is typical of a centrally illuminated source but with a central band of anomalously oriented polarization believed to occur in an optically thick disk.

Key words: : high angular resolution imaging — near infrared – young stellar objects — polarization — circumstellar matter

1. Introduction

Pre-main sequence T Tauri stars are usually surrounded by nebulous remnants of the 'prenatal' material and their radiation is often polarized. According to Bastien & Landstreet (1979), this polarization is due to Mie scattering by asymmetrically distributed material around the star. We therefore selected sources having a high degree of polarization at optical wavelengths and observed them with the SHARP (System for High Angular Resolution Pictures) camera at the ESO-NTT (New Technology Telescope). This speckle camera works at near-infrared wavelengths (1–2.5 μm), which are well-suited to investigate circumstellar matter. SHARP has been designed to be diffraction limited in the K band (2.2 μm) at the NTT. To complement our study, we recently installed a polarization mode on this camera in order to get diffraction limited images of the distribution of the polarized radiation. In this contribution we present both diffraction limited speckle images of the pre-main sequence star V536 Aql and a high spatial resolution two-dimensional polarization map of the reflection nebula IRN Cha.

2. V536 Aql

This young stellar object has recently (Ageorges *et al.* 1994) been discovered to be a binary (separation 0.52″ at 17°) with circumstellar matter in its close neighbourhood, i.e. at a maximum distance of ≈ 200 AU. The spectral energy distribution of this variable, optically highly polarized (6–8 %; Ménard & Bastien 1992) source presents an infrared excess that couldn't be explained before.

In April 1994, we reobserved this source and confirmed the presence of extended structures similar to those we observed in July 1993 (see Fig. 1). Although it is now unlikely that the presence of these structures is due to seeing calibration problems, the exact shape might be affected by them.

Astrophysics and Space Science **224**: 411–414, 1995.

Assuming that the observed structural differences are real, we can think of two possible explanations: either the 'circumstellar' material is an independent cloud, or simply gas or dust in front of the system in our line of sight, but close enough to still be illuminated by the binary; or it is linked to the system. It could then be some accreted material: the system might be in an intermediate phase, between phases b and c of Shu, Adams & Lizano's star formation model (1987) where it is about to finish accreting but where no outflows have yet formed (as yet, no outflow or jet has been found in this system). Ageorges & Duschl (in preparation) develop a consistent model based on a 'torch effect' to explain the observed variable structures. New observations are planned to get better constraints on those models.

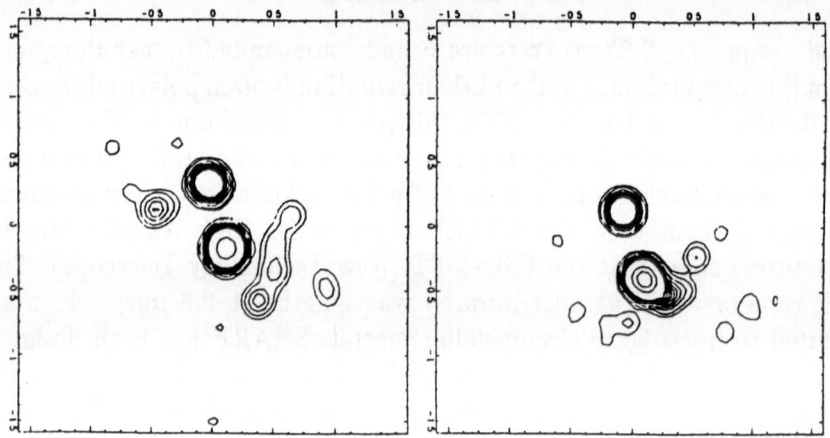

Fig. 1. Linear deconvolved speckle images of V536 Aql in the H (1.65μm) band. Left: Observations of July 1993; Right: Observations of April 1994. The intensity contours have been plotted at 0.5, 1, 2, 3, 4, 5, 6, 7, 8, 9, 10, 40, 70 and 100% of the maximum intensity. North is up, East to the left. Offsets are in arcseconds from the zero position.

3. IRN Cha

3.1 POLARIZATION MEASUREMENTS

In April 1994, we installed a polarization mode on the SHARP camera. Our grid polarizer is made of barium fluoride and characterised by 93 % transmission in the 1–5μm range. The polarizer was installed on a remotely controlled rotating table. We repeatedly observed the source and calibrator at six angles of the polarizer separated by 30°. After calibration and applying a simple shift-and-add algorithm, we fitted in selected apertures a cosine function to the data. In Fig. 2, we present the first results of bidimensional polarization measurements obtained on the InfraRed Nebula (Cha IRN) of Chamaeleon at near infrared wavelengths.

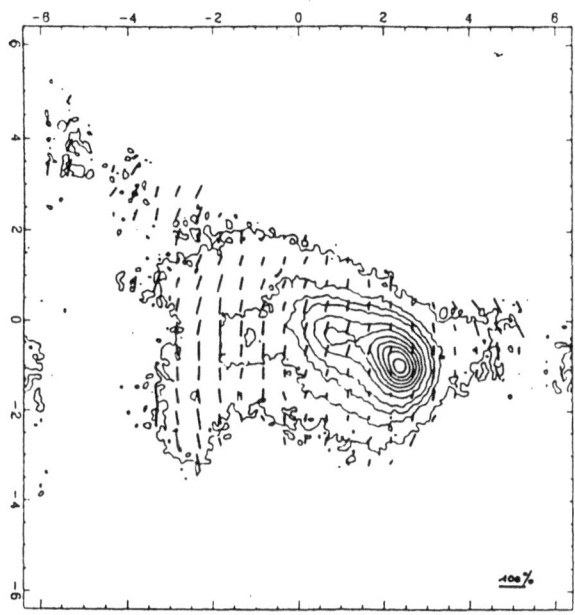

Fig. 2. Intensity contour and polarization map (12.8″ × 12.8″) of IRN Cha at 1.65 μm. The degree of polarization has been estimated every 0.5″ for this map. North is up, East to the left.

3.2 RESULTS

IRN Cha is a bipolar reflection nebula illuminated by an IR source, surrounded by an extensive disk of gas and dust. Its overall dimensions are about 125″ × 35″ (Schwartz & Henize, 1983). Cohen & Schwartz (1984) propose a possible identification of the bright knot, observed in the nebula, with the young IR star Cha T C9–2 discovered by Hyland *et al.* (1982). The structure of IRN Cha suggests that this is an outflow source. Like McGregor *et al.* (1994) we find that in H band (1.65μm) the core of IRN Cha has an integrated degree of polarization of 32% at a position angle of the order of 165°, in an 8″ aperture. Within a smaller aperture (≈ 1.4″), the knot presents a polarization of 22% (also measured by Scarrott *et al.* (1987) at optical wavelengths).

Maps of polarization show a high level of polarization (30–40 %) in both lobes of the nebula with an overall pattern of vector orientations typical of a reflection nebula illuminated from the inside (the central source). However, it is important to notice a systematic deviation of this pattern from complete centro-symmetry, particularly in the central region. The geometry of the source and a smaller optical depth explain that we observe the parallel band of poalrization with a nearly north-south axis centered on the embedded source and not between the two lobes (Scarrott *et al.* 1987). We explain the central band of anomalously oriented polarization by multiple Mie scattering in the optically thick disk (model of Bastien & Ménard (1990) = BM90).

The presence of a dark band between the two lobes is due to obscuration by the overlying disk in a system that is tilted to the line of sight. A comparison with theoretical models of protostellar disk (like BM90) suggests a disk inclination of the order of 70°. This angle combined with the apparent size of the gap between the two lobes suggests a minimun disk size of 800 AU.

A paper with complete detailed interpretation is in preparation.

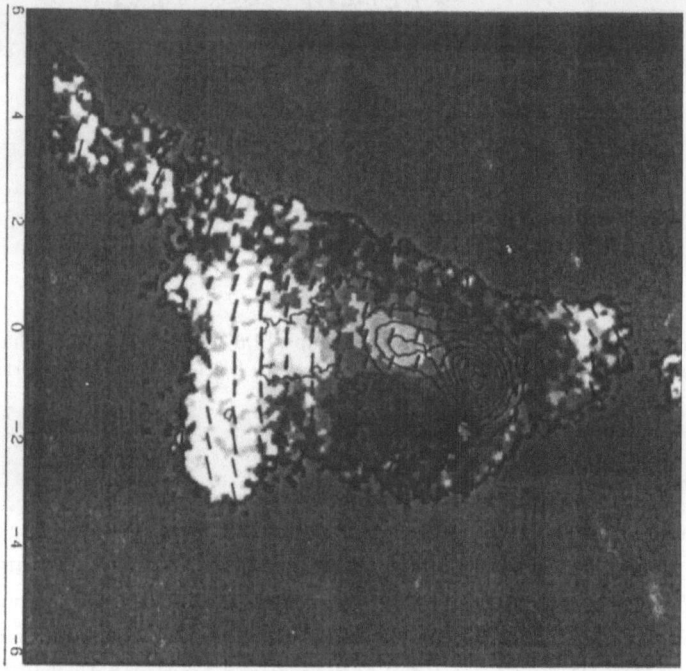

Fig. 3. Map of the degree of polarization of IRN Cha (calculated for each pixel) on which we plottted polarization and contour map (12.8″ × 12.8″). North is up and East to the left.

References

Ageorges, N. *et al.*: 1994, *A&A* **283**, L5
Bastien, P., Landstreet, J.D.: 1979, *ApJ* **229**, L137
Bastien, P., Ménard, F.: 1990, *ApJ* **364**, 232
Cohen, M., Schwartz, : 1984, *ApJ* **89**, 277
Hyland, A.R., Jones, T.J., Mitchell, R.M.: 1982, *MNRAS* **201**, 1095
Mc Gregor, P.J. *et al.*: 1994, *MNRAS* **267**, 755
Ménard, F., Bastien, P.: 1992, *AJ* **103**, 564
Scarrott, S.M. *et al.*: MNRAS, *1987* **228**, 827
Schwartz, R.D., Henize, K.G.: 1983, *AJ* **88**, 1665
Shu, F.H., Adams, F.C., Lizano, S.: 1987, *ARA&A* **25**, 23

CIRCUMSTELLAR MATTER IN DIRECT IMPACT ALGOL SYSTEMS

GEARY E. ALBRIGHT and MERCEDES T. RICHARDS

Department of Astronomy, University of Virginia

Abstract. Full-orbit Hα observations have been analyzed to determine the two-dimensional distribution of the circumstellar gas in the four short-period Algols, U Sge, U CrB, RS Vul, and SW Cyg. In these systems, the gas stream resulting from Roche-lobe overflow directly impacts the mass-gaining star and feeds material into a structure known as a 'transient accretion disk'. The accretion regions observed in these systems include a transient accretion disk that at least partially surrounds the mass gainer, the gas stream, and possibly a chromospheric component associated with the magnetically active cool star.

Key words: stars: Algol variables

1. Introduction

Algol-type binaries are interacting systems where the late-type G–K subgiant secondary star has filled its Roche lobe and is transferring material through the L_1 point onto the hot B5–A5 subdwarf primary star. The short-period Algols (P < 5–6 days) undergo direct-impact accretion where the gas stream strikes the primary star directly resulting in an asymmetric distribution of circumstellar gas known as a 'transient accretion disk' (Kaitchuck, Honeycutt, & Schlegel 1985; Richards 1993). We have obtained full-orbit Hα spectra of short-period Algol systems on the Coudé Feed (KPNO) to investigate the physical properties and two-dimensional distribution of the accretion regions which occurs in these systems. The entire orbit of each binary was covered within 3 to 4 weeks in an attempt to minimize the secular variability that is characteristic of these binaries. The analysis of Hα difference profiles calculated from these spectra reveals that variable circumstellar gas is almost always present in every system observed.

The detailed study of the circumstellar gas in the short-period Algol systems requires that the circumstellar components of the Hα line profile be deblended from the photospheric absorption components of the stars in the binary. This is accomplished by subtracting a composite theoretical photospheric line profile from the observed Hα line profiles, the resulting difference profile represents all the emission and excess absorption arising from circumstellar gas in the system. The phase variation of the difference profiles can be analyzed to determine the two-dimensional distribution of the accretion regions in each system.

Astrophysics and Space Science **224**: 415–416, 1995.
© 1995 *Kluwer Academic Publishers.*

2.   Results

A transient accretion disk was observed in U Sge (1993 May) during an epoch of enhanced mass transfer (Albright & Richards 1994). In addition, the strength of the transient disk was found to vary in only a few orbital cycles. This change in strength may result from an increase in the mass transfer rate or the deflection of the gas stream from the predicted path through interaction with gas in the disk. This deflection would result in a more oblique impact and presumably more gas fed into the transient disk. The transient disk is also the dominant feature in the difference profiles of SW Cyg. In this system, the gas stream impacts the mass gainer nearly tangential to the surface of the star facilitating the formation of a transient disk. The transient disk in SW Cyg is easily observed in the Doppler tomogram constructed from the same data set by Richards, Albright & Bowles (1994).

The difference profiles of U CrB display two blueshifted emission peaks from $\phi = 0.65 - 0.75$. During this phase interval, the gas stream is is aligned along the line of sight, our models identify the higher velocity peak as arising from emission due to the gas stream. Similar double-peaked blueshifted emission has been observed in another short-period Algol TX UMa (Albright & Richards 1993). The observations of U CrB also show an emission feature associated with the primary star at all phases around the orbit. This probably arises from a circumprimary 'bulge' of gas formed by the penetration of the gas stream deep into the photosphere of the mass gaining star. The nonsynchronous rotation of the primary star causes the bulge to be distributed symmetrically around the equator of the star. This type of bulge has been observed in other short-period Algols with high mass transfer rates (Olson 1980). The difference profiles of RS Vul are similar to those observed in U CrB displaying gas stream and circumprimary bulge emission.

Acknowledgements

We acknowledge observing grants from KPNO (NOAO), operated by AURA under cooperative agreement with the NSF. This work was partially supported by AFOSR grants F49620–92–J–0024 and F49620-94-1-0351, and NSF grants AST 91–114214 and AST 93–15108.

References

Albright, G. E., & Richards, M. T. 1994, *ApJ*, in press
Albright, G. E., & Richards, M. T. 1993, *ApJ*, **414**, 830
Kaitchuck, R. H., Honeycutt, R. K., & Schlegel, E. M. 1985, *PASP*, **97**, 1178
Olson, E. C. 1980, *ApJ*, **241**, 257
Richards, M. T. 1993, *ApJ Suppl.*, **86**, 255
Richards, M. T., Albright, G. E., & Bowles, L. M. 1994, *these proceedings*

FORMATION AND PHOTODESTRUCTION OF PAHS

T. ALLAIN* and E. SEDLMAYR
Institut für Astronomie und Astrophysik, Technische Universität Berlin, Sekr. PN 8-1, Hardenbergstraße 36, D-10623 Berlin, Germany

and

S. LEACH
D.A.M.Ap, Observatoire de Paris-Meudon, 5, Place Jules Janssen, 92195 MEUDON Cédex, France

Abstract. Photodissociation processes of PAHs are studied in the interstellar medium (ISM) and in the envelopes of carbon-rich C–stars.

Key words: ISM : molecules — molecular processes — circumstellar shells — carbon stars

1. Introduction

Since Léger and Puget (1984) first proposed free Polycyclic Aromatic Hydro-carbons (PAHs) as carriers of the so–called 'unidentified infrared bands' observed in the emission spectra of astronomical objects associated with strong UV–radiation fields, the question of the formation and destruction of PAHs is still open. We here discuss the process of destruction that is probably the most effective, namely photodestruction, and we consider its influence on the chemistry of C–star envelopes which represent one of the most important sites for PAH formation.

2. Photodestruction of PAHs

After absorption of a UV–photon PAH molecules have four main channels to dissipate the energy: **(i)** ionization, **(ii)** IR radiative relaxation from vibrational modes, **(iii)** electronic transitions, **(iv)** photoejection of H, H_2 or C_2H_2 (Jochims *et al.* 1994). Integrating the cross–sections of these processes over the energies available in the interstellar medium (ISM) provides the rates of the various processes (Allain *et al.* 1994). These rates show that only PAHs having more than 50 carbon atoms have a characteristic time for photodestruction greater than 10^9 years and can therefore surely survive the interstellar UV–field. On the other hand, PAHs smaller than coronene ($N_C = 24$) are destroyed within some 10^3 years.

* D.A.M.Ap, Observatoire de Paris–Meudon, 5, Place Jules Janssen, 92195 MEUDON Cédex, France

3. Formation of PAHs in C–star envelopes

Studies of PAH formation in circumstellar shells have paid so far no attention to the influence of the UV–radiation field coming from the ISM. Since
PAH formation constitutes the first step in the chemical pathway to dust, it
is however worth considering whether the UV interstellar field may disturb
the whole process of PAH and dust growth, and the stellar injection rate of
PAHs into the ISM. We therefore compare the characteristic timescales for
PAH photodestruction, PAH chemical formation and dynamical changes in
the envelope for a typical cool C–star with a stationary dust–driven wind
described by the self–consistent model of Winters *et al.* (1994). In the inner
part of the envelope PAH formation is the dominant process until the envelope region is reached where temperature and density have dropped to such
values that the chemistry becomes frozen. At distances greater than 10^3
stellar radii $R_\star$ from the star, photodestruction dominates. Photodestruction has clearly no influence on the formation of PAHs and does not disturb
the formation and growth of dust. PAHs can however effectively be destroyed
in the outer part of the envelope. In order to evaluate precisely up to what
distance from the star the formed molecules resist the penetrating interstellar UV field, we use the more complete and recent network of reactions
describing PAH formation from Cherchneff *et al.* (1992) and extend it by
the photoreactions of Sect. 2. The results show that small PAHs ($N_C \leq 20$)
are all destroyed at a distance $R \leq 10^4 \, R_\star$ before reaching the transition
region between stellar wind and interstellar medium ($\simeq 10^5 \, R_\star$). C–stars are
therefore not in the position to inject small PAHs into the ISM.

4. Conclusion

Small PAHs cannot survive the UV radiation field of the ISM and may
therefore exist in the ISM only if another mechanism (such as sputtering
of dust particles by shock waves or UV radiations) is effective enough to
balance the photodestruction processes. However, larger PAHs ($N_C \geq 50$)
resist the UV–field and can be produced in C–star envelopes. An extension
of the current network of PAH formation reactions is therefore required to
have a better estimate of their densities in the ISM.

References

Allain T., Leach S. and Sedlmayr E.: 1994, *Astron. & Astrophys.* in preparation.
Cherchneff I., Barker J.R. & Tielens A.G.G.M.: 1992, *Astrophys. J.* **401**, 269.
Jochims H.W., Ruhl E., Baumgartel H., Tobita S. & Leach S.: 1994, *Astrophys. J.* **420**,
    307.
Léger A. & Puget J.L.: 1984, *Astron. & Astrophys.* **137**, L5.
Winters J.M., Dominik C. & Sedlmayr E.: 1994, *Astron. & Astrophys.* **288**, 255

SILICATE DUST EMISSION IN SR VARIABLES[*]

B. ARINGER and J. HRON

Institut für Astronomie, Universität Wien, Austria

Abstract. We have investigated the IRAS LRS spectra of a sample of 56 Semiregular (SR) variables with silicate dust features (class 2 n) and existing near infrared photometry. We compare our results to similar work that has been done on Mira variables by Little-Marenin & Little (1990).

Key words: stars: AGB — stars: variable — stars: mass-loss

1. Introduction

One of the most interesting aspects of the circumstellar shells of AGB stars are the properties of dust grains. The IRAS LRS spectra provide the most complete and uniform database for investigations of that type (Jourdain de Muizon 1992). Laboratory experiments and theoretical calculations show that the optical depth of the shell, the grain size distribution, the dominant crystalline form of the grains and the degree of processing all affect the shape of the characteristic dust features observed in the LRS wavelength range (Simpson 1991, Borghesi *et al.* 1985, Nuth & Hecht 1990).

2. Analysis of the Spectra

We have defined a sample of 56 oxygen-rich SR stars with near infrared photometry (Kerschbaum & Hron 1994) and high quality fluxes in the IRAS Point Source Catalog. The form of the stellar and dust continuum for these objects has been calculated by fitting two blackbodies to the photometric data. Subsequently, the resulting function has been adjusted to the mean flux of the LRS spectra between 8 and 8.6 μm. After the subtraction of this continuum we studied the remaining dust spectra of the SR stars by using the classification criteria from Little-Marenin & Little (1990, LM): a) class Sil: Single emission feature with maximum at 10 μm; b) class Sil$^+$: Emission feature with maximum at 10 μm, but with a weak "bump" near 11.2 μm; c) class Sil^{++}: Emission feature with maximum at 10 μm, but with a strong "bump" near 11.2 μm; d) class 3C: Feature of three emission peaks around 10, 11 and 13 μm; Since we have only included objects of the LRS type 2 n in our study, we did not cover the other classes defined by LM. Nevertheless, we also found some stars with no obvious emission feature.

In Table 1 we present the statistics of our classification and compare it with the results from LM. It is evident that the variety of spectral classes

[*] This work is supported by the Austrian Science Fund Project P9638–AST.

Table 1: Results of our classification: Number and percentage of SR objects of different types and comparison with the statistics for Mira stars.

type	number	percentage	Miras (LM)
Sil	27	48%	53%
Sil+	15	27%	22%
Sil++	4	7%	12%
3C	6	11%	13%
no feature	4	7%	
all objects	56	100%	

is similar for SR and Mira variables. Even the distribution of the different classes seems to be very similar for those two groups. The numbers listed for the Miras in Table 1 are calculated by excluding all stars, which are not of type 2 n. The average main emission feature of the SR objects peaks at 10.0 μm. As it has been already noted by LM, this wavelength is a little bit longer than for Miras (9.7 μm).

3. Discussion

Contrary to the results obtained by LM for Miras, we found no correlation between the IRAS [12]–[25] colors, the amount of emission from the dust envelope and our classification. Although this difference could be partly due to our smaller sample size, we still believe that it is real.

While there are some subtle differences between the characteristics of the dust emission from SR variables with regard to Miras, the overall properties of the dust particles seem to be rather similar. A better understanding of the differences between the dust envelopes of Miras and SR objects, also with regard to possible evolutionary differences, needs more realistic models for dust formation and mass loss in AGB stars (e.g. Winters et al., this conference). Furthermore, we plan to extend our study also to 1 n LRS spectra and to the 18 μm feature.

References

Borghesi, A., Bussoletti, E., Colangeli, L., De Blasi, C., 1985, A&A 153, 1

Jourdain de Muizon, M., 1992, Proc.: Infrared Astronomy with ISO, Les Houches Summer School 1992, eds. Th. Encrenaz, M.F. Kessler, Nova Science Publ., Inc.

Kerschbaum, F., Hron, J., 1994, A&AS 106, 397

Little-Marenin, I.R., Little, S.J., 1990, AJ 99, 1173

Nuth, J.A., Hecht, J.H., 1990, Ap&SS 163, 79

Simpson, J.P., 1991, ApJ 368, 570

SIO OBSERVATIONS OF MASS-LOSING AGB STARS*

B. ARINGER

Institut für Astronomie, Universität Wien, Austria

G. WIEDEMANN and H.U. KÄUFL

European Southern Observatory, Garching, FRG

and

J. HRON

Institut für Astronomie, Universität Wien, Austria

Abstract. We have observed the first overtone rotation-vibration absorption bands of SiO near $\lambda = 4\,\mu\text{m}$ for a sample of 23 oxygen-rich Mira and Semiregular variables using the ESO NTT and IRSPEC. We discuss the strength of the SiO absorption in terms of the near infrared, IRAS and pulsational properties of the stars. Especially among the Miras there are big differences in the strength of the SiO bands between individual objects, which are probably due to pulsational variability.

Key words: stars: AGB — stars: variable — stars: mass-loss

1. Introduction

Among the molecules found in oxygen-rich AGB stars SiO is particularly interesting. It is very abundant in the atmospheres of such cool giants and represents the prime condensate for the circumstellar dust (Gail & Sedlmayr 1986), the formation of which is crucial for the mass loss as well as for the observable properties of the objects. In the radio range, thermal and maser emission from the rotational lines of SiO can be observed (Olofsson 1988), the latter being located at the base of the stellar wind (Elitzur 1992), where dust formation takes place. The same applies for the rotation-vibration bands in the near infrared. Therefore they can provide important additional constraints for theoretical models of dust formation, mass loss and pulsation in AGB stars (Fleischer *et al.* 1991, Dorfi & Feuchtinger 1991). SiO may also be an efficient cooling agent for stellar winds (Muchmore *et al.* 1987).

2. Results

In order to study the first overtone rotation-vibration bandheads of SiO we observed the spectral range from $3.95\,\mu\text{m}$ to $4.10\,\mu\text{m}$ for a sample of 23 oxygen-rich AGB objects, most of which are Mira or Semiregular (SR) variables. Also two Irregular variables have been included. The stars cover a large variety of pulsation periods and infrared colors. The spectra have been observed in June 1993 with the cooled grating spectrograph IRSPEC

* This work is supported by the Austrian Science Fund Project P9638–AST.

at the ESO NTT. They are composed of 5 overlapping parts corresponding to different grating positions, each with a resolution of approximately 4000. The band-heads of the three very distinct transitions V(2,0), V(3,1) and V(4,2) of the main isotope $^{28}Si^{16}O$ are situated in the observed wavelength range. To eliminate atmospheric features and effects of detector response, similar spectra of O and B stars were taken.

We found that most of the observed AGB stars have strong SiO bandheads. However, there are also several objects showing only a very weak or no SiO absorption. In the spectra of the Miras RR Sgr and R Oph, which belong to this group, Brackett-α emission at 4.05 μm can be observed. In order to evaluate the strength of the SiO absorption the equivalent widths of the V(2,0), V(3,1) and V(4,2) bandheads have been determined.

3. Discussion

Objects with weak or no SiO absorption are only found among the Mira stars. Since there are also some Miras showing strong bandheads, a large scatter of the equivalent widths appears for them. This may be due to the spectral variability found by Rinsland & Wing (1982), who noticed that the SiO bands become very weak or disappear about the time of the visual light maximum. On the other hand, all SR objects have a strong SiO absorption, which also indicates that a weakening of the bands is caused somehow by the intense pulsations of Miras.

We have also investigated the behavior of bandheads as a function of the ([12]-[25]) color, the (K-[12]) color and blackbody temperatures derived from IRAS and near infrared photometry. In the corresponding diagrams we find some correlations, which are mainly caused by the different infrared properties of the Miras and SR variables in our sample. We conclude that the spectral variability of the Miras may affect our whole discussion of the strength of the SiO absorption. Therefore it will be necessary to observe the objects again in order to study their spectral changes.

Until now we have only compared our observations to very simple synthetic spectra. More sophisticated models including realistic stellar atmospheres for AGB stars are needed for a correct interpretation of the data. The comparison with such models will be the next step in this investigation.

References

Dorfi, E.A., Feuchtinger, M., 1991, A&A 249, 417
Elitzur, M., 1992, *Astronomical Masers,* Kluwer Academic Publishers
Fleischer, A.J., Gauger, A., Sedlmayr, E., 1991, A&A 242, L1
Gail, H.P., Sedlmayr, E., 1986, A&A 166, 225
Muchmore, D.O., Nuth, J.A., Stencel, R.E., 1987, ApJ 315, L141
Olofsson, H., 1988, Space Science Reviews 47, 145
Rinsland, C.P., Wing, R.F., 1982, ApJ 262, 201

EFFECTS OF A CLUMPY CIRCUMSTELLAR MEDIUM ON THE DYNAMICS OF STELLAR WINDS

S. J. ARTHUR

Instituto de Astronomía, UNAM, 04510 México D.F., México.

Abstract. Mass pickup from clumps resulting from previous mass loss phases by a fast stellar wind can substantially modify its behaviour, even leading to a highly supersonic wind becoming transonic before leaving the mass-loading region. The observational consequences of mass-loading on these astrophysical flows is discussed.

Key words: hydrodynamics — stars: mass-loss — ISM: kinematics and dynamics — planetary nebulae

1. Introduction

The circumstellar medium (CSM) surrounding many stars is not homogeneous and often contains clumps and filaments on a variety of scales. Some planetary nebulae have clumpy cores and filamentary haloes, while some Wolf-Rayet ring nebulae possess filamentary structure. The clumps can come from previous evolutionary phases of the star (e.g. the breakup of a dense wind shell due to Rayleigh-Taylor instabilities, or simply an inhomogeneous dense wind phase) or can be clouds already present in the interstellar medium (ISM).

The environment in which these clumps find themselves is hostile. They are subject to the strong ionizing radiation of the central star and also to hydrodynamic ablation by strong stellar winds. Both of these processes will lead to a gradual erosion of the clumps, and clump material will be released into the surrounding flow (*mass-loading*) (Hartquist *et al.* 1986).

2. Mass-Loading in a Clumpy Planetary Nebula Core

The traditional picture of a planetary nebula consists of several distinct shells representing the various phases of the star's evolution. An important feature of the "standard" model is that the winds from the different phases do not mix and that material in the core is due to the current fast stellar wind, while the halo is composed of (shocked) gas from the AGB and red giant phases. In this work we propose an alternative picture, in which at the onset of the fast wind/ionization front stage of evolution, instabilities induced in the dense shell left over from the superwind phase cause it to break up (Breitschwerdt & Kahn, 1990). This shell break-up both leads to the formation of clumps (that can be ablated) and allows the fast wind to leak out of the core region.

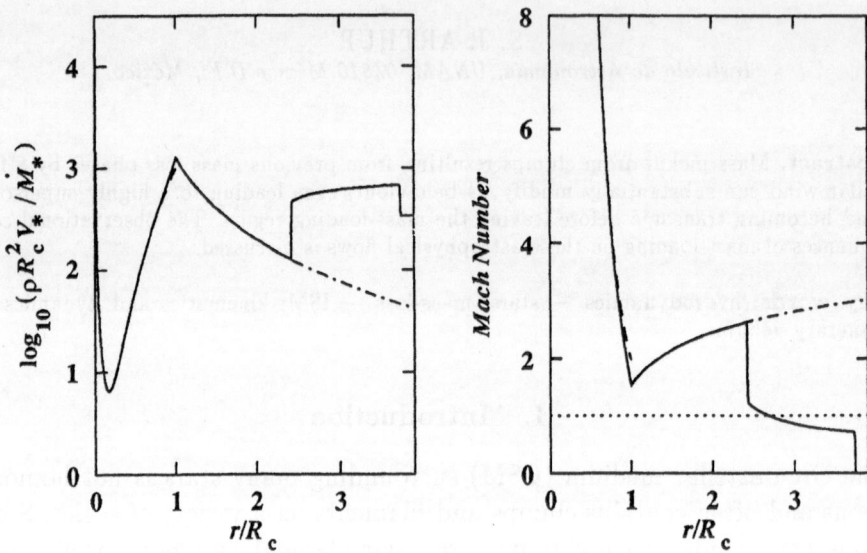

Fig. 1. Log(density) and Mach number for the numerical model (solid line) and analytic model (dashed line in core, dot-dashed line in halo).

We present a numerical model of the flow of a fast stellar wind through a clumpy core region. We assume that cooling of the wind/ablated material mixture behind the clumps is so efficient that the flow can be treated as isothermal, with the temperature being maintained at 10000K by photoionization. In Figure 1 we present the results of such a calculation for a mass-loading rate that is not quite high enough to reduce the Mach number of the flow to unity before it exits the core. It can be seen that the Mach number of the flow is reduced without the flow going through a shock wave. For higher mass-loading rates the flow can go subsonic within the core, but always adjusts to go through a sonic point at the edge of this region (Arthur, Dyson & Hartquist, 1994). As it flows into the halo this mass-loaded wind accelerates gently since the pressure in such transonic flows is non-negligible. Shocking of this transonic flow (Mach number < 3) around filaments in the halo could produce the low velocity dispersion "hot" (15000K) [OIII] that has been observed in some planetary nebulae (Middlemass et al. 1989; Bryce et al. 1992).

References

Arthur, S. J., Dyson, J. E. & Hartquist, T. W.: 1994, *MNRAS* **269**, 1117
Breitschwerdt, D. & Kahn, F. D.: 1990, *MNRAS* **244**, 521
Bryce, M., Meaburn, J., Walsh, J. E. & Clegg, R. E. S.: 1992, *MNRAS* **254**, 477
Hartquist, T. W., Dyson, J. E., Pettini, M. & Smith, L. J.: 1986, *MNRAS* **221**, 715
Middlemass, D., Clegg, R. E. S. & Walsh, J. E.: 1989, *MNRAS* **239**, 1

WHICH SIZE DISTRIBUTION IS BEST FOR IRC +10°216?

S. BAGNULO and J. G. DOYLE

Armagh Observatory, College Hill, Armagh BT61 9DG Northern Ireland

and

I. P. GRIFFIN

Armagh Planetarium, College Hill, Armagh BT61 9DB Northern Ireland

Abstract. A generalized two-streamed Eddington aproximation is applied to the problem of radiative transfer in a spherically symmetric dust shell. We investigate whether our technique allows us to draw conclusions about the size distribution of the grains.

Key words: stars: IRC +10° 216 — grain size distribution

In recent years, several different size distribution have been proposed to describe the nature of grains in the circumstellar dust envelope of the mass losing carbon star IRC +10° 210. Three such distributions for spherical grains with radius a are

$$\xi(a) = \delta(a - a_0) \tag{1}$$

$$\xi(a) = Aa^{-p} \text{ where } a_1 \leq a \leq a_2 \tag{2}$$

$$\xi(a) = Ba^{-q}e^{-a/b_0} \text{ where } a \geq a_1 . \tag{3}$$

Eq. (1) with a_0 ranging from $5 \times 10^{-2}\mu$ to $10^{-1}\mu$ has been used by Martin & Rogers (1987) to describe amorphous carbon dust in the circumstellar envelope of IRC +10° 216. Eq. (2) with p ranging from 3.3 to 3.6, and with a range of sizes depending upon chemical composition is the well known Mathis, Rumpl and Nordsieck (1977) distribution, which has been theoretically justified by Biermann & Harwitt (1980). Griffin (1990) adopted this distribution with $p = 3.5$, $a_1 = 5 \times 10^{-3}\mu$ and $a_2 = 5 \times 10^{-2}\mu$ for amorphous carbon and silicon carbide grains in a two component model of the circumstellar envelope of the star. Eq. (3) with $p = 3.5$ and $b_0 \sim 10^{-1}\mu$ has been suggested in a recent paper by Jura (1994).

In this paper we utilise the technique described by Haisch (1979) to examine which of these three grain size distributions best fits high quality photometric observations of IRC +10° 216 derived from the literature. The technique uses the generalised two-stream Eddington approximation to solve the equation of radiative transfer in a spherically symmetric medium.

Preliminary results, illustrated by Fig. (1), indicate that distribution described by Eq. (3) provides a poor fit to the data. However our modelling technique can not, at present, discriminate between the size distributions described by Eq. (1) and (2), both of which give better fits than Eq. (3).

Astrophysics and Space Science **224**: 425–426, 1995.
© *1995 Kluwer Academic Publishers.*

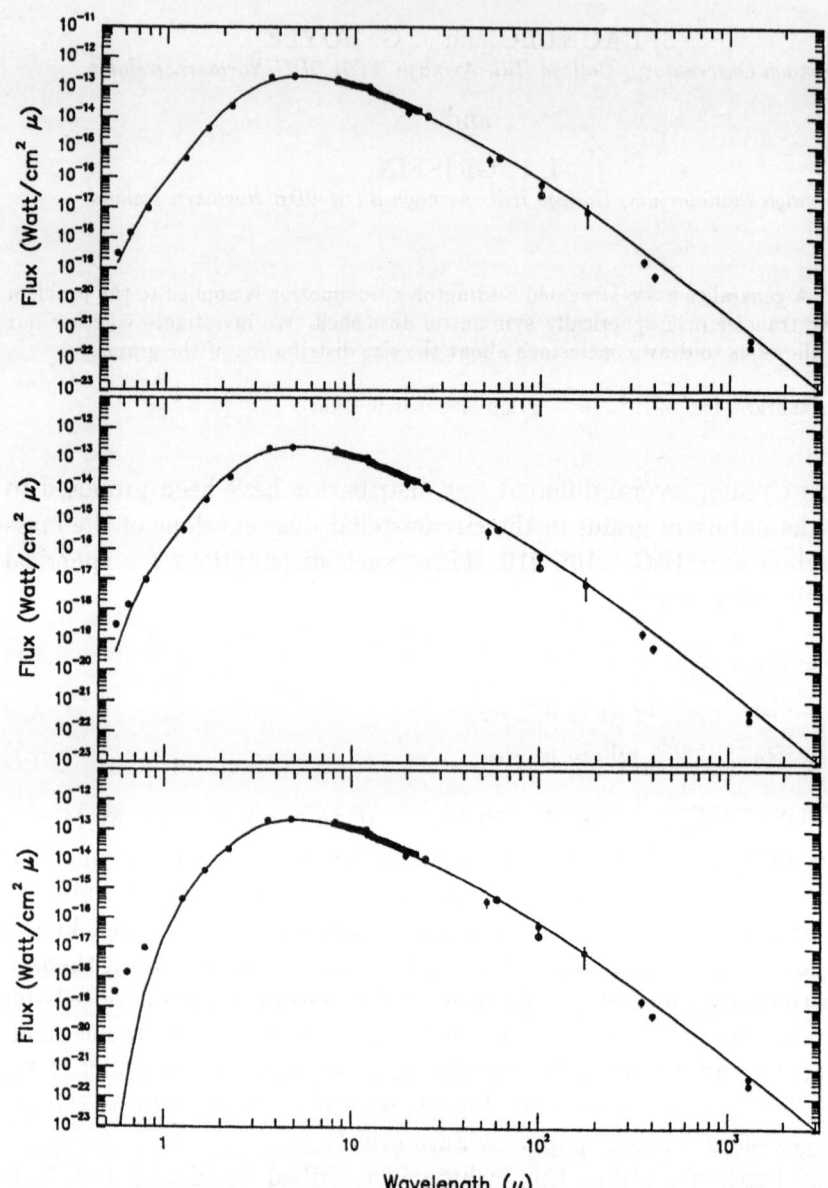

Fig. 1.   From the top to the bottom: fit obtained with Eq. (1) ($a_0 = 0.02\mu$); fit obtained
with Eq. (2) ($a_1 = 0.005\mu$, $a_2 = 0.05\mu$), fit obtained with Eq. (3) ($a_1 = 0.005\mu$, $b_0 = 0.05\mu$)

References

Biermann P. and Harwitt M.: 1980, *ApJ*, **241**, L105
Griffin I.P.: 1990, *MNRAS*, **247**, 591
Jura M.: 1994, *ApJ* (in press)
Haisch B.M.: 1979, *A&A* **72**, 161
Martin P.G. & Rogers C.: 1987, *ApJ*, **322**, 374
Mathis J.S., Rumpl W., and Nordsieck K. H.: 1977 *ApJ*, **217**, 425

EVOLUTION OF THE CIRCUMSTELLAR SHELL OF

PLEIONE (1981-94)

D. BALLEREAU and J. CHAUVILLE

Observatoire de Paris-Meudon - 92195 Meudon Cedex, France

and

J. ZOREC

Institut d'Astrophysique - 98bis Boulevard Arago, 75014 Paris, France

Abstract. Three spectra of Pleione (1981, 1987–88, 1994), obtained in the blue wavelength range, show strong long-term changes which indicate that the surrounding shell, first rather cool and close to the central star, then expands gradually in the equatorial plane of the star and towards the poles. From 1981 to 1987–88, the spectrum of Pleione is that of the central star, with a cool shell which globally mimics a late-type star. The year 1994 is marked by a pure and stronger Be spectrum, with no shell line.

Key words: stars : Be — stars: individual : Pleione

1. Introduction and observations

Since 1888, Pleione (HD 23862, B7-8IV-Ve, V sin i $=$ 320 km s^{-1}) has undergone several spectroscopic B, Be and Be-shell phases. The last evolving episode, which began in 1972 (Delplace & Hubert 1973), reached its maximum in 1982 (Sharov & Lyutyi 1988) and entered a new Be phase in 1984 (Ballereau *et al.* 1988), corresponds to the formation, stabilization and extension of the circumstellar gas. We present three sets of spectroscopic data, obtained in 1981 Sep. 25, 1987–88 and 1994 Aug. 9 with the T1.52m of Haute-Provence Observatory. The first two are photographic plates (IIaO, 12 Å mm^{-1}), the last one an AURELIE record (R $=$ 16400). Figure 1 gives an intensity tracing of the Hβ line profile at each epoch.

2. Discussion and conclusions

Very pronounced differences exist between these optical spectra. A shell spectrum is prominent in 1981 on Balmer and metallic ion lines; it disappears in 1987–88 except on FeII (mult. 42) and H lines (up to $\simeq$ H8); it vanishes completely in 1994. A weak, double-peaked emission spectrum is seen in 1981 on Hβ (V/R $\simeq$ 1); it is stronger in 1987–88 (V/R < 1), and present on metallic ion lines, with P Cyg-like profiles; it is stronger again in 1994 on Hβ (V/R $\ll$ 1) and FeII λ4924 Å (double-peaked, V/R $\ll$ 1). On Hβ, th emission peak separations Δe are respectively 273, 202 and 155 km

The spectral variations of Pleione result from the dynamic evolution of its shell and the successive radiative transfers from the star surface to the ISM.

Astrophysics and Space Science **224**: 427–428, 1995.

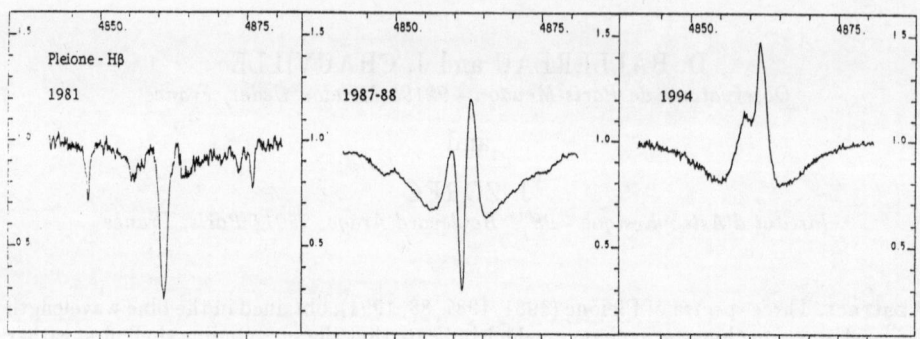

Fig. 1.   Intensity line profile of Hβ of Pleione at three observational epochs.

The central star has not necessarily changed its spectral type (Zorec 1994). If we admit that the H-emitting shell has an axisymmetric structure, its outer limit r_2, expressed in stellar radii, is given by Huang's (1972) formula : $r_2 = (2V \sin i / \Delta e)^{1/j}$ (j = 1/2, Keplerian motion). For the three values of Δe on Hβ, we obtain : 1981 : $5r_*$; 1987–88 : $10r_*$; 1994 : $17.0r_*$.

Several conclusions can be stressed :

- Before 1972, the H-emitting shell is in constant contraction. From 1972 to 1982, this limit has rather low values ($r_2 = 2$ to $5\ r_*$). Since 1982, beginning of the transition to a Be phase, this limit is in constant expansion.

- In 1981, the shell is compact and mimics a line spectrum of a late-type star. The global flux in the whole wavelength range decreases, which indicates that the shell is flattened (non-symmetrical diffusion of photons). This is supported by the strong decrease of the far-UV continuum observed at the same epoch by Doazan et al. (1988).

- In 1987–88, the Be phase is well developed and the compact screen of the shell dissipates radially in the equatorial plane and perpendicularly to it, and becomes more spherical and hotter.

- In 1994, the shallowing of the photospheric profile of Hβ can be explained by the veiling of an increasing continuum emission, filling-in by line emission, and/or electron scattering.

We thanks Dr. R. Hirata for very fruitful comments.

References

Ballereau, D., Chauville, J. & Mekkas, A.: 1988, A&AS **75**, 139
Delplace, A.-M. & Hubert, H.: 1973, C.R.Acad.Sc.Paris **277**, 575
Doazan, V., Thomas, R.N. & Bourdonneau, B.: 1988, A&A **205**, L11
Huang, S.-S.: 1972, ApJ **171**, 549
Sharov, A.S. & Lyutyi, V.M.: 1988, Astron. Zh. **65**, 593
Zorec J.: 1994, 'The photosphere of Pleione' in L.A. Balona et al., ed(s)., Pulsation, Rotation and Mass Loss in Early-Type Stars, Kluwer Academic Publishers, 362

INHOMOGENEITIES IN THE CIRCUMSTELLAR
ENVELOPE OF THE A0E HERBIG STAR AB AUR

NINA BESKROVNAYA and MIKHAIL POGODIN

Pulkovo Observatory, 196140 St. Petersburg, Russia

and

IVAN NAJDENOV and IOSIFF ROMANYUK

Special Astrophysical Observatory, 357147 Nizhnij Arkhyz, Russia

Abstract. Simultaneous high-resolution spectroscopy in Hα and *UBVRI* polarimetric observations are proposed as an effective method for the search for circumstellar inhomogeneities in A0-type Herbig stars. The new results for AB Aur are presented as a successful example of the use of this method. The analysis of about 100 CCD Hα profiles ($R = 30\,000$) and more than 150 polarimetric measurements obtained in January, 1994 allowed to discover a long-lived stream-like inhomogeneity in the circumstellar gaseous envelope.

Key words: line profiles – polarization – stellar wind – inhomogeneity

AB Aur (B9ve, $m_\mathrm{V} = 7^\mathrm{m}1$) is one of the brightest pre-main sequence Ae/Be stars. It belongs to the subclass of B8–A2 stars showing PCyg profiles at Hα . The existence of azimuthal inhomogeneities in the gaseous envelope of AB Aur was first reported by Praderie et al. Our observing programme included a search for rapid variability ($\tau \sim$ hours$\div$days) of the Hα line profile and of the *UBVRI* parameters of linear polarization in AB Aur . The choice of just these characteristics was connected with our intention to investigate a major part of circumstellar gas where hypothetical inhomogeneities could originate. The spectroscopic observations of AB Aur were carried out in the CrAO in January, 1994. The 2.6m telescope with a coudé spectrograph was used to obtain CCD spectra. Simultaneously AB Aur was observaed polarimetrically with two telescopes: 1.25m of the CrAO, equipped with the *IBVRI*-polarimeter and 1m telescope of the SAO with a photoelectric polarimeter. The results presented in Fig. 1–2 allow us to conclude that since January, 4 up to January, 9 in the envelope of AB Aur there existed a large-scale long-lived azimuthal inhomogeneity, rotating with the period equal to the rotational period of the star. The origin of this inhomogeneity is likely to be due to the stream-like outflow from the stellar surface, having essentially non-stationary (may be discrete) character, with relative constance of the kinematical law inside the stream during the whole period of observations.

References

Praderie, F., Simon, T., Catala, C., Boesgaard, A.M.: 1986, *ApJ* **303**, 311

Astrophysics and Space Science **224**: 429–430, 1995.
© 1995 *Kluwer Academic Publishers.*

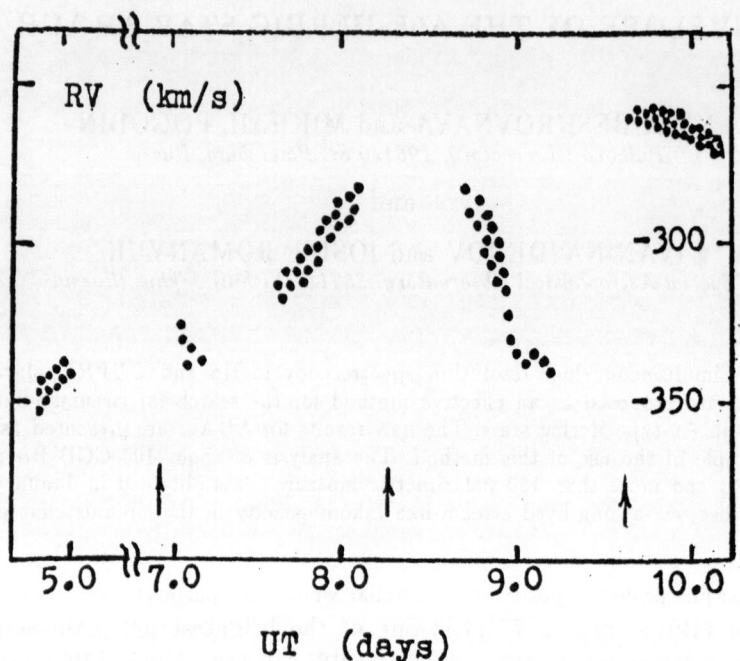

Fig. 1.

Variations of the RV position of the blue edge of the additional emission
connected with the inhomogeneity on the $0.75F_c$ level. The arrows mark
expected moments of the stream crossing the line of sight for $P_{rot} = 32^h3$.

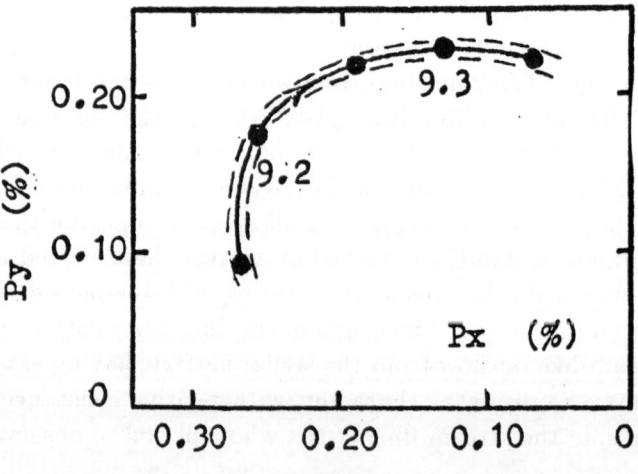

Fig. 2.

Trend of the polarization parameters in B-filter during the night January,
9/10 1994 on the (P_x, P_y)–plane. The UT-intervals of 0^d05 are marked on the
trace. The dashed lines indicate $\pm\sigma$-interval of the observational uncertainty

DUST AND GAS IN W49A

H. BUCKLEY

Institute for Astronomy, University of Edinburgh, Blackford Hill, Edinburgh, EH9 3HJ

and

D. WARD–THOMPSON

Royal Observatory, Blackford Hill, Edinburgh, EH9 3HJ.

Abstract. We have mapped the high-mass star-forming region W49A at 450, 800, and 1100 microns with the JCMT. Spectral index measurements suggest an increase in temperature towards the emission peaks, consistent with previous data. We derive the gas masses associated with the central and extended emission from each of the three components, and find a deficit of gas around W49SW. The mass found for the core of W49N is in good agreement with the value previously derived from $C^{34}S$ (5–4) maps (Serabyn *et al.*, 1993), and similar morphologies are found in the line and continuum maps.

Key words: circumstellar matter – stars:formation – ISM: dust

Introduction

W49A is one of the most luminous star formation regions in our galaxy. It lies around 11 kpc distant and is completely obscured in the optical. The three main subcomponents, W49N, SE and SW (in order of prominence), are each strong mid-infrared sources containing HII regions and masers. Our continuum maps allow a determination of the mass distribution and temperature of the dust, lying in the frequency regime where free-free emission is small and the emission is optically thin.

Results

Table I gives the masses of the components of W49A derived from our continuum maps. For each component a bright central region and more diffuse extended region can be distinguished. From the $450\mu m/800\mu m$ spectral index we infer a temperature difference between the central and extended regions in W49N and SE, consistent with the temperatures found by Ward Thompson *et al.* 1992 (hereafter WBR). The low flux from W49SW resulted in a noise dominated spectral index measurement. Gas masses were derived using the standard procedure (Hildebrand, 1983) with $\beta = 2$, and the temperatures given in WBR. We estimate a total uncertainty of around a factor of 2 in the masses of W49N and SE, and a factor of 3 for W49SW.

The derived masses for W49N (Table 1) are consistent with those found by Sievers *et al.* (1991) from 870 and 1300 μm maps. The W49SW mass quoted is a factor of three lower than their measurement, which may be partly due to the low signal to noise ($\simeq 3$) for this source in our maps. Estimating the luminosities using the IRAS fluxes (WBR), we find luminosity to mass ratios of 14, 22, and 160 times solar for the extended regions around W49N, SE

Astrophysics and Space Science **224**: 431–432, 1995.
© *1995 Kluwer Academic Publishers.*

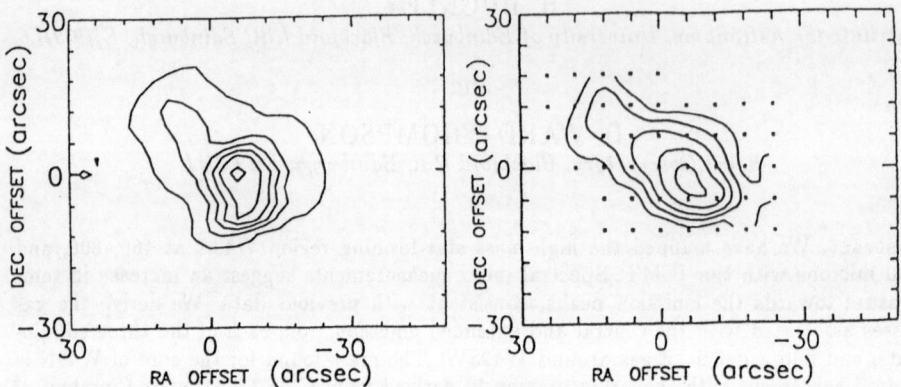

Fig. 1. (Left) 450μm continuum map of W49N. Contours at 15%, 35%, 50%, 65%, 80% and 95% of peak brightness. Map centre is at $(\alpha, \delta) = (19^h07^m50^s.0, +09°01'20")$. (Right) Same region mapped in $C^{34}S$ 5–4 with similar beamsize. Contours as before except that here the 15% contour is omitted (Serabyn et al., 1993).

TABLE I
Derived masses for W49A.

Source	Extended		Central		
	T (K)	M_g $(10^3 M_\odot)$	D_{ap} (")	T (K)	M_g $(10^3 M_\odot)$
W49N	12	400	48	50	54.0
W49SE	12	150	24	50	5.4
W49SW	12	19	24	50	2.5

and SW respectively. If our mass estimates are correct there is a deficit of cold molecular material around W49SW, which may indicate that it is in a more evolved state. Figure 1 shows a comparison between the the $C^{34}S$ 5-4 map of Serabyn et al. (1993) over the core of W49N, and the corresponding region of our 450μm continuum map. The morphologies are very similar, and the mass derived by Serabyn et al. for the core of W49N ($\simeq 10^5 M_\odot$) is about a factor of two greater than our measurement, showing reasonable consistency between the two methods used.

References

Hildebrand R.H., 1983, *QJRAS*, 24, 267

Serabyn E., Gusten R., Schulz A., 1993, *ApJ*, 413, 571

Sievers A. W., Mezger P. G., Gordon M. A., Kreysa E., Haslam C. G. T, Lemke R., 1991, *A&A*, 251, 231

Ward-Thompson D., Berry D. S., Robson E. I., 1992, *MNRAS*, 257, 180

CO INTERFEROMETRIC OBSERVATIONS OF M1-92
(THE MINKOWSKI'S FOOTPRINT)

V. BUJARRABAL and J. ALCOLEA
Centro Astronómico de Yebes
Apartado 148, 19080 Guadalajara, SPAIN

and

R. NERI and M. GREWING
IRAM
300 rue de la Piscine, 38406 St Martin d'Hères, FRANCE

Abstract. We present high spatial resolution observations of the CO emission at 2.6mm in the protoplanetary nebula M1-92 (Minkowski's Footprint), obtained with the Plateau de Bure interferometer. The total mass corresponding to the CO measured intensity is estimated to be large, of about 0.1-0.2 solar masses. The cartography is particularly rich and revealing. From the measured position–velocity distribution of brightness, we conclude that a strong dynamical interaction between the AGB and post-AGB winds is present.

Key words: Circumstellar matter – Stars: AGB and post-AGB – Planetary nebulae: individual: M1-92 – Wind interaction in protoplanetary nebulae

We present high spatial resolution observations of the CO emission at 2.6mm in the protoplanetary nebula M1-92 (Minkowski's Footprint), obtained with the Plateau de Bure interferometer. The results represent the best molecular image obtained of one of such objects; the main features of the structure and kinematics of the molecular envelope are clearly shown, over a total emitting region of about 12 arcsec. Three components are distinguished in the maps (Fig. 1). A bipolar outflow, with a (deprojected) velocity of about 65 km/s, appears at the line wings. A hollow prolate structure, with axial velocity increasing with the distance to the star up to about 60 km/s, is also identified from our maps by the presence of relative minima of intensity at intermediate velocities. Finally, a compact condensation, with a size of about 2 arcsec and velocities smaller than 10 km/s, is present in the central channels. The total mass corresponding to the CO emission is estimated to be 0.1-0.2 solar masses. Therefore, the CO observations probe the bulk of the nebular material, other (more excited) components representing a negligible contribution to the total mass.

The gradient in axial velocity can be measured along the whole structure. In fact, a remarkable continuity in position and velocity is found between the different components. We think that this property indicates that an important dynamical interaction between the AGB and post-ABG winds is active now and affects most of the nebular material. Such an interaction would consist in a significant momentum transport from the bipolar fast flow, characteristic of protoplanetary objects, to the rest of the nebula.

Astrophysics and Space Science **224**: 433–434, 1995.
© 1995 *Kluwer Academic Publishers.*

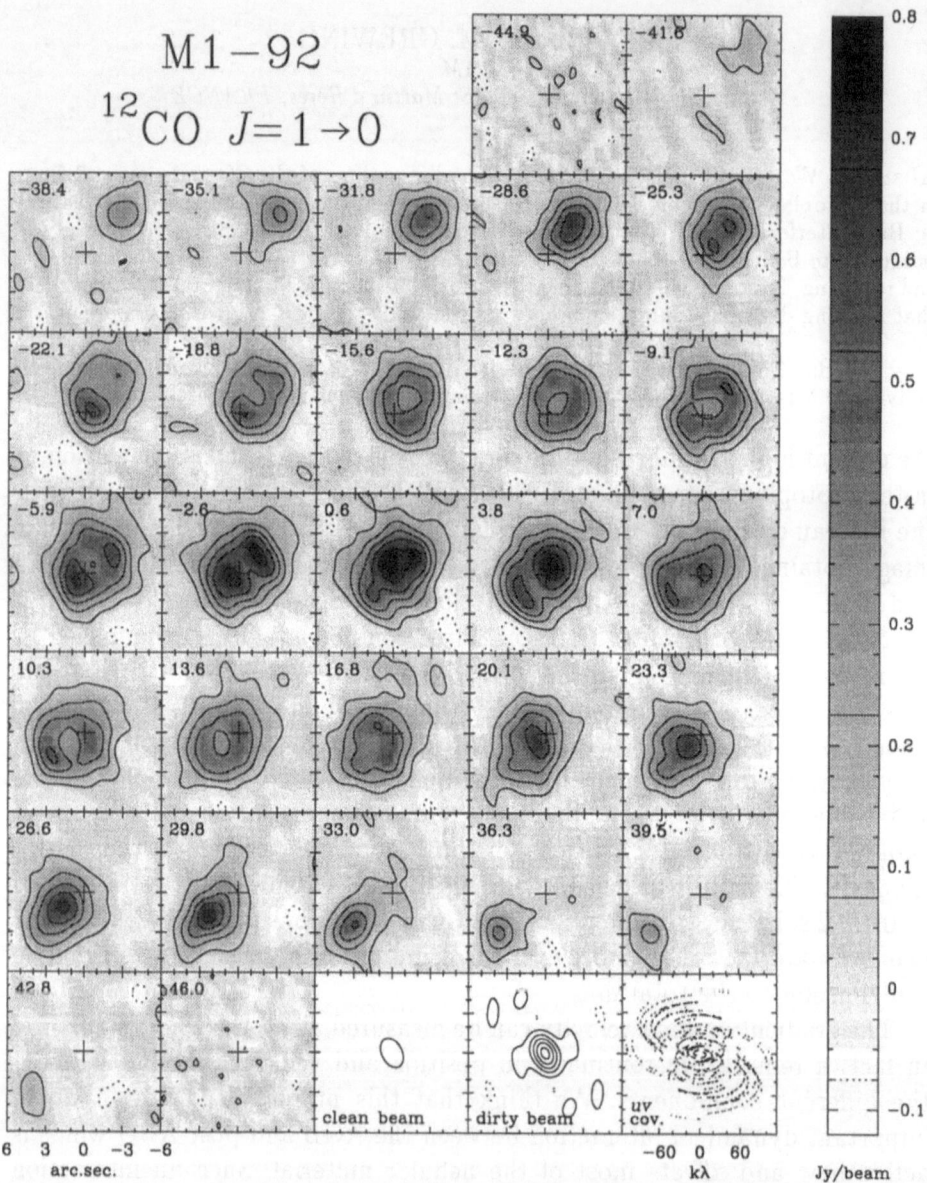

Fig. 1.   Maps of the CO 1-0 intensity for the *lsr* velocities indicated in the upper left corners. See the scale in the column, dashed lines represent negative contours. We also include the CLEANed beam (at half power), the dirty beam (contours from -10% to 90% by intervals of 20%) and the *uv* coverage.

NEW CANDIDATES FOR OBJECTS
WITH A 21 MICRON FEATURE

S. J. CHAN, TH. HENNING and B. BEGEMANN
Max Planck Society
Research Group "Dust in Star Forming Regions"
Schillergaesschen 3, D-07745 Jena, Germany

Abstract. We report the preliminary result of studying the nature of objects with a 21 μm feature. New candidates are presented and possible carriers of the 21 μm feature are discussed.

Key words: circumstellar matter — infrared spectra

1. Introduction

Recently, nine IRAS sources have been reported to show an unidentified emission feature at 21 μm in their Low Resolution Spectrometer (LRS) spectra and in their UKIRT spectra (Kwok *et al.* 1993 and references therein). Possible carriers of the 21 μm feature have been suggested by several research groups, such as PAHs (Buss *et al.* 1990, Hrivnak *et al.* 1994), SiS_2 (Goebel 1993), and iron oxides (Cox 1990). Currently, we are searching for additional objects with a 21 μm emission feature. Moreover, we are studying the possible carriers of this feature in our infrared spectroscopy laboratory.

2. Searching for Objects with a 21 μm Feature

At present, we are searching for the LRS spectra of all possible sources with a 21 μm feature and an IRAS flux at 12 μm greater than 5 Jy. The selection criterion is the ratio $\lambda F_\lambda(25\ \mu m)/\lambda F_\lambda(12\ \mu m)$. It has to be between -0.10 and 0.55 which is the colour (with uncertainties) of already known sources with a 21 μm feature. Up to now, we have found 14 candidates from our list (Figure 1) whose spectra probably show the 21 μm feature including two sources (IRAS 2000+3239 and IRAS 22574+6609) which were detected by Kwok and his colleagues from UKIRT observations.

3. General Characteristics of the LRS Spectra and General Properties of the Candidates

The general characteristics of the LRS spectra of the candidates are: a peak around 19 – 22 μm; a flat plateau between $\approx$ 12 μm and 18 μm; a drop between $\approx$ 7 to 11 μm which may be an absorption feature; some

Astrophysics and Space Science 224: 435–437, 1995.
© 1995 *Kluwer Academic Publishers.* .

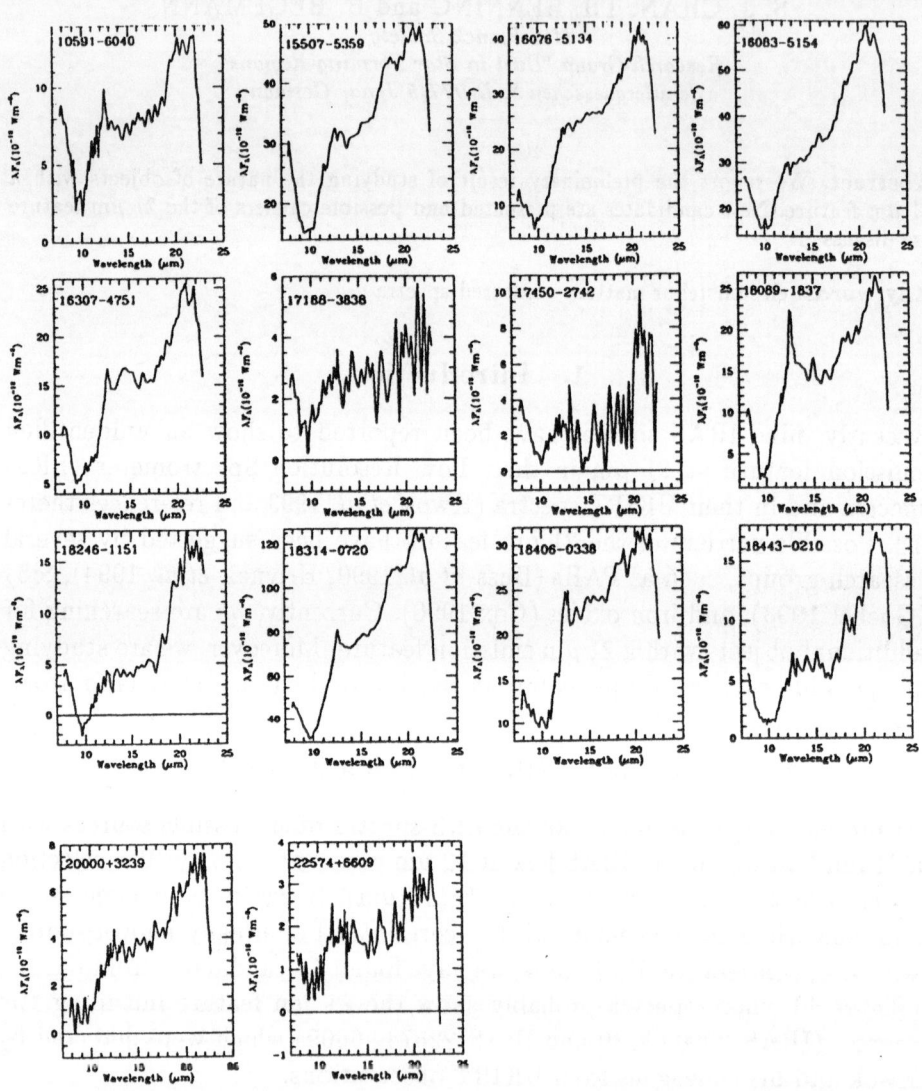

Fig. 1.   New candidates with a 21 μm feature

objects show a shoulder at $\approx$ 18 μm and some objects show a weak peak between $\approx$ 11 – 13 μm.

Those objects with a 21 μm feature are suggested to be C-rich bipolar proto-planetary nebulae (PPNe) (Meixner *et al.* 1994 and references therein). Cox (1990), however, proposed that they are not only evolved objects but also objects associated with H II regions. Furthermore, he also proposed that the 21 μm feature is ubiquitous in the infrared spectra of H II regions. A study of the nature of our candidates with a 21 μm feature shows results partly consistent with Cox's suggestions because most of our candidates are associated with radio H II regions, but some are associated with evolved objects. However, to find out if the 21 μm feature is really ubiquitous in the infrared spectra of H II regions needs more detailed studies.

4.   Possible Carriers of the 21 μm Feature

Three possible carriers of the 21 μm feature in C-rich PPN are proposed. Infrared spectra (transmission, polyethylene pellets) of PAHs such as Naphtacene and Pyrene show vibrational bands which really lie in the right wavelength region, but they are much too narrow in comparison with the observations. Inorganic substances as SiS_2 and iron oxides show much broader infrared vibrational bands. The spectrum of SiS_2 shows a 22 μm band, which was only examined in the 15 – 100 μm region. γ-Fe_2O_3 and Fe_3O_4 have two broad bands in the wavelength region from 13 – 33 μm. FeO (reflection, UV – FIR) shows only one broad band from 16 – 23 μm, with a peak at 20 μm (normalized absorption coefficent calculated for spheres from optical constants). In spite of the good correspondence, the existence of FeO as the main component of C-rich PPNe is unrealistic.

Acknowledgements

The IRAS data were obtained using the IRAS data base server of the Space Research Organisation of the Netherlands and the Dutch Expertise Centre for Astronomical Data Processing.

References

Buss, R.H. Jr., Cohen, M., Tielens, A.G.G.M., Werner, M.W., Bregman, J.D., Witterborn, F.C., Rank, D., Sandford, S.A.: 1990 *ApJ* **365**, L23

Cox, P.: 1990 *A&A* **236**, L29

Goebel, J.H.: 1993 *A&A* **278**, 226

Hrivnak, B.J., Kwok, S., Geballe, T.R.: 1994 *ApJ* **420**,783

Kwok, S., Hrivnak, B.J., Geballe, T.R., Langill, P.P.: 1993 *Astronomical Infrared Spectroscopy: Future Observational Directions, ASP Conf. Series* **41**, p123, ed. S.Kwok

Meixner, M., Graham, J.R., Skinner, C.J., Hawkins, G.W., Keto,E.: 1994, *Experimental Astronomy 3*, Kluwer Academic Publishers: Dordrecht, p53

GAS-GRAIN PROCESSES IN THE ENVELOPES OF LATE-TYPE STARS

S.B. CHARNLEY
Astronomy Department, University of California at Berkeley
and Space Science Division, NASA Ames Research Center

Abstract. Some results from studies of the gas-grain chemistry in oxygen-rich circumstellar envelopes (CSEs) are presented.

Key words: stars: late-type – stars: circumstellar matter

1. Introduction

The massive molecular envelopes of AGB stars (e.g. M giants, OH/IR stars etc.) are understood to form by radiation pressure on dust grains which then stream through, and drag along, the circumstellar gas. Observations of the refractory molecules SiO and SiS show that molecule-grain collisions lead to accretion and depletion of gas phase species (Jura & Morris 1985; Omont 1993). Water ice is directly observed in the outflows from several oxygen-rich stars (e.g. Smith et al. 1988). In this paper some connections between the gas-grain interaction and circumstellar chemistry are briefly described.

Modelling of a well-studied sample of OH/IR stars has been performed to determine how common circumstellar ice mantles may be, and how their formation depends upon the physical conditions in the envelope (Figure 1a; Charnley & Smith 1994). Removal of water from the gas, the major molecular coolant, affects the thermal balance in the flow and also reduces the number of hydroxl molecules photoproduced in the outer envelope. The influence of ice formation on the oxygen and sulphur chemistries of O-rich outflows has been investigated. For example, hotter gas leads to more efficient release of atomic S from H_2S and the S-chemistry is initiated at smaller radii (Charnley 1994).

An outstanding problem in circumstellar chemistry is the presence of C-bearing molecules (HCN & H_2CO) in ostensibly O-rich envelopes. Nejad & Millar (1988) suggested that a source of methane existed in the warm inner layers of the envelope. Kress et al. (1994) have proposed that reaction of CO and H_2 on warm metallic grains is this source. Charnley et al. (1994) showed that the this excess CH_4 drives a rich organic chemistry in the photodissociation layer which produces detectable abundances of CH_3OH and C_2H in some objects (Figure 1b).

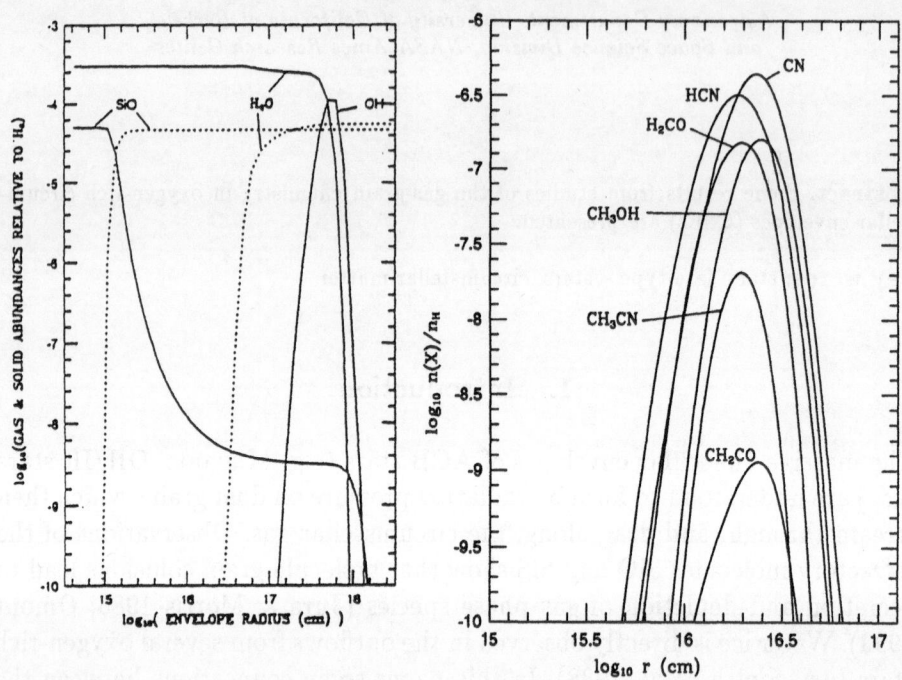

Fig. 1. (a) The depletion of SiO and water by accretion onto dust in an O-rich wind with physical conditions similar to OH26.5+0.6. (b) Radial distributions of organic molecules produced in the photochemistry driven by grain-catalysed methane.

References

Charnley, S.B. 1994, in preparation.
Charnley, S.B. & Smith, R.G. 1994, *MNRAS*, submitted.
Charnley, S.B., Tielens, A.G.G.M. & Kress, M.E. 1994, *A & A*, submitted.
Kress, M.E., Tielens, A.G.G.M. & Charnley, S.B., 1994, in preparation.
Jura, M. & Morris, M. 1985, *Ap. J.*, **292**, 487.
Nejad, L.A.M. & Millar, T.J. 1987, *MNRAS*, **230**, 79.
Omont, A. 1993, *J. Chem. Soc. Faraday. Trans.*, **89**, 2137.
Smith, R.G. et al. 1988, *Ap. J.*, **334**, 209.

THE INTERSTELLAR CHEMISTRY
OF PROTOSTELLAR DISKS

S.B. CHARNLEY

Astronomy Department, University of California at Berkeley
and Space Science Division, NASA Ames Research Center

Abstract. A study has been undertaken of the gas-grain chemistry of protostellar disks which are sufficiently cool that in the outer regions, where the gas density is less than $\sim 10^{13} \, cm^{-3}$ and the ionization rate highest, a bimolecular chemistry resembling that of dark clouds can occur. Since the gas-grain collision rate is so high, outgassing mantle molecules effectively determine the gas phase composition at any position in the disk. In contrast to previous work, a detailed gas phase chemistry is considered along with the accretion and desorption of mantle species which is controlled locally by the dust temperature.

Key words: ISM: molecules – solar system: formation

1. Introduction

The ionization structure of cool protostellar disks plays a key rôle in their physical evolution since it controls the degree to which the gas and magnetic field are coupled. This coupling determines how efficient differential rotation can be in amplifying the magnetic field, thus removing angular momentum from the disk. Previous work on this problem has included the charging effects of gas-grain collisions, but not the accretion and retention of neutral molecules, the balance of which depends upon the local dust temperature (Umebayashi & Nakano 1988). On the other hand, those models that have considered mantle outgassing in regions of comet formation have ignored its effect on gas phase chemistry and have used evaporation rates appropriate to the condensation temperatures of the pure substances (Yamamoto 1985). This (steady-state) gas-grain chemistry fixes the ionization structure in the disk as well as the molecular distributions and predicts, for example, that the dominant gas phase ion is that of the representative refractory metal; a conclusion difficult to justify given the high binding energies for physisorption of such species (Jura & Morris 1985)

A study was undertaken for a particular physical model, using a more consistent treatment of interstellar gas-grain chemistry, to explore how the molecular distributions in the disk are influenced by various assumptions involving thermal binding energies for physisorption; the gas phase abundances of refractory metal atoms; and the degree of dust sedimentation. The model adopted was that of Hayashi (1981) for the protosolar nebula : a one solar mass disk in Keplerian rotation with most of the dust grains precipitated to the midplane. From the gas temperature the dust temper-

Astrophysics and Space Science **224**: 441–442, 1995.

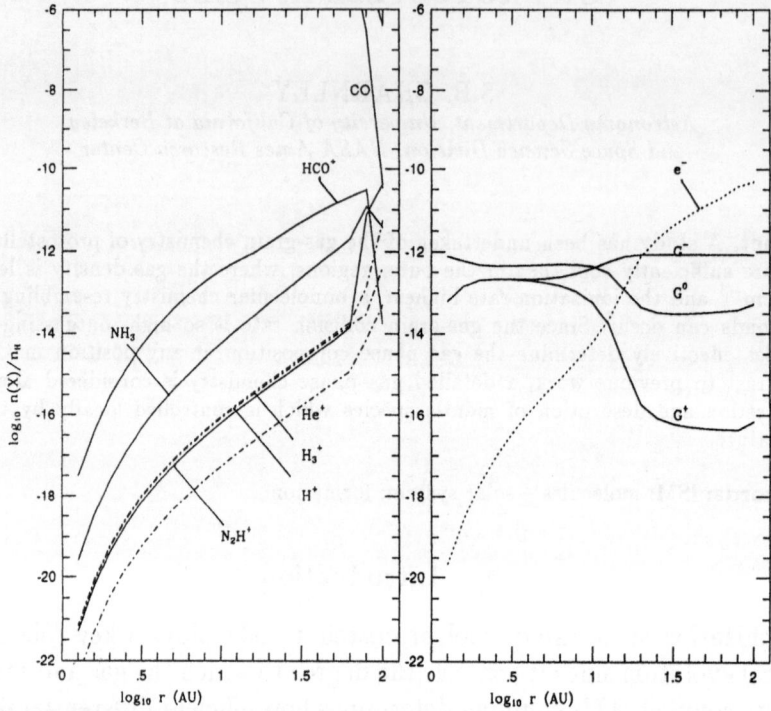

Fig. 1.   The steady-state abundance distributions of the principal species derived from
CO, N$_2$ and O$_2$. This model has an interstellar gas/dust number density ratio with all
refractory metals depleted in the grains. The volatile surface binding energies are those
corresponding to the pure substance (e.g. CO on CO).

ature can be calculated and hence the evaporation rate of the volatiles at
any radius.

Briefly, the models show that the most marked differences occur when
binding energies are employed appropriate to volatiles physisorbed on H$_2$O
ice (Tielens & Allamandola 1987). In evolved disks, significant grain sedimentation lowers the gas-grain collision rate and leads to 'inflated' abundances of several potentially detectable molecules (Charnley 1994). For the
model most closely representing that of Umebayashi & Nakano (i.e. Figure
1) there is, however, general agreement for the overall ionization structure.

References

Charnley, S.B. 1994, in preparation.
Hayashi, C. 1981, *Prog. Theor. Phys. Suppl.* **70**, 35.
Jura, M. & Morris, M. 1985, *Ap. J.*, **292**, 487.
Tielens, A.G.G.M. and Allamandola, L.J. 1987, in *Interstellar Processes*, eds. D.J. Hol-
    lenbach & H.A. Thronson Jr., Reidel, p.397.
Umebayashi, T. and Nakano, T. 1988, *Prog. Theor. Phys. Suppl.* **96**, 188.
Umebayashi, T. and Nakano, T. 1990, *MNRAS*, **243**, 103.
Yamamoto, T. 1985, *A & A*, **142**, 31.

MOLECULAR SIGNATURES OF MHD WAVES

S.B. CHARNLEY

*Astronomy Department, University of California at Berkeley
and Space Science Division, NASA Ames Research Center*

and

H. BUTNER

Carnegie Institute of Washington, Department of Terrestrial Magnetism

Abstract. The small ion-neutral drift speeds attained in MHD waves lead to a significant increase in the rates of several important ion-molecule reactions. The effect on the deuterium chemistry of dark clouds could allow MHD wave motion to be observed directly. The results of some recent observations to test this theory are outlined.

Key words: ISM: molecules — ISM: kinematics and dynamics

1. Introduction

The relative streaming of ions and neutral particles (ambipolar diffusion) is an important physical process in the partially-ionized plasma that comprises dark interstellar clouds (Zweibel & Josafatsson 1983). Ambipolar diffusion plays a key rôle in the initial stages of low-mass star formation and in the propagation of interstellar magnetohydrodynamic (MHD) waves.

MHD waves may be responsible for the large superthermal linewidths observed in cold molecular clouds and so, as a first attempt to determine observable diagnostics for them, Charnley & Roberge (1992, 1994) explored the chemistry of molecular gas that is swept by (linear) Alfvén waves. In cold (10K) gas, ion-neutral streaming induced by these waves can drive reactions with small activation barriers or small endothermicities. In particular, the reaction of H_2D^+ with H_2 can proceed at a non-thermal rate much larger than the expected rate at 10K. The destruction of deuterated ions in the waves affects the D/H ratios in certain observable species (N_2D^+/N_2H^+ and DCO^+/HCO^+).

If MHD waves play a part in interstellar chemistry, then the theory of Charnley & Roberge makes several predictions. The conservative ion-neutral drift speeds produced (< 1 km s^{-1}) in linear MHD models are sufficient to alter the deuterium fractionation ratios of several molecules over distances of 0.02–0.12 pc, depending upon the fractional ionization (and density) of the gas. The theory predicts that R(XD) ($=n(XD)/n(XH)$) is inversely proportional to Δv, at least on the scale of the ion-neutral collision length, and tentatively suggests that superthermal linewidths and the observed small-scale chemical abundance gradients in dark clouds (*e.g.* Olano *et al.* 1988; Swade & Schloerb 1992) could be connected. High resolution mapping of cold

Astrophysics and Space Science **224**: 443–444, 1995.

molecular clouds to seek spatial variations of $R(DCO^+)$ and $R(N_2D^+)$ on these scales offer the best prospect of detecting low-amplitude MHD waves by molecular spectroscopy.

To test this theory we have observed several molecules in dark clouds, as well as their deuterated forms, to measure relative spatial distributions of emission and linewidth (Butner et al. 1994). The results of this study are outlined and compared with the predictions.

2. Observations

In December 1993 we observed the TMC-1 core along the DCO^+/NH_3 ridge which runs SE $\longrightarrow$ NW. We used the NRAO 12m telescope and observed the J = 1–0 transitions of DCO^+, N_2D^+, $H^{13}CO^+$ and N_2H^+, as well as several transitions in deuterated neutral species such as DCN, DNC and C_3HD. A preliminary analysis of the data shows two trends. First, apart from N_2D^+, the linewidths and emission of the observed molecules both display approximately the same spatial variation along the ridge. Second, as one moves away from the 'DCO$^+$ peak', the N_2D^+/N_2H^+ ratio decreases and the N_2D^+ linewidth increases whereas that of N_2H^+ is almost unchanged. We also see that the DCO^+ line widens slightly as $DCO^+/H^{13}CO^+$ decreases, and as DCN decreases in intensity, it widens also.

The fact that DCO+ is less affected than N_2D^+ is consistent with the theory since deuterated ions other than H_2D^+ can form it whereas N_2D^+ is only deuterated by H_2D^+ (Millar et al. 1989). Thus, in TMC-1 most of the H_2D^+ present is at the 'DCO$^+$ peak', where linewidths are narrowest; it is destroyed by the effects of ambipolar diffusion as one moves down the ridge. More observations are required before claiming a definitive confirmation of the theory, but the fact its key predictions appear to have been borne out is encouraging.

References

Butner, H. et al. 1994, in preparation.
Charnley, S.B. & Roberge, W.G. 1992 in Astrochemistry of Cosmic Phenomena, Ed. P.D. Singh, (Kluwer), p155.
Charnley, S.B. & Roberge, W.G. 1994, in preparation.
Millar, T.J., Bennett, A. and Herbst, E. 1989, Ap.J **340** 906.
Olano, C.A., et al. 1988, A & A, **196**, 194.
Swade, D.A. & Schloerb, F.P. 1992, Ap.J, **392**, 543.
Zweibel, E.G. & Josafatsson, K. 1983, Ap.J **270** 511.

VLBI STUDY OF THE SIO MASER EMISSION AT 43 GHZ IN THE CIRCUMSTELLAR ENVELOPE OF R CAS

FRANCISCO COLOMER*

Centro Astronómico de Yebes. Apartado 148. E - 19080 Guadalajara. Spain

Abstract. We discuss the European effort in the study of millimeter–wave circumstellar masers with very long baseline interferometry. We show some results of the first observations, and discuss present and future work.

Key words: Stars: Radio Interferometry — Masers: SiO — stars: individual: R Cas

We have undertaken the study of the silicon monoxide (SiO) maser emission in the circumstellar envelopes of late–type stars with European VLBI telescopes since 1990. We have obtained the first detection of compact SiO maser features with interferometric baselines up to 1750 km, providing fringe spacings as small as $0\rlap{.}''0004$ (Colomer *et al.* 1992).

We observed the Mira star R Cas for 6 hours with the Onsala–Effelsberg baseline. The fringe-rate map we obtain in Figure 1 is consistent with previous observations of this source (Lane *et al.* 1982; McIntosh *et al.* 1989). We clearly distinguish four regions of emission, distributed within 40 mas. Amplitude visibilities for this source yield Gaussian sizes in the range $1-2$ mas (4 to $8 \cdot 10^{12}$ cm at 270 pc), demonstrating the existence of structure at milliarcsecond level.

We have performed a new set of observations with 14 VLBI telescopes (the VLBA telescopes in the USA, Onsala, Effelsberg, Yebes, and the IRAM 30–m on top of Pico Veleta) to be processed soon at the VLBA correlator. We expect to produce high quality results thanks to the much improved u–v coverage obtained.

Acknowledgements

This project is being developed in cooperation with D.Graham, T.Krichbaum, A.Witzel, R.Booth, B.Rönnäng, P.de–Vicente, A.Barcia, J.Gómez–González, V.Bujarrabal, J.Alcolea, A.Baudry, and N.Brouillet. It would not have been possible without the expertise and help of the staff crews at all participating observatories.

* Previously at Onsala Space Observatory. S - 439 92 Onsala. Sweden

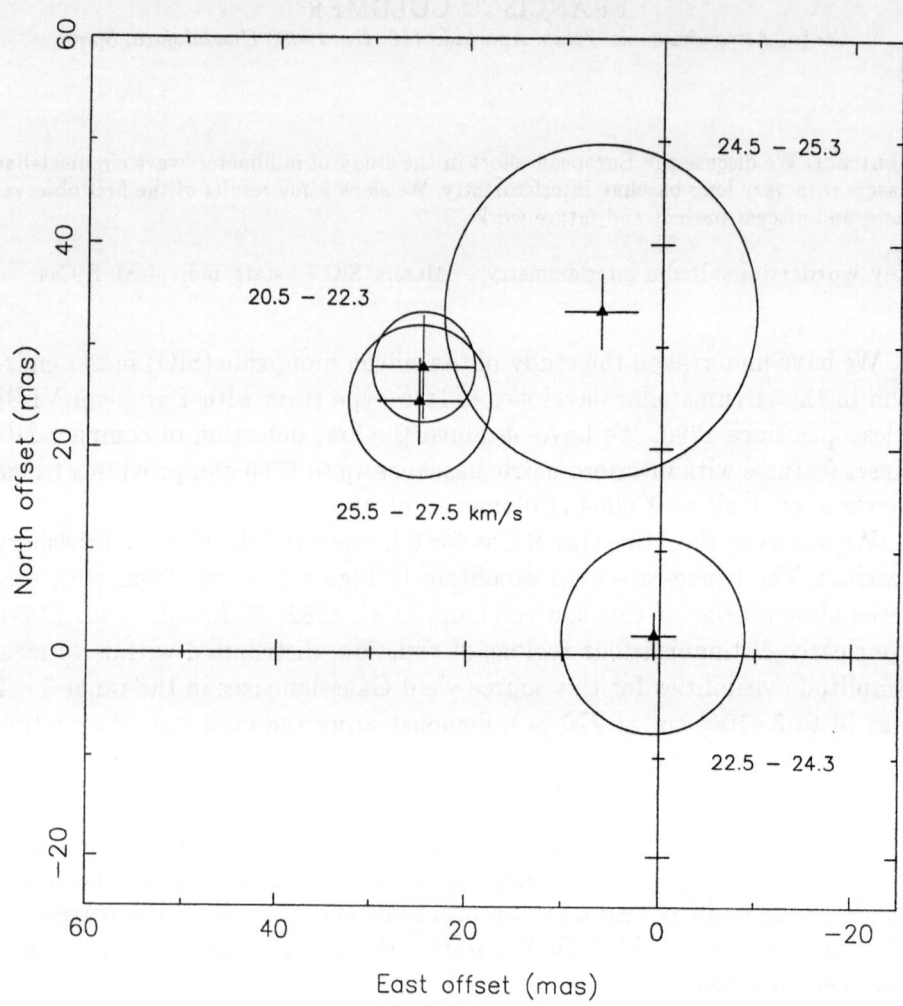

Fig. 1.   Fringe-rate map of the SiO masers in R Cas.

References

Colomer, F., et al. 1992, *A&A* **254**, L17.

Lane, A.P., 1982, *Observations of SiO masers in the circumstellar envelopes of late-type stars, PhD Thesis*, University of Massachusetts.

McIntosh, G.C., Predmore, C.R., Moran, J.M., Greenhill, L.J., Rogers, A.E.E., Barvains, R., 1989, *Ap J* **337**, 934.

INNER RADII OF DUST SHELLS AND STELLAR
DIAMETERS OBTAINED BY AN INFRARED STELLAR
INTERFEROMETER AT 11 MICRON WAVELENGTH

W.C. DANCHI, M. BESTER, L.J. GREENHILL; C.G. DEGIACOMI;**
N. GEIS, D. HALE;*** B. LOPEZ‡ and C.H. TOWNES
Space Sciences Laboratory, University of California, Berkeley, CA 94720-7450 USA

Abstract. Visibility data and analyses are discussed for 16 late-type stars observed with the Infrared Spatial Interferometer (ISI) of the University of California, Berkeley.

Key words: interferometry — mass loss — dust shells — stellar diameters — masers

Visibility data and analyses are discussed for 16 late-type stars observed with the Infrared Spatial Interferometer (ISI) of the University of California at Berkeley. The ISI is a two-element heterodyne interferometer operating in the mid-infrared (9–12 μm) and is located at Mt. Wilson. The present status of the instrument and current upgrades are discussed in Bester *et al.* (1994). Initial data and modeling of IRC +10216, α Ori, and o Cet were reported in Danchi *et al.* (1990) and Bester *et al.* (1991). Visibility data from the ISI were compared with visibility curves calculated from a radiative transfer model. A χ^2 fitting procedure was used to estimate the inner radii, optical depths, and temperatures at the inner radii of the dust shells. For stars in which the dust is completely resolved, estimates of the stellar diameter and temperature were also made.

An initial survey of 13 of 16 program stars including details of radiative transfer models was reported in Danchi *et al.* (1994a). An important conclusion from the survey is that the inner radii of the dust shells varies markedly with spectral type. Approximately half of the stars surveyed have dust far from their photospheres. Among these are the supergiants α Ori, α Sco, and α Her, which have a mean distance from the photosphere of 38 stellar radii (R_*). The S stars χ Cyg and W Aql have dust relatively far from the star but somewhat closer than the supergiants; the mean inner radius for these stars is 12 R_* from the photosphere. The remainder of the stars have dust less than 6 R_* from their photospheres, and these include the carbon star IRC +10216, the Mira variables R Leo, o Cet, and NML Tau, the supergiants VX Sgr and VY CMa, and the symbiotic star R Aqr. The mean distance between the inner radii of the dust shells and the photospheres of these stars

* Present Address: Center for Astrophysics, 60 Garden St., Cambridge, MA 02138 USA
** Universität zu Köln, I. Physikalisches Institut, D-50937, Köln, GERMANY
*** Physics Department, University of Pittsburgh, Pittsburgh, PA 15260 USA
‡ Observatoire de la Côte d'Azur, 06304, Nice cedex 4, FRANCE

Astrophysics and Space Science **224**: 447–448, 1995.
© 1995 *Kluwer Academic Publishers.*

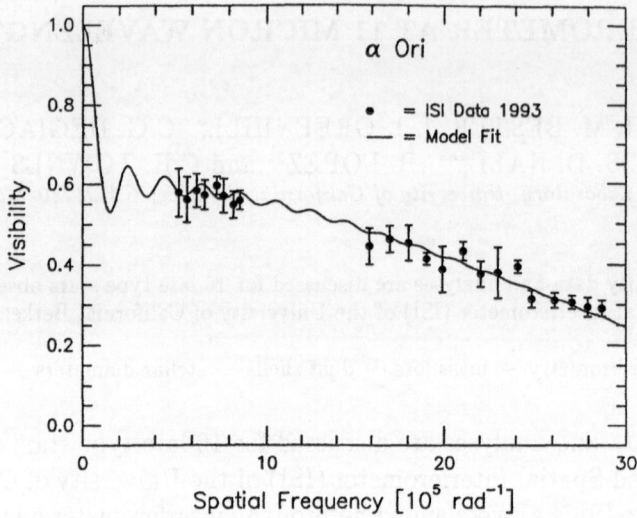

Fig. 1.   Visibility data and model for α Ori.

is 3.5±1.8 R_*. An anomaly is the oxygen-rich Mira variable, U Ori, which has an inner radius of 10 R_* from the photosphere, and which would be expected to have dust closer to the star.

Preliminary visibility data for the stars NML Cyg and IRC +10420 have recently been obtained using a 10 m baseline (Danchi *et al.* 1994b). In addition, data have been obtained using a 32 m baseline (Fall 1993 & Spring 1994) for α Ori, α Sco, o Cet, α Her, IRC +10216, R Leo, VY CMa, R Aqr, VX Sgr, χ Cyg, W Aql, W Hya, and IRC +10420. The visibility data for α Ori cover a sufficient range of spatial frequencies to make a determination of its diameter, which is 53±3 milli-arcsec (mas) (95% confidence limit, CL) (cf. Fig. 1). A preliminary estimate of the diameter of α Sco is 42±6 mas (95% CL). For the other stars the recent 32 m baseline data are compared with previous models and further constrains some of their parameters (cf. Danchi *et al.* 1994b).

References

Bester M., Danchi, W.C., Degiacomi, C.G., Townes, C.H., & Geballe, T.R. 1991, ApJ, 367, L27.
Bester, M., Danchi, W.C., Degiacomi, C.G., & Bratt, P.R. 1994, Proc. SPIE, 2200, 274.
Danchi, W. C., Bester, M., Degiacomi, C. G., McCullough, P. R., & Townes, C. H. 1990, ApJ, 359, L59.
Danchi, W.C., Bester, M., Degiacomi, C.G., Greenhill, L.J., & Townes, C.H. 1994a, AJ, 107, 1469.
Danchi, W.C., Bester, M., Greenhill, L.J., Degiacomi, C.G., & Townes, C.H. 1994b, Proc. SPIE, 2200, 286.

THE EQUATORIAL EXPANSION OF BE STARS
A THIN-LINE WIND?

F. X. DE ARAÚJO

Observatoire de la Côte d'Azur, B.P. 229 - 06304 Nice cedex 4, France
Observatório Nacional, R. Gal. Bruce 586, Rio de Janeiro, Brazil

Abstract. Numerical solutions for the expansion motion equation in the equatorial plane of a rotating early type Be star are presented. Decreasing terminal velocities are obtained as lower values of the radiative parameter α are used.

Key words: : Stars:Be – circumstellar matter –hydrodynamics

1. Introduction

Be stars are supposed to have a two-component envelope: a fast and not very dense wind and a much more dense and slowly expanding equatorial "disc". Radiative models can describe quite well the polar physical characteristics but not the equatorial ones. In particular the terminal expansion velocities computed from the models are always much higher than those estimated from observations. In order to solve this discrepancy we have considered an equatorial expansion driven mainly (or uniquely !) by optically thin lines. This is done by decreasing the value of the radiative parameter α, which gives the relative contribution of optically thick and optically thin lines. If $\alpha = 0$ is adopted the radiative acceleration has a simple r^{-2} functional dependence. It leads to a momentum equation that is similar to the well known solar wind equation.

2. Stellar and wind parameters

It was assumed $M = 15 M_\odot$, $R = 10 R_\odot$, $T_{eff} = 25000 K$ and a rotation rate which is 2/3 of critical ("break-up") velocity. Concerning the envelope itself we have adopted $T_{env} = 0.8 T_{eff}$, Keplerian rotation and a boundary condition given by the density at the basis of the wind (r = R) $\rho_0 = 1.0\, 10^{-11} g cm^{-3}$.

3. Results

Our computations have shown that a shallowing expansion is obtained as lower α_s are used (see Fig. 1). Moreover, this result is independent of stellar and wind parameters as rotation rate, photospheric density, envelope temperature etc. On the other hand reasonable mass loss rates require large values for the radiative parameter k.

Astrophysics and Space Science **224**: 449–450, 1995.
© *1995 Kluwer Academic Publishers.*

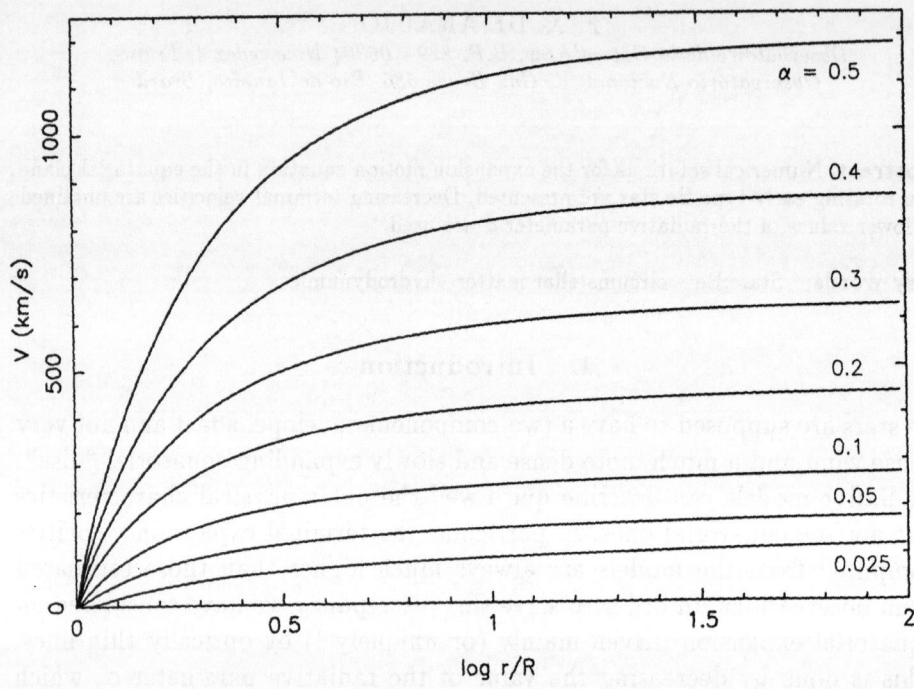

Fig. 1.   Expansion velocity for several values of α

4.   Conclusions

It seems that a radiatively force due basically to optically thin lines may lead to to an outflow with the desired properties, but a indeed large number of these lines is need. Whether or not this is the major mechanism acting in the equatorial plane of Be stars remains to be seen. More sophisticated models should be developed: the radiative force must be self-consistently calculated rather than simply adopting values for the parameters α and k.

THE EXTENDED DUST EMISSION AROUND THE NOVA
GK PERSEI

S.M. DOUGHERTY
Liverpool John Moores University, UK

L.B.F.M. WATERS
University of Amsterdam, The Netherlands
SRON, Groningen

M.F. BODE
Liverpool John Moores University, UK

and

D.J.M. KESTER and TJ.R. BONTEKOE
SRON Groningen, The Netherlands

Abstract. We have applied maximum entropy reconstruction methods to the IRAS observations of the nova GK Persei to examine the spatial distribution of the far-IR emission.
We have discovered discrete regions of emission in a co-linear structure extending to 17
arcmin on either side of the binary system, supporting a stellar origin for the structure.
We postulate that the evolved secondary is the progenitor of the circumbinary envelope.

The classical nova GK Per is one of the more unusual cataclysmic binaries:
it has the longest known orbital period for a CV (1.997 days), and it contains
a white dwarf with an evolved K0–2 sub-giant secondary. Most remarkably, it
exhibits an extended region of far-IR emission, shown in the co-added maps
of the IRAS Sky Survey Atlas. The IRAS data contain information at higher
resolution than presented in the co-added maps of the IRAS Sky Survey
Atlas. The higher intrinsic resolution and the multi-scan observing strategy
allow image construction with a resolution approaching the diffraction limit
of the telescope, which is 1 arcmin at 60 μm and 1.7 arcmins at 100μm
(Bontekoe *et al.* 1991). The 100μm maximum entropy reconstruction shown
in Figure 1 was obtained with the HIRAS software package developed at
SRON Groningen (Bontekoe, Koper & Kester 1994).

The maximum entropy reconstructions show that the emission at both
60 and 100μm is resolved into several discrete emission regions that form
an almost linear feature extending $\sim$ 17 arcmins to the SE and NW of GK
Per, which is located in a saddle at the centre of this feature (Dougherty *et
al.* 1994). The discrete regions are evident in both bands on each side of the
nova, with their positions being remarkably symmetric about GK Per. The
highly symmetric distribution of the emission features about the position of
GK Per is very unlikely in the context of superimposed interstellar cirrus,
and strongly supports a stellar origin for the far-IR emitting matter around
GK Per. In addition, observations of ^{12}CO line emission within 2 arcmins of

Astrophysics and Space Science **224**: 451–452, 1995.

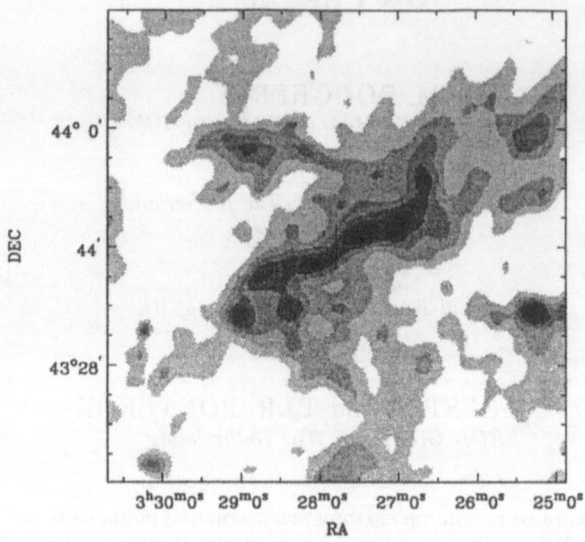

Fig. 1. The maximum entropy reconstruction of the 100μm IRAS data. The cross denotes
the position of the CV.

GK Per support a stellar origin for the molecular gas (Scott *et al.* 1994).

Bode *et al.* (1987) propose that the emission is from the remnant of a
PN. However, there are a number of issues that require addressing in this
model, in particular the age of the envelope. The secondary is of low mass
(~ 0.5 M$_\odot$) and has evolved to a sub-giant. This evolutionary state gives a
minimum age of the binary at $\sim$ a few 10^{10} years. The 1 M$_\odot$WD primary
had a main sequence progenitor of $6-7$ M$_\odot$ (Weidemann 1992), which would
have evolved through the AGB to a WD in $\sim 10^8$ years. Since the dynamical
age of the IR emitting envelope is $\leq 10^6$ years (Dougherty *et al.* 1994), the
stellar envelope that formed the AGB star associated with WD could not
have been the source of the circumbinary envelope. We conclude that the
secondary is the source of the material in the recently ejected extended
envelope.

References

Bode M.F., Seaquist E.R., Frail D.A., Roberts J.A., Whittet D.C.B., Evans A., Albinson
    J.S. 1987, Nature, 329, 519
Dougherty S.M., Waters L.B.F.M., Bode M.F., Kester D.J.M., Bontekoe Tj. R., 1994, in
    preparation
Bontekoe Tj. R., Koper E., Kester D.J.M. 1994, Astron. & Astrophys. , 284, 1037
Scott A.D., Rawlings J.M.C., Evans A. 1994, Mon. Not. of the RAS ; in press
Weidemann V., 1992, in *Proceedings of the Second ESO/CTIO workshop on Mass loss
    on the AGB and Beyond*, La Serena, Chile, June 1992, ESO Conference/Workshop
    Proceedings No. 46, ESO, Garching, p. 55

HIGH RESOLUTION RADIO IMAGES OF THE SYMBIOTIC SYSTEM R AQUARII

S.M. DOUGHERTY, M.F. BODE and H.M. LLOYD
Liverpool John Moores University

and

R.J. DAVIS and S.P. EYRES
University of Manchester, NRAL

Abstract. We present MERLIN images of the symbiotic system R Aquarii obtained at 1.7 and 5 GHz. We identify the emission from the binary system and derive a mass loss rate for the Mira that is commensurate with that in typical Miras. We show that variations in the radio emission over the last decade are consistent with the jet model for the system originally presented by Solf (1992). In addition, we calculate the spectral index distribution and find that there is no evidence of non-thermal emission.

R Aquarii is a D-type symbiotic containing a M7 Mira long period variable (LPV) and a suspected white dwarf companion. The region within $10''$ of the LPV has been observed extensively at optical and radio wavelengths, revealing several discrete emission regions in a "jet-like" structure. Several models have been advanced to explain these features, and it is widely thought that the discrete features are condensations associated with an expanding nebula which have been shock excited by a highly collimated, supersonic wind ($\sim 300 - 500$ km s^{-1}) from the binary system (*e.g.* Solf 1992).

We observed R Aquarii at 5 and 1.7 GHz using the Nuffield Radio Astronomy Laboratory MERLIN array. The final synthesized images are shown in Fig. 1. Paresce & Hack (1994) have established the position of the LPV in R Aquarii using HST images at 550 nm and 190 nm. Comparison of Fig. 1a with the HST images show that feature C1c is associated with the LPV. We can estimate the mass loss rate from the LPV, free of contamination by emission from other nearby emitting features, namely C1a and C1b. To allow for the possibility that the Mira wind is partially ionized, we use the model of Seaquist, Taylor & Button (1984) for the radio emission in symbiotic stars. We have solved the radiative transfer equation for the free-free emitting, ionized region around the hot component and find that the observed 5 GHz flux from the binary system implies a mass loss rate from the LPV of $\sim 5 - 15 \times 10^{-7}$ M$_\odot$ yr^{-1}, similar to that expected from typical Miras.

The emission in features C1 and C2 have positive spectral indices, indicative of partially optically thick thermal free-free emission, an interpretation which is supported by the brightness temperatures of the features being $\sim 10^3 - 10^4$ K, typical of free-free emitting regions. The spectral index of A′ is 0.12 ± 0.34, consistent with optically thin bremsstrahlung. The spectral

Astrophysics and Space Science **224**: 453–454, 1995.

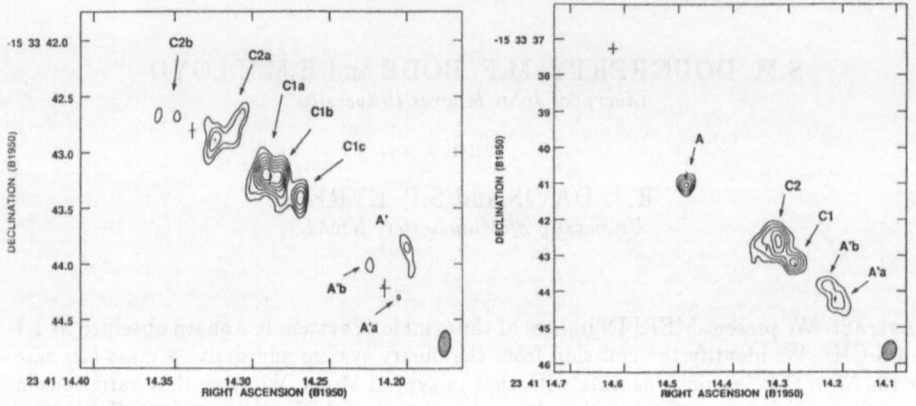

Fig. 1.   a) C band and b) L band image of R Aquarii. Synthesized beam sizes are 0.24×0.11 arcseconds and 0.55×0.38 arcseconds respectively. Crosses indicate the positions of previously identified radio features C1, C2, A, A' and B, deduced from the proper motions of Lehto & Johnson (1992).

index value for feature A is more difficult to establish. Since A is resolved out at 5 GHz in these observations, we have a lower limit to the total flux at 5 GHz of ∼ 1.6 mJy from Lehto & Johnson (1992). This gives a lower limit to the spectral index of 0.18 ± 0.13, suggesting partially optically thick thermal emission. Our observations show that the 5 GHz flux of A' has increased by a factor of 17 ± 3 over a decade. This flux increase could be due to cooling. The derived densities and temperatures imply a cooling time of the order of tens of days (Dougherty *et al.* 1994). Thus, changes in flux could occur on very short timescales but are probably dominated by the timescale governing whichever process is responsible for the heating. If this is due to shock heating in a hydrodynamic interaction with an outflow from the central binary, as proposed by Solf (1992), then a ten year timescale for variation would imply a lower limit to the outflow velocity of ∼ 300 km s^{-1} (based on the size of A' derived from these observations). Solf suggests that the "jet" components are shock heated by a collimated wind at a velocity of 300 − 500 km s^{-1}, so any variation in the conditions in this wind will naturally produce flux variations on time-scales of a few to ten years.

References

Dougherty S.M., Bode M.F., Lloyd H.M., Davis R.J., Eyres S.P, 1994, Mon. Not. of the RAS , submitted
Lehto H.J., Johnson D.R.H. 1992, Nature, 355, 705
Paresce F., Hack W. 1994, Astron. & Astrophys. , 287, 154 (PH)
Seaquist E.R., Taylor A.R., Button S. 1984, Astrophys. J. , 284, 202
Solf J. 1992, Astron. & Astrophys. , 257, 228

A FLARE IN THE MIRA X OPH

S. ETOKA and A.M. LE SQUEREN

Observatoire de Meudon F-92195 Meudon Principal cedex, France.

Abstract. Before the observation of the 1974 U Ori eruption, it was considered that the Mira stars had only some regular OH variations. With this eruption, we realized that sometimes flares can occur in this type of star. In the course of an OH Mira star monitoring programme with the Nançay radio telescope, we have discovered a new eruptive type of OH maser emission in several sources. Especially, in early 1992, we observed a quickly rising 1665 Mhz emission in the Mira X Oph. The main characteristics of this flare were: large flux variations independent of the light curve; large degree of circular polarization; radial velocity emission close to the stellar velocity.

Key words: Flare – stars : individual : X Oph

X Oph is a cool oxygen-rich M type Mira star of 334 days period. It's a binary star; its companion is located at 64 AU with a systemic period of about 310 years according to Fernie (1959). This Mira has a weak [25-12] color, therefore a thin envelope. From 1978 to 1992 only a weak fleeting 1667 Mhz feature was detected but suddenly, in 1992 May, X Oph flared up. Strong 1665 Mhz emission arose in less than four months, followed by a slow decline lasting about 1.5 years. The decline was in phase with the light curve with a degree of polarization and an amplitude slowly declining in time as a damped oscillator. But no important variations were seen in the light curve before or during the flare. Taken as a whole, the decaying of the emission is associated with a drift in velocity.

Frequency=1665 Mhz   polarization=left and right   date=92.06.28   source=X OPH

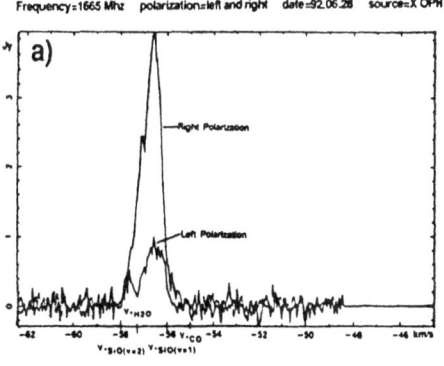 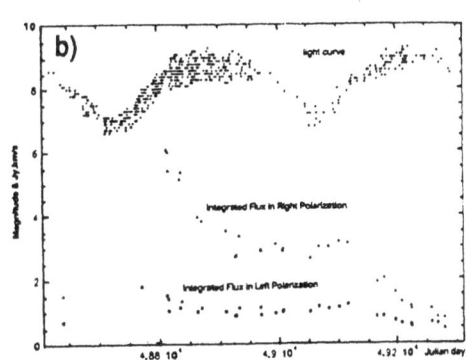

Fig. 1.   **a)** Spectra at 1665 Mhz in left and Right polarization. Velocities as estimated from CO line (Young, K., private communication, 1994), as well as from H_2O (Bowers & Hagen, 1984) and from SiO (Bujarrabal *et al.*, 1987) lines are displayed. **b)** Integrated flux at 1665 Mhz in left and right polarization with the corresponding part of the light curve (Mattei, J., 1993 and Schweitzer, E., 1994, private communications).

The spectrum displays a rather narrow emission spread over about 2.5

km/s. It consists of a group of blended multi-components. It has a strong degree of polarization, slowly decaying during time. The flare emission is located near the stellar velocity as estimated from CO line (-55 ± 0.5 km/s, Young, K., private communication, 1994), likely coming from the front side of the shell. The extinction of several components took place from the negative velocity values toward the positive ones. Thus this extinction starts in the farthest part of the shell.

U Ori underwent an OH flare at 1612 Mhz in May 1974 (Reid *et al.*, 1977). This flare lasted more than 11 years (it was still emitting in May 1985 (Bowers & Johnston, 1988)). The similarities with the X Oph eruption are the following : OH variations were in phase with the light curve (Cimerman, 1979) slowly decaying in strength (Cimerman, 1979; Jewell *et al.*, 1979). Small IR radiation excess indicating the existence of a thin dust shell (Cimerman, 1979). The emission seemed to originate from the front part of the shell (Reid *et al.*, 1979). OH masers coming from a shell located at about 5.10^{14} cm from the star, i.e. 3 or 10 time shorter than the standard distance (Fix *et al.*, 1980). Changes in spectral shape and in the velocity emission were observed (Jewell *et al.*, 1981, Chapman & Cohen, 1985).

Flare up events are rather rare phenomena. We have observed only six in the course of a Nançay monitoring concerning about 40 Mira stars started 10 years ago. These events take place in the inner part of the Mira envelope, near the grain condensation zone. They are probably triggered by a perturbation coming from the deep part of the star and are likely a signature of a transition in the stellar evolution. From our OH observations we shall attempt to elaborate a model, but in our opinion, an interferometric map as well as infrared photometry especially at 2.8 μm would be of great help to study in more detail this type of events.

References

Bowers, P. F., Hagen, W.: 1984, *ApJ* **285**, 637
Bowers, P. F., Johnston, K. J.: 1988, *ApJ* **330**, 339
Bujarrabal, V., Planesas, P., del Romero, A.: 1987, *A&A* **175**, 164
Chapman, J. D., Cohen, R. J.: 1985, *MNRAS* **212**, 375
Cimerman, M.: 1979, *ApJ (letters)* **228**, L79
Fernie, J. D.: 1959, *publication of the Goethe Link Observatory, Indiana Univ.* **33**, 611
Fix, J. D., Mutel, R. L., Benson, J. M., Claussen, M. L.: 1980, *ApJ (letters)* **241**, L95
Jewell,P. R., Elitzur, M., Webber, J. C., Snyder, L. E.: 1979, *ApJ Suppl.* **41**, 191
Jewell,P. R., Webber, J. C., Snyder, L. E.: 1981, *ApJ* **249**, 118
Reid, M. j., Muhleman, D. O., Moran, J. M., Johnston, K. J., Schwartz, P. R.: 1977, *ApJ* **214**, 60
Reid, M. J., Moran, J. M., Leach, R. W., Ball, J. A., Johnston, K. J., Spencer, J. H., Swenson, G. W.: 1979, *ApJ (letters)* **227**, L92

DIRECT OBSERVATIONS OF BIPOLAR OUTFLOW FROM HM SAGITTAE

S.P.S. EYRES, R.J. COHEN and R.J. DAVIS
University of Manchester, NRAL, Jodrell Bank, Macclesfield, Cheshire, SK11 9DL, England

H.T. KENNY
Royal Military College of Canada, Kingston, Ontario K7K 5LO, Canada

and

H.M. LLOYD, M.F. BODE and S.M. DOUGHERTY
School of Chemical and Physical Sciences, Liverpool John Moores University, Byrom Street, Liverpool, L3 3AF, England

Abstract. We observed HM Sagittae with the *Multi-Element Radio Linked Interferometer Network* (MERLIN*) at 6 cm and 18cm. We find non-thermal bipolar outflow in the east-west direction, associated with optical emission lines, and thermal ridges to the north and south associated with the UV nebulosity detected by the HST.

Key words: binaries: symbiotic — radio continuum: stars — stars: individual: HM Sge.

1. Observations

HM Sagittae is a symbiotic novae, which underwent an optical outburst of over 6 magnitudes in 1975 has since developed into the strongest radio-emitting symbiotic, with an extensive and evolving circumstellar structure.

We observed HM Sagittae at both 6 cm and 18 cm with MERLIN. The former observations were taken in December 1992, while the latter were made in October 1993. At both observing wavelengths, we have resolved new structure (see Fig. 1). At 6 cm, we resolve the double-peaked structure seen by Kwok, Bignell & Purton (1984) into the two ridges N and S, to the north and south of a central minimum. The ridges have peak brightness temperatures of around 45,000 K, which exceeds that expected for radiatively excited hydrogen plasma. The later 18 cm map is elongated in the E-W direction, and has a far greater extent than the 6 cm structure. Radio spectral indices calculated using our 18 cm image, and a VLA 6 cm image from Kenny *et al.* (1993) are consistent with non-thermal emission from the features E and W.

---

* MERLIN is a national facility operated by the University of Manchester on behalf of PPARC.

Astrophysics and Space Science **224**: 457–458, 1995.

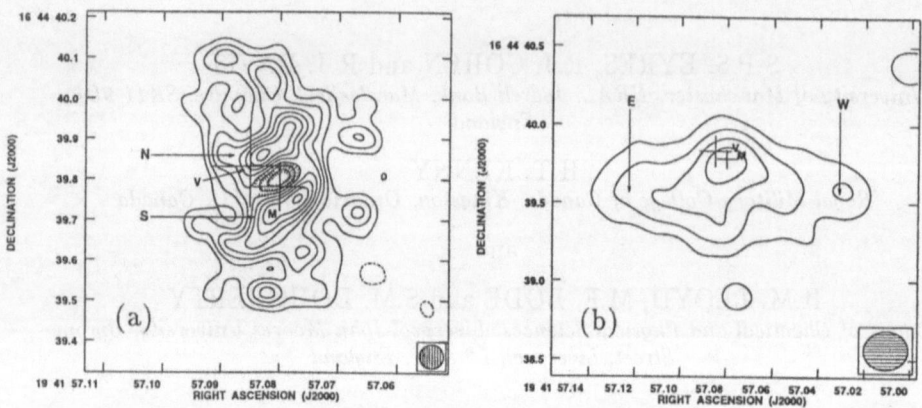

Fig. 1.   Radio maps of HM Sagittae made using MERLIN. (left) At 6 cm, the contours
are −3, 3, 6, 9, 12, 15, 18, 21, 24 times the rms noise of 90 μJy beam^{-1}. The ridges are
marked N and S. (right) At 18 cm, the contours are −3, 3, 6, 9, 12 times the rms noise of
240 μJy beam^{-1}. The approximate positions of the non-thermal emission are marked E
and W. M is the mean optical position, and V the position of the VLA 6 cm radio peak,
from Kenny *et al.* (1993). The size of each cross indicates the 1σ error.

2.   Correlations with other Wavelengths.

Of particular interest are the correlations between our observations and
observations at other wavelengths. Hack & Paresce (1993) obtained an ultraviolet image with the HST only 3 weeks before our 6 cm observations. Features N and S are clearly present, with the only major discrepancy being the
central peak in the UV image, which we attribute to the presence of a white
dwarf in the system. Optical spectroscopy carried out by Solf (1984, 1993)
has allowed the determination of the approximate positions of five separate
radial velocity features. The highest velocity components correspond spatially to features E and W at 18 cm. This indicates that we have directly
observed bipolar outflow from this source.

References

Dokuchaeva O. D., 1976. *Inf. Bull. Variable Stars*, No. 1189
Hack W. J., Paresce F., 1993, *PASP*, **105**, 1273
Kenny H. T., Taylor A. R., Davis R. J., Pavelin P. E., Bode M. F., Bang M., 1993: in
    Sub-arcsecond Radio Astronomy ed. Davis R. J., Booth R. S., (Cambridge: Cambridge
    Univ. Press), p. 18
Kwok S., Bignell R. C., Purton C. R., 1984, *ApJ*,**279**, 188
Lloyd H. M., O'Brien T. J., Bode M. F., Kahn F. D., 1993, *MNRAS*, **265**, 457
Lloyd H. M., Bode M. F., O'Brien T. J., Kahn F. D., 1994, in *Kinematics and dynamics
    of diffuse astrophysical media* ed. Dyson J. E., Carling E. B., (Kluwer:Dordrecht)
Solf J., 1984, *A & A*, **139**, 296
Solf J. 1993. Max-Planck-Institut für Astronomie Preprint, to appear in *Stellar and
    Circumstellar Astrophysics*, Astronomical Society of the Pacific Conference Series.

STRUCTURE AND EVOLUTION OF CIRCUMSTELLAR
SHELLS FORMED BY MASSIVE STAR WINDS

GUILLERMO GARCÍA-SEGURA
University of Illinois at Urbana-Champaign
and Instituto de Astrofísica de Canarias

MORDECAI-MARK MAC LOW
University of Chicago
and University of Illinois at Urbana-Champaign

and

NORBERT LANGER
Max-Planck-Institut für Astrophysik, Garching

Abstract. We follow the interaction of massive stars with their circumstellar gas over their entire life-times by combining hydrodynamic stellar evolution calculations for 35 and 60 $M_\odot$ stars and one- and two-dimensional gas dynamical calculations for the circumstellar medium.

Key words: massive stars — mass loss — LBV-nebulae — WR nebulae

Circumstellar structures around a 60 $M_\odot$ star. The computed 60 $M_\odot$ model, which loses 56 $M_\odot$ in the form of stellar winds during the evolution, evolves through a Luminous Blue Variable (LBV) stage into a Wolf-Rayet (WR) star (*cf.* Langer *et al.* 1994). Two observable circumstellar shells form, the first during the LBV stage, and the second at the beginning of the WR phase. The total mass of the LBV wind is 8 $M_\odot$, and the succeeding fast WR wind captures 75 % of this mass (6 $M_\odot$) in the form of a ring nebula. We obtain a large number of growing radial filaments in the LBV ring nebula, with an almost steady angular spacing. They are produced by the ram–ram pressure instability referred to as the "nonlinear thin shell instability" in Vishniac (1994), which allows clumps to be formed with shorter angular wavelengths than the Vishniac (1983) instability. Comet-like tails and jets appear as this instability becomes non-linear. Broad spectral lines with FWHM of 40–50 % of the expansion velocity — far greater than thermal widths — are predicted in LBV nebulae. Our model is in excellent agreement with HST observations of the nebula surrounding the LBV AG Carinae (Nota 1994). P Cygni is another direct test of our computations since this LBV lies on our 60 $M_\odot$ evolutionary track (*cf.* Langer *et al.* 1994). We successfully predict its recently discovered fragmented circumstellar nebula (Clampin *et al.* 1994). There is no circumstellar shell at the end of the star's life, except the large shell of interstellar material swept up by the main sequence wind.

Astrophysics and Space Science **224**: 459–460, 1995.

Circumstellar structures around a 35 M$_\odot$ star. We have also computed the circumstellar structure produced by a star of 35 M$_\odot$, using the same method. This star evolves through a long-lived red supergiant phase before it transforms into a WR star. A filamentary ring nebula is produced when the WR wind hits the red supergiant wind shell. Our results support the Three-Wind Model of García-Segura & Mac Low (1994). We find that pure Rayleigh-Taylor instabilities in thick shells do not form confined clumps. For clump formation to occur, shells must be thin. Otherwise the gas cannot move transversely to create clumps via the Vishniac instability. The star WR 136 fits to our evolutionary track. The ring nebula NGC 6888 surrounding this star agrees well with the computed nebula if we assume a red supergiant wind velocity of ~15 km s^{-1}. The ring nebula Sh 308 (Chu *et al.* 1982) and its central star WR 6 also agree with our model, if the preceding red supergiant wind had a moderate velocity of < 75 km s^{-1}. The H II regions Sh 303 and Sh 304 are interpreted as parts of the fossil shell of the main-sequence bubble swept up by WR 6. The quasistationary flocculi of Cas A can be understood as clumps formed in the WR shell before the supernova explosion. However, the clumps in our 35 M$_\odot$ model have moved out to a radius of 20 pc at the time of the supernova explosion, suggesting that the progenitor of Cas A must have had a smaller mass and so a smaller shell at the time of explosion.

We thank M. L. Norman and the Laboratory for Computational Astrophysics for the use of ZEUS-3D. The calculations were performed on a Cray Y-MP C90 at the Pittsburgh Supercomputing Center, and visualized at the National Center for Supercomputing Applications. This work was partially supported by NASA grants NAG5-2245 and NAGW-2379, and by the Deutsche Forschungsgemeinschaft through grant La 587/8-1.

References

Chu Y.-H., Gull T.R., Treffers R.R., Kwitter K.B., Troland T.H.: 1982, *ApJ*, **254**, 562
Clampin et al. : 1994, *ApJ*, in press
García-Segura G., Mac Low, M.-M.: 1994, in: *Circumstellar Media in the Late Stages of Stellar Evolution* R.E.S. Clegg, P. Meikle, & I. Stevens, eds., Cambridge University Press: Cambridge, p.85
Langer N., Hamann W.-R., Lennon M., Najarro F., Pauldrach A., Puls, J.: 1994, *A&A*, **290**, 819
Nota, A. : 1994, *Science*, **264**, 1668
Vishniac, E. T.: 1983, *ApJ*, **274**, 152
Vishniac, E. T.: 1994, *ApJ*, **428**, 186

THE NEY-ALLEN NEBULA: AN UNAMBIGUOUS EXAMPLE OF CIRCUMSTELLAR DUST EMISSION

DANIEL Y. GEZARI

NASA/Goddard Space Flight Center, Code 685, Greenbelt, MD 20771

Abstract. Diffraction-limited array images of the Trapezium/Ney Allen infrared nebula have been obtained at six wavelengths between 7.8 and 12.4 microns, including the 9.7 micron silicate feature. Extended emission from warm dust shows significant differences in structure around each of the four Trapezium stars. The most dramatic infrared source is associated with θ^1 Orionis D, where the bright mid-infrared emission is found to be a distinct crescent-shaped ridge or shell, concentric with the O star. This unambiguous relationship between a known type stellar luminosity source and a distinct circumstellar dust cloud of known distance and dimensions provides a unique opportunity to test the predictions of dust grain emission models for circumstellar infrared sources.

Key words: Orion — circumstellar dust

Astrophysics and Space Science **224**: 461, 1995.
© 1995 *Kluwer Academic Publishers.*

4.8 – 20 MICRON IMAGING OF ORION BN/KL:
LUMINOSITY SOURCES AND THE ROLE OF IRC2

DANIEL Y. GEZARI

NASA/Goddard Space Flight Center, Code 685, Greenbelt, MD 20771

and

DANA E. BACKMAN

Dept. of Physics and Astronomy, Franklin and Marshall College, Lancaster, PA 17604

Abstract. Mid-infrared imaging photometry of the Orion BN/KL infrared cluster at eight wavelengths between 5 and 20 μm using a 58 x 62 pixel imaging array camera has revealed new compact sources and the large-scale structure of the region in diffraction-limited (1 arcsec) detail. Several new objects have been detected within a few arcsec of IRc2, widely thought to be the principal luminosity source for the entire BN/KL complex. Detailed color temperature and emission opacity images are derived from the 7.8, 12.4 and 20.0 μm observations, and the 9.8 μm image is used to derive an image of "silicate" dust extinction for the region. The color temperature, opacity, and extinction images show that IRc2 may not be the single dominant luminosity source for the BN/KL region; substantial contributions to the luminosity could be made by IRc7, BN, KL, and five new compact 10 μm sources detected within a few arcseconds of IRc2. We suggest that a luminous, early-type star near IRc2, which is associated with the compact radio source "I" and the Orion SiO maser, is the dominant luminosity source in the BN/KL region, hidden from view by cool dust material with at least Av $\sim$ 60 mag of visible extinction.

Key words: : Orion BN/KL — Orion IRc2

Astrophysics and Space Science **224**: 463, 1995.
© 1995 *Kluwer Academic Publishers.*

NGC 7027: ANALYSIS OF 7.8–20 MICRON CONTINUUM, SILICATE AND PAH GRAIN DISTRIBUTIONS

D. Y. GEZARI, M. D. THORNLEY AND F. VAROSI

NASA/Goddard Space Flight Center, Infrared Astrophysics Branch, Code 685, Greenbelt, MD 20771

Abstract. Images of mid-infrared (5 - 20 μm) circumstellar dust sources have been obtained with a new 58 x 62 pixel infrared array camera system. A seven-color imaging study of the bright planetary nebula NGC 7027 challenges the assertion that polycyclic aromatic hydrocarbons (PAH) may extend further from the center of the nebula than the continuum emission from silicate dust grains. It appears that the overall distributions are nearly identical, ruling out differences in intensity between the PAH emission and the general "silicate" dust material. A rigorous comparison between the infrared image data and new visible CCD images of NGC 7027 is made.

Key words: PAH grains — NGC 7027

Astrophysics and Space Science **224**: 465, 1995.
© 1995 *Kluwer Academic Publishers.*

HH24–26: STRUCTURE, DYNAMICS AND CHEMISTRY

A.G. GIBB AND L.T. LITTLE

Electronics Laboratory, University of Kent, Canterbury, Kent, UK. CT2 7NT
Email contact: agg@star.ukc.ac.uk

Abstract. New J=1-0 HCO$^+$ and J=2-1 C^{18}O observations of HH24–26 reveal striking differences between these and previous maps of higher-J transitions of HCO$^+$ and CS. The high-J HCO$^+$ and CS emission traces the densest portions of the cloud, while C^{18}O traces the more tenuous envelope. This is also evident in the velocity structure. HCO$^+$ and C^{18}O appear to be depleted from the gas phase by amounts which vary with position.

Key words: Molecular clouds — HH24–26 — abundances — structure

1. Introduction

HH24–26 is located within the Lynds 1630 molecular cloud at a distance of 400pc. HCO$^+$ J=4-3 emission (Gibb & Heaton 1993) shows that the region consists of a collinear array of at least six clumps, some of which have infrared sources associated with them, with at least two large-scale bipolar molecular outflows. A NE–SW temperature gradient is observed in ^{12}CO and CI.

2. New results

The CI spectra tend to be brighter and exhibit broader profiles towards HH24. This broadening suggests that the line is saturated. Near HH24 the peak intensities imply a temperature of 16K, while near HH26 the temperature is 11K. These values are lower than those derived from ^{12}CO indicating that the CI is tracing material deeper within the cloud, and not just at the cloud edge.

A C^{18}O J=2-1 map made at the JCMT (Figure 1a) shows a gradient in intensity in the same manner as the ^{12}CO and the CI. HCO$^+$ J=1-0 and 4-3 emission is shown in figure 1b. Despite the differences in beamsize (55 vs 14″) the transitions are clearly tracing different density regimes. The C^{18}O emission follows more closely the 1-0 HCO$^+$ rather than the 4-3 HCO$^+$. 1100μm continuum photometry at the clump positions reveals that the clumps are real entities.

A distinct, systematic north–south velocity gradient of magnitude 1.4 km s^{-1} pc^{-1} is apparent in the 4-3 HCO$^+$ data, but absent from the HCO$^+$ $1-0$ and C^{18}O 2-1.

Assuming excitation temperatures of typically 10–13K, derived from ammonia (1,1) and (2,2) observations reinforced by estimates based on the

Astrophysics and Space Science **224**: 467–468, 1995.
© 1995 *Kluwer Academic Publishers.*

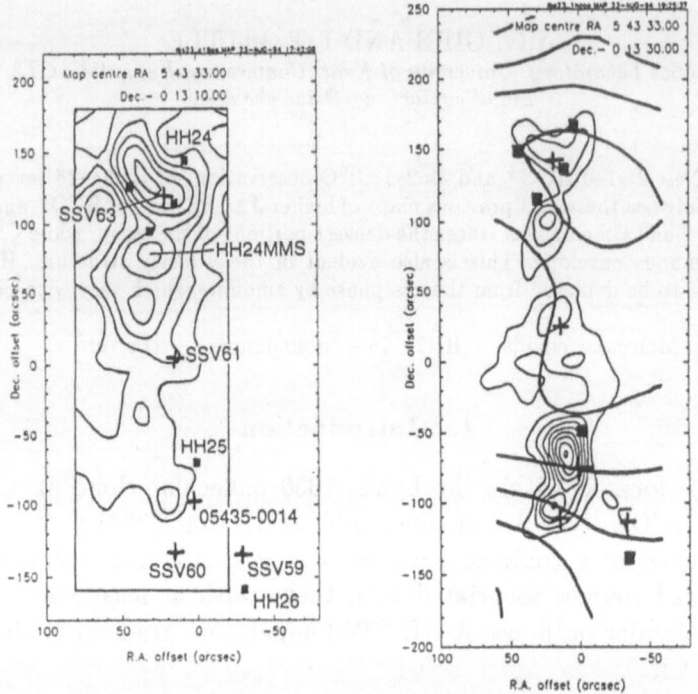

Fig. 1.   a) $C^{18}O$ 2–1 emission b) HCO^+ 1–0 emission (bold contours) with 4–3 contours overlaid.

HCO^+ 4–3 line brightness, LTE column densities for HCO^+ and $C^{18}O$ have been calculated at the location of all the clumps and some intermediate positions. In addition, the H_2 column density has been calculated from the $1100\mu m$ data as in Hildebrand (1983), assuming a uniform sphere of radius equal to that seen in 4–3 HCO^+ emission. Comparison of these column densities show that the HCO^+ and the $C^{18}O$ are invariably underabundant by factors of up to 100.

Acknowledgements

We wish to acknowledge Dr. K. Mattila and coworkers at the University of Helsinki, Finland for the allocation of time at Metsähovi.

References

Hildebrand R.H. (1983) *QJRAS* **24**, 267
Gibb A.G. and Heaton B.D. (1993) *A&A* **276**, 511

THE SIO MASERS AND DUST SHELL OF VX SGR

L.J. GREENHILL, F. COLOMER* and J.M. MORAN
Harvard-Smithsonian Center for Astrophysics
60 Garden St, Cambridge, MA 02138 USA

D.C. BACKER
Department of Astronomy, University of California
Berkeley, CA 94720 USA

and

W.C. DANCHI and M. BESTER
Space Sciences Laboratory, University of California
Berkeley, CA 94720 USA

Abstract. The circumstellar envelope of the late-type supergiant VX Sagittarii has been studied with VLBI observations of SiO maser emission ($J = 1 \rightarrow 0$ $v = 1$) and interferometric observations of mid-infrared dust emission.

Key words: VX Sgr — interferometry — masers — mass loss — dust shells

Observations with the Very Long Baseline Array (VLBA) shows that the SiO ($J = 1 \rightarrow 0$ $v = 1$) maser emission lies in the extended stellar atmosphere of the semi-regular variable star VX Sagittarii, projected in a ring-like region of radius 1.4 R_* (Fig. 1). The measurement of the stellar radius is discussed below. Ring-like distributions have also been observed for TX Cam, U Her, and W Hya (Diamond *et al.* 1994, Miyoshi *et al.* 1994). The velocity field of the emission in VX Sgr is asymmetric, and there are no systematic gradients (Greenhill *et al.* 1995). The velocity structure may be consistent with turbulence of a few km s^{-1}, which is on the order of both the sound speed and the Alfvén speed. The diameters of the masers that lie along the limb of the star are $\sim$ 0.4 AU (0.02 R_*), for an adopted distance of 1.7 kpc (Humphreys, Strecker & Ney 1972). The fraction of the total power that is imaged is substantially less than unity for all spectral features, which suggests that some emission arises on scales of at least $\sim$ 30 AU). The diameter of the inner edge of the dust shell is probably the upper limit to the angular scale on which SiO maser emission can occur.

Emission at 11.15 μm from the dust shell has been resolved by the U. C. Berkeley Infrared Spatial Interferometer (Danchi *et al.* 1994, Greenhill *et al.* 1995). The dust emission lies well outside the radius of the maser shell. We estimate dust shell parameters by matching the visibility function predicted by a radiative-transfer model to the observed visibility function. At an optical phase of about 0.9, the stellar radius is 12 ± 2 mas and the inner radius of the shell is 6.6 R_*. At phase 0.4, which corresponds to the time of

* Current address: Centro Astronómico de Yebes, Apartado 148, 19080 Guadalajara, Spain

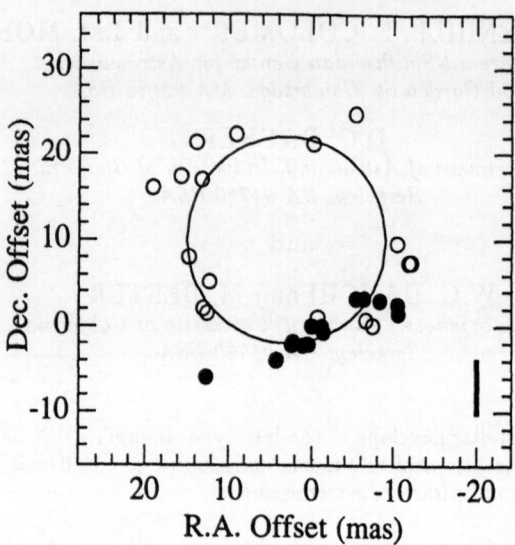

Fig. 1. Ring-like distribution of maser features around VX Sgr. Unfilled circles represent emission within 4.5 km s^{-1} of the stellar velocity of 5.3 km s^{-1} (Chapman & Cohen 1986). The disk of the star is shown and is assumed (in the absence of precise optical and radio astrometry) to be roughly centered within the maser ring. The scale bar corresponds to 10 AU.

the VLBA observations, the inner radius is 4.6 R$_*$ for the same stellar radius. For an exponential stellar atmosphere, the location of the maser shell, and the gas densities in the dust shell, maser shell, and photosphere may be used to constrain scale heights. These heights are consistent with models describing how radial pulsations of the stellar photosphere create shocks that, over time, cause the atmosphere to be greatly extended compared with those of non-variable stars (cf. Bowen 1988, and references therein). The region between the photosphere and the shocks has a scale height of about 0.02 R$_*$. The scale height within the shocked atmosphere is about 1 R$_*$. A pulsation amplitude of 2 – 4 km s^{-1} would cause a scale height of 0.6 to 0.7 R$_*$.

References

Bowen, G. H. 1988, *Ap. J.*, **329**, 299

Chapman, J. M. & Cohen, R. J. 1986, *M.N.R.A.S.*, **220**, 513

Danchi, W. C., Bester, M., Degiacomi, C. G., Greenhill, L. J., & Townes, C. H. 1994, *A. J.*, **107**, 1469

Diamond, P. J., Kemball, A. J., Junor, W., Zensus, A., Benson, J., & Dhawan, V. 1994, *Ap. J. (Letters)*, **430**, L61

Greenhill, L. J., Colomer, F., Moran, J. M., Backer, D. C., Danchi, W. C., & Bester, M. 1995, *Ap. J.*, submitted

Humphreys, R. M., Strecker, D. W., & Ney, E. P. 1972, *Ap. J.*, **172**, 75

Miyoshi, M., Matsumoto, K., Kameno, S., Takaba, H., & Iwata, T. 1994, *Nature*, **371**, 395

CO MODELLING OF CW LEO

MARTIN GROENEWEGEN

Institut d'Astrophysique de Paris, 98 bis Boulevard Arago, F-75014 Paris, France

Abstract. Preliminary results are presented of fitting the spectral energy distribution and newly obtained ^{12}CO and ^{13}CO J =1–0, 2–1 and 3–2 line profiles of CW Leo. This is done with a model which includes a self-consistent calculation of the gas temperature.

Key words: CO observations — stars: AGB — stars: individual: CW Leo — mass loss

1. Introduction

Fitting observed molecular line profiles with an appropriate radiative trans-fer model allows a detailed study of the mass loss rate or the molecular abundances. It has been recognised that such a radiative transfer model should include a self-consistent calculation of the gas temperature. Several such models have recently been published (Kastner 1992, Justtanont *et al.* 1994 and Groenewegen 1994). The main difference between my model and the other two is that the cooling by CO molecules is treated exactly and that photoelectric heating is taken into account.

However, there still is freedom in the modelling due to uncertainties in the dust-to-gas ratio, the luminosity, etc. These uncertainties can be greatly reduced by fitting the observed spectral energy distribution (SED) and the CO line profiles *simultaneously*. The basic idea is that the grain properties influence both the SED and the shape and intensity of the CO line profiles as the main heating rate is dust grain-gas collisions. In practice this means one has the following free parameters for the CO model: luminosity (L), dis-tance (D), mass loss rate ($\dot{M}$), dust-to-gas ratio (Ψ), grain size (a), effective absorption coefficient (Q), dust opacity at a reference wavelength (κ) and the CO abundance, with the constraints L/D^2 = constant and $\dot{M} \, \Psi \, Q_\lambda/a$ = constant (Q being the wavelength-integrated, flux-weighted value of Q_λ). The latter relation comes from the dust optical depth which is determined by fitting the SED.

2. CW Leo

I have fitted the SED of CW Leo with a dust radiative transfer model (Groe-newegen 1993) taking photometry from the literature. This fixes the quan-tities $\dot{M} \, \Psi \, Q_\lambda/a$ and L/D^2, as well as the shielding by dust of external UV radiation (needed for the calculation of the photoelectric heating rate in the CO model).

Astrophysics and Space Science **224**: 471–472, 1995.

Regarding the CO calculations, the 8-dimensional parameter space has been scanned at the moment for $L = 15\,000$, $20\,000$ and $25\,000\ L_\odot$. The models are fitted to recently obtained high-quality ^{12}CO and ^{13}CO J = 1-0, 2-1 and 3-2 line profiles (Groenewegen *et al.* 1994; IRAM and JCMT data). The ^{13}CO abundance is determined from a ^{12}CO/^{13}CO ratio of 44 as determined from optically thin lines (Kahane *et al.* 1992). The best-fitting models also predict the observed ^{12}CO(4-3) intensity (Williams & White 1992) and are in agreement with previous J = 1-0, 2-1 on-source observations of CW Leo with different telescopes. The solutions that fit the data best tend to have high dust opacities (130-200 cm^{2}gr^{-1}), large grain sizes (0.2-0.3 μm) and high CO abundances (12-17 10^{-4} relative to H$_2$).

From the present modelling, luminosities below $15\,000\ L_\odot$ can be excluded and the best-fitting model with $L = 15\,000\ L_\odot$ fits the observations just at the 3σ level. The best-fitting models for the higher luminosities tend to be equally good. For each of the luminosities investigated, the best-fitting models have about the same mass loss rate ($\dot{M} = 1.9 \pm 0.3$ 10^{-5} M$_\odot$/yr) and dust-to-gas ratio ($\Psi = 0.0027$-0.0042).

3. Conclusions

Since the dust properties (opacity, grain size, dust-to-gas ratio) influence both the observed spectral energy distribution and the CO line profiles, it is argued that, by fitting both simultaneously, maximum use is made of the available constraints. Although numerically expensive, since many models have to be run, it is the most powerful method to determine interesting quantities like the total mass loss rate and the dust-to-gas ratio.

On-source spectra obtained with telescopes of different beam sizes, and especially mapping data (e.g. Huggins *et al.* 1988) can constrain the mass loss history. Based on the preliminary models so far, the data for CW Leo appear to be consistent with a constant mass loss rate. A model where the mass loss rate was increased by a factor of 10 relative to its present-day value, 720 years ago can be excluded.

References

Groenewegen M.A.T.: 1993, *Ph.D. thesis*, University of Amsterdam
Groenewegen M.A.T.: 1994, *A&A* , in press
Groenewegen M.A.T., Baas F., de Jong T., Loup C.: 1994, *A&A* , submitted
Huggins P.J., Olofsson H., Johansson L.E.B.: 1988, *ApJ* **332**, 1009
Justtanont K., Skinner C.J., Tielens A.G.G.M.: 1994, *ApJ* , in press
Kahane C., *et al.*: 1992, *A&A* **256**, 235
Kastner J.H.: 1992, *ApJ* **401**, 337
Williams P.G., White G.J.: 1992, *A&A* **266**, 365

IS C/O ENHANCED IN NGC 253?

ANDREW HARRISON AND PETER BRAND

Institute for Astronomy, University of Edinburgh, Blackford Hill, Edinburgh, EH9 3HJ

and

ADRIAN RUSSELL AND PHIL PUXLEY

Royal Observatory, Blackford Hill, Edinburgh, EH9 3HJ.

Abstract. We have mapped the nuclear region of the starburst galaxy NGC 253 in the $^3P_1 \rightarrow ^3P_0$ line of neutral carbon using the JCMT. Carbon is widespread across the nuclear region with a similiar distribution to CO as expected. Previous studies of Galactic star-forming regions showed that carbon emission is enhanced in photon-dominated regions (where UV photons impinge upon molecular clouds). Previous observations of other PDR tracers such as ionized carbon and FIR continuum constrain the physical conditions in the PDR gas of NGC 253. The carbon we have observed is far brighter than predicted by theoretical models of PDRs with solar elemental values. This indicates that carbon emission is not a reliable diagnostic of the physical conditions in the nuclear region of a galaxy undergoing a burst of star formation.

Key words: ISM: carbon — PDRs

1. Introduction

NGC 253 is a nearby, highly inclined spiral galaxy and, along with M 82, has become an archetypal starburst, e.g. Carral *et al.* (1994). We have mapped the central 500pc of NGC 253 in [CI]. Theoretical models of photon dominated regions (PDRs) predict that the majority of [CI] emission should arise from a thin layer of gas at the interaction between HII regions and molecular gas, typically at a few A_v (Hollenbach *et al.* 1991). The level of [CI] emission, according to the models, is fairly insensitive to both the density of the molecular gas and the impinging UV field. Observations of [CI] from Galactic star-forming regions indicate that although the majority of [CI] does arise from a layer away from the ionization front, there is also [CI] found deep within the cloud at high exctinction, e.g. Keene (1990). Suggestions which have been proposed to explain the "deep" carbon include clumpy clouds allowing the permeation of UV photons through the cloud, non-equilibrium chemistry and shocks. Thus, although PDRs may not account for all the [CI], the fact that they can account for virtually all the [CI] means that [CI] should be a good estimate of the number of PDRs within a single-beam observation of the nucleus of a galaxy.

Astrophysics and Space Science **224**: 473–474, 1995.
© 1995 *Kluwer Academic Publishers.*

2. Existing PDR models cannot explain the level of [CI] emission from the nucleus of NGC 253

Carral *et al.* (1994) observed [OI] (63μm) and [CII] (158μm) from the nuclear region of NGC 253, and, using the FIR continuum measurements of Telesco & Harper (1980) modelled the PDR emission from NGC 253 as arising from a population of n=10^4cm^{-3}, G_0=10^4 PDRs filling the KAO beam with a factor of order unity. The intensity of [CI] in such a PDR is predicted by Hollenbach *et al.* (1991) to be $\sim 2 \times 10^{-6}$erg cm^{-2} s^{-1} sr^{-1}. However we find across the $40'' \times 10''$ CO bar in NGC 253, $< I_{CI} > \sim 5 \times 10^{-5}$erg cm^{-2} s^{-1} sr^{-1}, a factor of $\sim$25 greater than predicted.

It is possible to make [CI]/[CII] and [OI]/[CII] consistent with a PDR origin for [CI] in two ways, either by reducing the density of the PDR gas or by lowering the UV field. Both ways result in much less [OI] and [CII] being produced, but have little effect on the [CI]. However, [OI] drops off much more dramatically than the [CII]. Since shocks result in [OI]/[CII]>1, Hollenbach & McKee (1989), by having an ensemble of PDRs and shocks, with large filling factors, $\sim$25, we can make I[CII]/I[CI] and I[OI]/I[CII] consistent with observations and the PDR models of Hollenbach *et al.* (1991). Unfortunately, changing either the density or the UV field of the PDR results in conflicts with the PDR model predictions and other observations.

Reducing the UV field results in large pressure differentials between different components of the ISM, too much CO(1→0) emission and requires that the UV flux from OB stars only plays a minor role in heating the dust.

Reducing the density of the PDRs results in more H^0 than is observed, as well as producing large pressure differentials between different components of the gas. There is also little observational evidence for such a major component of diffuse gas.

It is only at low densities or at low UV fields that I[OI] and I[CII] are low enough to make the O^0, C$^+$ and C^0 consistent with each other. Hence intermediate densities and UV fields are also ruled out. The C^0 emission we have detected is inconsistent with the models of Hollenbach *et al.* (1991) which would indicate that either the assumption that the majority of [CI] emission arises from PDRs in NGC 253 is flawed, or the elemental values used to model the PDR gas are incorrect.

References

Carral,P., Hollenbach,D.J., Lord,S.D., Colgan,S.W.J., Haas,M.R., Rubin,R.H. & Erickson,E.F. 1994, *ApJ*, **423**, 223

Hollenbach, D.J. & McKee,C.F. 1989, *ApJ*, **342**, 306

Hollenbach,D.J., Takahashi,T. & Tielens, A.G.G.M. 1991, *ApJ*, **337**, 192

Keene,J. 1990, in *Carbon in the Galaxy: Studies from Earth & Space*, ed. Tarter,J.C., Chang,S. & Defrees,D.J. (NASA CP-3061), 181

Telesco,C.M. & Harper,D.A. 1980, *ApJ*, **235**, 392

FAR-INFRARED OBSERVATIONS OF MAIN SEQUENCE
STARS SURROUNDED BY DUST SHELLS

P. M. HARVEY, B. SMITH and J. DI FRANCESCO

Astronomy Dept., University of Texas
Austin, TX 78712 USA

Abstract. We present 50 and 100 μm photometry and size information for several main sequence stars surrounded by dust shells. The observations from NASA's Kuiper Airborne Observatory include the "Vega-like" stars, Beta Pic, Fomalhaut, as well as four stars suggested by Walker and Wolstencroft to belong possibly to the same class. The results of our observations are best interpreted as upper limits to the far-infrared sizes of the dust clouds around all of the stars except Fomalhaut and Beta Pic. We have also fit simple, optically thin models to the Beta Pic data to explore the range of shell parameters consistent with our limits and with previous observations.

One of the most exciting phenomena discovered by the IRAS sky survey was the presence of dust surrounding several nearby main-sequence stars. These dust clouds appear to be composed of grains, substantially larger than typical interstellar grains, in orbit around the central star, and almost certainly replenished because of their short lifetime due to a variety of effects (e.g. Backman & Paresce 1993). The angular resolution of the KAO is critical to determine accurately the sizes of the shells in order to place the strongest limits on the grain properties and spatial distribution of the circumstellar material. Our observations were made on several deployments of the KAO to the southern hemisphere in 1992 – 1994. We used the University of Texas 20-channel bolometer array described briefly by Smith *et al.* (1994) operating at 50 and 100μm with diffraction limited beam sizes of $15 \times 25''$ and $25 \times 35''$ respectively for each pixel.

Table 1 shows the photometry in the KAO beam sizes with both the statistical errors of the measurements and the total uncertainties including absolute calibration errors; also indicated is the IRAS measured flux density in its much larger beam. The simplest conclusion from these results is that the dust clouds around all the stars except for Fomalhaut and β Pic are quite compact since the KAO photometry gives fluxes close to but less than the IRAS results. For β Pic and Fomalhaut the discrepancy between the KAO and IRAS fluxes as well as spatial measurements (not shown) suggest that the circumstellar disks are resolved.

Models essentially identical to Backman, Gillett, & Witteborn's (1992) numbers "10" and "11" do an excellent job of reproducing the spatial and spectral data for β Pic. The important features of these models are: a much lower grain density inside a radius of order 100 a.u. than outside this radius; a small inner radius and/or small grain size for the inner grains in order to

Astrophysics and Space Science **224**: 475–476, 1995.
© 1995 *Kluwer Academic Publishers.*

TABLE I
Photometric Results

Star	KAO F_ν (Jy)	± stat.	± total	λ	IRAS F_ν (Jy) @ λ
β Pic	12.9	± 1.0	± 1.6	50μm	18.8 ± 0.9 @ 60 μm
	8.5	± 0.6	± 1.0	100μm	11.2 ± 1.0 @ 100 μm
Fomalhaut	5.6	± 0.65	± 0.95	50μm	9.8 ± 0.5 @ 60 μm
	6.7	± 0.6	± 1.0	100μm	9.8 ± 0.5 @ 60 μm
HD 135344	24.3	± 1.8	± 3.5	50μm	26.3 ± 1.5 @ 60 μm
HD 139614	14.0	± 1.3	± 2.5	50μm	18.3 ± 1.2 @ 60 μm
HD 142527	98	± 2.0	± 15	50μm	106 ± 6 @ 60 μm
	84	± 1.2	± 12	100μm	82 ± 5 @ 100 μm
HD 169142	22.9	± 2.1	± 3.5	50μm	28.9 ± 2 @ 60 μm

reproduce the shorter wavelength infrared emission from the hottest grains; and an $r^{-1.7}$ surface density gradient in the outer disk to be consistent with optical surface brightness measurements.

The basic conclusions of this study are that the high spatial resolution KAO data for β Pic are consistent with existing models for the dust emission from this star which require a two-component disk with a strong dust "depletion" in the inner disk. Fomalhaut is clearly resolved by its much smaller flux density in the KAO beam than in the IRAS beams as well as spatial measurements not shown here. For the four stars from the list of Walker & Wolstencroft (1988) we find that the far infrared emission is barely resolved, if at all, based on the ratio of KAO to IRAS measured flux density. Because of the much greater distance to these stars, this result does not place strong constraints on the grain sizes, but it does call into question the conclusion that these stars' dust shells were resolved by IRAS.

Acknowledgements We thank the staff of the KAO for their superb support during these deployments to the southern hemisphere which made possible these observations. This research was supported by NASA grant NAG 2-67 to the University of Texas.

References

Backman, D.E., Gillett, F.C., & Witteborn, F.C.: 1992, *Ap.J.* **385**, 670
Backman, D.E. & Paresce, F.: 1993, in *Protostars and Planets III*, eds. E.H.Levy & J.I. Lunine, University of Arizona Press:Tucson, 1253.
Smith, B.J., Harvey, P.M., Colome, C., Zhang, C.Y., & Di Francesco, J.: 1994, *Ap.J.* **425**, 91
Walker, H.J. & Wolstencroft, R.D.: 1988, *Pub.A.S.P.* **100**, 1509

DETECTION OF DETACHED DUST ENVELOPES AROUND OXYGEN-RICH AGB STARS

O. HASHIMOTO
Department of Applied Physics, Seikei University, Musashino, Tokyo, Japan

H. IZUMIURA* and D. J. M. KESTER
SRON Laboratory for Space Research Groningen, Groningen, The Netherlands

and

TJ. R. BONTEKOE
Astrophysics Division, Space Science Department of ESA, ESTEC, Noordwijk, The Netherlands

Abstract. Extended emission components are clearly found in the IRAS scan data of optically visible oxygen-rich AGB stars which show no 10 μm silicate band feature in the IRAS LRS spectra but a strong infrared excess in the IRAS photometric data. It is most likely that these stars really have their circumstellar dust envelopes, which are detached from the central stars, indicating a halting of mass loss for a significant period.

Key words: circumstellar dust — stars : AGB — Mass loss — stars: individual : R Hya

1. Introduction

A number of optically visible M-type AGB stars show no or a very weak silicate band feature around 10 μm in their IRAS low resolution spectra (LRS) while they show strong infrared excesses in their IRAS photometric data (Hashimoto 1994). It is very difficult to understand such infrared properties in terms of a circumstellar dust envelope of silicate grains surrounding a central star which is undergoing continuous mass loss. Such a dust envelope with a strong infrared excess should show a significant 10 μm band feature at the same time.

2. Analysis of IRAS image data

Using IRAS survey data at 60 μm, we examined the circumstellar structure of thirteen of those red M giant stars: T Lep, R Aur, R Cnc, BK Vir, Y UMa, R Hya, g LC Her, IRC+20326, X Oph, SU Sgr, S Pav, V1943 Sgr, and T Mic. Extended emission components are clearly found in the individual scan data for most of the stars (9 stars), which are consistent with the infrared excess. Some of the brightest sources also show the extended circumstellar emission components in the high resolution 60 μm images reproduced by the pyramid

* On leave from Department of Astronomy and Earth Sciences, Tokyo Gakugei University, Koganei, Tokyo, Japan

Astrophysics and Space Science **224**: 477–478, 1995.
© *1995 Kluwer Academic Publishers.*

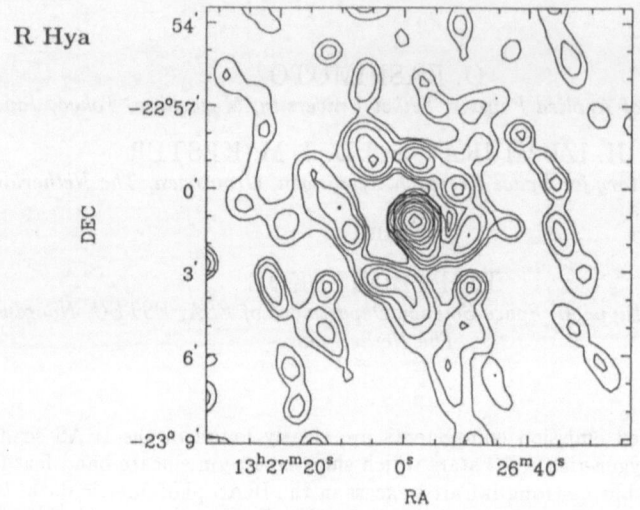

Fig. 1.   Infrared image of R Hya at 60 μm, reconstructed by the PME techniques. Contour lines are 1, 2, 4, 8, 16, 32, 64, 128, 256, 512, and 1024 MJy str⁻¹

maximum entropy (PME) image reconstruction techniques developed for the IRAS survey data (Bontekoe *et al.* 1991, 1994, Waters *et al.* 1994). The image of R Hya reconstructed by the PME techniques is presented in Fig.1, for example, where the spatial resolution is about 1 arc minute. A possible detached shell structure is found in the image of R Hya.

3.   Discussion

It is most likely that those M stars really have their circumstellar dust envelopes. Our model analyses (Hashimoto 1994) together with the present results on R Hya further suggest that those envelopes should be detached from the central stars so that they do not show any significant 10 μm feature while showing the strong infrared excess in the IRAS photometric data. Those M-type AGB stars are likely to be experiencing a halting of mass loss which has lasted for about 50–200 years, and which has made the inner boundary of the dust envelope farther than the region warm enough to produce the 10 μm feature (Hashimoto 1994).

References

Bontekoe Tj. R., Kester D. J. M., Price S. D., De Jonge A. R. W. and Wesselius P. R.: 1991, *A&A* **248**, 328
Bontekoe Tj. R., Koper E. and Kester D. J. M.: 1994, *A&A* **284**, 1037
Hashimoto O.: 1994,*A&AS*, in press
Waters L. B. F. M., Loup C., Kester D. J. M., Bontekoe Tj. R. and de Jong T.: 1994, *A&A* **281**, L1

UNVEILING THE FORMATION OF STARS AND PLANETS WITH A LARGE INFRARED-OPTICAL TELESCOPE

SAEKO S. HAYASHI

SUBARU Telescope Project Office
National Astronomical Observatory of Japan
Osawa, Mitaka, Tokyo 181, JAPAN

Abstract. Diffraction-limited imaging in the near-IR (with the aid of adaptive optics) to middle-IR wavelengths will be a major breakthrough with 8m IR-optical telescopes. Special interest is on the scale of planet formation, that is, within 100 AU of the embedded sources in the nearby Taurus or Ophiuchus clouds. In this report, we demonstrate the sensitivity of a telescope with a virtual instrument that covers a wide spectral range.

Key words: star formation — planetary formation — large IR-optical telescope

Characteristics of the spectral energy distributions (SEDs) of protostars, classical and weak-line T Tauri stars have been sampled by recent infrared through mm observations, revealing gaseous and dusty compact structures. However, the very vicinity of the central source is still difficult to resolve. The peak of the SED directs one to pursue the near to middle infrared wavelength regime for such a detailed study. In order to understand the dynamical accretion and other important mechanisms in star formation, as well as the formation process of planetary systems, high resolution observations at infrared wavelengths are essential. It is also useful in identifying major constituents of the circumstellar matter and protoplanetary systems.

Motivated by these requirements, several new 8m class telescopes are under construction. Essentials of the Japanese one are as follows:

- Image size 0″.23 FWHM with a monolithic 8.2m primary mirror.
- Active optics of the primary, supported by 261 mechanical actuators.
- Overall telescope emissivity 8% or less.
- Careful control of thermal environment with flushing enclosure.
- Accurate and stable pointing and tracking.

Started in 1991, construction of SUBARU (Pleiades in ancient Japanese) Telescope is proceeding on Mauna Kea in Hawaii, USA, amongst a cluster of front-line optical through millimetre telescopes. Polishing the primary mirror blank is starting. The telescope pier and the enclosure foundation are completed, and the upper enclosure assembling started. The truss, mounting structure and primary mirror cell are finishing. Mass production of actuators is catching up with the shop erection of the mirror cell. Primary, Cassegrain, and two Nasmyth foci will accommodate varieties of astronomical instruments after the first light at the end of fiscal 1998.

Astrophysics and Space Science **224**: 479–480, 1995.
© 1995 *Kluwer Academic Publishers.*

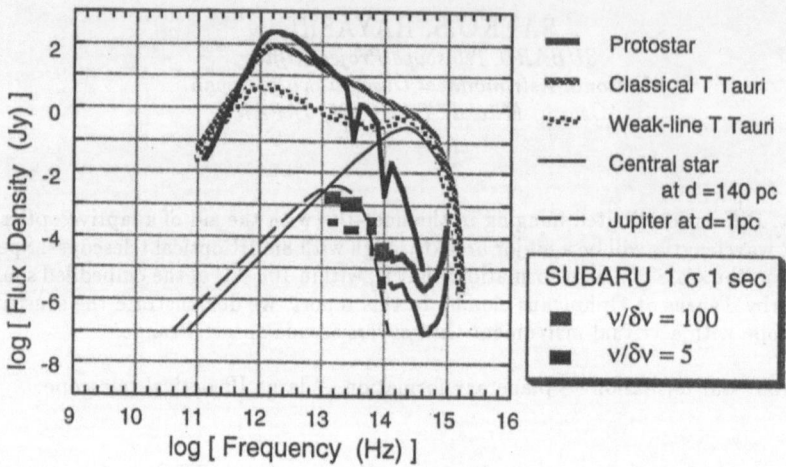

Fig. 1. Sensitivity with SUBARU Telescope, superposed onto the spectral energy distributions of young stellar objects in different evolutionary stages.

The expected sensitivity is evaluated for two resolutions representing photometric and spectroscopic measurements; R $(\nu/\delta\nu)$ = 5 and 100. The image size is 0″.2 or diffraction limited where larger. A virtual detector of pixel size 20 micron is adopted; the optical response is with TI CCD efficiency, readout 5 e⁻, and IR performance is 10/50 e⁻ readout, quantum efficiency 0.2. One sigma level in 1 second integration is shown in Jy. Other details are based on Hayashi et al. (1992). In the optical range, micro-Jansky level is measurable. Though thermal background is dominating in the infrared, the sensitivity level will dramatically be improved by the Adaptive Optics.

Acknowledgements

This paper is dedicated to Hikaru of 21 month old who suffered from the author's absence during the conference.

References

Hayashi, S. S., Okamura, S., and Shibai, H.: 1992, *Publ. Nat. Ast. Obs. Japan*, **2**, 547.

EMISSION LINE WIDTHS OF AN ASYMMETRIC
BOWSHOCK INSIDE A CLOUD OF DUST

W. J. HENNEY

Instituto de Astronomía, UNAM, Apartado Postal 70-264, 04510 México D. F., México

Abstract. A model is presented that explains the anomalous emission line widths in the Herbig-Haro object HH 1 in terms of emission from a lop-sided bowshock and scattering by surrounding dust.

Key words: ISM: individual (HH 1) – ISM: jets and outflows – reflection nebulae

The Herbig-Haro object HH 1 is often considered to be a bowshock, moving almost in the plane of the sky, formed by the action of a jet from a young stellar object. While simple bowshock models have been successful in reproducing observed characteristics of HH 1, such as line ratios and gross morphology, certain discrepancies remain, such as the spatial distribution of the emission line widths (Solf *et al.* 1991), which do not follow the pattern expected from simple dynamic models (i.e. maximum velocity dispersion just behind the bowshock head with the line widths decreasing towards the wings). In fact, observations show a more complicated pattern, with the line widths decreasing along one wing of the bowshock and increasing along the other, in addition to a second maximum of velocity dispersion just in front of the leading condensation. This discord between theory and observations has been tentatively attributed to the effects of surrounding dust, which can scatter and hence broaden the emission lines from the bowshock. This idea is investigated here using Monte Carlo simulations of emission line scattering. Details will be published elsewhere; see Henney (1994) for the analytical background to the problem.

The model adopted is of a bowshock moving inside a spherical cloud of scatterers. The ambient medium is inhomogeneous such that the density decreases from top to bottom of the cloud as seen by an observer. The shape of the bowshock is calculated by assuming pressure balance between the thermal pressure of the shocked jet gas and the ram pressure of the ambient gas due to the bowshock motion. It is the inhomogeneity in the ambient density that leads to the skew shape of the bowshock. The emission from the bowshock is calculated assuming a line (with a strength scaling as the cube of the normal shock velocity) that is allowed to scatter from ambient dust, whose distribution follows that of the ambient gas. The dust is assumed destroyed by the shock so that there are no scatterers inside the bow. Multiple scattering is fully taken into account. Figure 1 shows contours of constant width of the emission lines from the bowshock (solid lines) superimposed on surface brightness contours (dashed lines). Comparison

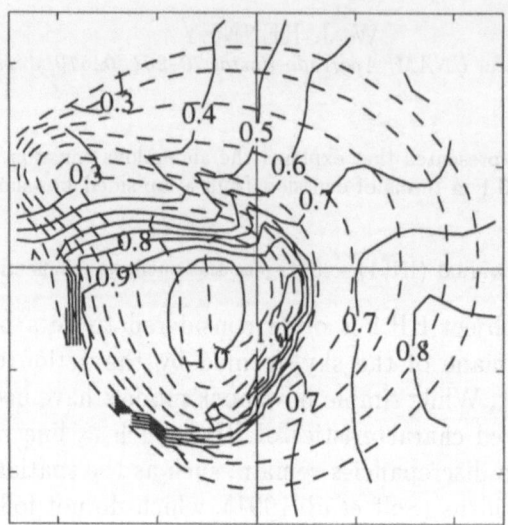

Fig. 1. Theoretical map of velocity dispersion in HH 1. Solid lines are contours of emission line full-width-half-maximum and the emission line surface brightness contours are shown by dashed lines. The line width contours are labelled in units of the bowshock speed.

with Figure 4 of Solf *et al.* (1991) shows that the two observed maxima are in qualitative agreement with simulations, the first just upstream of the head of the bowshock, and the second just behind the head and displaced towards the fainter wing. The upstream maximum is due to scattering and occurs where the direct light from the bowshock, which is centred on zero velocity, and the scattered light from upstream dust, which is highly blueshifted (Henney, Raga & Axon 1994), have comparable brightness and hence produce a broad blended line. The downstream maximum is due entirely to the direct light from the bowshock since the scattered light is unimportant in this region, even for the optical depth $\tau = 1$ that has been adopted here. The displacement of this maximum from the bowshock axis is caused by the ambient density gradient and, although observations show this maximum to be shifted further downstream than is indicated by the simulations, this is probably due to the model's neglect of the finite cooling distance of the postshock gas. (However, Eislöffel *et al.* 1994 suggest that this region may be contaminated by a separate flow superimposed on HH 1).

References

Eislöffel, J., Mundt, R. and Böhm, K.-H.: 1994, *AJ* **108**, 1042
Henney, W. J., Raga, A. C. and Axon, D. J.: 1994, *ApJ* **427**, 305
Henney, W. J.: 1994, *ApJ* **427**, 288
Solf, J., Raga, A. C., Böhm, K. H. and Noriega-Crespo, A.: 1991, *AJ* **102**, 1147

HIGH RESOLUTION OBSERVATIONS OF THE RADIO EMISSION FROM LUMINOUS YSOS

M. G. HOARE* and J. E. DREW

Dept of Physics, Astrophysics, University of Oxford, Keble Road, Oxford, OX1 3RH, UK.

S. T. GARRINGTON

Nuffield Radio Astronomy Laboratory, Jodrell Bank, Lower Withington, Cheshire, SK11 9DL, UK.

and

M. J. MCCAUGHREAN

Max-Planck-Institut für Astronomie, Königstuhl 17, D-69117 Heidelberg, Germany.

Abstract. New, high-resolution 1.66 GHz MERLIN maps of Cep A2 and LkHα101 are presented and discussed.

Radio continuum emission from luminous young stellar objects (YSOs) has a spectral index close to the $S_\nu \propto \nu^{0.6}$ expected for constant velocity outflow and the presence of a wind is indicated by the broad (few 100 km s^{-1}) wings on near-IR H I lines. These winds are predicted to recombine at distances of around 100 AU which can be resolved with the upgraded MERLIN array and shed light on the wind morphology and driving mechanism.

Observations of two luminous YSOs were made with the MERLIN array plus the Lovell 76m telescope at Jodrell Bank at 1.66 GHz for a full track. For Cep A2 (Fig 1) we see the elongated structure of the main 2(ii) component at PA≈45° in agreement with previous observations (Hughes 1991, Rodríguez et al. 1994). The patches of emission to the SW may well be an extension of this structure which would lend support for a jet interpretation for the radio morphology. Possible further support for this comes from a new wide-field K-band image which reveals a cone of scattered light emission extending to the SW of Cep A2 which could be related to an outflow in this direction. A new radio component 2(iii) is also revealed clearly in our MERLIN map although its relation to 2(ii) is not clear.

LkHα101 is a more evolved object and Fig 2 shows that it is more symmetric although there is a bright bar of emission at PA≈135° and as in our higher resolution 5 GHz map (Hoare et al. 1994) there is a generally clumpy appearance. A wider sample needs to be mapped to see which of the structures: disk-like in S106IR (Hoare et al. 1994), perhaps jet-like in Cep A2,

* Current address: Max-Planck-Institut für Astronomie, Königstuhl 17, D-69117 Heidelberg, Germany.

Astrophysics and Space Science **224**: 483–484, 1995.

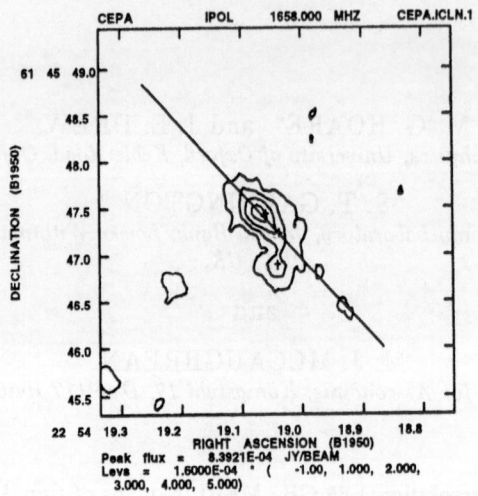

Fig. 1.   1.66 GHz map of Cep A2 at 0.25″resolution. Sources (+) N to S are 2(ii), 2(i) and 2(iii).

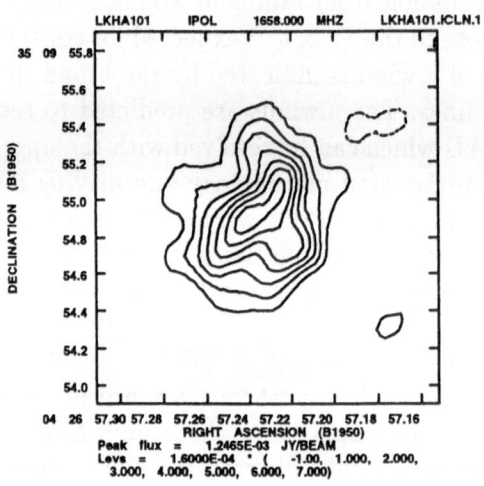

Fig. 2.   1.66 GHz MERLIN map of LkHα101 at 0.23″ resolution.

and more symmetric in LkHα101 is the norm and whether the determining factor is evolutionary stage, initial mass, angular momentum etc.

References

Hoare M. G., Drew J. E., Muxlow T. B., Davis R. J., 1994, Astrophys. J. 421, L51.
Hughes V. A., 1991, Astrophys. J. 383, 280.
Rodríguez L. F., Garay G., Curiel S., Ramírez S., Torrelles J. M., Gómez, Velázquez A., 1994, Astrophys. J. 430, L65.

NONLINEAR MODELS OF PULSATION AND MASS LOSS

S. HÖFNER and M.U. FEUCHTINGER

Inst. f. Astronomie d. Univ. Wien, Türkenschanzstr. 17, A-1180 Wien, Austria

Abstract. We briefly discuss the current status of our radiation-hydrodynamical models of pulsation in various stellar objects and of dust-driven mass loss in LPVs. We emphasize the importance of a future combined modelling of pulsation and mass loss in AGB stars which has to be based on reliable physical and numerical methods. *

Key words: pulsation — mass loss — dust — long period variables — AGB stars

1. Stellar Pulsation

Pulsation plays an important role in the heavy mass loss observed in Miras and other LPVs by creating an extended atmosphere which allows efficient dust formation. Subsequent acceleration of the newly formed dust grains by radiation and momentum transfer to the gas leads to slow but massive stellar winds and the formation of a circumstellar dust shell.

Due to the complicated nonlinear interactions of gas dynamics, radiation and dust, only a combined self-consistent modelling of all components and phenomena can lead to a realistic picture of these objects. The aim of our project is the construction of a radiation-hydrodynamical model for LPVs which includes both the pulsation zone and the extended atmosphere where the dust driven mass loss is initiated.

As a first step to develop a deeper understanding of stellar pulsation we have investigated classical variables (RR Lyr stars, Feuchtinger & Dorfi 1994) and found that the light curves and limit cycle characteristics obtained with our models closely match the observations. Unfortunately, there exist several physical and numerical problems concerning the pulsation calculations for LPVs and we are currently working to overcome these difficulties.

2. Dust-Driven Mass Loss

At the same time we have started to explore the dynamics of the circumstellar envelope with radiation-hydrodynamical models which include additional equations describing time-dependent dust formation, growth and evaporation (Gail & Sedlmayr 1988, Gauger *et al.* 1990). Presently, the pulsation of the star is simulated by a piston below the photosphere and the chemistry is restricted to C-rich objects. More details about the input physics can be found in Höfner & Dorfi (1992) and Höfner (1994). The models exhibit the

* This work is supported by the *Österreichischer Fonds zur Förderung der wissenschaftl. Forschung* under project numbers P8758 and P9694

Astrophysics and Space Science **224**: 485–486, 1995.

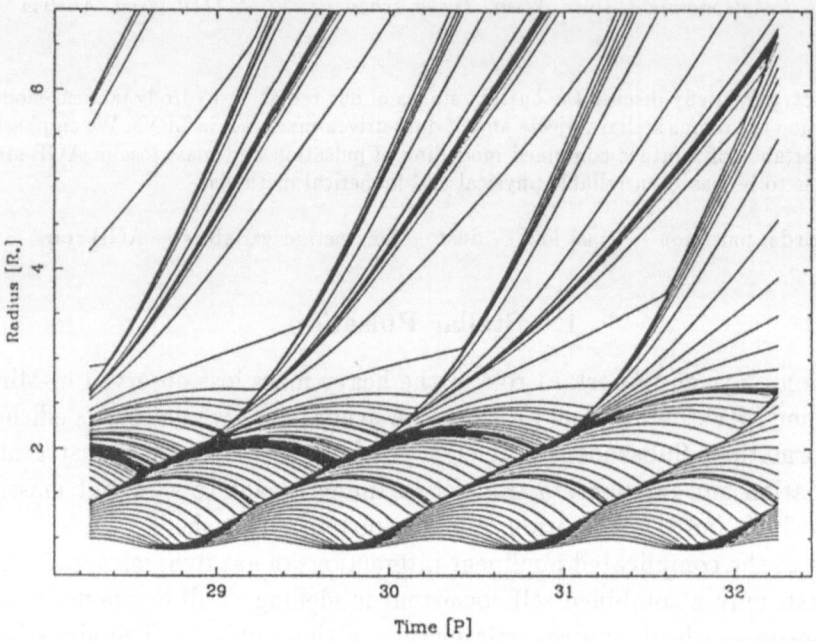

Fig. 1. Positions of selected mass shells as a function of time: Each line corresponds to the path of a given element of matter (test particle). The dust formation is triggered by the shock waves caused by the piston. The propagation of the shocks is indicated by the sharp bends of the lines. As soon as a new dust layer has formed (around phase 0.0 of the pulsation cycle near $2R_*$) it is accelerated by the radiation pressure. The dust-free material behind decelerates and falls back until it is overrun by the next shock wave.

typical discrete dust layers associated with strong shock waves already found in the models of Fleischer *et al.* (1992).

Fig. 1 shows the dynamics of the circumstellar envelope for a model with the following parameters: $M_* = 1M_\odot$, $L_* = 10^4 L_\odot$, $T_{\text{eff}} = 2700$ K (hydrostatic initial model); carbon/oxygen abundance $\varepsilon_C/\varepsilon_O = 1.77$; piston: period $P = 650$ d, velocity amplitude $\Delta u = 2.0$ km/s; The resulting terminal velocity and mass loss rate are $u_\infty = 24.7$ km/s and $\dot{M} = 1.0 \cdot 10^{-5} M_\odot$/yr.

References

Feuchtinger, M.U., Dorfi, E.A.: 1994, *A&A*, in press
Fleischer, A.J., Gauger, A., Sedlmayr, E.: 1992, *A&A* **266**, 321
Gail, H.-P., Sedlmayr, E.: 1988, *A&A* **206**, 153
Gauger, A., Gail, H.-P., Sedlmayr, E.: 1990, *A&A* **235**, 345
Höfner, S.: 1994, Ph.D.-thesis, Univ. Wien
Höfner, S., Dorfi, E.A.: 1992, *A&A* **265**, 207

PROBING THE MAGNETIC FIELD AROUND PROTOSTARS

WAYNE S. HOLLAND and JANE S. GREAVES

Joint Astronomy Centre, 660 N. A'ohoku Place, Hilo, HI 96720, USA

and

DEREK WARD-THOMPSON

Royal Observatory, Blackford Hill, Edinburgh EH9 3HJ, UK

Abstract. We present recent JCMT polarimetric observations of the ρ Oph A cloud core and the S106 HII region. We infer the magnetic field structure in these regions.

Key words: polarisation – stars: formation – ISM: magnetic fields

1. Background and Observations

Magnetic fields are believed to play an important role in the evolution of interstellar molecular clouds. For example, magnetic support can prevent cloud collapse on free-fall timescales, as the field lines resist compression. Massive stars are thought to form when the initial mass of the core is too great to be supported (*supercritical collapse*), whilst low mass objects form when this support is lost more gradually (*ambipolar diffusion*). Some of the first direct observational evidence for this bi-modal scenario is presented in Greaves, Murray and Holland (1994 – hereafter GMH). The magnetic field still plays a role after the initial collapse, as it may become twisted-up in a rotating disk, and/or channel bipolar outflows. Models predict that toroidal fields may be observed in the accretion disk, while the field in the outflowing gas may be helical, or perpendicular to the disk. By using submm polarimetry we hope to obtain a clearer understanding of the role of magnetic fields in the star formation process. The observations were mainly carried out in June 1994 at JCMT, using the Aberdeen/QMW Polarimeter with the bolometer UKT14 at $800\mu m$. Photometry was performed at a number of equally spaced waveplate positions, and the resulting sinusoidal modulation used to deduce the source degree of polarisation and position angle.

2. Results and Discussion

Fig 1 shows 800 μm isophotal contour maps of the ρ Oph A region (map from André, Ward-Thompson and Barsony, 1993), and the S106 HII region (map from Richer et al., 1993). Superposed on the maps are the measured polarisation vectors, with the circles representing the 13.5" FWHM beam. The polarisation arises from aligned nonspherical dust grains. The alignment mechanism is not fully understood, but all models have the inferred magnetic

Astrophysics and Space Science **224**: 487–488, 1995.
© 1995 *Kluwer Academic Publishers.*

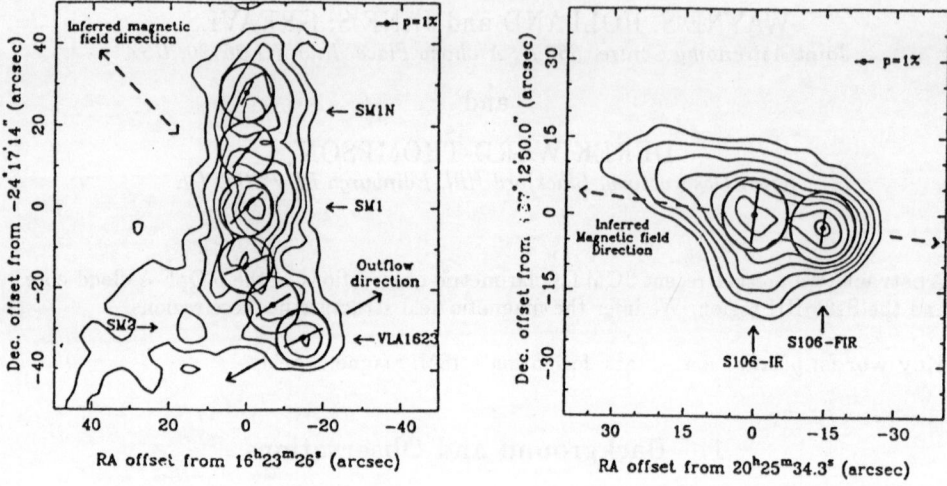

Fig. 1. (a) The ρ Oph A cloud core; (b) S106 HII region

field direction perpendicular to the grain long axis, and hence perpendicular to the submm polarisation vectors. In the case of the low mass ρ Oph A core, the magnetic field is roughly parallel throughout the cloud core, lying approximately NE-SW. There is no evidence for any change in direction of the field from the pre-protostellar SM1, to the Class 0 protostar VLA1623. Interestingly, since the magnetic field does not deviate in the vicinity of VLA1623, and is almost exactly *perpendicular* to the bipolar CO outflow, we conclude that the large-scale field *cannot* be collimating the outflow. Alternative mechanisms will be discussed in a future paper.

The S106 HII region has a bipolar morphology, with a dark lane running approximately E-W, and has been shown to contain two fragments at submm wavelengths, one associated with the young star S106IR and the other with a candidate protostar, S106FIR. The magnetic fields in the two objects appear to be closely aligned, and lie roughly along the dark lane. Our measured core field is perpendicular to the large scale N-S field deduced from Zeeman data (Roberts et al. 1994), and may be interpreted as twisting of the field during the cloud collapse to form the two stars. Similar morphologies have been found around other high mass regions (GMH).

References

Andre, P., Ward-Thompson, D., Barsony, M: 1993, *Astrophysical Journal* **406**, 122
Greaves, J., Murray, A., Holland, W: 1994, *Astronomy & Astrophysics* **284**, L19 (GMH)
Richer, J.S., et al: 1993, *Monthly Notices of the RAS* **262**, 839
Roberts, D.A., Crutcher, R.M., Troland, T.H: 1994, *Astrophysical Journal* , in press

POTENTIAL PROTOSTARS IN CLOUD CORES
H_2CO Observations of Serpens

ROBERT L. HURT and MARY BARSONY

University of California, Riverside

and

ALWYN WOOTTEN

National Radio Astronomy Observatory

Abstract. We present H_2CO observations of young protostar candidates in the Serpens Cloud Core. We find evidence for dense molecular gas in the cores of these objects that is warmer than the surrounding dust. The strong emission and gas properties support the premise that many of these sources may be very young protostars.

Key words: star formation — protostars — formaldehyde

In studying the processes of cloud collapse and subsequent star formation, one wishes to detect protostellar sources at increasingly earlier evolutionary stages. One expects the youngest observable protostars to lie at the core of clouds of high extinction and would escape detection at visible and even near infrared wavelengths. A proposed category for such young 'Class 0' objects defines them based on their undetectability at $\lambda < 10~\mu$, their high ratios of submm to bolometric luminosity, their narrow spectral energy distributions resembling a blackbody of T < 30 K, and the presence of a molecular outflow (Andre *et al.* 1993). The Serpens Cloud Core is a rich hunting ground for potential Class 0 protostar candidates. At a distance of about 310 pc (de Lara *et al.* 1991), it is notable for its near infrared clusters and its strong far infrared source FIRS1 (or S68N) with associated radio jet. Various FIR and mm studies have found gas and dust temperatures in the region around 30 K (e.g. Zhang *et al.* 1988; McMullin *et al.* 1994). Of particular interest are several submm continuum sources with no NIR counterparts (Casali *et al.* 1993) which are potential Class 0 sources.

We have observed the submm continuum sources in the Serpens Cloud Core in several transitions of the formaldehyde molecule in an effort to probe the gas properties in the cores of these objects and to determine whether they are indeed young protostars. H_2CO is particularly well-suited for studying gas properties in protostellar cores; its low optical depths and high critical density of $\sim 10^6$ cm^{-3} make it a better probe of temperature and density in dense cores than other common molecular species.

We observed up to four different H_2CO transitions for the Serpens sources SMM1-6 and S68N. The spectra of the strongest line are presented in Fig. 1. The two weakest sources (SMM5 & SMM6) have NIR counterparts, indicat-

Astrophysics and Space Science **224**: 489–490, 1995.

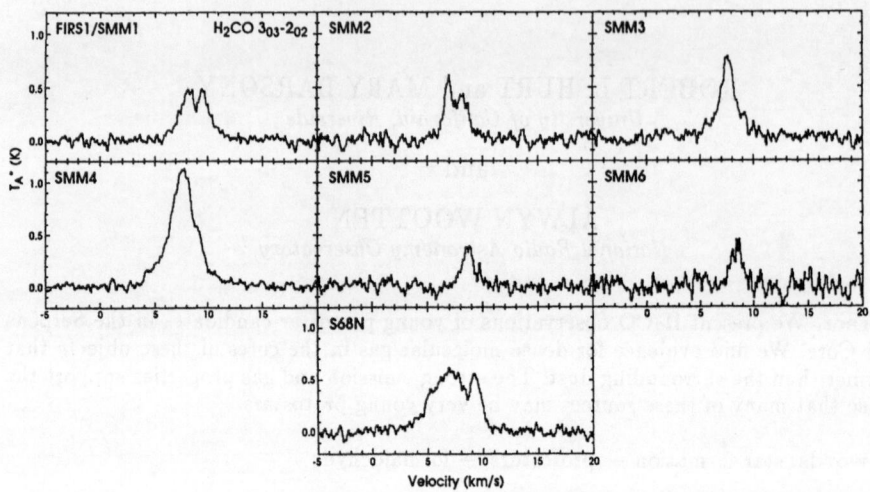

Fig. 1.   H$_2$CO (3_{03}–2_{02}) lines for the Serpens Cloud submm continuum sources.

ing possible gas depletion around older star-forming cores. All of the stronger lines are broad and exhibit non-Gaussian line wings, suggestive of molecular outflows. There is also clear evidence for self-absorption in sources SMM1, SMM2, and S68N indicating that the gas is probably optically thick.

By comparing the integrated line ratios with LVG calculations we derive properties of the gas in some of these protostellar cores, although potentially large optical depths can limit our accuracy. The two strongest potential outflow sources (SMM4 & S68N) exhibit interestingly high temperatures (T = 70–85 K) for gas at relatively high densities (n = $10^{6.5}$ cm^{-3}). These temperatures are higher than the ~ 30 K inferred for the dust and other molecular species but are consistent with other sources observed in H$_2$CO (Mangum & Wootten 1993). The fits for sources SMM1 and SMM3 are not so well-constrained but are consistent with the results for SMM4 and S68N.

We conclude that there are substantial dense, warm cores within the Serpens submm sources lacking NIR counterparts. Their strong H$_2$CO emission, possible outflows and inferred gas properties make them excellent candidate Class 0 protostellar sources.

References

Andre, Ward-Thompson, & Barsony, 1993, ApJ, 406, 122
Casali, Eiroa, & Duncan, 1993, A&A, 275, 195
de Lara, Chavarria-K., & Lopez-Molina, 1991, A&A, 243, 139
Mangum & Wootten, 1993, ApJS, 89, 123
McMullin, Mundy, Wilking, Hezel & Blake, 1994, ApJ, 424, 222
Zhang, Laureijs, & Clark, 1988, A&A, 196, 236

FORMATION RATE OF TRIPLE STARS RELATIVE TO DOUBLE STARS

SYUZO ISOBE

National Astronomical Observatory, Mitaka, Tokyo, Japan

and

MIKINORI NI-INO

Department of Physics, Chuo University, Bunkyo-ku, Tokyo, Japan

Abstract. A series of our and McAlister *et al.*'s speckle observations of spectroscopic binaries show that certain fractions of them are triple systems. A typical configuration of those systems is that of two stars separated by about 0.1 AU with a third component in an orbit at a distance about 10 AU from the central stars. These results suggest that the third star was formed from the outer circumstellar envelope after forming the central double.

Key words: Triple stars — Circumstellar envelope — Speckle observations

Stars are formed through contraction processes of interstellar clouds. Following theoretical evolution models (Shu, Adams, & Lizano 1987), a protostellar cloud falls onto a disk with a radius less than 100 AU, and three different systems such as a binary star system, a planetary system surrounding a star, and a single star are formed depending on its angular momentum.

Isobe *et al.* (1992, 1993) found an excess of triple stars with distances, 10 to 100 AU, of the third component from the central double stars by their speckle observations. Because of short distances of the third component from the central double stars all the triple systems of this type can not be formed by capture mechanisms, but are interpreted as being formed in a protostellar cloud. If this statement is true, we are able to give some constraint to the presently developed star formation theory which is how a gas disk with a radius less than 100 AU breaks into three stars.

To study this problem further, we examine rate of triple stars relative to double stars from presently available data.

As our working catalogue, a list of spectroscopic binaries compiled in Sky Catalogue 2000 is adopted. There are 532 stars in the list, out of which 11 stars have some relation with the other stars shown as A, B or Aa, Bb in the list and form each system. Therefore, the total number of stars studied to detect multiple systems is 521.

All the reference used in this study are: Sky Catalogue 2000 (Spectroscopic binaries, Visual Binaries, and Double stars); Yale Bright Star Catalogue; 2nd Catalog of Interferometric Measurements of Binary Stars (McAlister *et al.* 1988); ICCD Speckle Observations of Binary Stars I, II, IV, and V (McAl-

Astrophysics and Space Science **224**: 491–492, 1995.
© 1995 *Kluwer Academic Publishers.*

ister *et al.* 1987, 1987, 1989, 1990) and Speckle Observation of Spectroscopic Binaries I, II, III, IV, and V (Isobe *et al.* 1990a, 1990b, 1992, Miura *et al.* 1992, 1993).

83 objects were resolved by speckle observations. 56 objects out of 83 are multiple systems. Additionally to these, we find 14 multiple systems by comparing catalogues given above (HD 26961, 35411, 36695, 37021, 37041, 60178/9, 68273, 74874, 76644, 108248, 116656/7, 131041, 156015, 157978/9, 173648, Rigel B). In total, we have 70 multiple systems which correspond to 13% of 521 spectroscopic binaries. Combining this result and that by Isobe *et al.* (1993), 7% of spectroscopic binaries are triple systems with the configuration: two stars separated by about 0.1 AU and the third component star in an orbit with a distance about 10 AU from the central pair of stars.

7% is not small fraction. Therefore, we need some explanation to make this configuration during its formation process. 10 AU is too short a distance for the system to be formed by a capture mechanism. One should study a formation mechanism of the third component star in an envelope of spectroscopic binary stars.

References

Hirshfeld, A. & Sinnott, R. W. 1985, *Sky Catalogue 2000.0 Volume 2* (Cambridge University Press, Cambridge).

Hoffleit, D. 1982, *Bright Star Catalogue* (Yale University Observatory, New Haven).

Isobe, S., Norimoto, Y., Noguchi, M., Ohtsubo, J., Baba, N., Miura, N., Yanaka, H. & Tanaka, T. 1990a, Publ. Natl. Astron. Obs. Japan, 1, 217.

Isobe, S., Norimoto, Y., Noguchi, M., Ohtsubo, J., Baba, N., Miura, N., Yanaka, H. & Tanaka, T. 1990b, Publ. Natl. Astron. Obs. Japan, 1, 318.

Isobe, S., Noguchi, M., Ohtsubo, J., Baba, N., Miura, N., and Ni-ino, M. 1992, Publ. Natl. Astron. Obs. Japan, 2, 459.

Isobe, S., and Baba, N. 1992, in *Complementary Approaches to Double and Multiple Star Research, IAU Colloquium No. 135*, eds. H. A. McAlister and W. I. Hartkopf (Astronomical Society of Pacific, San Francisco), p. 555.

Isobe, S., Baba, N., Miura, N., and Ohtsubo, J. 1993, in *New Frontiers in Binary Star Research*, eds. K. C. Leung and I.-S. Nha (Astronomical Society of Pacific, San Francisco), p. 81.

McAlister, H. A., Hartkopf, W. I., and Hutter, D. J. 1987, Astron. J., 93, 688.

McAlister, H. A., Hartkopf, W. I., Hutter, D. J., Shara, M. M., and Franz, O. G. 1987, Astron. J., 93, 183.

McAlister, H. A., and Hartkopf, W. I. 1988, 2nd Catalog of Interferometric Measurements of Binary Stars (Center for High Angular Resolution Astronomy, Atlanta).

McAlister, H. A., Hartkopf, W. I., Dombrowski, E. G., and Franz, O. G. 1989, Astron. J., 97, 510.

McAlister, H. A., Hartkopf, W. I., and Franz, O. G. 1990, Astron. J., 99, 965.

Miura, N., Baba, N., Ni-ino, M., Ohtsubo, J., Noguchi, M., and Isobe, S. 1992, Publ. Natl. Astron. Obs. Japan, 2, 561.

Miura, N., Ni-ino, M., Baba, N., Iribe, T., and Isobe, S. 1993, Publ. Natl. Astron. Obs. Japan, 3, 153.

Shu, F. H., Adams, F. C., and Lizano, S. 1987, Ann. Rev. Astrophys., 25, 23.

APERTURE SYNTHESIS OBSERVATIONS OF HCN J=1-0 EMISSION FROM Y CANUM VENATICORUM

H.IZUMIURA and A.FUJIYOSHI
Department of Astronomy and Earth Sciences, Tokyo Gakugei University
4-1-1 Nukui-kita, Koganei, Tokyo 184, Japan

and

N. UKITA
Nobeyama Radio Observatory, National Astronomical Observatory
Minami-maki, Minami-saku, Nagano 384-13, Japan

Abstract. We have made high resolution observations of HCN (1–0) emission from the carbon star Y Canum Venaticorum using the Nobeyama Millimeter Array. We find that the emission region is not well resolved by the synthesized beam of 3.7"×4.6" over the entire velocity range (V_{LSR} =10 to 35 km s^{-1}). We find that the true brightness temperature probably exceeds 200 K at many velocity channels as well as at the 26 km s^{-1} maser spike. The broad emission component may be the result of superimposed maser spikes. The high brightness requires an unreasonably high HCN fractional abundance if LTE is assumed. It is likely that the HCN abundance previously reported for the star is considerably affected by the maser action. A new maser spike has been found at V_{LSR} = 29 km s^{-1}

Key words: carbon stars — HCN — abundances — masers — Y CVn

Although it has been widely detected from both optically bright and infrared luminous carbon stars, the HCN J=1–0 emission is still a matter of controversy (Olofsson *et al.* 1993; Izumiura, Ukita & Tsuji 1995). The reason is the existence of maser action in the line, which may affect the HCN abundance measure. For correct understanding, direct measurements of the outer radius and brightness temperature of the emission region are essential. We therefore made high resolution observations of the HCN emission toward Y CVn using the Nobeyama Millimeter Array. The star has been recognized as a source of both the H^{12}CN and H^{13}CN ground state masers (Izumiura *et al.* 1987; Izumiura *et al.* 1995). The observations were made in February and April 1993, with each of 1024 spectral channels having 0.53 km s^{-1} velocity width. The synthesized beam was 4.7"×3.7"(FWHM).

The extent of the emission was not well resolved even in the individual channel maps. We detected a new maser component at V_{LSR} = 29 km s^{-1} in the reconstructed spectrum. The spectrum was almost identical to that observed with the 45m telescope on 6 March 1994 confirming the new component. We fitted a one-component 2-dimensional Gaussian to the emission in each channel map (IMFITtwo, and deconvolved the restoring beam from the Gaussian to obtain a size estimate of underlying true emission region. The possible maximum sizes of the the emitting region in the various velocity channels ranged around 2" (FWHM). We then derived the intrinsic brightness profile of the emission (Figure 1).

Astrophysics and Space Science **224**: 493–494, 1995.

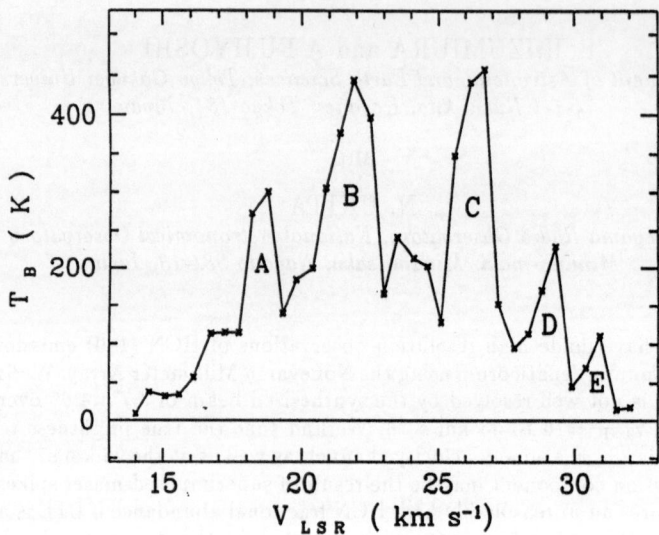

Fig. 1. Derived profile of the intrinsic brightness of the HCN emission against the line of sight velocity. We identified five brightness peaks (A-E). Peaks C and D correspond to the formerly known maser spike and a new one detected in this study, respectively.

The figure shows that the intrinsic brightness probably exceeds 200 K at many velocity channels. Moreover, there are several brightness peaks (A–E). The peaks C and D precisely correspond to the 26 and 29 km s^{-1} maser spikes. The peaks A, C, and E correspond to the hyperfine components of F=0–1, 2–1, and 1–1, respectively, in terms of their radial velocities. Furthermore, the separation between the peaks B and D is just the same as that of the F=0–1 and 2–1 hyperfine components. These results imply that the HCN emission from Y CVn mostly consists of a series of high brightness peaks which are related to the hyperfine pattern.

Our results indicate that the extent of the emission region is likely compatible with the photodissociation radius derived by Olofsson *et al.* (1993). If we assume LTE, we would have to introduce an unrealistic HCN fractional abundance ($\sim 10^{-3}$) as discussed in Izumiura *et al.* (1995). Special attention should be paid to the excitation condition of HCN molecules when the fractional abundance is derived through the mm-wave emission.

References

Izumiura, H., Ukita, N., Kawabe, R., Kaifu, N., Tsuji, T., Unno, W., & Koyama, K.: 1987, *ApJL* **323**, L81
Izumiura, H., Ukita, N., & Tsuji, T.: 1995, *ApJ* **20 Feb 1995 issue**, in press
Olofsson, H., Eriksson, K., Gustafsson, B., & Carlström, U.: 1993, *ApJS* **87**, 305

EXTENDED DUST SHELLS AROUND CARBON STARS
RESOLVED BY "HIRAS"

H. IZUMIURA *, D. J. M. KESTER and T. DE JONG
SRON Laboratory for Space Research Groningen
Landleven 12, P.O.Box 800, 9700 AV Groningen, The Netherlands

C. LOUP
Institut d'Astrophysique de Paris, France

L. B. F. M. WATERS
Astronomical Institute Anton Pannekoek, University of Amsterdam, The Netherlands

and

TJ. R. BONTEKOE
Bontekoe Data Consultancy, J.Bergmanstraat 3, 2221 BM Katwijk, The Netherlands

Abstract. We have examined forty-two carbon stars which show excess emission at 60 and/or 100 μm by applying maximum-entropy image reconstruction techniques to the IRAS 60 μm survey data. Thirteen stars are found to be extended in the reconstructed images. Four of them show a detached ring centered on the stellar position. In particular, U Ant may have a double detached dust shell. The implications of our results are discussed concerning the variation of mass loss on the AGB evolution.

Key words: HIRAS — carbon stars — dust shells — thermal pulses

There is growing evidence that some carbon stars possess detached dust shells that produce excess 60 and/or 100 μm emission observed with IRAS (Waters *et al.* 1994). The geometry of these dust shells allow us to investigate the history of mass loss on relatively long time scales.

We examined forty-two carbon stars with 60 and/or 100 μm excess and a 60 μm flux greater than 5 Jy by applying maximum-entropy image reconstruction techniques (HIRAS, Bontekoe *et al.* 1994) to the IRAS 60 μm survey data to examine the spatial distribution of the excess emission. Thirteen stars were found to be extended in the reconstructed images. In Fig. 1, U Hya, Y CVn and X Tra show detached rings centered on the stellar positionand U Ant a well resolved central plateau. This probably corresponds to the detached shell seen in CO emission in this star (Olofsson *et al.* 1990). Furthermore, U Ant possibly possesses an outer faint shell. The shell parameters are given in Table 1. These results suggest that mass loss rates on the AGB vary considerably on time scales compatible with those of thermal pulses, and that the higher mass loss phase may be repeated several times among a certain kind of carbon stars (e.g. Vassiliadis & Wood 1993). Moreover, U Ant may imply a possibility of an interpulse period less than 10^4 years.

* On leave from Dept. of Astronomy and Earth Sciences, Tokyo Gakugei University

Astrophysics and Space Science **224**: 495–496, 1995.
© 1995 Kluwer Academic Publishers.

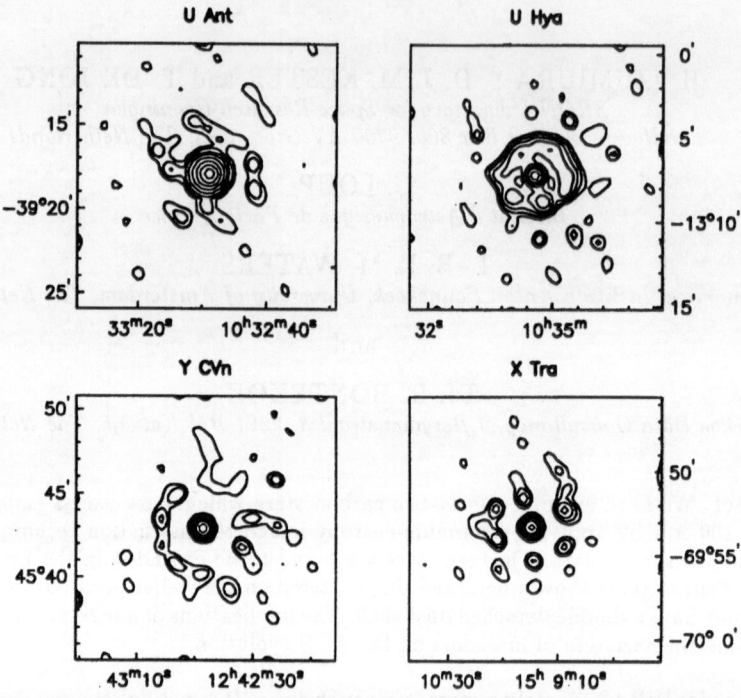

Fig. 1. Reconstructed images of U Ant, U Hya, Y CVn, and X Tra. Each map shows a square area of $16' \times 16'$. The resolution is about $1'$. Contours are in steps of powers of 2 MJy sr^{-1} from 2 MJy sr^{-1}

TABLE 1
Dust shell parameters.

	IRAS PSC		Radius	Age	Ve[1]	Distance[1]
Star	F60 (Jy)	Structure	(')	($\times 10^4$yrs)	(km s^{-1})	(kpc)
U Ant	27.1	semi-detached	<1.4	<0.6	20	0.32
...	...	detached	2.8	1.3	...	...
U Hya	17.2	detached	1.8	1.0	15	0.29
X Tra	14.8	detached	2.3	1.3	15	0.30

1) Values assumed for the fifth column.

References

Bontekoe, Tj. R., Koper, E., & Kester, D. J. M.: 1994, *A&A* **284**, 1037

Olofsson, H., Carlström, U., Eriksson, K., Gustafsson, B., & Willson, L.A.: 1990, *A&A* **230**, L13

Vassiliadis, E. & Wood, P. R.: 1993, *ApJ* **413**, 641

Waters, L.B.F.M., Loup, C., Kester, D. J. M., Bontekoe, Tj., R., & de Jong, T.: 1994, *A&A* **281**, L1

THE CIRCUMSTELLAR STRUCTURE OF LYNDS 379 IRS1

M.L. KELLY and G.H. MACDONALD

Electronics Laboratory, University of Kent, Canterbury, Kent.

Abstract. ^{12}CO J=2–1 maps of L379 IRS1 show a molecular outflow seen almost end-on while C^{18}O J=2–1 emission covers a smaller central region, tracing virially bound material deeper within the cloud. Continuum maps at 450, 800 and 1100μm all trace an identical double peaked arc west of IRS1 and VLA NH$_3$ (1,1) & (2,2) integrated intensity maps reveal the same double-peaked structure. An identical velocity gradient is seen in ^{12}CO, ^{13}CO, C^{18}O and NH$_3$ (1,1) & (2,2) following the arc-like structure of the continuum emission.

Key words: YSO — molecular cloud — L379 — IRAS 18265–1517

1. Observations and Background

Lynds 379 is a sharp-edged dark molecular cloud at a distance of ~2kpc. IRS1 (IRAS 18265–1517) is the brightest of three deeply embedded IRAS sources in L379. In April 1994, L379 IRS1 was mapped in the J=2–1 transition of ^{12}CO, ^{13}CO and C^{18}O and 450, 800 and 1100μm continuum using the JCMT. This project aimed to clarify the nature of IRS1 and improve upon previous ^{12}CO maps of the molecular outflow associated with it (Hilton *et al.* 1986, Wilking *et al.* 1990). Results from these observations and previous NH$_3$ (1,1) and (2,2) VLA maps of IRS1, are presented here.

2. Results

The blue and red shifted ^{12}CO emission (Fig. 1) are virtually coincident, but the peaks are separated by 26″ (0.25pc) suggesting that the outflow is orientated almost end-on to the observer. The apparent centre of the outflow is 20″ (0.2pc) north of IRS1, so it is not likely that IRS1 is the source of the outflow.

The C^{18}O map (not shown) peaks ~0.1pc W of IRS1 with a FWHM radius of 0.3pc. The average column density N(C^{18}O) is 9.8×10^{16} cm^{-2} and this gives a mass of ~2600M$_\odot$ within a radius of 1.55pc. The corresponding virial mass is only ~1600M$_\odot$ assuming a constant density sphere, hence the cloud core is virially bound.

All the continuum maps delineate the same arc-like structure with one peak northwest of IRS1 and a second peak southwest of IRS1 (Fig.2a). Integrated NH$_3$ emission in both mainline (Fig. 2b) and hyperfine maps has the same appearance and this arc is also seen on a larger scale in the molecular outflow traced by CO.

Astrophysics and Space Science **224**: 497–498, 1995.

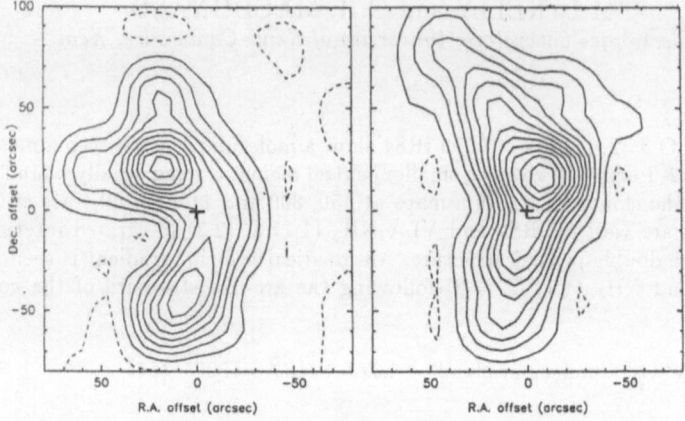

Fig. 1. ^{12}CO 2–1 maps: (a) Red–shifted emission integrated from 25 to 40 kms^{-1} (b) Blue–shifted emission integrated from -10 to 10 kms^{-1}.

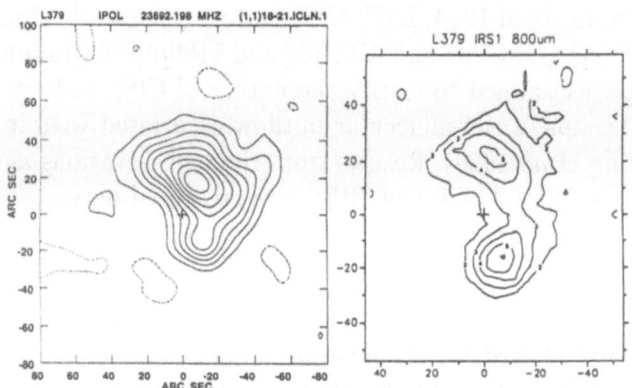

Fig. 2.   (a) NH$_3$ (1,1) integrated emission (b) 800μm emission (Jy/beam)

NH$_3$ channel maps (not shown) demonstrate the migration of the peak intensity along the arc from north to south with decreasing velocity. This is measured at 14 kms^{-1}pc^{-1} and can be compared with equivalent velocity gradients of 13 kms^{-1}pc^{-1} from C^{18}O, 11 kms^{-1}pc^{-1} from ^{13}CO and 12 kms^{-1}pc^{-1} from ^{12}CO. Therefore gas near the north dust clump is receding and gas near the south clump is approaching the observer, both in the inner and the outer regions of the molecular cloud.

References

Hilton, J., White, G.J., Cronin, N.J., Rainey, R.: 1986, *A&A* **154**, 274.
Wilking, B.A., Blackwell, J.H., Mundy, L.G.: 1990, *ApJ* **100**, 758.

CO OBSERVATIONS OF SEMIREGULAR VARIABLES*

F. KERSCHBAUM
Institut für Astronomie, Wien, Austria

H. OLOFSSON
Stockholms Observatorium, Saltsjöbaden, Sweden

and

J. HRON and B. ARINGER
Institut für Astronomie, Wien, Austria

Abstract. O-rich SRa and SRb variables have been observed in the ^{12}CO(1–0) and (2-1) lines. A total of 40 stars were observed and 19 detections can be reported. The majority of the objects are weak in CO and have envelopes with small expansion velocities between 1.9 to 15.6 km/s. No correlation is found between period and expansion velocity for the short period objects whereas a trend may be present for the long period ones.

Key words: stars: asymptotic giant branch — stars: variable — stars: mass-loss

1. Introduction

Kerschbaum & Hron (1992, 1994) proposed that the semiregular variables of type SRas are a mixture of Miras and SRbs. O-rich SRbs consist of a short period, 'blue' group without mass-loss and a 'red' group similar to Miras but with shorter periods. The 'blue' SRVs seem to be on the early AGB, the 'red' SRVs could precede the Mira phase. Jura & Kleinmann (1992) arrived at similar conclusions. In order to supplement our knowledge about dust mass-loss (IRAS photometry) with information about gas mass-loss a sample of O-rich SRVs were selected for observations in thermal ^{12}CO (1–0) and (2–0) lines. The observations were obtained with the 15 m Swedish-ESO Submillimeter Telescope (SEST), Chile and the 20 m dish at Onsala Space Observatory (OSO), Sweden.

2. First results

For our 40 stars 19 detections (16 new) and 2 tentative ones can be reported. The majority of the detected objects, covering short and long periods, are weak in CO, i.e. they are low mass-loss rate objects, and have envelopes with small expansion velocities ranging from 1.9 to 15.6 km/s. This is in agreement with the results of Kahane & Jura (1994). Moreover, the velocities are similar to those derived by Netzer & Elitzur (1993) from models of O-rich

* Supported by the Austrian *FWF*, project number P 9638-AST. Based on observations at ESO, Chile and Onsala Space Observatory, Chalmers Tekniska Högskola, Sweden. This research has made use of SIMBAD, CDS, Strasbourg, France.

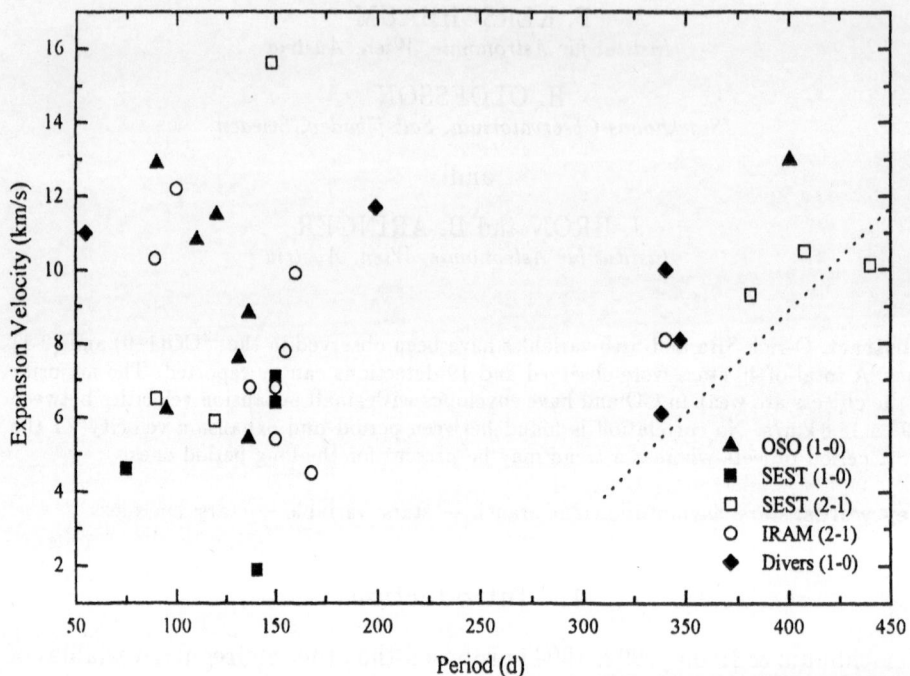

Fig. 1. Expansion velocity as a function of period for O-rich SRVs. OSO and SEST represent our new observations, IRAM (Kahane & Jura 1994) and Divers (Loup et al. 1993) come from the literature. The dashed line is taken from Wood (1990).

objects with mass-loss rates of a few $10^{-7}\ M_\odot \mathrm{yr}^{-1}$. In Fig. 1 the expansion velocity is plotted versus the period for SRVs with $P < 450$ days. To increase the number of objects data of Kahane & Jura (1994) and Loup et al. (1993) are included. The velocity ranges are similar for stars with periods below 200 and above 350 days. No correlation is found between period and expansion velocity for short period objects whereas a trend may be present for the long period stars. From studies of Mira variables Wood (1990) proposed a correlation of expansion velocity with period. Wood's P/v_{exp} relation is indicated by a dashed line in Fig. 1. We will perform additional observations in order to better characterize the mass loss properties of O-rich SRVs, and comparisons with Miras as well as carbon stars will be made.

References

Jura M., Kleinmann S.G.: 1992, *ApJS* **79**, 105
Kahane C., Jura M.: 1994, *A&A*, in press
Kerschbaum F., Hron J.: 1992, *A&A* **263**, 97 (Paper I)
Kerschbaum F., Hron J.: 1994, *A&AS* **106**, 397 (Paper II)
Loup C, et al.: 1993, *A&AS* **99**, 291
Netzer, N., Elitzur, M.: 1993, *ApJ* **410**, 701
Wood, P.R.: 1990, Proc.: From Miras to Planetary Nebulae, eds. M.O. Mennessier, A. Omont, Editions Frontières, Gif sur Yvette, p. 67

INTERFEROMETRIC OBSERVATIONS OF
HCN AND CN TOWARDS CARBON STARS

M. LINDQVIST

Sterrewacht Leiden, Postbus 9513, 2300 RA Leiden, The Netherlands

R. LUCAS

IRAM, 300 rue de la Piscine, F-38406 St Martin d'Heres Cedex, France

H. OLOFSSON

Stockholm Observatory, S-13336, Saltsjöbaden, Sweden

A. OMONT

Institut d'Astrophysique de Paris, CNRS, 98 bis bd Arago, 75014 Paris, France

and

K. ERIKSSON and B. GUSTAFSSON

Astronomical Observatory, Box 515, S-75120 Uppsala, Sweden

Abstract. Using the IRAM interferometer we have observed four carbon stars (U Cam, CIT6, Y CVn, IRC+40540) in the $HCN(J = 1 \rightarrow 0)$ and $CN(N = 1 \rightarrow 0)$ lines. Here we present some results for CIT6 and U Cam.

Key words: stars: AGB — stars: carbon — stars: individual: CIT6, U Cam

Carbon-rich circumstellar envelopes (CSEs) have proven to be rich in various molecular species. However, observational estimates of molecular abundances, such as for HCN and CN, often rely on calculations of the sizes of the emitting regions (see, e.g., Olofsson *et al.* 1993). This is very unsatisfactory and we have therefore initiated a project aimed at direct measurements of the sizes. Furthermore, the molecular line brightness distributions provide information about the geometrical structures of the CSEs, and hence about the spatial properties of the stellar mass loss.

The observations were made using the IRAM interferometer on Plateau de Bure during 1993 and 1994. The data were calibrated, mapped and deconvolved using the IRAM software. Prior to mapping we averaged the hyperfine components of CN not affected by blending.

The brightness distributions of $CN(N = 1 \rightarrow 0)$ and $HCN(J = 1 \rightarrow 0)$ towards CIT6 are shown in Fig. 1a. The CN emission outlines a hollow shell surrounding and centred on the HCN emission, which peaks, within the errors, at the stellar position. This confirms the expectation that, in the case of carbon stars, HCN originates in the stellar photosphere and CN is a photodissociation product of HCN (Huggins & Glassgold 1982). However, the spatial extents of both HCN and CN appear to be significantly larger

Astrophysics and Space Science **224**: 501–502, 1995.

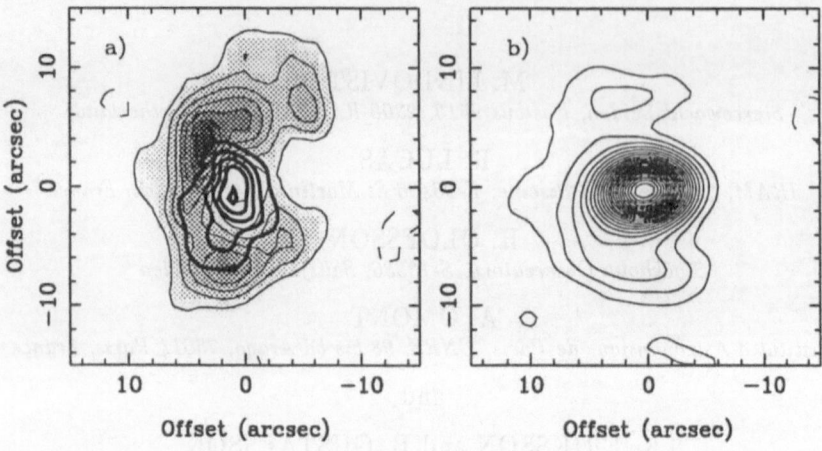

Fig. 1.   Velocity channel maps of *a)* CN($N = 1 \rightarrow 0$) (grey-scale+thin contours) and HCN($J = 1 \rightarrow 0$) (thick contours) towards CIT6 at $v_{\mathrm{LSR}} = -1.0\,\mathrm{km\,s^{-1}}$, roughly the systemic velocity, and *b)* of HCN($J = 1 \rightarrow 0$) towards U Cam at $v_{\mathrm{LSR}} = 0.6\,\mathrm{km\,s^{-1}}$, close to the systemic velocity ($\Delta v \approx 2.1\,\mathrm{km\,s^{-1}}$ in all three cases). The synthesized beams for CIT6 are (using uniform weighting) $\approx 4'' \times 3''$ and $\approx 3'' \times 2''$ for CN and HCN, respectively. The synthesized beam for U Cam is (using natural weighting) $\approx 6'' \times 4''$.

than was calculated by Olofsson *et al.* (1993) using a photodissocation model. They found a radius of $\approx 1''$ for the HCN envelope, and a radius of $\approx 3''$ for the CN shell. This could be due to several things, e.g., a weak interstellar UV-field, an efficient dust screening, and/or a clumped medium. The CN data also show the existence of a clear departure from spherical symmetry.

The HCN($J = 1 \rightarrow 0$) brightness distribution towards U Cam shows evidence for an outer HCN envelope, possibly a hollow shell with a radius of $\approx 8''$, that surrounds the bright central HCN emission peak (Fig. 1b). We estimate that the inner envelope and the outer shell have expansion velocities of 13 and $26\,\mathrm{km\,s^{-1}}$, respectively, though these might be a slight overestimate due to the hyperfine structure of HCN. These data, together with the $60\,\mu\mathrm{m}$ excess inferred from the IRAS data, suggest that U Cam has experienced a phase of episodic mass loss.

Acknowledgements

ML acknowledges the Leidsch Kerkhoven-Bosscha Fonds for travel support. ML is supported by an ESA external fellowship.

References

Huggins P.J., Glassgold A.E., 1982, ApJ, 252, 201
Olofsson H., Eriksson K., Gustafsson B., Carlström U., 1993, ApJS, 87, 305

THEORETICAL VISIBILITY CURVES

OF LATE TYPE STARS

S. LORENZ MARTINS

Observatoire de la Côte d'Azur, B.P. 229 - 06304 Nice cedex 4, France
Observatorio Nacional, R. Gal. Bruce 586, Rio de Janeiro, Brazil

and

J. LEFEVRE AND D. MEKARNIA

Observatoire de la Côte d'Azur, B.P. 229 - 06304 Nice cedex 4, France

Abstract. Theoretical visibility curves are presented for one carbon and one oxygen star. We aim to show the importance of measurements at wavelengths across the band.

Key words: Stars: individual: Y Tau, α Ori — circumstellar matter — visibility curves

1. Introduction

Late type stars represent one of the most interesting types of object to be studied by high resolution methods. These stars have extended atmospheres and high luminosities in red and near-infrared wavelengths. Carbon stars present SiC band emission at 11.3 μm and oxygen stars present silicate band emission at 9.7 and 18 μm.

Infrared interferometry with a baseline of few tens of metres allows for the bright, and nearest, sources to be resolved in the shells but not the stars. This circumstance is favourable for separating in the visibility curve the relative contributions of stellar and shell radiation to the total flux. Across an absorption band (SiC or silicate) this relative contribution varies rapidly: the extinction of stellar radiation is enhanced in the band whereas the shell emission is at maximum. In some cases the resultant variation of the visibility is large.

2. Theoretical visibility curves

The Monte Carlo method was employed for representing the propagation of radiative energy photon by photon, solving the radiative transfer in the envelope (Lorenz Martins & Lefèvre 1994). The models provide, among other quantities, the brightness distribution as a function of wavelength.

We have calculated theoretical visibility curves at several wavelengths and for several baselines. For a fixed value of the baseline, the variation of the visibility with the wavelength is maximum when the relative contribution of the star and the shell to the total flux are equal - either in the band or in the nearby continuum (Fig. 1).

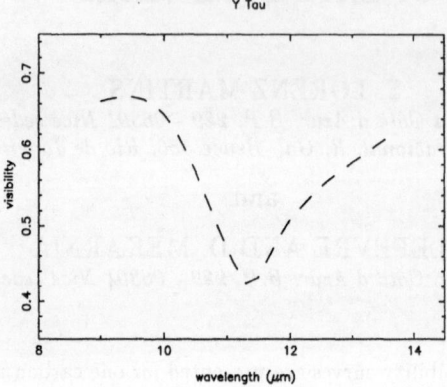

Fig. 1. Variation of the visibility with wavelength for (a) Y Tau and (b) α Ori

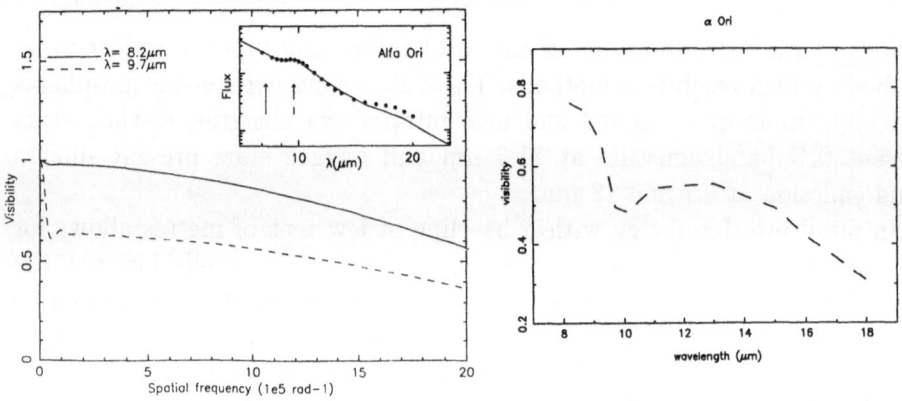

Fig. 2. Theoretical visibility for α Ori: visibility in the continuum (full line) and in the silicate band (dashed line). The inset shows the best fit for the silicate band.

The theoretical visibility curve (Fig. 2) reveals the importance of measurements in several wavelengths across the band. This result should encourage interferometrists to develop infrared interferometry in the 10 μm band with spectral resolution. The main physical result from such measurements will be a significant improvement of our knowledge of optical properties of silicate and SiC grains in this wavelength range.

References

Lorenz-Martins, S. and Lefèvre, J. (1994) A&A in press

INFRARED VELOCITY MAPPING OF COMPACT HII REGIONS

STUART LUMSDEN

Anglo-Australian Observatory, PO Box 296, Epping, NSW2121, Australia

and

MELVIN HOARE

Max-Planck-Institut fur Astronomie, Konigstuhl, D-6900 Heidelberg 1, Germany

Abstract. We present initial results of a programme to map the velocity field of a sample of compact HII regions using near infrared HI Brγ emission. Data are shown in particular for the well studied region G29.96–0.02.

Key words: Compact HII Regions

1. Introduction

Compact HII regions are the most visible manifestation of young hot OB stars which are still deeply embedded within their natal molecular clouds (e.g. Churchwell 1990). Although heavily obscured, these objects can be seen even on the far side of the galaxy at IR and radio wavelengths. Hence they are extremely useful probes of the earliest stages of massive star evolution and the star formation rate (SFR). Although this has been realised for ~ 30 years now it was only relatively recently that Wood & Churchwell (1989: WC89) discovered a huge excess in the number of compact HII regions. They found over an order of magnitude more compact sources in a VLA survey of star formation regions than expected from the predictions of a simple Strömgren sphere analysis.

New theoretical models have been advanced to explain the WC89 result. Any mechanism that severely constrains the growth of the nebula may be sufficient to explain the observed overabundance. In this context, the bow-shock model of Van Buren & MacLow (1992: VB) has received much attention recently. Here, supersonic *stellar* motions trap the ionization front, which is continually fed with fresh neutral gas from the stellar motion through the natal molecular cloud. Wood & Churchwell (1991: WC91) have used high spatial resolution radio recombination line (RRL) data to map the kinematics of a single compact HII region, G29.96–0.02. VB have analysed this data within the context of their model and claim good agreement. However, there are problems with relying on radio data alone. RRL's are intrinsically very weak, and prone to pressure broadening effects at high densities. One way to circumvent all of these problems is to map the HII regions using near infrared HI lines.

2.   Observations

We have obtained near-IR HI Brγ maps with the common user cooled grating spectrometer CGS4 on UKIRT. The effective spectral resolution is $\sim$ 20kms^{-1}, and the spatial resolution is $\sim$ 2″. The slit was aligned with the assumed symmetry axis of the HII region. The data shown here are for G29.96-0.02. Eleven separate spectra were required to make the velocity maps. Our data are in excellent agreement with WC91, but cover the whole nebula and not just the the bright core (a difference of 20–30 in area), illustrating the difference in sensitivity between radio and IR methods.

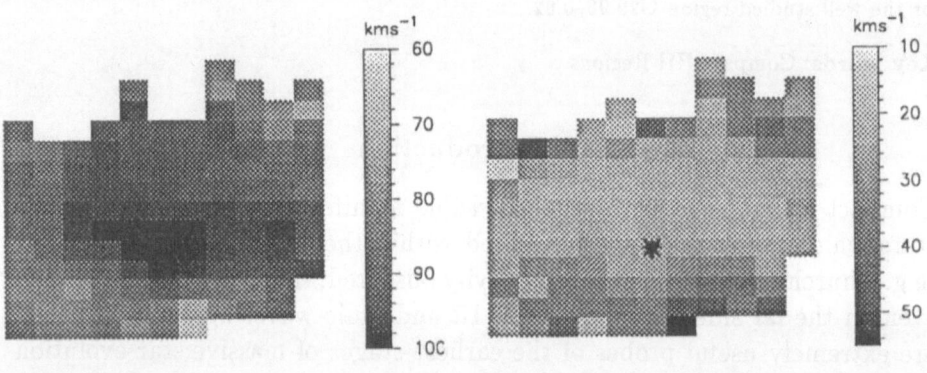

Fig. 1.   Velocity centroid and line width grey scale maps of G29.96. 'Up' is 60° east of north. The position of the exciting star is indicated (*).

The bow-shock model predicts the line FWHM will be largest near the 'head' of the region, where a stagnation point is created and supported by the shock that the stellar wind drives through the molecular cloud. As noted by VB, the data are consistent with viewing angle for G29.96 of $\sim$ 135°. Therefore, we also see shifts in the line centre velocity, since our line of sight intersects the 'tail' at different distances from the core, and hence we are observing different velocity gas on the near and far sides of the region. VB show that the most redshifted gas should be near the position of the star, in agreement with our data, and that as the sight-line intersects more of the tail, the gas velocity should tend back towards that seen at the head. Although a fuller analysis of our data is required, it seems likely that G29.96–0.02 is the best candidate for an HII region bow-shock currently available.

References

Churchwell, E.: 1990, *Astronomy & Astrophysics Review* **2**, 79
Van Buren, D. and MacLow, M.-M.: 1992, *ApJ* **394**, 534
Wood, D.O.S. and Churchwell, E.: 1989, *ApJS* **69**, 831
Wood, D.O.S. and Churchwell, E.: 1991, *ApJ* **372**, 199

THE CHEMISTRY OF SILICON IN HOT MOLECULAR CORES

D.D.S.MACKAY

Electronic Engineering Laboratories, The University of Kent, Canterbury, Kent

Abstract. The gas phase chemistry of hot molecular cores is thought to depend upon the composition of evaporated ice mantle material and its subsequent gas phase reactions. We have modelled the silicon chemistry in these conditions from an SiH_4 precursor and find substantial fractional abundances of SiO, H_2SiO and HNSi on a timescale of a few 10^4 yr.

1. Introduction

The evidence for grain surface hydrogenation of accreting atomic species and incorporation of the molecular products into mantle ices in cold dense clouds continues to grow. A number of studies have been made of the chemistry of gas phase molecular material in hot core conditions associated with high mass protostars, assuming this material has been seeded with evaporating grain mantles. Charnley *et al.* (1992) have modelled the O- and N-rich chemistries observed in the Orion hot cores, assuming only a mixture of simple C, N and O molecular species are released. This modelling has been extended to incude the simpler alcohols (Charnley *et al.* 1993), sulphur (Charnley, 1993) and phosphorus (Charnley & Millar 1994).

The chemistry of silicon is similarly of interest in this connection, having a cosmic abundance approximately twice that of sulphur. Observed abundances of this element in the gas phase in diffuse clouds suggests that it is heavily depleted due to incorporation into silicate grains. Just a few percent of the solar abundance remains (Whittet 1992), although this may still amount to a fractional abundance of perhaps 10^{-8} or more. We assume that monosilane, SiH_4, is the likely reservoir for this remaining silicon accreted within dense clouds.

2. Results

We have calculated results for a range of densities and temperatures appropriate to hot core conditions. Detailed results will be published elsewhere (MacKay 1994), but Fig.1 shows results from a representative model in which $n(H_2) = 2 \times 10^7$ cm^{-3} and the kinetic temperature is 300K. SiH_4 is not readily destroyed but, from the 10% or so that is, significant fractional abundances of SiO, H_2SiO and HNSi are generated(Fig.1a). In any significant absence of molecular oxygen, the fractional abundances of atomic silicon, HCSi and $SiCH_2$ are much enhanced (Fig.1b).

Astrophysics and Space Science **224**: 507–508, 1995.

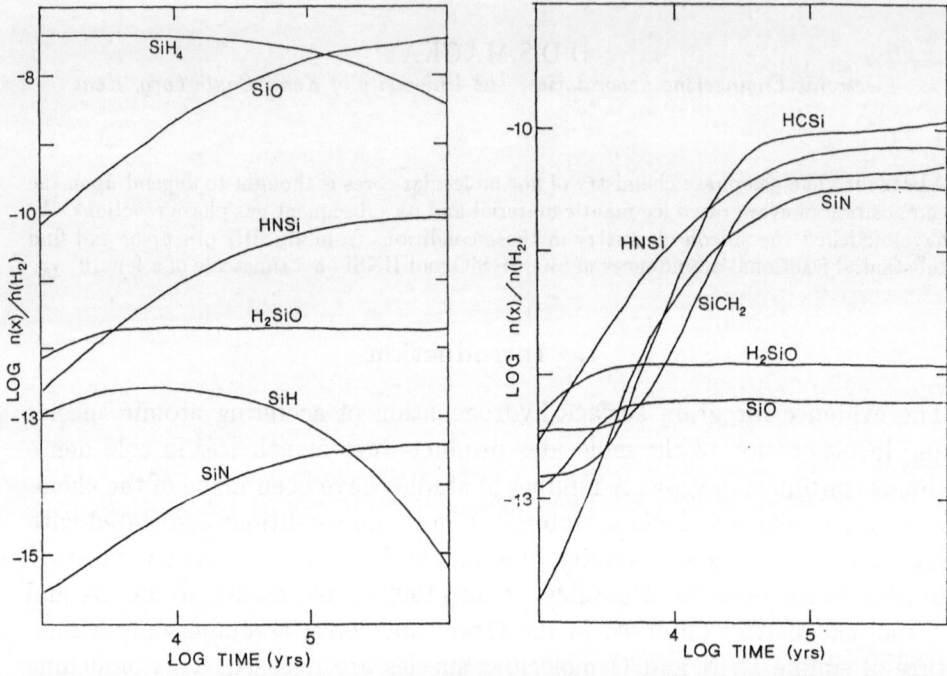

Fig. 1. (a)Fractional abundances as a function of time at a gas kinetic temperature of 300K and particle density of $2 \times 10^7 cm^{-3}$.(b) As for (a) but with zero O_2 initially.

A recent IR absorption non-detection of SiO against GL2591 has been reported at these proceedings by Evans (Carr *et al.* 1994) in which the column density limit is deduced to be $1.6 \times 10^{16} cm^{-2}$ at 200K. This represents a sensitivity in fractional abundance relative to H_2 down to a few 10^{-7}, which is only just above the fractional abundance predicted by our model for the reported conditions.

References

Carr, J.S., Evans, N.J., Lacy, J.H., Zhou, S., 1994, ApJ submitted
Charnley, S.B., Tielens, A.G.G.M., Millar, T.J., 1992, ApJ 399, L71
Charnley, S.B., 1993, ApJ submitted
Charnley, S.B., Kress, M.E., Tielens, A.G.G.M., Millar,T.J., 1993, ApJ submitted
Charnley, S.B., Millar, T.J., 1994, MNRAS 270, 570
MacKay, D.D.S., 1994, MNRAS submitted
Whittet, D.C.B., 1992, *Dust in the Galactic Environment*, IOP Publishing Ltd., Bristol

INFRARED POLARIMETRY AND IMAGING OF ULTRACOMPACT PARTIALLY IONIZED OPTICAL SOURCES IN THE ORION NEBULA

BRUCE MCCOLLUM
Astronomy Programs, Computer Sciences Corporation

and

MICHAEL W. CASTELAZ
East Tennessee State University

Abstract. Hubble Space Telescope images of the Orion nebula taken with the Wide-Field Camera have revealed subarcsecond structure in several dozen objects which are apparently ionized externally from nearby stars. We have obtained near-IR images and IR polarimetry of the Orion region to search for correlations with the WFC objects. We find that all of the ultracompact WFC objects are associated with IR features of some sort, and that some are associated with strongly polarized IR emission. The object with strongest polarization also shows small IR "lobes". In addition, we find some previously unreported sources, showing polarized IR emission, outside the field of the HST images, which we believe may be the same sorts of object. We note that the object with strongest polarization has a double-lobed appearance in the K band image.

Key words: infrared — PIGs — proplyds — Orion nebula

Hubble Space Telescope Wide-Field Camera (WFC) images have revealed several dozen subarcsecond optical structures in the Orion Nebula having diameters of 100 to 500 AU (O'Dell, Wen & Hu 1993; O'Dell & Wen 1994), many of which correspond to the $H\alpha$ emission sources discovered by Laques & Vidal (1979) and subarcsecond sources seen at radio wavelengths (Churchwell *et al.* 1987; Garay, Morgan & Reid 1987). The WFC images were taken with narrow-band filters centered on $H\alpha$, S[II], O[III], and N[II]. O'Dell & Wen (1994) give a comprehensive list of all such Orion sources discovered so far. These objects are sometimes called "PIGs" (partially ionized globules) or "proplyds" (protoplanetary disks), and are similar to what was predicted by Dyson (1968). Vidal (1982), Garay (1987), Meaburn (1988) and O'Dell *et al.* (1993) have modelled these as having high-density neutral interiors and outer parts which are ionized by nearby stars.

Using the Simultaneous Quad-Color Imaging Device ("SQIID") at KPNO, we have obtained simultaneous IR images in J, H, and K, with polarimetry in K, covering an area of about 5 x 5 arc minutes centered on the Trapezium, at a resolution of 1.2 arc sec per pixel, in order to compare with archival HST WFC images and look for correlation between optical and IR features.

Astrophysics and Space Science **224**: 509–510, 1995.
© 1995 *Kluwer Academic Publishers.*

Since our data were taken, O'Dell & Wen (1994) obtained additional WFC images covering another part of Orion which is not completely included in our IR field of view. We find that all of the proplyds in the area covered by our images correspond to IR features of some sort. Outside the WFC coverage, we find some polarized IR emission sources which do not correspond with previously known objects, which we believe may be additional PIGs/proplyds. We find no clear correlation between optical emission intensity as seen in WFC images and intensity of IR emission or polarization.

Our data show that 25 of 28 proplyds have 2μ emission counterparts. Eight of these show extended 2μ emission. All eight WFC sources having extended 2μ emission also have considerably larger polarization vectors than their surrounding areas, with vectors of up to 15% as compared with the approximately 5% average polarization of the background. We also note that these eight sources are associated with dark patches approximately 5" across in the WFC images, which show an absence of emission at the WFC filter wavelengths, and that these features are the ones with the largest IR polarization vectors. We believe that the strong IR polarization is due to scattering off dust which is probably physically associated with the WFC objects.

We call particular attention to object 183-405 (as listed in O'Dell & Wen 1994), which has an IR polarization of almost 15% at 2μ, the largest in our sample. In our IR images, this object appears bipolar, with the optical source in the middle of the IR "lobes" about 3 arc sec across. Prosser *et al.* (1994) find the central star to have V = 19.46 and V–I_c = 3.54, with a position on the color-magnitude diagram corresponding to an age of 3×10^5 – 10^6 years. In the WFC image, this object shows absorption at all filters. New post-repair WFC images (O'Dell and Wen 1994) show that 183-405 has a clearly resolved oblong shape (0.9" × 1.2"), and that it appears to be silhouetted against the nebular background.

References

Churchwell. W., Felli, M., Wood, D.O.S., & Massi, I. 1987, ApJ, 321, 516.
Dyson, F. 1968, Ap&SS, 1, 388.
Garay, G. 1987, Rev.Mex.Astron.Astrof., 14, 489.
Garay, G., Moran, J.M & Reid, M.J. 1987, ApJ, 147, 471.
Lacques, P. & Vidal, J.L. 1979, A&A, 73, 97.
Meaburn, J. 1988, MNRAS, 233, 791.
O'Dell, C.R, Wen, Z. & Hu, X. 1993, ApJ, 410, 696.
O'Dell, C.R. & Wen, Z. 1994, ApJ, to appear in November 20 issue.
Prosser, C.F., Stauffer, J.R., Hartmann, L., Soderblom, D.R., Jones, B.F., Werner, M.W. & McCaughrean, M.J. 1994, ApJ, 421, 517.
Vidal, J.L. 1982, in *Symposium on the Orion Nebula to Honor Henry Draper*, Ann. N.Y. Acad. Sci., 395, 176.

OBSERVATIONS WITH A MID-INFRARED CAMERA
OF AGB STAR ENVELOPES

M. MARENGO, M. BUSSO and L. ORIGLIA
Osservatorio Astronomico di Torino, 10025 Pino Torinese, Italy

P. PERSI and M. FERRARI-TONIOLO
Istituto di Astrofisica Spaziale CNR, 00044 Frascati, Italy

G. SILVESTRO
Istituto di Fisica Generale, Università di Torino, 10125 Torino, Italy

and

M. TAPIA
Istituto de Astronomia, UNAM, Ensenada, Baja California, Mexico

Abstract. We observed 17 AGB and post-AGB circumstellar envelopes wuth the mid-infrared camera TIRCAM. The collected photometry is compared with the IRAS LRS spectra and a two color diagram for the chemical classification of the dust is developed.

Key words: Stars: circumstellar shells — Infrared: sources

1. Introduction

Mass loss processes in the late evolutionary stage of intermediate and low mass stars known as Asymptotic Giant Branch (AGB) stars, cause the formation of optically opaque circumstellar envelopes of dust and gas, which will later evolve into planetary nebulae (Habing, 1989). The dust component of the envelopes dictates the global optical properties of the circumstellar matter, and is responsible for the observed infrared emissions of these sources (continuum emission from dust, emission/absorption bands from silicates at 9.7 μm in oxygen-rich envelopes, and from SiC at 11.3 μm in carbon-rich ones). A transition from O-rich to C-rich envelopes is expected to occur in the AGB as a consequence of nucleosynthesis and mixing (third dredge-up, Willems & deJong 1988): it can be analized by the IRAS [12]-[25], [25]-[60] color-color diagram (van der Veen & Habing 1988), which locates in separate regions AGB envelopes in different evolutionary stages (although a complete separation between O-rich and C-rich sources requires higher spectral resolution than available with the satellite).

A mid-infrared camera with good sensitivity and angular resolution appears to be a suitable instrument to investigate AGB circumstellar envelopes in the transition stage.

Astrophysics and Space Science **224**: 511–512, 1995.
© 1995 *Kluwer Academic Publishers.*

2. Observations and Discussion

The mid-infrared camera TIRCAM (Tirgo InfraRed CAMera, Persi *et al.*, 1994) equipped with a 10 × 64 Si:As array sensitive in the 5–25 μm wavelength range, was used in a few observing runs (1992–1994) at the National Mexican Observatory of San Pedro Martir (SPM, Baja California) and at the Italian Infrared Telescope (TIRGO) for imaging and photometry of 17 envelopes around AGB stars (9 O-rich, 7 C-rich and one post-AGB star).

For each source we collected the images in the available filters (see Table 1). Good agreement was found between the photometry thus obtained and the Low Resolution Spectra (LRS, 1986) measured by IRAS.

A color-color diagram in the TIRCAM photometric system has been reconstructed in order to chemically classify the sources, being the [8.8]–[12.5] color related to the chemical nature of the dust (O-rich showing larger color with respect to C-rich) and the [8.8]–[9.8] color associated with the evolutionary stage of the envelopes (redder colors imply more evolved envelopes).

TABLE I

Filters	Band	λ_{eff}	$\Delta\lambda$	Chemical characteristics
F3	8.4– 9.2	8.81	0.87	Dust continuum
F4	9.3–10.3	9.80	0.99	Silicate emission features (9.7 μm)
F5	9.8–10.8	10.27	1.01	Dust continuum
F6	11.2–12.3	11.69	1.11	SiC emission features (11.3 μm)
F7	11.9–13.1	12.49	1.16	Dust continuum

References

Habing, H., J.: 1989, 'The Evolution of Red Giant to White Dwarfs, a review of the Observational Evidence' in M. O. Menessier and A. Omont, ed(s)., *From Miras to Planetary Nebulae: Which Path for Stellar Evolution?*, Editions Frontiéres:Gif sur Yvette Cedex - France, 16

IRAS Catalogue and Atlases, Atlas of low-resolution spectra, Iras Science Team: 1986, *A&A Suppl* **65**, 607

Persi, P., Shivanandan, K., Busso, M., Bonazzola, G., Corcione, L., Ferrari-Toniolo, M., Nicolini, G., Racioppi, F., Robberto, M., Tofani, G.: 1994, *Experimental Ap.* , submitted

van der Veen, W., E., C., J. and Habing, H., J.: 1988, *A&A* **194**, 125

Willems, F., J. and deJong, T.: 1988, *A&A* **196**, 173

POLARIMETRIC PROPERTIES OF THE REFLECTION NEBULA R MON/NGC 2261

MASAFUMI MATSUMURA

Faculty of Education, Kagawa University, Takamatsu 760, Japan

and

MUNEZO SEKI

Astronomical Institute, Tohoku University, Sendai 980-77, Japan

Abstract. The linear polarization of the reflection nebula R Mon/NGC 2261 has been observed since November 1991 with a multi-channel polarimeter at Dodaira Observatory, one of the branches of the National Astronomical Observatory of Japan. Results of the observations are reported.

Key words: Polarization – Circumstellar grains – R Mon/NGC 2261

1. Introduction

Large linear polarizations are observed at optical and near infrared regions in the young stellar object R Mon, and in the nebula NGC 2261 which surrounds R Mon (Scarrott *et al.* 1989 for a review). Mapping observations reveal that polarization vectors make a 'centrosymmetric' pattern at the outer region. This can be easily explained by single scattering of light.

In the region close to R Mon, another pattern is observed. The polarization vectors are aligned along a certain direction. At least two interpretations are proposed for this pattern, namely, (1) dichroic extinction by aligned grains (e.g. Aspin *et al.* 1985), and (2) multiple scattering of light by circumstellar grains (e.g. Bastien & Ménard 1988, Fischer *et al.* 1994). Theoretical calculations of multiple scattering can explain the aligned pattern (Fischer *et al.* 1994), although we still cannot exclude the possible effect of dichroic extinction.

Observationally, temporal variations are reported both in brightness and in polarization of R Mon/NGC 2261. The properties of these variations are still not clear, although many observations have been made.

2. Results and Discussion

We observed the linear polarization of R Mon, i.e., the region of 'aligned pattern', with a multi-channel polarimeter at Dodaira Observatory, to discuss the mechanism of polarization. Since November 1991, we have made observations on several nights a year. A diaphragm of 18" has been used, because the 'aligned pattern' of polarization extends over 20" or 30" around R Mon.

Astrophysics and Space Science **224**: 513–514, 1995.
© 1995 *Kluwer Academic Publishers.*

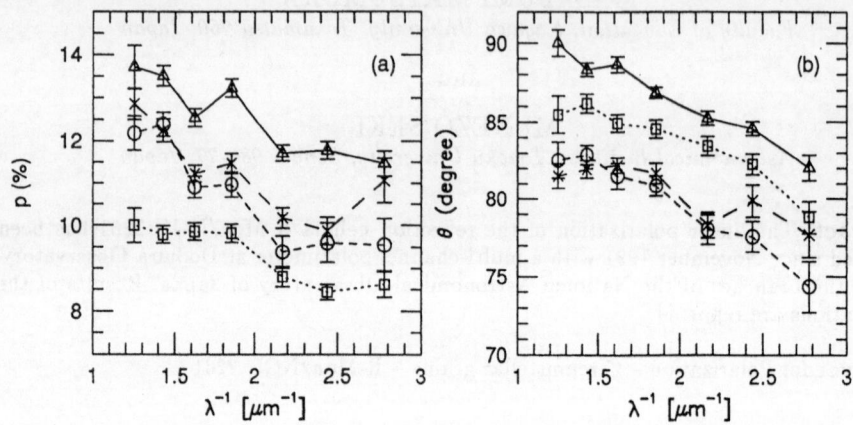

Fig. 1.   (a) Dependence of fractional polarization p on λ^{-1} and (b) that of position angle θ. The triangles indicate the results in November 1991, the squares November 1992, the circles January 1994, and the crosses February 1994.

Results of the observations are shown in Fig.1. Though temporal variations are apparent, permanent properties can be recognized. Fractional polarization p decreases monotonically with λ^{-1} (Fig.1a), where λ is wavelength. This behavior is completely different from the Serkowski law which is observed in the interstellar polarization. The position angle θ decreases with λ^{-1} all the time (Fig.1b), and this cannot be explained by the assumption that scattering particles are distributed in axial symmetry.

During the series of observation, we saw a variation in the flux from R Mon amounting to about 0.5 mag. This variation is found to be positively correlated with that of p. The increase of the magnitude, i.e. the decrease of flux, may indicate that the direct light from the star is extinguished by opaque dust clouds which happen to cross the line of sight. If the scattered light, which is strongly polarized, is less extinguished than the direct light, the observed net polarization will increase.

If this interpretation is correct, the circumstellar matter may be clumpy around R Mon, and the process of light scattering may be much more complicated than the currently proposed models.

References

Aspin, C., McLean, I. S., & Coyne, G. V.: 1985, *Astron. & Astrophys.* **149**, 158
Bastien, P. and Ménard, F.: 1988, *Astrophys. J.* **326**, 334
Fischer, O., Henning, Th., & Yorke, H. W.: 1994, *Astron. & Astrophys.* **284**, 187
Scarrott, S. M., Draper, P. W., & Warren-Smith R. F.: 1989, *Mon. Not. of the RAS* **237**, 621

NEW MERLIN OBSERVATIONS OF OH/IR STARS:

OH357.31–1.33 AND OH0.9+1.3

V. MIGENES

ATNF, PO Box 76, Epping, NSW 2121, Australia

E. LÜDKE, R.J. COHEN

Manchester University, NRAL, Jodrell Bank, Cheshire, SK11 9DL, UK

M. SHEPHERD

California Institute of Technology, Pasadena, CA, 91125, USA

and

P.F. BOWERS

Naval Research Laboratories, Code 7220, Washington DC. 20375, USA

Abstract. We present recent radio interferometer measurements of the OH 1612 MHz maser emission from the OH/IR sources OH0.9+1.3 and OH357.31 obtained with the enhanced MERLIN. Some preliminary results are briefly discussed. These results are part of an on-going observational campaign to obtain the best radio maps of bright OH/IR stars with MERLIN, VLA and VLBA, in order to understand the nature and dynamics of their circumstellar matter and evolution.

Key words: OH/IR Stars — MERLIN — VLA — VLBA

OH/IR stars are long-period variable stars which lie on the Asymptotic Giant Branch (AGB), with mass loss rates $\sim 10^{-6}$ $M_\odot$ yr^{-1} and are surrounded by oxygen-rich circumstellar envelopes with strong far-infrared emission from dust and maser emission from OH, H_2O and SiO molecules. These stars are also useful probes in the study of stellar galactic dynamics and AGB stellar evolution, since they are possibly precursors of planetary nebulae (Iben & Renzini, 1983).

The two sources were observed in the OH 1612 and 1667 MHz lines during December 1993 and January 1994 with spectral resolution of 3.9 kHz. Only the 1612 MHz results will be discussed in detail. Figure 1 shows results for OH0.9, which is a transition object with radio continuum emission as well as OH maser emission (Pottasch *et al.* 1987). Previous MERLIN observations of the source by Shepherd *et al.* (1990) indicated shell structure. With the improved signal-to-noise ratio on the longest baselines the shell is now properly resolved. The data confirm the steady increase in source flux. The strong blue-shifted component at −124 km s^{-1} has continued to increase at about 1Jy per year ($S_{1994.1} \sim 53.4$ Jy).

The double-peaked 1612 MHz spectrum of OH357.31–1.33, is given in Figure 2, together with a map of the emission at −39.5 km s^{-1}. This is one of the largest OH shells known. In order to map it accurately the MERLIN

Astrophysics and Space Science **224**: 515–516, 1995.

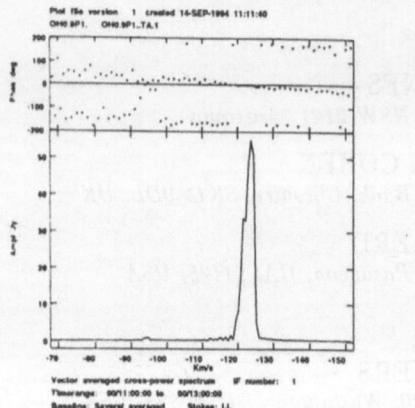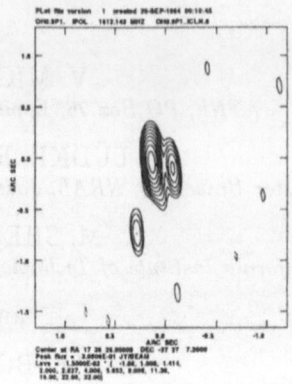

Fig. 1.   Total spectrum of OH0.9+1.3 and map at 1612 MHz at a radial velocity of −95 km s⁻¹. The shell is resolved with a beam of 0.35 × 0.1 arcseconds.

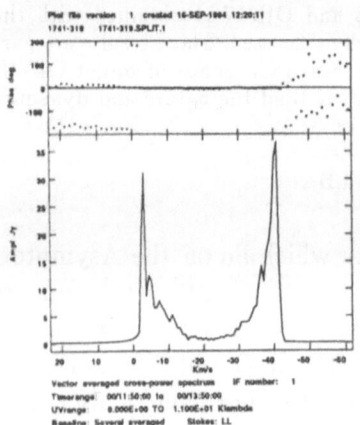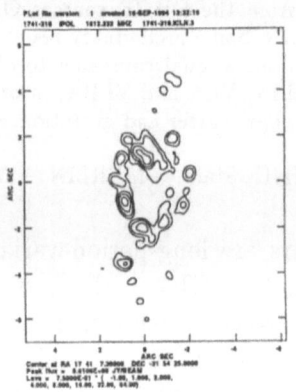

Fig. 2.   Total spectrum of OH357.3−1.3 and map at 1612 MHz at a radial velocity of −39.5 km s⁻¹, at 350 mas resolution. The spectrum at 1612 MHz is more symmetric than the 1667 MHz (not shown) which is possibly due to sudden changes in the excitation conditions at different positions in the circumstellar envelope.

data, which show the fine-structure in the shell, will be combined with VLA data which trace the large-scale structure. The final data set will consist of combined array maps which will allow us to study the fine structure of the shells, to estimate both shell size and thickness which is a direct indicator of the mass loss rate in post AGB stars.

References

Iben, I. & Renzini, A., 1983, *Ann Rev A &A*, **21**, 271-342.
Pottasch, S.R., Bignelli, C. & Zijlstra, A., 1987, *A &A*, **177**, L49-L52.
Shepherd M.C., et al., 1990, *Nature*, **344**, 522.

LONG-SLIT SPECTROSCOPY OF THE COMPACT
PLANETARY NEBULA HU 2-1

L.F. MIRANDA
Dpto. Astrofísica, Facultad de Ciencias Físicas, Universidad Complutense de Madrid,
E-28040 Madrid, Spain

M.T. EIBE
Armagh Observatory, College Hill, Armagh BT61 9DF, N. Ireland

C. EIROA AND E. ORTIZ
Dpto. Física Teórica CXI, Universidad Autónoma de Madrid, Cantoblanco, E-28049
Madrid, Spain

and

J.M. TORRELLES
Instituto de Astrofísica de Andalucía CSIC, Ap. Correos 3004, Sancho Panza s/n,
E-18080 Granada, Spain

Abstract. Long-slit spectra of high spectral and spatial resolution of the compact planetary nebula Hu 2-1, are presented. The analysis of the [NII] 6583 emission line detected in the spectra allows us to identify the kinematical components present in the nebula and to deduce their basic geometry. We use position-velocity maps of the [NII] 6583/Hα line intensity ratio in order to identify nebular regions in which shock-excitation and/or overabundace of N exist.

Key words: ISM: kinematics and dynamics — planetary nebulae: individual (Hu 2-1)

1. Introduction

Hu 2-1 is a planetary nebula (PN) which presents characteristics of a young PN: stellar image, low expansion velocity and relatively high electron density (Maciel & Pottasch 1980; Martin 1981; Sabbadin *et al.* 1984). In addition, the central star is located in the HR diagram at the begining of the evolutionary tracks of PNe (Martin 1981). At 5 and 8.4 GHz (Kwok 1985; Kwok & Aaquist 1993), Hu 2-1 exhibits an elongated structure with the major axis oriented at PA≈320°. In order to deduce the spatio-kinematical structure we have taken long-slit spectra of high spectral and spatial resolution of Hu 2-1. Here we summarize the results obtained.

2. Observations and Results

The spectra were obtained on 1993 July with the f/12 camera of the coudé spectrograph of the 2.2 m telescope at Calar Alto Observatory. The spectra cover the range 6540-6593 Å at a dispersion of 2.2 Å mm^{-1}. The slit was centered on the object and oriented at position angles (PAs) 229°, 245°,

Astrophysics and Space Science **224**: 517–518, 1995.
© 1995 *Kluwer Academic Publishers.*

310°, 320° and 330°. The spectral resolution (FWHM) is ≈12 km s^{-1} and the spatial resolution (FWHM) is ≈ 0".9.

The analysis of the data indicates that Hu 2–1 is a bipolar PN consisting of an equatorial toroid and two faint bipolar lobes. The polar axis of the nebula is oriented at PA ≈320° and tilted by ≈55° with respect to the line of sight. Equatorial expansion velocity is ≈19 km s^{-1}, polar expansion velocity is ≈39 km s^{-1}. The kinematical age of these structures is ≈500 yr (assumed distance 2.35 kpc, Maciel & Pottasch 1980, Martin 1981), a value compatible with the idea that Hu 2–1 is a very young PN. Two highly collimated bipolar condensations have been detected. They move along the polar axis at ≈100 km s^{-1} and their kinematical age is ≈450 yr. The [NII] 6583 line also shows the existence of two small components located antisymmetrically with respect to the center of the emission feature. The angular separation between these components varies from ≈ 0".34 at PA 320° up to ≈ 0" at PA 245°. The separation in radial velocity is ≈8.5 km s^{-1} and does not vary with PA. The spatio-kinematical properties of these components suggest that they represent a bipolar flow oriented at PA ≈320° and tilted by >90° with respect to the line of sight. This bipolar flow could be younger than the rest of nebular components. All nebular components just described are embedded in an extended circular halo. The halo can be traced from ≈ 5" up to ≈ 11".5 from the central star and its expansion velocity is ≈12.5 km s^{-1}. The kinematical age of the halo, which probably represents the remnant envelope of the red giant progenitor of Hu 2–1, ranges from ≈4000 yr at its inner edge up to ≈10000 yr at is outer edge.

The [NII] 6583/Hα intensity ratio indicates that shock excitation occurs in the bipolar condensations, although overabundance of N may also be present. In addition, the position-velocity [NII]/Hα maps reveal the existence of a toroidal region where [NII] emission is relatively strong. This region surrounds the equatorial toroid identified in the [NII] emission. Its expansion velocity is ≈25 km s^{-1} and its kinematical age is ≈1400 yr. The low expansion velocity suggests that shock-excitation does not occur in this region and that the relatively high [NII]/Hα ratio is due to overabundance of N.

References

Kwok S.: 1985, *AJ* **90**, 49
Kwok S., Aaquist O.B.: 1993, *PASP* **105**, 1456
Maciel W.J., Pottasch S.R.: 1980, *A&A* **88**, 1
Martin W.: 1981, *A&A* **98**, 328
Sabbadin F., Gratton R.G., Bianchini A., Ortolani S.: 1984, *A&A* **136**, 181

MASSIVE EARLY-TYPE EMISSION-LINE STARS:
AN ATTEMPT TO DISTINGUISH A NEW GROUP OF
STELLAR OBJECTS

A.S. MIROSHNICHENKO, YU.K. BERGNER
Central Astronomical Observatory of the Russian Academy of Sciences at Pulkovo,
196140, Saint-Petersburg, Russia, e-mail: anat@gaoran.spb.su

and

D.B. MUKANOV, T.A. SHEJKINA
Fessenkov Astrophysical Institute of the National Academy of Sciences of Kazakhstan
Republic, 480068, Almaty, Kamenskoe plato, Kazakhstan

Abstract. A new sample of possibly massive early-type emission-line stars (METELS) based on the previous lists of peculiar Be stars is presented. It consists of 36 objects divided amongst supergiants, possible binaries, and candidates to the list. The central stars are probably more massive than $\sim 10 M_\odot$. Two new relations allowing idientification of possible binaries among the objects are proposed.

Key words: emission line stars — supergiants — binaries

Peculiar early-type stars show strong emission-line spectra which points to strong mass loss. Most of those known today were identified by Allen & Swings (1976) and Carlson & Henize (1979). The purposes of our study are (1) to analyse published data and studies of Bep stars, (2) to present an extended sample of such objects, (3) to separate possible subgroups in the sample and (4) to report relationships amongst observational characteristics of these objects. A new impulse for investigations of early-type stars with circumstellar (CS) envelopes came from the papers of Dong & Hu (1991) where the emission-line stars with strong IR excesses were separated on the basis of the catalogues of Wackerling (1970) and IRAS Point Sources, and of Thé et al. (1994) where the new catalogue of Herbig Ae/Be (HAEBE) and related objects was presented. From our and published results, we have formed a list of objects with the following characteristics: (1) estimated masses more than $10 M_\odot$, (2) possible supergiants with strong Hα and IR excesses, (3) a strong emission-line spectrum and early-B types but (4) not considered as Planetary Nebulae (PN) or symbiotic systems.

Our list consists of 36 objects divided into three subgroups:
1. Supergiants: P Cyg, η Car, AG Car, HR Car, HD 160529, WRA 751, He3-519, HD 87643, HDE 316285, CPD$-52°9243$, CPD$-57°2874$, HDE 326823, HDE 327083, 3 Pup, MWC 300, MWC 314, MWC 349,
2. Possible binaries: GG Car, HD 89249, XX Oph, MWC 84, MWC 297, MWC 342, MWC 623, MWC 930, MWC 1080,

3. Candidates: MWC 137, WRA 1484, CD$-$42°11721, HD 45677, HD 50138, He3-759, WRA 966, $L_k H_\alpha 101$, AS 78, LS II +22 8.

We argue that this group has to be studied separately from other groups of early-type emission-line stars, because it possibly represents an evolutionary sequence of stars. Most massive HAEBEs and symbiotics at early stages of their evolution can be contained in the sample.

During our study we obtained two interesting results for the objects of our list. Studying the diagram $Log S_{12}/S_{25} \sim Log S_{12}/S_V$ we have found that several METELS with CS dust are situated in the region of late- type stars (MWC 84, MWC 623, CPD -52°9243, CPD -57°2784, HD 89249). MWC 623 and HD 89249 were suspected in binarity earlier (Zickgraf & Stahl 1989 and Stahl & Leitherer 1987, respectively). We have detected absorptions of a late-type star in the spectrum of MWC 84 (Miroshnichenko 1994). Other stars are considered to be single objects.

The diagram of Balmer decrements $(Log(I_{H_\alpha}/I_{H_\beta}) \sim Log(I_{H_\gamma}/I_{H_\beta}))$ shows that single stars with strong CS envelopes and binaries with dominating late stars (symbiotics) are situated near the reddening line. Only several objects lie beyond the common relationship (MWC 84, MWC 930, XX Oph, HD 89249, CPD $-$52°9243). MWC 930 was suspected in binarity because it showed a quasi-periodic variability and absorptions of a late-type star (Miroshnichenko 1994).

In our opinion, the following steps must be made to accelerate investigations of METELS: (1) obtain homogeneous data on all METELS : photometry from UV to far IR, at least optical polarimetry, spectra of medium and high resolution, (2) to use as many as possible methods to determine interstellar reddening, (3) to organize a program to investigate METELS variability, (4) to make a search for new candidates to the list of METELS and (5) to model observational characteristics of METELS (emission-line profiles, SEDs, wavelength-dependence of polarization) in the framework of non-spherical envelope models (for example, Efstathiou & Rowan-Robinson 1990 and de Araujo et al., 1994).

References

Allen D.A. & Swings J.P.: 1976, *A&A* **47**, 293
Carlson E.D. & Henize K.G.: 1979, *Vistas in Astronomy* **23**, 213
de Araujo F.X., de Freitas Pacheco J.A. & Petrini D.: 1994, *MNRAS* **267**, 501
Dong Yi-sun & Hu Jing-yao: 1991, *Chin. A&A* **15**, 275
Efstathiou A. & Rowan-Robinson M.: 1990, *MNRAS* **245**, 275
Miroshnichenko A.S.: 1994, *Astronomical and Astrophysical Transactions* **5**, 1
Stahl O. & Leitherer C.: 1987, *A&A* **177**, 105
Thé P.S., de Winter D. & Pérez M.R.: 1994, *A&AS* **104**, 315
Wackerling L.R.: 1970, *Mem.RAS* **73**, 153
Zickgraf F.-J. & Stahl O.: 1989, *A&A* **223**, 165

MID-IR IMAGING POLARIMETRY AND CIRCUMSTELLAR MAGNETIC FIELDS

T. J. T. MOORE, C. H. SMITH, D. K. AITKEN and T. FUJIYOSHI

Physics Department, University College, ADFA
Canberra, ACT, Australia 2600

Abstract. We have obtained diffraction-limited images of the mid-infrared emission and polarization patterns in a number of southern-sky objects. By mapping the polarization produced in absorption or emission by aligned aspherical dust grains, we have been able to trace the detailed spatial structure of magnetic fields in the warm circumstellar material of young molecular-outflow sources and HII regions, in the expanding dust shell of the mass-losing star Eta Carinae, and in the inner parsec of the Galactic Centre.

Key words: circumstellar matter — magnetic fields — polarization

1. Introduction

Polarization observed at mid-infrared wavelengths is due to absorption by or emission from aligned aspherical dust grains. The alignment may be produced directly by the presence of an ordered magnetic field via the classical or modified Davis-Greenstein mechanisms or by other processes such as gas-grain streaming. In the latter case, the Barnett effect should induce rapid precession of the grain spin about any ambient magnetic field and the local field direction will still determine the mean orientation of the alignment. Mid-infrared polarization patterns, therefore, yield important information on the spatial distribution of any magnetic fields present in dusty circumstellar material around objects of many different types.

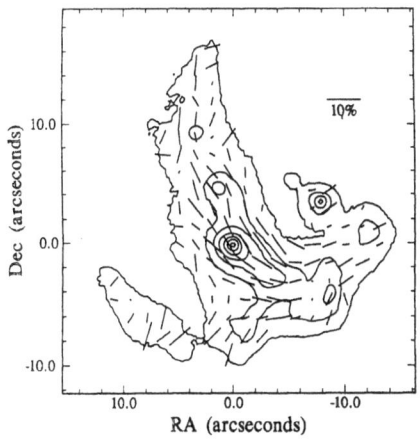

Fig. 1. The Galactic Centre

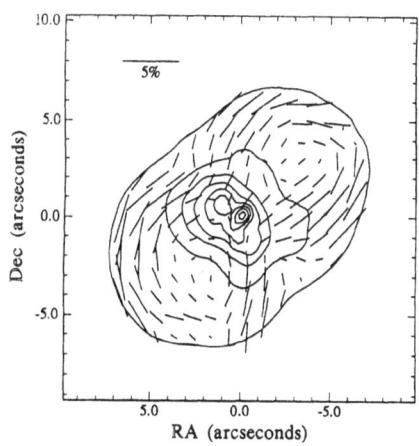

Fig. 2. η Carinae

Astrophysics and Space Science **224**: 521–522, 1995.

2.   Observations and results

Observations were made at the Anglo-Australian Telescope in 1994 May, using the new mid-IR imaging polarimeter (NIMPOL) built at University College, UNSW, Canberra (Smith *et al.* 1994). The results reveal the distribution of polarization with diffraction-limited ($\sim 0.7''$) resolution in a number of sources of key astrophysical interest. All images were taken using a 10% $\lambda 12.5\,\mu$m filter except that of η Carinae, which is broad-band N. Fig. 1 shows the central parsec of the Galactic Centre. The distribution of emission from warm dust is closely correlated with the radio continuum (e.g. Killeen & Low 1989). Polarization vectors are drawn normal to **E** and hence parallel to a magnetic field that closely follows the curve of the 'Northern Arm', along which material is streaming towards the Galactic Centre. In Fig. 2, emissive polarization in η Carinae forms a strikingly organised pattern (vectors are drawn parallel to the aligning magnetic field) which traces the morphology of the expanding dust shell ejected from this massive, unstable star. Fig. 3 shows absorptive polarization in G298.6–0.2, a distant (12 kpc) HII region. The vectors are drawn parallel to **E** and their alignment is close to the direction of the interstellar magnetic field as determined from observations of field stars out to 5 kpc (Wright 1994). Fig. 4 is the southern HII region G333.6–0.2. Here the polarization is due to a combination of absorption, which dominates the outer regions, and emission in the bright inner zone.

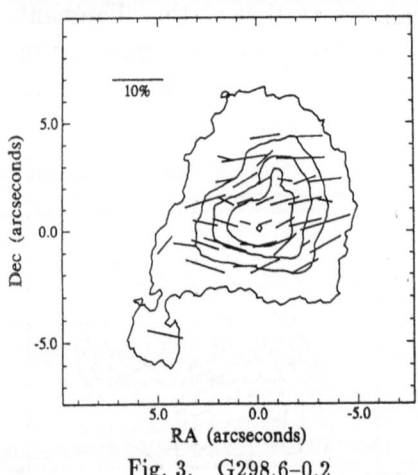

Fig. 3.   G298.6–0.2

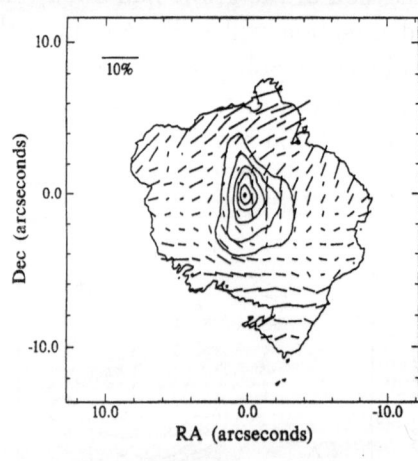

Fig. 4.   G333.6–0.2 at 12.5μm

References

Killeen, N.E.B., & Low, K.Y.: 1989, *The Center of the Galaxy*, ed. Morris, M. Proc. IAU 136 p. 453

Smith, C.H., Aitken, D.K. & Moore, T.J.T.: 1994, *Instrumentation in Astronomy VIII*, eds Crawford, D.L. & Craine, E.R. Proc. SPIE 2198, in press

Wright, C.M.: 1994, *PhD Thesis*, The University of New South Wales

THE MASS OF THE STAR FORMED IN A CLOUD CORE

TAKENORI NAKANO

Nobeyama Radio Observatory, National Astronomical Observatory, Japan

TETSUO HASEGAWA

Institute of Astronomy, University of Tokyo

and

COLIN NORMAN

Department of Physics and Astronomy, Johns Hopkins University

Abstract. The forming star grows by mass inflow from the parent cloud core, mainly through the accretion disk. However, the core matter which has not yet contracted much is seriously disturbed by the activities of the forming star. We consider mass outflow and emission of ultraviolet radiation as such activities and determine the stellar mass as a function of the physical quantities of the parent cloud core.

A high-velocity mass outflow is expected around a forming star with a rate $\dot{M}_O$ about 0.1 times the inflow rate $\dot{M}_I$ from the core (Shu *et al.* 1988). The mass outflow gradually pushes out the surrounding matter forming a thin, dense shell. When the shell radius grows nearly to the initial radius of the cloud core, a considerable fraction of the core matter is blown off and supply of matter to the disk and thence the star effectively stops because the remaining core matter is no longer gravitationally bound and disperses unless the core is initially much more massive than the generalized Jeans mass M_J. In this way the mass of the forming star is fixed to, say, $M_*^{(O)}$.

As the central star evolves to have a high luminosity and high effective temperature, a compact HII region grows and pushes out the surrounding matter and finally stops accretion fixing the stellar mass to $M_*^{(HII)}$. Although the mass outflow and the HII region may work simultaneously, the stellar mass M_* is approximately given by the smallest among $M_*^{(O)}$, $M_*^{(HII)}$, and the cloud core mass M_c. The stellar mass determined in this way is a function of the core density n_c, $\dot{M}_I$, $\dot{M}_O/\dot{M}_I$, and $f_M \equiv M_c/M_J$, or a function of n_c, M_c, $\dot{M}_O/\dot{M}_I$, and f_M. Fig.1 shows the results for the case of $\dot{M}_O/\dot{M}_I = 0.1$. The solid lines represent the contours of constant stellar mass for $f_M = 1$. The lower, steeper parts of the lines for $M_* \geq 10 M_\odot$ represent the stellar mass determined by the HII region, $M_* = M_*^{(HII)}$, and the rest $M_* = M_*^{(O)}$. The dashed lines represent $M_* = M_*^{(HII)}$ for the case of $f_M = 3$ though the deviation from the case of $f_M = 1$ is very small ($M_*^{(O)}$ does not depend on f_M). The dots in this figure represent the cloud cores in the Orion A molecular cloud (Tatematsu *et al.* 1993). It is to be noticed that the stellar mass is determined by the outflow for all the cores.

Astrophysics and Space Science **224**: 523–524, 1995.
© 1995 *Kluwer Academic Publishers.*

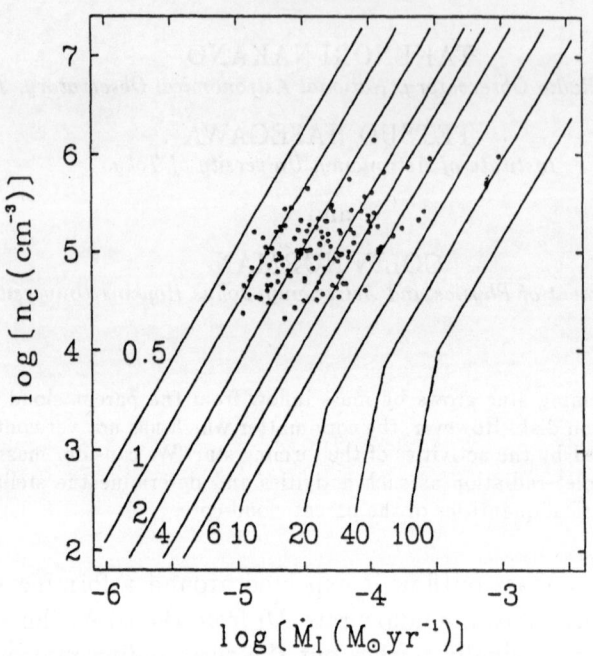

Fig. 1.   The stellar mass M_* as a function of the cloud core density n_c and the mass inflow rate $\dot{M}_I$ for the case of $\dot{M}_O/\dot{M}_I = 0.1$. The solid lines represent the contours of constant M_* labeled by the values of $M_*/M_\odot$ for the case of $f_M = 1$, and the dashed lines for $f_M = 3$. The dots represent the cloud cores in the Orion A molecular cloud.

When the mass outflow is dominant in determining the stellar mass, we have $M_* \propto M_c^{7/6} n_c^{1/12}$ with the proportional coefficient dependent on $\dot{M}_O/\dot{M}_I$, and then M_* is almost independent of n_c. Therefore, the star formation efficiency M_*/M_c is mainly determined by $\dot{M}_O/\dot{M}_I$ and is only weakly dependent on the core parameters M_c and n_c. We have $M_*/M_c \sim 0.04$ for $\dot{M}_O/\dot{M}_I = 0.1$ at $M_c \approx 100 M_\odot$.

We have calculated the initial mass function (IMF) of stars for Orion A from the stellar masses estimated for 125 cores in Fig.1. The IMF at $M_* \geq 4 M_\odot$ can be approximated by a power law $dN_*/d\log M_* \propto M_*^{-1.7}$, which is in reasonable agreement with the IMF of field stars $\propto M_*^{-1.5}$ at $3 \leq M_*/M_\odot \leq 60$ (Scalo 1986), though the turnover of the IMF occurs at $M_* \sim 2 M_\odot$ which is considerably higher than in the IMF of field stars. The discrepancy at low stellar mass must be partly due to the possible existence of subcondensations in the cores.

References

Scalo, J.M.: 1986, *Fundam. Cosmic Phys.* **11**, 1
Shu, F.H., Lizano, S., Ruden, S.P., and Najita, J.: 1988, *Ap. J.* **328**, L19
Tatematsu, K., Umemoto, T., Kameya, O., *et al.*: 1993, *Ap. J.* **404**, 643

A SEARCH FOR PLANETARY SYSTEM CANDIDATES

Northern Nearby A Stars With Dust Disks And Circumstellar Gas

JAMES E. NEFF

Astronomy & Astrophysics, Penn State University, University Park, PA 16802

and

KWANG-PING CHENG

Physics Dept., California State University, Fullerton, CA 92634

Abstract. We report on our search for possible planetary system candidates in a volume-limited sample of 62 nearby A stars. Since the evolutionary lifetimes of A stars ($\leq 10^9$ yrs) roughly correspond to the era of planet formation and subsequent "heavy bombardment" in our solar system, our study could provide valuable insight into the origin of our own Solar System. From our ground-based visual and IUE high-resolution spectroscopy of all the northern nearby A stars, we have identified at least 12 stars with circumstellar gas. Combining these results with our previous IRAS survey we are probing the link between stars with circumstellar gas and those showing circumstellar dust disks. Our aim is not just to identify stars with gas, or stars with both gas and dust, but to identify systems with dynamic spectral activity similar to β Pic, a well known proto-planetary system candidate. By measuring the gas dynamics in the disks of these β Pic-like stars, we can begin to study the physics of accretion disks of young evolving systems.

Key words: Protoplanetary Systems — Circumstellar Gas — A Stars — Ca II

1. IR/IRAS Survey

We first performed a careful analysis of the Faint Source Survey database assembled from IRAS infrared data for a large sample of A stars. Specifically, we examined all the 62 A stars in the Woolley catalogue of stars within 25 pc of the Sun (Cheng et al. 1992). Among these program stars, we found 11 stars having [12]–[25] and [25]–[60] colors consistent with circumstellar dust. Any newly identified dusty A stars provide additional comparisons for β Pic or Vega, the two well-known proto-planetary system candidates having very different IR colors. If a dusty system has a gaseous disk similar to β Pic and viewed approximately edge-on, it should have a similar Ca II spectrum, with a narrow absorption feature near the bottom of the photospheric line. The detection of strongly redshifted, variable absorption in the circumstellar lines of ionized elements such as Ca II, Mg II, Fe II, and Al III might be explained (Beust & Tagger 1993) by the evaporation of large bodies falling onto the star.

Astrophysics and Space Science **224**: 525–526, 1995.
© 1995 *Kluwer Academic Publishers.*

2. Visual/Ground-based Survey

Using high-resolution ($\lambda/\Delta\lambda \approx 125,000$), high S/N ($> 300$) spectra obtained at the National Solar Observatory's McMath–Pierce telescope, we are able to detect narrow absorption features and measure the amount of gas in the line of sight. This could be either circumstellar (CS) or interstellar (IS) gas, but we can discriminate between them in several ways. For example, ratios of the equivalent widths of various lines can be used to discriminate between CS and IS origin. Different line ratios can arise from abundance differences or from differences in density or temperature. For example, the Ca II/Na I equivalent width ratio for IS clouds is between The CS gas around β Pic has a Ca II/Na I ratio of $\sim$9.5 (Lagrange-Henri et al. 1990).

We have completed seven observing runs (42 nights) at the McMath-Pierce telescope for our ground-based survey. All of the 42 A stars in the northern hemisphere part of our survey have been observed in both Ca II K line and Na I D line regions. At least 12 of these stars show discrete, probably circumstellar absorption features in their Ca II K lines. We have measured the the velocity and equivalent width of these features.

3. UV/IUE Survey

Using IUE observations of targets identified in our ground-based survey, we are able to resolve any ambiguity between CS and IS lines. The fine-structure lines of Fe II of the resonance UV1 multiplet near 2600 Å are excellent diagnostics of CS gas. The narrow Fe II absorption line seen in the core of the photospheric feature arising from the zero-volt level, $\lambda2599.396$, could have contribution from both the IS and the CS gas. However, absorption from excited J-levels in the ground configuration ($\lambda2607.0876$ from 667.683 cm^{-1} and $\lambda2611.8743$ from 384.79 cm^{-1} above ground) would only be seen in the CS gas. For densities in excess of 10^3 cm^{-3}, the excited levels in the ground configuration of Fe II should be significantly populated, approaching LTE values. Therefore, a fairly pronounced sharp absorption feature should be apparent for circumstellar Fe II absorption comparable to that seen in β Pic (Boggess et al. 1991; Cheng et al. 1994). We have examined all the existing IUE high-dispersion spectra of the 12 candidate stars identified to date by our ground-based survey.

References

Beust, H., Tagger, M.: 1993, *Icarus* **106**, 42
Boggess, A. et al.: 1991, *ApJ* **377**, L49
Cheng, K.-P. et al.: 1994, *ApJ* , submitted
Cheng, K.-P., Bruhweiler, F.C., Kondo, Y., Grady, C.A.: 1992, *ApJL* **396**, L83
Lagrange, A.M. et al.: 1990, *AApS* **85**, 1089

MOLECULAR ABUNDANCES IN CARBON-RICH CIRCUMSTELLAR ENVELOPES

L.-Å. NYMAN

ESO, Casilla 19001, Santiago 19, Chile

and

H. OLOFSSON

Stockholms Observatorium, S-13336 Saltsjöbaden, Sweden

Abstract. Millimetre observations of three southern carbon stars, IRAS 07454–7112, IRAS 15082–4808 and IRAS 15194–5115 detected 14 molecular species and some of their isotopomers. The $^{12}C/^{13}C$ ratio was found to vary between sources.

Key words: stars: carbon — molecules

1. Introduction

We have performed molecular radio line surveys of the three southern carbon stars IRAS 07454–7112, IRAS 15082–4808, and IRAS 15194–5115 using the Swedish–ESO Sub–mm Telescope (SEST). IRC+10216 was also observed as a comparison object. The IRAS 15194–5115 data have been presented in Nyman *et al.* 1993.

The purpose of the observations is to obtain molecular abundances from a consistent set of data obtained with the same telescope. The observations were done in the 3 mm and 1.3 mm bands. 14 molecular species and some of their isotopomers were detected. Table I lists some of the parameters for the objects: adopted distances together with expansion velocities and mass loss rates determined from the CO (1-0) and (2-1) observations.

TABLE I

Source parameters

Source	Distance (pc)	V_{exp} (km s^{-1})	Mass loss rate ($M_\odot$ yr^{-1})
IRAS 07454-7112	1800	13.6	$3.5 \ 10^{-5}$
IRAS 15082-4808	1500	20.2	$6.1 \ 10^{-5}$
IRAS 15194-5115	1200	22.2	$5.0 \ 10^{-5}$
IRC+10216	200	14.4	$2.0 \ 10^{-5}$

2. Results and Abundances

The following species were observed:
CO, CN, CS, HCN, HNC, SiS, SiO, SiC_2, C_3N, C_2H, C_3H, C_3H_2, C_4H, HC_3N, and some of the ^{13}C isotopomers of the carbon bearing species.

Abundances were calculated for species where the emission was supposed to be optically thin (all except CO and HCN) following the methods in Nyman *et al.* 1993. Since the observations of the different species were made with the same telescope (same calibration etc.), and the abundances were calculated in the same way for each source, we estimate that the relative abundances are accurate to within a factor of 3.

The abundances of most species are similar to within a factor of 3. The exceptions are C_2H and C_4H which can vary by more than a factor of 10 between different sources. However, their abundance ratio is relatively constant. The $^{12}C/^{13}C$ ratio also varies between the different sources. The ratios listed in Table II were calculated using abundance ratios of CN, ^{13}CN, CS, ^{13}CS, HC_3N, $HCC^{13}CN$, and $HC^{13}CCN$.

TABLE II

$^{12}C/^{13}C$ ratios

Source	$^{12}C/^{13}C$ ratio
IRAS 07454-7112	15
IRAS 15082-4808	>30
IRAS 15194-5115	5
IRC+10216	25

Due to optical depth effects in some of the lines the $^{12}C/^{13}C$ ratio should be regarded as a lower limit (e.g. the ratio in IRC+10216 has been estimated to be about 50 using optically thin lines). IRAS 15194–5115 has a surprisingly low $^{12}C/^{13}C$ ratio considering the fact that it is a star with a high mass loss rate.

References

Nyman, L.-Å., Olofsson, H., Johansson, L.E.B., Booth, R.S., Carlström, U., Wolstencroft, R.,: 1993, *A&A* **269**, 377.

LONG-SLIT SPECTROSCOPY OF HERBIG-HARO OBJECTS

E. ORTIZ and C. EIROA

Dpto. Física Teórica, C-XI, Facultad de Ciencias,
Universidad Autónoma de Madrid, Cantoblanco, E-28049 Madrid, SPAIN

and

L.F. MIRANDA

Dpto. Astrofísica, Facultad de Ciencias Físicas,
Universidad Complutense de Madrid, E-28040 Madrid, SPAIN

Abstract. Long-slit spectra of intermediate spectral and high spatial resolution of new Herbig-Haro objects in AFGL 5142, AFGL 5157 and IRAS 2237+7455 and of the already known GGD17/HH1 were obtained. The results are shown in the form of position-velocity diagrams. Radial velocities and electron densities were measured.

Key words: Star Formation — HH Objects

1. Introduction

In the last years, the known number of Herbig-Haro (HH) objects has greatly increased. They present a variety of morphologies, in many cases difficult to interprete within the various models. Spectroscopic studies provide information about physical parameters, which can be compared with models in order to distinguish between general properties and particularities of each HH object.

We have carried out long-slit spectroscopy of new HH objects AFGL5142, AFGL5157, GGD17/HH1 and IRAS2237+7455 (Carballo & Eiroa 1992; Torrelles *et al.* 1992; Eiroa *et al.* 1994).

2. Observations

The spectra were taken in 1993 December at the red channel of the Twin Cassegrain Spectrograph of the 3.5m telescope on Calar Alto (Almería, Spain). The spectra cover the spectral range 6175 – 7045 Å. The spectral and spatial resolution (FWHM) are $\approx$65 km/s and $1''$– $2''$, respectively.

3. Results

We show position-velocity diagrams in H_α and [SII] emission lines of IRAS 2237+7455 in Fig 1. In Table I the measurements of radial velocity in H_α and [SII] emission and mean electron density are presented.

Astrophysics and Space Science **224**: 529–530, 1995.

References

Carballo, R. & Eiroa,C.: 1992, *A&A* **262**, 295

Eiroa, C., Torrelles, J.M., Miranda, L.F. & Anglada, G.: 1994, *A&A* **in press**,

Torrelles, J.M., Eiroa, C., Mauersberger, R., Estalella, R., Miranda, L.F. & Anglada, G.: 1992, *ApJ* **384**, 528

TABLE I

Spectrosopic Results

Object	Knot	[SII] V_{LSR} km/s	[SII] FW0.1 km/s	H_α V_{LSR} km/s	H_α FW0.1 km/s	$I_{[SII]}/$ I_{H_α}	N_e (cm^{-3})
AFGL 5142		-3	170	-17			10
AFGL 5157	HH1	-84	206	-89	213	1.89	265
	HH2	-76	173	-89	216	1.36	265
	HH3	-35	188	-50	198	3.56	445
	HH4	-77	211	-89	305	2.37	250
GGD17	HH1	-180	150				150
IRAS 2237	A	-56	150	-56	274	0.27	560
+7455	B	-46	183	-16	293	0.44	120
	C	-4	188	-14	224	0.99	335

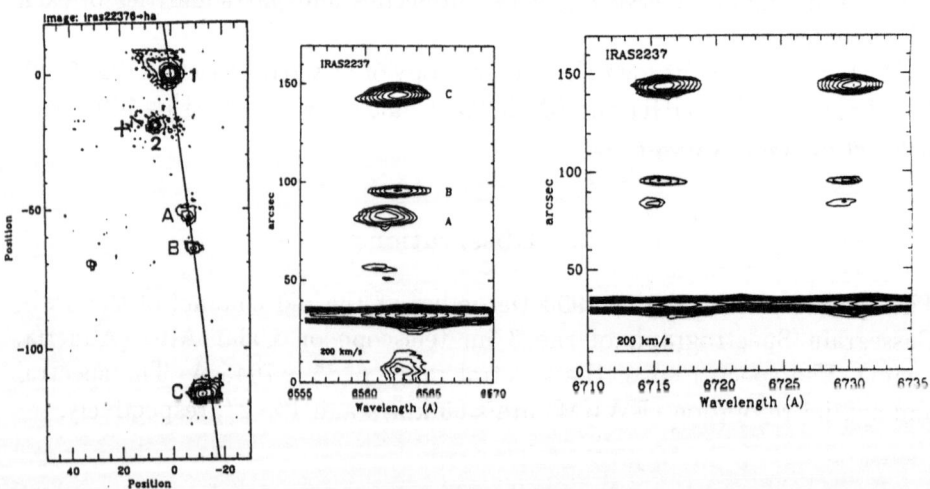

Fig. 1. (a) H_α image adapted from Eiroa *et al.* (1994). (b) PV diagrams of the H_α and [SII] emission lines. Contours are logarithmic separated by a factor $2^{1/2}$ in intensity. Data deduced from the spectrum are presented in Table I. The radial velocity is different in each of the three condensations. Knot A is of high excitation whereas knot C is of low excitation.

SILICON CARBIDE CLUSTERS IN CIRCUMSTELLAR ENVIRONMENT

G. PASCOLI and M. COMEAU

Faculte des Sciences, Department of Physics, Amiens, France

Abstract. The presence of small clusters of silicon carbide (SiC) in circumstellar dust shells surrounding late-type stars is inferred from a broad emission feature peaking at around 11 micrometre in infrared spectra (Little–Marenin ,1986 ApJ Lett. 307, L15). These clusters are expected to condense from molecular arrangements composed of a few carbon and silicium atoms which are present in stellar winds surrounding carbon–rich late–type stars. we have searched for all the possible geometric structures of SiC_n^+ radicals ($n <= 5$) with help of ab initio calculations ($T = 0$ K). Vibrational frequencies of the most stable species have then been determined . the destabilizing influence of a finite temperature effect on these structures has also been studied by using general considerations of thermal statistics. We show that for $n >= 3$ linear structures are energetically favored compared to the planar and three–dimensional ones. A comparison with other results published in this context is also made.

Key words: stars:Silicon Carbide – stars:stellar winds

Astrophysics and Space Science **224**: 531, 1995.
© 1995 *Kluwer Academic Publishers.*

FORMATION OF A FILAMENTARY STRUCTURE IN PLANETARY NEBULAE

G. PASCOLI and J. LECLERCQ

Faculte des Sciences, Department of Physics, Amiens, France

Abstract. Intricate filamentary structure and multiple shell–like appearance are very common phenomena in Planetary Nebulae.

In addition, recent observations also indicate that the individual filaments present in these objects can have larger velocities than the adjacent smooth background (Pascoli, 1992 PASP 104, 350 and paper quoted therein).

We have hypothesized that non linear hydrodynamical processes existing within the nebular gas are, possibly, responsible for these structures. As a matter of fact, it is argued that such a characteristic morphology, reinterpreted as a intermingled network of solitary waves or solitons, can be spontaneously generated in Planetary Nebulae as soon as one assumes that the nebular gas is permeated by a weak magnetic field whose strength is about 10^{-5} to 10^{-4} gauss.

Main results of this work and further comments will be subsequently published in Ap&SS.

Key words: stars:planetary nebulae – stars:stellar winds – stars:filaments

DIRECT MID–IR IMAGES OF GL 2591,W51–IRS2 AND S140

P.PERSI, M.FERRARI–TONIOLO and A.R.MARENZI
Istituto Astrofisica Spaziale,CNR, C.P.67, 00044 Frascati, Italy

M.BUSSO, L.CORCIONE and M.MARENGO
Osservatorio Astronomico di Torino, Pino Torinese, Italy

and

M.TAPIA
Instituto de Astronomia, UNAM, Ensenada,B.C., Mexico

Abstract. We present the results of direct mid–IR images of the luminous young stellar objects GL 2591, S140–IRS1, and W51–IRS2. The sources show an extended mid–IR emission at the limit of our spatial resolution indicating the presence of dense circumstellar dust disks.

Key words: Young stellar objects – Molecular outflows – mid–IR images

1. Introduction and Observations

It is well known that the process of star formation is accompanied by dynamic and energetic activity. The interactions of these outflows with the surrounding medium on arcsec spatial scales can be well studied with the newly available infrared array detectors. We have employed the mid–IR camera TIRCAM (Persi et al. 1994) based on 10×64 Si:As array from Hughes Co., to image the vicinity of three massive star–forming regions GL 2591,S140, and W51–IRS2 associated with outflows. The observations were carried out with an image scale of 1.34 arcsec/pix at the 2.1 m telescope of S.Pedro Martir (Baja California, Mexico).

2. Results

2.1 GL 2591

The observed spatial distribution of H_2 and $Br\gamma$ lines and 2.2 μm continuum is consistently explained by a dust disk model around GL 2591 (Tamura & Yamashita, 1992). The image at 11.7 μm (Fig.1a), shows a slight extended structure with a size of $\sim$4 arcsec (10^{17}cm)(FWHM) that seems to confirm the presence of a dense dust disk surrounding the YSO.

2.2 S140

The K-image of the source S140 IRS–1 associated with a bipolar molecular outflow (Forrest & Shure 1986) shows an elongated N–S structure around

Astrophysics and Space Science **224**: 535–536, 1995.
© *1995 Kluwer Academic Publishers.*

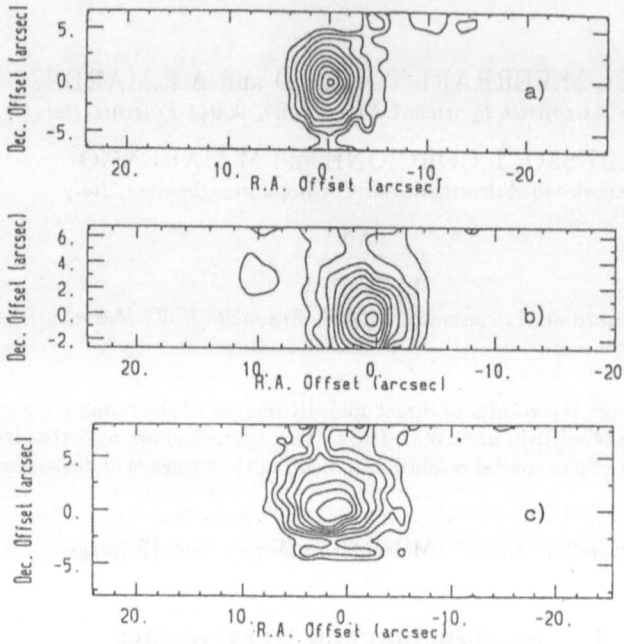

Fig. 1. Contour maps of:(a) GL 2591 11.7μm,(b) S140 12.5μm, and (c) W51–IRS2 12.5μm.
The lowest contours are 20% of the peaks surface brightness.

the YSO. Similar morphology is observed in our 12.5 μm image (Fig.1b).
This morphology could be explained by the presence of a dust circumstellar
disk.

2.3   W51–IRS2

Two centers of activities have been identified in this region. The direct 12
μm image of IRS2 obtained by Bally et al. (1987), suggests the presence
of a massive disk (100–200M$_\odot$) surrounding a luminous protostellar object.
The 12.5 μm image here reported (Fig.1c) is consistent with the previous
observation, showing an extended E–W emission of 024 pc (D=7.5 Kpc).

In conclusion, mid–IR images of the luminous YSOs GL 2591, S140 and
W51–IRS2 show extended emission. This result seems to confirm the model proposed for luminous YSOs with molecular outflows, in which a dense
circumstellar dust disk is responsible for the observed mid–IR radiation.

References

Bally et al.: 1987, *Ap.J.Lett.* **323**, L73

Forrest, W.J. and Shure, M.A.: 1986, *Ap.J.Lett.* **311**, L81

Persi et al., 1994, in: I.S. McLean (ed), *Infrared Astronomy with Arrays: the next genera-*
    tion, Dordrecht: Kluwer, p. 331

Tamura, M. & Yamashita, T.: 1992, *Ap.J.* **391**, 710

EXCITATION OF INTERSTELLAR WATER BY ORTHO-
AND PARA-HYDROGEN

TIMOTHY R. PHILLIPS
Hughes STX Corporation, New York, NY USA

and

SHELDON GREEN
NASA/Goddard Institute for Space Studies, New York, NY USA

Water (H_2O) is fairly abundant in some regions of circumstellar and interstellar space. Maser emissions from a number of millimeter wave and microwave transitions are thought to be associated with interstellar regions near newly-forming stars, and near certain classes of late-type stars, such as Mira variables (Cohen 1989). Many other rotational transitions of water are difficult or impossible to detect from the Earth's surface due to atmospheric H_2O, and must be observed from high altitudes or must await the the availability of orbital telescopes such as the planned International Submillimeter Observatory and Submillimeter Wave Astronomical Satellite (SWAS). In any case, analysis of observations of H_2O rotational transitions requires some knowledge of the behavior of the H_2O molecule on collision with the most abundant interstellar and circumstellar species, hydrogen (H_2) and helium (He). Theoretical studies of H_2O–He collisions have been presented by Green (1980), and more recently by Palma *et al.* (1988, 1989) and by Green *et al.* (1993). Studies of H_2O–H_2 collisions are more complicated and have therefore been fewer, but one recent contribution has been made by Balasubramanian *et al.* (1993). Here we present the first results of our theoretical investigation of the H_2O–H_2 collision problem. We have calculated cross sections for the collisional $1(01) \rightarrow 1(10)$ excitation of ortho-H_2O by ortho- and para-H_2O using quantum-mechanical molecular collision theory as implemented in the MOLSCAT program (version 12, Hutson & Green 1993). The H_2O and H_2 molecules are considered to be rigid. Their potential energy of interaction is described by a global expansion in angular functions originating at the H_2O center-of-mass :

$$V(R\alpha\beta\gamma\theta_2\phi_2\theta\phi) = \sum_{p_1 q_1 p_2 p} V_{p_1 q_1 p_2 p}(R)$$
$$\times \sum_{r_1 r_2 r} \begin{pmatrix} p_1 & p_2 & p \\ r_1 & r_2 & r \end{pmatrix} D^{p_1}_{q_1 r_1}(\alpha\beta\gamma) + (-1)^{p_1+q_1+p_2+p} D^{p_1}_{-q_1 r_1}(\alpha\beta\gamma)$$
$$\times Y^{r_2}_{p_2}(\theta_2\phi_2) Y^r_p(\theta\phi) [1 + \delta_{q_1 0}]^{-1}$$

with radial coefficients $V(p_1, q_1, p_2, p; R)$, which were calculated by a least-squares fit of the above expansion to a set of ab-initio potential energies

Astrophysics and Space Science **224**: 537–538, 1995.
© 1995 *Kluwer Academic Publishers.*

TABLE I

Rate (cm^{-3}/sec) of collisional excitation of $1(01) \rightarrow (10)$ transition of water by ortho- and para-H_2 at selected temperatures. Numbers in parentheses indicate powers of 10.

collision partner	40K	60K	80K	100K
ortho-hydrogen	1.5(-10)	1.9(-10)	2.1(-10)	2.2(-10)
para-hydrogen	1.2(-11)	1.6(-11)	1.9(-11)	2.1(-11)

calculated for 722 relative configurations of H_2O and H_2 (Phillips *et al.* 1994). The rotational basis set consisted of a single (j=1) H_2 rotational state in the case of excitation by ortho-H_2, and two (j=0,2) H_2 rotational states in the case of excitation by para-H_2. Enough H_2O rotational states were included to produce cross sections converging to within a few percent. In most cases this meant rotational states with j * 4. The $1(01) \rightarrow 1(10)$ collisional excitation rate coefficient was calculated by Boltzmann average of the cross sections calculated at 13 H_2O–ortho H_2 collision energies, and at 5 H_2O–para-H_2 collision energies. The results are shown in Table 1. Excitation by ortho-hydrogen is found to have a rate coefficient which exceeds that for excitation by para-hydrogen by nearly an order of magnitude.

Acknowledgements

This work was supported in part by NASA Headquarters, Office of Space Science and Applications, Astrophysics Division, Infrared and Radio Astrophysics Program, and in part by NSF Grant CHE-920166C. T.P. extends thanks to Dr. George Flynn of Columbia University for workstation time, and to Lilly del Valle of Hughes STX Corporation for help in preparing the original conference poster.

References

Balasubramanian V., Balint-Kurti G.G. & van Lenthe J.H. 1993, J. Chem. Soc. Faraday Trans. 89, 2239
Cohen R.J. 1989, Rep. Prog. Phys. 52, 881
Green S. 1980, Ap. J. (Suppl.) 42, 103
Green S., Maluendes S. & McLean A.D. Ap.J. (Suppl.) 85, 181.
Hutson J.M. & Green S. 1993, MOLSCAT version 12, CCP6, UK Science and Engineering Research Council
Palma A., Green S., DeFrees D.J. & McLean A.D. 1988, Ap.J. (Suppl.) 68, 287
Palma A., Green S., DeFrees D.J. & McLean A.D. 1989, Ap.J. (Suppl.) 70, 681
Phillips T.R., Maluendes S. Mclean A.D. & Green S. 1994, J. Chem. Phys. 101, 5824

STRUCTURAL PECULIARITIES OF CIRCUMSTELLAR ENVELOPES OF THE YOUNG A0E HERBIG STARS

MIKHAIL POGODIN

Pulkovo Observatory, 196140 St. Petersburg, Russia

Abstract. Circumstellar peculiarities of the A0-type Herbig stars are analyzed on the basis of high-resolution CCD spectroscopic data, obtained in 1991-1994 at the ESO and the Crimean Astrophysical Observatory (about 250 spectrograms). The results of investigation of the rapid line profile variability in $H\alpha$, $H\beta$, HeI 5876 and DNa lines are presented for AB Aur, HD 163296, HD 36112 and HD 100546. The conclusion is made that the behaviour of these lines can be explained in the framework of the model containing an equatorially concentrated and azimuthally inhomogeneous stellar wind and an external cool shell occasionally losing matter in the form of infall on to the star.

Key words: emission line profiles – stellar wind – remote shell – inhomogeneity

AB Aur and HD 163296 are the most wellknown A0e Herbig stars from PCyg-subclass. HD 36112 was included to the new catalogue of Herbig stars by The et al. (1994), and HD 100546 was proposed by Hu et al. (1989) as a candidate Herbig star.

1. Global envelope structure

1.1 EQUATORIALLY CONCENTRATED NON-STABLE WIND

Occasional disappearance of PCyg-profile structure was observed in AB Aur (March–October 1988 and January 1993) and in HD 36112 (Jan. 1994). This phenomenon can be explained in the framework of envelope model by Pogodin (1992). According to this model, the envelope near the star is formed by an equatorially concentrated stellar wind with variable latitudinal distribution of the gas density.

1.2 FORMATION OF REMOTE SHELL

The shell blueshifted relatively stable absorption is superimposed on the variable emission profile (with PCyg-signs or not) originated in the inner active region of the envelope (HD 163296, $H\alpha$, see The et al. 1985 and Pogodin 1994). This shell can be formed by low energy part of the wind at velocities lower than the escape velocity (Mundt 1984).

1.3 MATTER INFALL FROM THE SHELL ONTO THE STAR

The $H\alpha$ and NaI D line profiles of HD 100546 show the matter infall onto the star as the main large-scale motion in the circumstellar kinematics. The observed HeI 5876 line profiles of HD 100546 can be interpreted as superposition of:

Astrophysics and Space Science **224**: 539–540, 1995.

- absorption component formed in a turbulent and rotating chromosphere;
- blue emission peak from transparent, relatively stable stellar wind; and
- variable redshifted component (sometimes containing both emission and absorption) originated in the azimuthally inhomogeneous infalling gas.

The angular momentum of gas in the shell formed by the low energy part of the wind should be rather small, and its orbital eccentricity is close to 1. In this case an active interaction is expected between the wind and gas of the shell moving in direction to the star. This interaction can lead to formation of discrete space distribution of circumstellar matter, which can be observed as moving inhomogeneities.

2. Circumstellar inhomogeneity

The main type of rapid line profile variability in this type stars has a form of local intensity changes of different profile components without any positional shifting (AB Aur , Jan.1994, HD 100546, March 1994 and HD 163296, see Pogodin 1994). Detailed investigation of the wellknown profile transformation from PCyg II to PCygIII as the appearance of a secondary emission peak at the blue edge of the PCyg-absorption (see, for example, Beskrovnaya et al. 1991) have shown that this type of variability is a particular case of a more common observational phenomenon.

This type of variability is likely to be connected with the motion of circumstellar inhomogeneities at some distance from the star with a specific kinematics of its envelope (see Beskrovnaya et al. 1994)

Extremely sharp bumps shifting positionally on the $H\alpha$ profile of HD 163296 can be explained as local short-lived condensations moving near the stellar surface (Pogodin 1994).

References

Beskrovnaya, N.G., Pogodin, M.A., Shcherbakov, A.G., Tarasov, A.E.: 1991, *Lett. to Sov.Astron.* **17**, 825
Beskrovnaya, N.G., Pogodin, M.A., Najdenov, I.D., Romanyuk, I.I.: 1994, *A&A*, submitted
Hu, J.Y., The, P.S., de Winter D.: 1989, *A&A* **208**, 213
Mundt, R.: 1984, *ApJ* **280**, 749
Pogodin, M.A.: 1992, *Lett. to Sov.Astron.* **18**, 1066
Pogodin M.A.: 1994, *A&A* **282**, 141
The, P.S., Felenbok, P., Cuypers, H., Tjin A Djie, H.R.E.: 1985, *A&A* **149**, 429
The, P.S., de Winter, D., Perez, M.R.: 1994, *A&A*, in press

THE GEOMETRY OF DECOUPLED LINE DRIVEN WINDS

JOHN M. PORTER & JANET E. DREW

Astrophysics,
Nuclear and Astrophysics laboratory,
Keble Road,
Oxford,
OX1 3RH.

Abstract. Bjorkman & Cassinelli (1993) have proposed a mechanism that is expected to produce strong equatorial focusing of the radiation-driven winds from rapidly-rotating B stars. Here the possibility of the decoupling of the stellar radiation field and the outflow is considered. We find that greater equator to pole density ratios may be generated than the standard model, bringing it more into line with results implied by observations.

Key words: Be stars – wind – mass loss

1. Introduction

As a consequence of a relatively high initially rotational component, B and Be wind streamlines approach the star's equatorial plane (Bjorkman & Cassinelli 1993) and generate a wind compression disc. If the radiative force is terminated, then further increase in the wind speed is inhibited. This enhances the effect of the rotational motion of the gas, leading to greater fraction of momentum being focussed onto the equatorial plane. The enhancement, as measured by the normal component of the momentum fed into the disc, is in the region of a factor of 2 (Porter & Drew 1994).

2. Wind Decoupling

We examine two mechanisms which can result in the wind and radiation field decoupling: ion-stripping (Springmann & Pauldrach 1992) and shock disruption of the wind ionization balance (Castor 1987). They are most effective in the fast, low density winds generated by B and Be stars.

Momentum is transferred to the outflow through scattering of stellar photons from line transitions of heavy metal ions. The rest of the outflow couples to these driven ions through a frictional process. At high drift velocities, the frictional drag decreases with increasing drift velocity between the driven ions and the rest of the plasma. When the wind enters this regime the driven ions are quickly stripped out of the outflow (ion stripping), and the wind then receives no further radiative acceleration from the stellar photons.

Radiatively driven winds are susceptible to the growth of instabilities – once a small scale perturbation forms, it may quickly steepen to form a shock (Owocki, Castor & Rybicki 1988). In the low density winds generated

Astrophysics and Space Science **224**: 541–542, 1995.

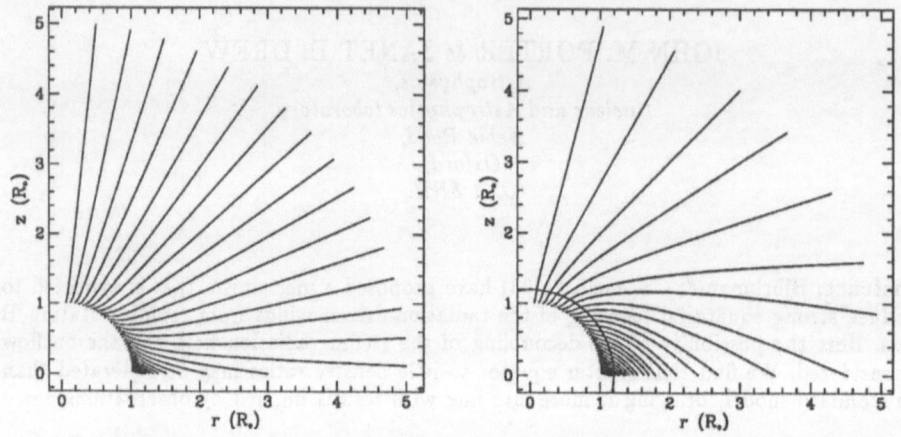

Fig. 1.   Streamlines for the representative Be star. v_{rot}/v_{crit} = 0.5. The left frame is the standard Bjorkman & Cassinelli model, the right frame includes decoupling. The solid line in the right frame marks the radius at which decoupling occurs, set here to be $1.2R_*$.

by B and Be stars, the higher ionization state postshock gas may not recombine and therefore, the preshock ionization state may not be regained. This depletion of the relevant ionization states leads to a substantial decrease or even elimination of the path through which a large fraction of the stellar radiative momentum is transmitted to the wind.

3.   Streamlines of a Decoupled Wind

Streamlines are now generated for the flow around a representative star - we use stellar parameters of $M_* = 13M_\odot$, $R_* = 7R_\odot$, $T_{eff} = 22100K$ and $\dot{M} = 10^{-9.9}M_\odot\,yr^{-1}$. Also the rotation is set at 50% of the critical rotation rate and the decoupling radius (for a typical B star) at $1.1R_*$.

The streamlines generated from the model with and without decoupling are striking in their difference. Truncation of the velocity profile at relatively low velocities leads to a higher fraction of mass loss from the star being focussed onto the equatorial plane. The decoupled material flows into the disc much less obliquely, enabling it to apply larger compressive forces. It is particularly effective at the lower rotations where it may be the dominant cause of compression enhancing the Bjorkman & Cassinelli effect.

References

Bjorkman, J. E., Cassinelli, J.P., 1993, Astrophysical Journal, 409, 429
Castor, J. I., 1987, in *Instabilities in Luminous Early-Type Stars*, eds. H. J. G. L. M. Lamers & C. W. H. de Loore, D. Reidel, The Netherlands, p159
Owocki, S.P., Castor, J.I., Rybicki G.B., 1988, Astrophysical Journal, 335, 914
Porter, J. M., Drew, J. E., 1994, *A&A*, in press
Springmann, U. W. E., Pauldrach, A. W. A., 1992, *A&A*, 262, 515

DUST PROCESSING IN NOVA WINDS

J.M.C. RAWLINGS
Department of Physics and Astronomy,
University College London,
Gower Street, London WC1E 6BT

and

A. EVANS
Department of Physics,
Keele University,
Staffordshire ST5 5BG

Abstract. We have radically re-assessed the conditions required for the formation and growth of carbon grains in the ejecta of novae. The stability and hence the ultimate fate of the grains is primarily determined by the degree to which they are annealed by the nova's ultraviolet radiation field.

Key words: Novae — Dust — Interstellar Medium

1. Summary

We consider the processes that govern the electrostatic charge of carbon grains in nova winds. Physical conditions in the nova ejecta, such as the opacity of the dust shell and the relative rates for the annealing and rehydrogenation of dust grains are continually changing. We have modelled the time-dependence of the nature of the dust grains in the nova environment. We find that the grains are positively charged unless the dust shell is extremely opaque ($\tau_{uv} \gtrsim 7$), and that the dust shell responds rapidly to the changing nature of the grains.

2. Ejecta Ionization and the Stability of Dust Grains

The observed timescale for dust formation is short, typically of the order of 1–20 days. The formation of dust will itself result in a sharp increase in the opacity to ionizing radiation, thus retarding the advance of the CII ionization front. These facts imply that, contrary to the findings of Mitchell & Evans (1984) and Mitchell *et al.* (1986), not only dust nucleation *but also growth* must take place in a CI region.

Substantial dispersion of the dust shell is required for carbon ionization to occur. With the increase in ionization a number of changes take place:

— The grains 'see' more UV flux and the grain temperature rises

— Shocks propagate into the ejecta, resulting in grain-grain collisions

Astrophysics and Space Science **224**: 543–544, 1995.
© 1995 *Kluwer Academic Publishers.*

– The grains become increasingly annealed and the sp^2:sp^3 ratio increases. One important effect of this is that the photoelectric properties of the grains and hence their stability changes

2.1 CHEMICAL CHARACTERISTICS

Earlier models of ionization in nova ejecta failed to take account of the *chemical* characteristics of the different ionization zones and the implications that these have on the growth/destruction balance of the grains: The onset of carbon ionization results in CO photodissociation/ionization to C^+ and O, and H_2 photodissociation. Basic molecular chemistry is supressed. The increase in ionization is accompanied by a rise in the gas temperature. Chemisputtering by oxygen is likely to be a dominant destruction mechanism in the CII region. The efficiency of monomeric accretion (of CII) will be determined by the grain charge and the gas temperature. If the grain charge is negative then net growth may occur. If the grains are positively charged then accretion will be inhibited and net destruction is more likely.

3.   Amorphous Carbon and Grain Charge

In the harsh environment of a nova the dust grains will have an electrostatic charge which, unlike the situation in interstellar clouds, may be large and significant with respect to their stability. Polymeric carbon (sp^3) has a larger band gap ($E_g \sim 2.5$V) and a much larger yield ($y_\infty \sim 0.5$) than graphitic (sp^2) carbon ($E_g \sim 0.3$V, $y_\infty \sim 0.05$). Thus, the efficiency of photoelectric emission is determined by the degree of annealing, which in turn is determined by the ionization conditions.

We have used an iterative model to obtain the equilibrium grain charge for several combinations of the gas density and temperature, the ionization and the time since outburst. Calculations were performed for both sp^2 and sp^3 carbon grains. A (preliminary) set of results as applicable to a CII region show that the grain charges on sp^2 and sp^3 grains differ substantially in regions of intermediate dust extinction ($\tau_{UV} \sim 4$–8) and that the annealing of dust grains may strongly enhance their prospects for survival at later times. We are currently assessing the implications that this has in the context of recent observations of hydrocarbon features in the ejecta of novae.

References

Mitchell, R.M. & Evans, A.: 1984, *MNRAS* **209**, 945.
Mitchell, R.M., Evans, A. & Albinson, J.S.: 1986, *MNRAS* **221**, 663.

THE CIRCUMSTELLAR ENVELOPE OF S PERSEI

A. M. S. RICHARDS and R. J. COHEN

NRAL Jodrell Bank, University of Manchester, England

and

J. A. YATES

School of Chemistry, University of Bristol, England

Abstract. We present MERLIN * observations of OH and H_2O masers in the circumstellar envelope of S Per. The results are consistent with a model of a thick shell of H_2O masers in a region which is still accelerating.

Key words: Stars : Late-type — Mass Loss — Masers — stars: individual : S Per

S Per is a long-period variable M-type supergiant with an estimated mass loss rate of $2.7 \times 10^{-5} M_{\odot}$ (Gehrz & Woolf 1971). It has strong circumstellar maser emission which has previously been mapped by Diamond *et al.* (1987) and Yates & Cohen (1994). We observed S Per in the OH transitions at 1665 and 1667 MHz in December 1993 with an angular resolution of 120 mas and a velocity resolution of 1.6 km s^{-1} and in the H_2O transition at 22 GHz in March 1994 with an angular resolution of 10 mas and a velocity resolution of 0.2 km s^{-1}. Further details of observational methods and data reduction procedures are in the *MERLIN User Guide* (Thomasson *et al.* 1994).

Figure 1 shows the integrated 22 GHz line emission from the envelope. It contains maser hotspots, such as were known from previous work, plus a weaker extended shell of emission. The hotspots have brightness temperatures of $\approx 10^9$ K, whereas the extended shell of emission has a typical brightness temperature of 3×10^6 K. There is almost continous H_2O emission from −14 to 11 km s^{-1} with respect to the stellar velocity, with a strong peak close to V_*. The most extended emission is close to the stellar velocity. The more highly blue-shifted features are closer to the centroid of emission. The kinematics are well fitted by a simple model of accelerating outflow, such as that applied to VX Sgr by Chapman & Cohen (1986). The acceleration in the shell can also account for the tangential beaming of the maser emission, seen as limb-brightening in Figure 1.

The OH 1665 and 1667 MHz mainline emission is shown to the right in Figure 1. It is different in detail at both frequencies and both polarisations, but in all cases it is twin peaked, with stronger blue shifted features and little or no emission close to the stellar velocity. It is distributed along a NNW–SSE axis, with the brightest features closest to the stellar position. The general appearance of the source is similar to that found ten years ago

* The Multi-Element Radio Linked Interferometer, a national facility operated by the University of Manchester on behalf of PPARC

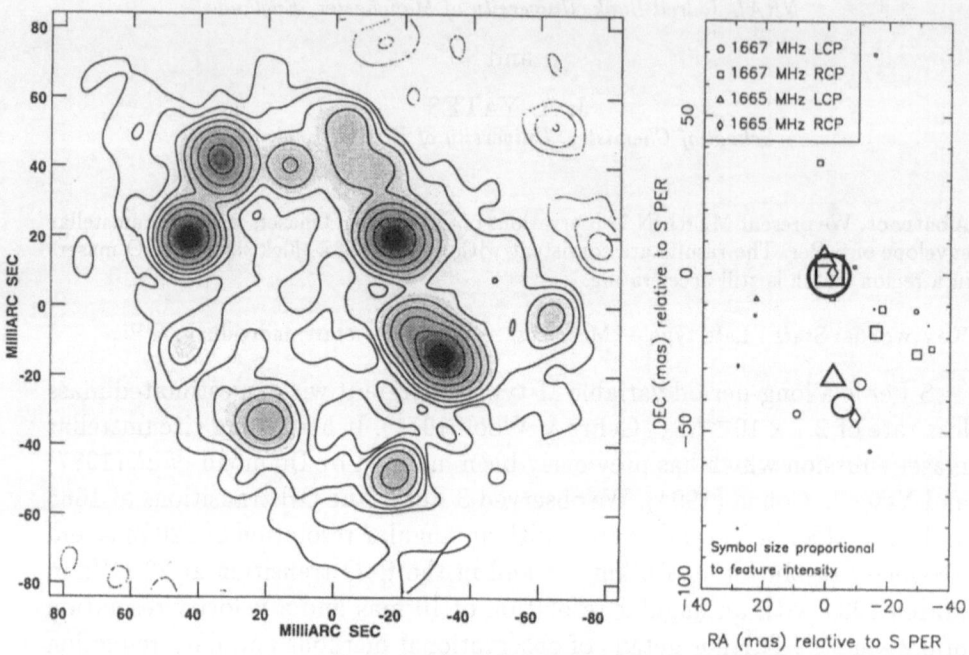

Fig. 1.   Left: H₂O 22 GHz maser emission in the envelope of S Per integrated over all velocity channels. Right: OH mainline emission features.

by Diamond *et al.* (1987). The OH kinematics are consistent with radial beaming from a shell with a smaller velocity gradient than found for the water masers. The kinematics of this region, and the contrasting behaviour of the thick shell giving rise to the 22 GHz emission, appear to have remained broadly similar over at least four stellar luminosity periods.

Further modelling of the water masers is in progress. We hope to use the inner cut-off to the maser distribution, where maser action is quenched, to determine an accurate value for the mass loss rate.

References

Chapman, J. S. & Cohen, R. J., 1986. *Mon. Not. R. Astr. Soc.*, **220**, 513.

Diamond, P. J., Johnston, K. J., Chapman, J. M., Lane, A. P., Bowers, P. F., Spencer, J. H. & Booth, R. S., 1987. *Astron. Astrophys.*, **174**, 95.

Gehrz, R. D. & Woolf, N. J., 1971. *Astrophys. J.*, **265**, 158.

Thomasson, P., Garrington, S. T., Muxlow, T. W. B. & Leahy, J. P., 1994. *MERLIN User Guide*, NRAL Jodrell Bank Cheshire SK11 9DL.

Yates, J. A. & Cohen, R. J., 1994. *Mon. Not. R. Astr. Soc.*, in press.

DOPPLER TOMOGRAPHY OF ACCRETION REGIONS

IN ALGOLS

MERCEDES T. RICHARDS, GEARY E. ALBRIGHT and
LARISSA M. BOWLES
Department of Astronomy, University of Virginia

Abstract. The technique of Doppler Tomography has been used to image the accretion regions in five short-period Algols. There is clear evidence of gas flows along the predicted free-fall path of the gas stream as well as asymmetric disk-like structures around the mass gainer. Another source of Hα emission is associated with the cool magnetically active star.

Key words: stars: binary — stars: individual: Algol, U Sge, U CrB, SW Cyg, RS Vul

Hα line profiles of five short-period ($P < 5$ days) Algols were obtained at closely-spaced positions around the entire orbit of each binary. These were obtained at the Kitt Peak National Observatory 0.9m Coudé Feed Telescope with a reciprocal dispersion of 6.9 Å mm^{-1} and a spectral field of 177 Å, during 1993 May and June (RS Vul) and 1994 June (U Sge, U CrB, SW Cyg). The data set also included earlier (Richards 1993) Hα spectra of β Per (Algol) with a dispersion of 16 Å mm^{-1}. These observed spectra were converted to difference profiles in the manner described in Richards (1993) and Albright & Richards (1993). The difference profiles represent the contributions of all non-photospheric gas flows in the binary, namely those produced by Roche lobe overflow or the chromosphere of the magnetically active cool secondary star, and typically show blends of emission and absorption, with either single or double-peaked emission at orbital phases outside of the eclipses.

A Back Projection Doppler tomography code (Horne 1992; Robinson, Marsh, & Smak 1993) was used to reconstruct the images the emission sources in β Per ($P = 2.87$ days), U Sge ($P = 3.38$ days), U CrB ($P = 3.45$ days), RS Vul ($P = 4.48$ days), and SW Cyg ($P = 4.57$ days). The tomograms of RS Vul, U Sge, U CrB and Algol all show distinct elongated emission along the gas stream trajectory from the L_1 point towards the primary star. The collimation of the gas stream is more pronounced in U CrB than in the other systems, where a more diffuse flow along the gas stream is observed. An emission feature that partially encircles the primary star is also present in the tomograms of these systems. This disk-like feature is enhanced in the tomograms when the difference profiles are analyzed during the orbital phases where the gas stream is expected to be occulted by the

Astrophysics and Space Science **224**: 547–548, 1995.

primary star. It is strongest in U CrB, which also displays the strongest gas stream emission of all the Algols in our sample. During the interval from $\phi =$ 0.30 − 0.69, the disk contours near the primary are found at higher velocities (in $-V_y$) and the strongest emission source is nearly centered on the primary star. Similar disk-like patterns are found for Algol, RS Vul, and U Sge. The disk emission has velocities extending out to several times the synchronous rotation rate of the primary. The part of the disk with the highest velocity is found along the line of centers on the side of the mass gainer facing the secondary. Another emission source in the vicinity of the L_1 point may arise from the chromosphere of the cool secondary star. Chromospheric contributions to the Hα difference profile have often been assumed to be negligible in the Algols, although they are a dominant source in the magnetically active RS CVn binaries (Richards & Albright 1993). The Doppler tomograms of the Algols show, for the first time, that the chromospheres of the secondaries provide a weak but important emission source.

The Doppler tomogram of SW Cyg is different from those of the other Algols described here because it displays a prominent almost Keplerian disk which is similar to those found in the cataclysmic variables. Work in progress on 1993 observations of U Sge, apparently during an epoch of enhanced mass transfer, shows a disk similar to that of SW Cyg. So it is apparent that the disk-like structures are dominant in the Doppler tomograms during epochs of enhanced mass transfer and in the longer period systems. These results may lead us to explore possible links between the classical disks in the cataclysmic variables and the transient disks in the Algols.

Acknowledgements

We thank K. Horne for the use of his Back Projection tomography code. We acknowledge observing grants from KPNO, National Optical Astronomy Observatories (NOAO), operated by AURA under cooperative agreement with the NSF. This work was partially supported by AFOSR grants F49620–92–J–0024 and F49620-94-1-0351, and NSF grants AST 91–114214 and AST 93–15108.

References

Albright, G. E., & Richards, M. T. 1993, *ApJ*, **414**, 830
Horne, K. 1992, private communication
Richards, M. T. 1993, *ApJ Suppl.*, **86**, 255
Richards, M. T., & Albright, G. E. 1993, *ApJ Suppl.*, **88**, 199
Robinson, E. L., Marsh, T. R., & Smak, J. I. 1993, in *Accretion Disks in Compact Stellar Systems*, ed. J. C. Wheeler (Singapore: World Scientific), p. 75

CIRCUMSTELLAR MATERIAL AROUND THE PROTO-PNE

IRAS 10178−5958 AND IRAS 11065−6026

A. RIERA

Departament de Física i Enginyeria nuclear, E.U.P.V.G., U.P.C. i Departament d'Astronomia i Meteorologia, Universitat de Barcelona, Av. Diagonal 647, E-08028 Barcelona (Spain)

P. GARCíA-LARIO

INSA; ESA IUE OBSERVATORY. Villafranca del Castillo, Apdo. de Correos 50727, E-28080 Madrid (Spain)

and

A. MANCHADO

Instituto de Astrofísica de Canarias, E-38200 La Laguna, Tenerife (Spain)

Abstract. Spectroscopy of Hen 401 (IRAS 10178−5958) and WRA 751 (IRAS 11065−6026) reveal strong HI, HeI, FeII and forbidden emission lines. Images of Hen 401 reveal the bipolar morphology of the nebula.

Key words: stars: post-AGB — stars: individual: WRA 751, Hen 401

1. Introduction

Hen 401 and WRA 751 are two of the stars included in a multiwavelength observational program of unidentified IRAS sources with far infrared colours similar to those of known PNe (Pottasch *et al.* 1988; Manchado *et al.* 1989). Recently, Loup *et al.* (1990) and Bujarrabal & Bachiller (1991) detected CO emission associated with Hen 401, which is commonly found in Proto-PNe. In order to study the spectral energy distributions and the optical properties of the nebulae associated with both sources, NIR photometry, optical long slit spectroscopy and CCD images (in the Hα and [OIII] filters) were obtained during May–June 1990 at the European Southern Observatory.

2. Infrared Emission

The strong near and far infrared excess shown by Hen 401 can only be explained by the presence of cold dust in a circumstellar envelope around the central star together with a hot dust component, which is interpreted as the result of recent mass loss episodes. The locations of Hen 401 and WRA 751 in the IRAS two-colour diagram correspond to objects which have left the AGB and are now evolving towards lower dust temperatures. The excess in J shown by Hen 401 can be due to HeI λ 1.083 μm emission, which may indicate that the star has begun to ionize its circumstellar envelope.

3. Optical Spectra

The cores of both sources show a rich emission line spectra characterized by strong HI Balmer lines, HeI recombination lines, and some forbidden and permitted iron emission lines. Hβ and Hα profiles show high-velocity non-Gaussian wings, where the redward edge implies LSR velocities $\geq$ 1100 km s^{-1} and $\geq$ 500 km s^{-1} in Hen 401 and WRA 751, respectively.

A very rich FeII, [FeII] and [FeIII] emission spectrum was observed from the Hen 401 core, whilst in WRA 751 the iron spectrum is dominated by [FeII] and much weaker FeII lines. Other emission lines also detected in both cores are [OI], [SII], [NII], strong [NiII] and weak SiII and OI.

From the analysis of forbidden emission lines arising from the core of Hen 401, we determine [NII], [SII] and [OI] emission is being formed in regions with $n_e \geq 10^5$ cm^{-3}, whilst [FeII] emission lines would probably be forming in regions with densities from 10^5 to 10^6 cm^{-3}. The rich forbidden [FeII] emission lines of WRA 751 might indicate the presence of an extended region with electron density $< 10^6$ cm^{-3}.

4. Optical Morphology

A bipolar nebula has been detected in the light of Hα and [OIII] around Hen 401. The object show in both images an inner bright bipolar nebula, with two lobes of unequal surface brightness, with the axis at a P.A. of 73°, and an equatorial waist. The difference in brightness between both lobes suggests that there is a dense circumstellar disk located in the equatorial waist which is obscuring the light coming from the NE–E lobe. Wether the optical emission is due to an ionization front advancing throught the red–giant remnant or due to shocks developed as the PNe fast wind interacts with the red–giant remnant is still unclear.

5. Conclusions

Hen 401 and WRA 751 have been identified as post–AGB stars in the transition phase between the end of the AGB and the formation of a new PN. The images of Hen 401 reveal the bipolar morphology of the nebula and suggest the presence of a circumstellar disk.

References

Bujarrabal, V., Bachiller, R., 1991, A&A, **242**, 247

Loup, C., Forveille, T., Nyman, L.Å., Omont, A., 1990, A&A, **227**, L29

Manchado, A., Pottasch, S.R., García-Lario, P., Esteban, C., Mampaso, A., 1989, A&A, **214**, 139

Pottasch, S.R., Bignell, C., Olling, R., Zijlstra, 1988, A&A, **205**, 248

THE SPATIAL STRUCTURE OF THE HH 30 JET

A. RIERA

Departament de Física i Enginyeria Nuclear, EU de Vilanova i la Geltrú (U. Politécnica de Catalunya). Av. Víctor Balaguer s/n, E-08000 Barcelona (Spain).

A. RAGA

Mathematics Department, UMIST, Manchester M60 1QD (UK)

and

R. LÓPEZ, G. ANGLADA, R. ESTALELLA

Departament d'Astronomia i Meteorologia (U. Barcelona). Av. Diagonal 647, E-08028 Barcelona (Spain)

Abstract. New, deep, wide-field [SII] images of the HL Tauri region show the extended spatial structure of the HH 30 jet and counter-jet. At an angular distance of ~ 300 arcsec toward the NE, the HH 30 jet ends in a group of scattered condensations. This previously undetected structure might correspond to a broken-up working surface. Our images also include HH 262, which is shown to have a previously undetected extended emission region.

Key words: HH objects: individual: HH 30, HH 262

1. Introduction

The region around HL Tauri (in the NE of the L1551 dark cloud) actually has two main jets, the spatially more extended one being ejected from the "HH30-star", the second one ejected from HL Tauri, and a third, less well defined, jet possibly arising from the VLA-HL Tau radio source (Mundt *et al.* 1987). A detailed study of the structure of the HH 30 and HL Tau jets can be found, e.g., in Mundt, Ray & Raga (1991). Even though these two jets have a structure of aligned knots that qualitatively resembles the well-known HH 111 and HH 34 jets, they differ from these objects in that they do not have clearly defined, bow-shaped "heads".

We have obtained a new, deep wide-field [SII] $\lambda\lambda$ 6717, 6731 CCD image of the HL Tauri region which shows unprecedent details of this region.

2. Observations

Observations were carried out on 1993 December 15–16 at the Prime focus of the 2.5 m Isaac Newton Telescope. We obtained eight images of 1800 s exposure each using a [SII] filter of central wavelength $\lambda_0 = 6730$ Å and bandpass $\Delta\lambda = 48$ Å . We also obtained a 1200 s exposure image through a "continuum" filter of central wavelenght $\lambda_0 = 6925$ Å and bandpass $\Delta\lambda = 50$ Å . The eight [SII] images were recentered and median averaged to obtain a final image with a total integration time of four hours.

Astrophysics and Space Science **224**: 551–552, 1995.

3. Results

In our deep [SII] image, the HH 30 system shows chains of aligned knots that extend more or less radially away the source in both directions. This image shows the detailed structure of the redshifted HH 30 counter-jet, which was only marginally visible in the images of Mundt *et al.* (1988, 1990). Toward the NE, the blueshifted HH 30 jet appears to end in a group of scattered condensations at an angular distance of $\sim$ 300 arcsec (0.24 pc at a projected distance of 160 pc). This previously undetected structure is invisible in our continuum image, thus indicating that the condensations are emission-line objects. These condensations form a structure that probably corresponds to a broken-up working surface, being somewhat reminiscent of the well known object HH 2. No similar structure was detected for the SW counterjet. The total angular extent of the HH 30 system is $\sim$ 8.2 arcmin or 0.38 pc (including the counterjet and the recently detected condensations toward the NE of the jet). It makes HH 30 one of the longest HH outflows known.

We also detect a spatially extended condensation (of $\sim$ 20 arcsec) that is precisely aligned with the NE extension of the jet from HL Tauri at an angular distance of $\sim$ 4.5 arcmin (or 0.16 pc) from the HL Tauri star position. This condensation might correspond to a working surface. Our image includes HH 262, a group of HH condensations that have been thought to be associated either with L1551–IRS5 or with the L1551NE source. In our image, HH 262 shows an "hour-glass" shaped structure that qualitatively resembles the morphology expected for a bipolar cavity blown by the wind from an undetected source located in the centre of HH 262.

Acknowledgements

The Isaac Newton Telescope is operated on La Palma (Spain) by the Royal Greenwich Observatory in the spanish Observatorio del Roque de los Muchachos of the Instituto de Astrofísica de Canarias. We thank our support astronomer R. Peletier for his help during the observations. This work has been supported by the *Acción Integrada Hispano-Británica* HB93-024 and by the spanish DGICYT grant PB92-0900.

References

Mundt, R., Brugel, E.W. and Buhrke, T. 1987, *ApJ*, **319**, 275
Mundt, R., Ray, T.P. and Raga, A.C. 1991, *A&A*, **252**, 740
Mundt, R., Ray, T.P. and Buhrke, T. 1988, *ApJL*, **333**, L69
Mundt, R., Ray, T.P., Buhrke, T., Raga, A.C. and Solf, J. 1990, *A&A*, **232**, 37

CIRCUMSTELLAR X-RAY EMISSION FROM SN1978K

ERIC M. SCHLEGEL,* E. COLBERT** and R. PETRE
NASA-GSFC, Code 668, Greenbelt, MD 20771 USA

Abstract. We present the X-ray light curve in the 0.2–2.4 keV band based on five *ROSAT* observations of SN1978K in NGC 1313. The X-ray emission is believed to arise from the interaction of the reverse shock and the expanding debris from the supernova. The reverse shock becomes established after the outgoing shock runs into circumstellar matter.

Key words: X-rays — circumstellar matter — supernovae

1. Introduction

The *Einstein* IPC first detected X-ray emission from a supernova, SN1980K, 35 days after discovery (Canizares *et al.* 1982). Since then, four more supernovae have been detected in X-rays (Schlegel 1994a): SN1987A in the LMC, SN1986J in NGC 891, SN1978K in NGC 1313, and SN1993J in NGC 3031. SN1980K has been re-observed and probably re-detected (Schlegel 1994b), as has SN1987A (Gorenstein *et al.* 1994; Beuermann *et al.* 1994).

X-ray emission is expected to arise by at least two processes: Compton-scattered γ-rays from radioactive decay, and the interactions of the shock from the detonation with any circumstellar matter. Of the five supernovae, the X-ray emission of only one may be attributed definitively to Compton-scattered γ-rays: SN1987A in the LMC. The other four almost certainly arise from circumstellar interaction. Two were detected just after the explosion (SN1980K, SN1993J); three were detected at roughly a decade post explosion (SN1980K, SN1978K, SN1986J).

If the X-ray emission arises from circumstellar interaction, then X-ray data can provide a measure of the envelope density distribution and the circumstellar matter distribution, if both are describable by power laws. The light curve of the interaction responds to both distributions. Radio observations also provide such a measure (e.g., Weiler *et al.* 1986).

2. The Observations and Implications

The observations were obtained with the *ROSAT* Position-Sensitive Proportional Counter (PSPC) and the High-Resolution Imager (HRI). A single point has been obtained using *ASCA* satellite (Petre *et al.* 1994).

The X-ray light curve is shown in the figure. It clearly shows an increase in the X-ray flux from the non-detection observation using the *Einstein* IPC.

* Research Scientist, Universities Space Research Association
** University of Maryland

Astrophysics and Space Science **224**: 553–554, 1995.

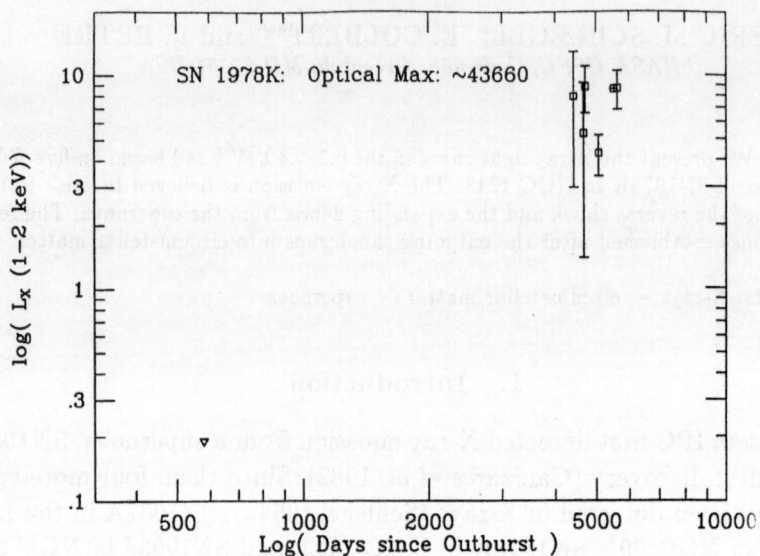

Fig. 1. The X-ray Light Curve of SN1978K to date.

The question to be addressed by future observations is whether the light curve has peaked. The supernova may soon enter the adiabatic phase where the light curve will decline as free-free radiation dominates. Observations to date do not indicate whether SN1978K has entered this phase (Schlegel et al. 1994). Testing for the early phase is easy: it is dominated by substantial absorption. A decrease in the absorption column tells us the first phase is ending. An *ASCA* re-observation has been requested for this purpose. Eventually, the light curve will measure the circumstellar density distribution. The decline rate, for example, is $\propto t^{-[(2s-3)n-5s+6]/(n-s)}$, where s is the circumstellar power law index ($\sim$2) and n is the index for the supernova ejecta ($\sim$8–12) (Chevalier & Fransson 1994; Fransson et al. 1994). A comparison can then be made between the values of n and s determined from X-ray observations and the values from radio observations.

References

Beuermann, K., Brandt, S., & Pietsch, W.: 1994, *Astron. & Astrophys.* **281**, L45
Canizares, C., Kriss, G., & Feigelson, E.: 1982, *Astrophys. J.* **253**, L17
Chevalier, R. & Fransson, C.: 1994, *Astrophys. J.* **460**, 268
Fransson, C., Lundqvist, P., & Chevalier, R.: 1994, *Astrophys. J.* , (submitted)
Gorenstein, P., Hughes, J., & Tucker, W.: 1994, *Astrophys. J.* **420**, L25
Petre, R., et al: 1994, *PASJ* **46**, L115
Schlegel, Eric M.: 1994a, *Reports on Prog. in Phys.* , (in preparation)
Schlegel, Eric M.: 1994b, *Astron. J.* **to appear**, Nov 1994
Schlegel, Eric M., Colbert, E., & Petre, R.: 1994, *Astrophys. J.* , (in preparation)
Weiler, K., et al.: 1986, *Astrophys. J.* **301**, 790

STATISTICAL RESULTS FROM A JCMT SURVEY OF MM AND SUB-MM EMISSION IN HERBIG AE/BE STARS

S. SKINNER

Inst. of Space & Astronautical Science, 3-1-1 Yoshinodai, Sagamihara 229 Japan

A. BROWN

JILA / Univ. of Colorado, Boulder, Co 80309-0440 USA

and

W. DENT

Joint Astronomy Center, 660 N. Aohoku Pl., Hilo, HI 96720 USA

Abstract. We have detected 1.1 mm continuum emission from 24 of 53 Herbig Ae/Be stars surveyed with the JCMT. Survival analysis shows that 1.1 mm luminosity is correlated with bolometric luminosity and with IRAS 25 μm luminosity. For those stars that were also detected at 0.45 or 0.8 mm we find a typical flux dependence of the form $S_\nu \propto \nu^3$, which is steeper than that of most classical T Tauri stars.

Key words: radio continuum: stars – stars: emission line, Be – stars: pre-main-sequence

1. Summary of Results

Herbig Ae/Be stars are intermediate mass (2 – 10 $M_\odot$) pre-main-sequence stars that typically show strong IR excesses due to circumstellar dust. However, the dust properties and geometry remain quite uncertain. Millimetre-submillimetre observations with instruments such as the UKT14 bolometer on the James Clerk Maxwell Telescope (JCMT) are now providing crucial information on the spectral energy distributions (SEDs) and dust properties of individual Herbig stars (cf. Mannings, this volume). It is our purpose to supplement these studies of specific stars with a broader picture of the mm properties of the class as a whole. In doing so, we can address questions of a statistical nature such as (i) how do the mm properties of Herbig stars compare with the lower mass T Tauri stars, and (ii) do fundamental relationships exist between the mm properties and other stellar parameters?

To this end, we have conducted a sensitive JCMT/UKT14 survey of 53 Herbig stars selected from Finkenzeller & Mundt (1984) (FM). We obtained 1.1 mm beam-switched photometry centered on the optical position of each star, yielding 24 detections. Typical upper limits for non-detections are $\approx$40 mJy (3σ). We did not map the sources, but information on spatial structure is being obtained in other JCMT programs (Sandell & Weintraub 1994). Using additional data acquired at 0.45 and/or 0.8 mm for 14 of the detections, we find typical spectral indices $\alpha \approx 3$ ($S_\nu \propto \nu^\alpha$). Two exceptions are

Astrophysics and Space Science **224**: 555–556, 1995.

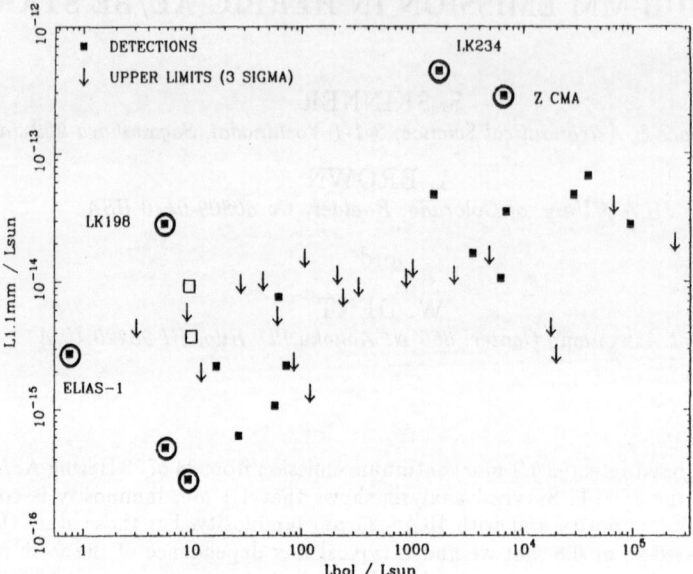

Fig. 1. Millimetre vs. bolometric luminosity for Herbig stars. Circled stars are binaries or have IR companions. Open squares are T Tauri stars DG and HL Tau.

Elias 1 and HD163296 which have flatter energy distributions with $\alpha \approx 2$, similar to those of classical T Tauri stars. Elias 1 has a close IR companion that may be dominating the mm-submm emission. High resolution IR observations of HD163296 to search for a close companion would be useful.

Using survival analysis methods that account for non-detections (Isobe et al. 1986), we find evidence for a correlation between 1.1 mm luminosity ($L_{1.1mm}$) and L_{bol}. Both Cox's and Kendall's tests give a high correlation probability, P(corr) ≥ 0.996, and a maximum likelihood (ML) fit gives $L_{1.1mm} \propto L_{bol}^{0.36\pm0.10}$. Fig. 1 shows that four stars with close IR companions are overluminous at 1.1 mm for their assigned L_{bol} (FM nos. 1,4,22,52). We thus suspect that the IR companions are contributing significantly to the mm flux. In addition, several stars with large L_{bol} were *not* detected (FM nos. 12,20,24,26,51). These are all early Be stars with small IR excesses (Group III objects). It is not yet known if they are classical Be stars, or perhaps young stars that have managed to shed most of their pre-natal dust. We also find a strong correlation between $L_{1.1mm}$ and IRAS 25 μm luminosity ($L_{25\mu m}$), with P(corr) ≥ 0.999 and $L_{1.1mm} \propto L_{25\mu m}^{0.89\pm0.10}$. This correlation provides quantitative support for the usual assumption that the far-IR and mm emission in Herbig stars arises from a common mechanism.

References

Finkenzeller, U., Mundt, R., 1984, *A&A Suppl.*, **55**, 109 (FM)
Isobe, T., Feigelson, E., Nelson, P., 1986, *ApJ*, **306**, 490
Sandell, G., Weintraub, D., 1994, in *The Nature & Evolutionary Status of Herbig Ae/Be Stars*, ed. P. Thé, M. Pérez, & E. van den Heuvel (San Francisco: ASP), 261.

CIRCUMSTELLAR MATTER IN SELECTED SYMBIOTIC STARS

A. SKOPAL

Astronomical Institute of the Slovak Academy of Sciences, 059 60 Tatranská Lomnica, Slovakia

Abstract. A group of symbiotic binaries exhibits periodic variation in the optical continuum during orbital motion. It is suggested that this variation can be produced by different projection of the circumstellar matter, located in major part between the components of the binary, into the line of sight at different orbital phases.

Key words: stars: symbiotic — stars: individual: EG And, V443 Her

1. Introduction

Periodic, wave-like modulation of the star's brightness along the orbital cycle, displaying a relatively large amplitude (~ 1 mag or higher) and depending on the wavelength ($\Delta U \geq \Delta B \geq \Delta V$), is a common feature of a group of symbiotic stars (e.g. EG And, BD-21.3873, RW Hya, SY Mus, V443 Her, He2-467, AS 338, AG Peg, V1329 Cyg, AG Dra). At present, in spite of some difficulties in explaining, mainly, the very large amplitude of the light curve, this variation is generally ascribed to a reflection effect.

In this contribution, it is suggested that the circumstellar matter located mainly between the components of the binary, is responsible for such the behaviour in the optical continuum. We consider whether the observed wave-like variation can be produced by different projection of this matter into the line of sight when viewing the system at different orbital phases.

2. Results and Discussion

We tested behaviour of the optical continuum of the symbiotic stars EG And and V443 Her by the model of the circumstellar material in the binary system suggested by Skopal *et al.* (1993). A schematic picture of this model is drawn in Fig. 2. The fitting of the theoretical curves to the observed data was performed by trial and error. The values of the resulting parameters are considered best by eye estimate, without the application of any formal fitting algorithm. Our preferred models of the U-light curves for EG And ($P_{\text{orb}} = 482$ days, mass ratio $q = M_{\text{cool}}/M_{\text{hot}} = 3$, inclination of the orbit $i = 45°$) and V443 Her ($P_{\text{orb}} = 600$ days, $q = 2$, $i = 18°$) are shown on Fig. 1. Differences between our idealized model and the real composition can be caused by an inhomogenously distributed material in the 'envelope'.

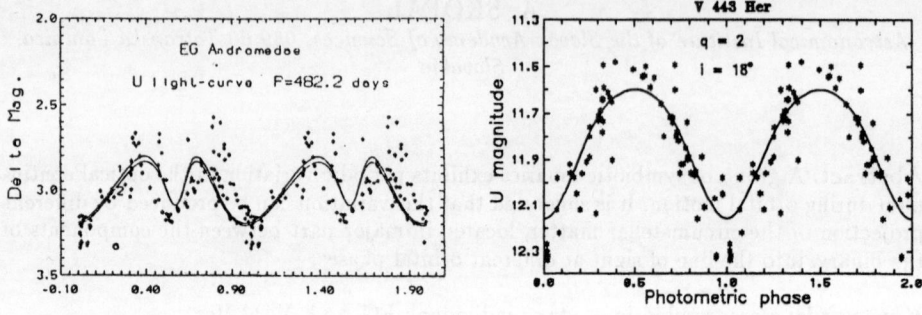

Fig. 1.   The U-light curves of the symbiotic binaries EG And (after Skopal *et al.* 1993) and V443 Her. Full lines represent model solutions.

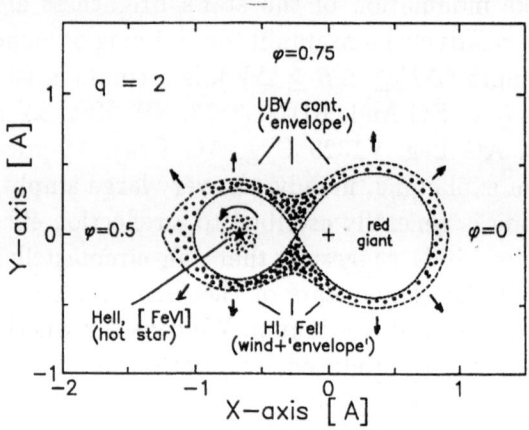

Fig. 2.   Sketch of the symbiotic binary exhibiting the wave-like variation in the light curve along the orbital cycle. It is suggested that the circumstellar matter located primarily between the components is responsible for such the behaviour. Separation of the components $A = 1$.

Acknowledgements

The author is grateful to the British Council for receipt of support within 'Promoting cultural, educational and technical co-operation between Britain and other countries', and Liverpool John Moores University for a travel grant.

References

Skopal A., Vittone A. & Errico L. 1993, Ap&SS, 209, 79

CIRCUMSTELLAR MATTER IN THE SYMBIOTIC BINARY CH CYG DURING ITS CURRENT OUTBURST

A. SKOPAL

Astronomical Institute of the SAV, 059 60 Tatranská Lomnica, Slovakia

M.F. BODE

Chemical & Physical Sciences, Liverpool JM University, Liverpool L3 3AF, U.K.

M. BRYCE and J. MEABURN

Dept of Physics and Astronomy, University of Manchester, Manchester M13 9PL, U.K.

and

S. TAMURA

Astronomical Institute, Tohoku University, Sendai 980, Japan

Abstract. We present optical high-resolution spectroscopy and UBV photometry of the symbiotic binary CH Cyg during the current outburst which began in 1992 February.

Key words: stars: individual: CH Cyg — binaries: symbiotic - spectra: line profiles

1. Introduction and Observations

The symbiotic star CH Cyg is an eclipsing binary with a 5700 d orbital period (e.g. Yamashita & Maehara 1979). It contains a late-type giant star and a compact companion. At the beginning of 1992 CH Cyg began a new outburst (Fig. 1). High-dispersion spectroscopy in the Hα region was secured at the La Palma and Okayama (OAO) Astrophysical Observatories. The UBV photometry was performed at the Skalnaté Pleso Observatory.

2. Results and Discussion

The blue continuum exhibits three types of variation: (i) A gradual increase from U~8.5 in the spring of 1992 to U~6.7 in the summer of 1994. (ii) Irregular variability on a time-scale of days to weeks. (iii) Short-term flickering variations running from minutes to hours. Development of hydrogen line profiles during the outburst is summarised in Fig. 2. (a) At the beginning of the outburst, simple P-Cygni profiles were present. (b) In the summer of 1992 a complex structure on the violet side of the hydrogen line profiles had developed. (c) During the period from 1993 June to 1993 September, only a single emission peak was observed in the Hα profile. (d) In 1993 October the Hα line profile again underwent large and abrupt changes.

The red giant in CH Cyg appears to be a semi-regular variable with a period of ~756 days (Skopal *et al.* 1994). It should thus lose mass at a rate of 10^{-5} to $10^{-4}\,M_\odot\,yr^{-1}$ (Schild 1989). The corresponding mass accretion rate is

Astrophysics and Space Science **224**: 559–560, 1995.
© 1995 *Kluwer Academic Publishers.*

approximately $1.3\,10^{-7}$ to $1.3\,10^{-6}\,M_\odot\,yr^{-1}$ (Livio & Warner 1984), which overlaps in part the range of those predicted to cause steady hydrogen burning on a white dwarf in such a system. Although the luminosity produced during the outburst of CH Cyg is below that predicted theoretically for such a case, it is obvious that it is changes in accretion rate direct from the wind of the red giant, as the binary separation changes during its orbit with the compact companion, that gives rise to outbursts.

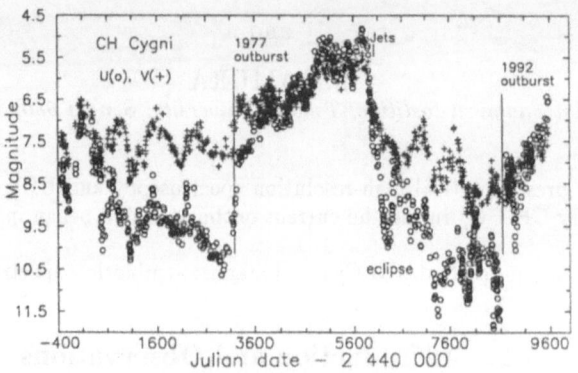

Fig. 1 The U (o) and V (+) light curves covering the period 1967–1994.

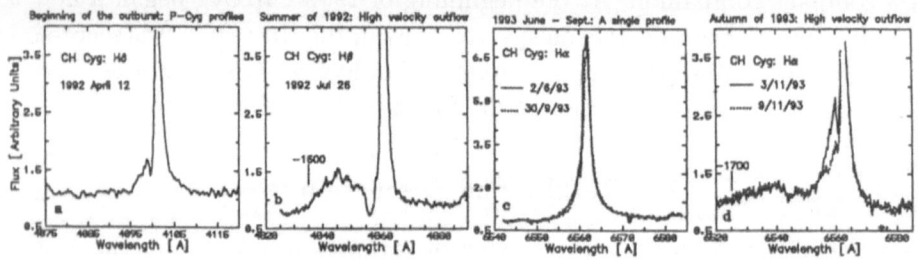

Fig. 2 Development of the hydrogen line profiles during the 1992 outburst.

Acknowledgements

We thank to A. Tajitsu for his help during reduction of the spectra taken at the OAO. AS is grateful to the British Council for receipt of support and Liverpool John Moores University for a travel grant.

References

Livio M. & Warner B., 1984, Observatory, 104, 152
Schild H., 1989, MNRAS, 240, 63
Skopal A., Bode M.F., Evans A., Errico L. et al., 1994, MNRAS, submitted
Yamashita Y., Maehara H., 1979, PASJ, 31, 307

COMPARISON BETWEEN A RADIATIVE WIND MODEL
FOR BE STARS AND HIGH ANGULAR RESOLUTION
DATA FROM THE GI2T INTERFEROMETER

PH. STEE

Observatoire de la Côte d'Azur,
CNRS URA 1361, Plateau de Calern,
06460 St Vallier de Thiey, France.
email: stee@rossini.obs-nice.fr

Abstract. I present theoretical line profiles and intensity maps from an axi-symmetric radiative wind model from a rapidly rotating Be star. The introduction of a viscosity parameter in the latitude-dependent hydrodynamic code enables us to consider the effects of the viscous force in the azimuthal component of momentum equations (Araújo et al. 1994). Both velocity field and density law derived from the hydrodynamic equations have been used for solving the statistical equilibrium equations. By adopting the Sobolev approximation, we could easily obtain a good estimate of both electronic density and hydrogen level populations throughout the envelope. The numerical calculation was performed for parameters characterisic of the Be star γ Cassiopeiae.

Key words: stars: Be — stars: individual: γ Cas

1. A radiative wind model for γ Cas

A challenging problem to solve for most radiative wind models resides in the Balmer emission line profiles which remain exessively wide. Recent quantitative computations by Stee & Araújo 1994 (hereafter SA) who found a separation of 2000 km/s between the V and R peaks of Hα reinforced this conclusion. In order to overcome this difficulty, SA suggest the possibility of a flow which is essentially driven by a large number of optically thin lines for the equatorial region (see also Lamers 1986; Marlborough 1987; Boyd & Marlborough 1991). Such a wind would have a lower terminal velocity and would produce emission line profiles less broadened.

We have built a model which reproduces both spectroscopic and interferometric data that have been measured with the GI2T (Grand Interferometre a 2 Telescopes) interferometer at the Observatoire de la Côte d'Azur (Mourard *et al.* 1994). We have obtained an Hα emission profile from our model that is in good agreement with the observed spectra (Fig.1). Our computed intensity maps, in both the continuum and Hα, agree reasonably well with maps derived from observed visibility (Fig. 2).

The model indicates that a radiative wind, driven mainly by optically thin lines at the equator, is a likely scenario for γ Cas. This is discussed in greater depth in Stee *et al.* (1994).

Astrophysics and Space Science 224: 561–562, 1995.
© 1995 *Kluwer Academic Publishers.*

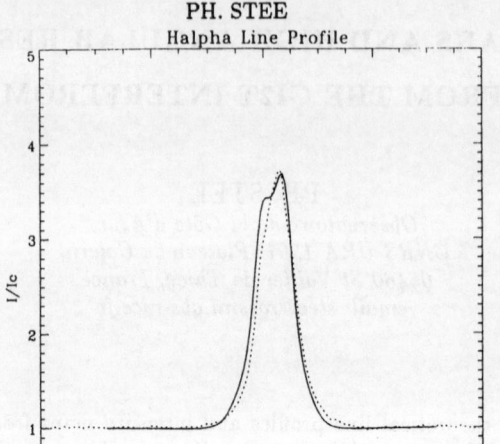

Fig. 1. Normalized $H\alpha$ line profile. Solid line: model, dotted line: GI2T spectrum

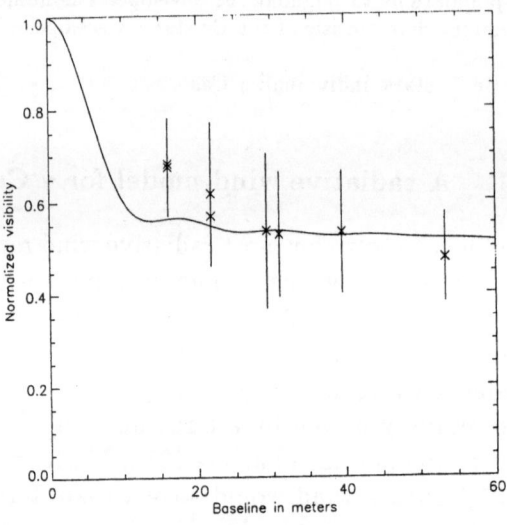

Fig. 2. Relative visibility curve (Visibility in the Hα line normalized to the visibility in the continuum). Solid line: model, Cross: GI2T data

References

Araújo, F.X., Freitas Pacheco, J.A., Petrini, D. 1993, MNRAS, 267, 501

Boyd, C.J. and Marlborough, J.M. 1991, ApJ, 369, 191

Lamers, H.J.G.L.M. 1986, A&A, 159, 90

Marlborough, J. M. 1987, IAU Colloquium 92, ed. A. Slettbak and T.P. Snow, Cambridge, Cambridge University Press

Mourard, D., Bosc, I., Blazit, A., Bonneau, D.,Merlin, G., Morand, F., Vakili, F., Labeyrie, A. 1994, A&A, 283, 705

Stee, Ph. and Araújo, F.X. 1994, A&A, in press

Stee, Ph., Araújo, F.X., Vakili, F., Mourard, D., Bonneau, D., Bosc, I. and Morand, F. 1994, A&A, in preparation

SILICATES IN EVOLVED STARS: THE LRS-MASER CHRONOLOGY REVISITED

ROBERT E. STENCEL

Department of Physics & Astronomy, University of Denver, Denver CO 80208 U.S.A

Abstract. Since the publication of our correlation (1990 Ap.J. 350, L45) suggesting a temporal relationship among: (a) shapes of 10 micron silicates seen in IRAS Low Resolution Spectra (LRS); (b) IRAS broad band colors; (c) light curve asymmetries, and (d) the types of masers associated with various Mira variable stars, the question has arisen whether the implied chronology refelcts the interval between thermal pulses, or the more lengthy ascent of the AGB. There is evidence in favor of the former interpretation. A strong implication of this idea is that variability among pre-silicate spectra should be significant. Several examples of this have been found in the extended LRS database, and are presented. Finally, plans are presented to continue monitoring of selected variables in the 10 micron region, to verify the variability suggested in the limited phase coverage provided in the LRS database. Monitoring will include use of our facilities at the Mt.Evans-Womble Observatory — highest in the world at 4313 meters altitude.

Key words: circumstellar material — IRAS — infrared spectroscopy

1. Introduction

Because we anticipate that dust formation in optically thin dust shell Miras may follow the pulsational driving cycle, we have been investigating 10 micron spectra of a variety of Mira variables in search of this effect.

Stencel *et al.* (1990) announced a series of correlations of LRS spectral shapes with IRAS broad band colors, light curve asymmetries, and the types of masers associated with various Mira variable stars. Originally, we argued that the sequence of silicate shapes represented the formation and annealing of silicate dust grains in the circumstellar space surrounding oxygen rich Mira stars. This interpretation is probably overly simplistic, given the ambiguities of grain formation history, and stellar variability on short- (Mira cycle, ~ 1 yr), intermediate- (thermal pulses, $\sim 10^3$ yrs), and long- (AGB ascent, $\sim 10^{4-5}$ yrs) scales. One diagnostic for which of these cycles is more predominant in the sample under study (see Little-Marenin, Staley & Stencel, 1993), is the variability level of features on the short period timescales.

2. New Observations

New post-IRAS data sources have emerged in the past couple of years that produce important spectral scans in the silicate region for moderately bright stars, including HiFOGS, CGS3 and GLADYS. Our Continuum Atmospheric Emission Spectro-Radiometer (CAESR, Creech-Eakman *et al.* 1994) is a high resolution ten micron spectrometer capable of recording 7–13 micron

Astrophysics and Space Science **224**: 563–564, 1995.

stellar spectra at ~250 resolution. A preliminary result of a recent observing run at WIRO is that while the shape and flux agreement, modulo airmass, among selected bright evolved stars are reasonable longward of 8.5 microns, we find that the shape diverges markedly from the LRS shape shortward of 8.25 microns. In many class 20+ spectra, LRS shows a general rise as one might expect from a blackbody distribution away from the silicate emission line. However, the falloff to shorter wavelengths seen in the CAESR data is consistent with a strong molecular SiO absorption that should be present in most red giants near 8 microns (Rinsland and Wing 1982). Although our reductions are preliminary and most instruments suffer from correcting for terrestrial methane obscuration shortward of 7.5 microns, the LRS calibration too has been shown to be suspect at short wavelengths. Because of the undeniable chemical pathway between SiO and silicates like $(MgFe)_2SiO_4$, in circumstellar outflows, further investigation of the true profile between 7 and 13 microns is appropriate. Another preliminary conclusion based on these new, higher resolution data is that there is little evidence for sharp spectral features superposed on the broad silicate profiles in all subclasses. As mentioned, we are in the process of upgrading our infrared observatory near the summit of Mt.Evans, Colorado (Mack *et al.* 1994), in order to pursue these and similar investigations regarding the process of planet forming around IRAS infrared excess main sequence stars, in part in support of an Infrared Space Observatory NASA Key Project investigation. Observing time will be granted to anyone citing this line in the paper, and we welcome inquires on the part of institutions interested in collaboratively developing a large aperture infrared telescope on the site. We thank the estate of William Herschel Womble for providing the seed funds needed to pursue these possibilities.

I am pleased to acknowledge support for these investigations from NASA grant NAGW-3680 to the University of Denver. I also thank Irene Little-Marenin, Greg Sloan, Fred Witteborn, Diane Wooden and Roger Sylvester for helpful information exchange.

References

Creech-Eakman, M., Stencel, R., Klebe, D., Williams, W.J. 1994 *Bull. Amer. Astron. Soc.* **26**, 948.

Kessler, M., Metcalfe, L. & Salama, A. 1992 *Space Sci. Rev.* **61**, 45.

Little-Marenin, I., Staley, S. & Stencel, R. 1993 in *Astronomical Infrared Spectroscopy: Future Observational Directions*, ed. S. Kwok, ASP Conf. Series **41**, p117.

Mack, J., Stencel, R. *et al.* 1994 *Bull. Amer. Astron. Soc.* **26**, 895.

Rinsland, R. & Wing, R. 1982 *Ap.J.* **262**, 201.

Stencel, R., Little-Marenin, I., Little, S. & Nuth, J. 1990 *Ap.J.* **350**, L45.

MAGNETICALLY-AIDED EVOLUTION OF PROTOPLANETARY DISKS

T. F. STEPINSKI and M. REYES-RUIZ *

Lunar and Planetary Institute **

3600 Bay Area Blvd.

Houston, Texas 77058 USA

Abstract. We investigate the global evolution of a turbulent protoplanetary disk incorporating the effects of Maxwell stress due to a large-scale magnetic field permeating the disk. A magnetic field is produced continuously by an $\alpha\Omega$ dynamo and the resultant Maxwell stress assists the viscous stress in p roviding the means for disk evolution. The most striking feature of magnetized disk evolution is the presence of the surface density bulge located in the "magnetic gap", the region of the disk where the degree of ionization is too low to allow for coupli ng between the magnetic field and the gas. The bulge persists for a time of the order of 10^5–10^6 yr. The presence and persistence of the surface density bulge may have important implications for the process of planet formation and the overall characteristics of resultant planetary systems.

Key words: protoplanetary disks – magnetic field – dynamo

1. Introduction

The physical conditions and evolution of protoplanetary disks (hereafter called PD) are determined mainly by the angular momentum transport mechanism acting in their interiors. In modeling PDs, it is usually assumed that turbulent viscosity, described by a dimensionless parameter α_{ss}, is solely responsible for the outward transport of angular momentum and the inward transport of mass. In addition to turbulence, a large-scale magnetic field embedded in a PD may also transport angular momentum. The presence of a large-scale magnetic field may introduce some changes in the overall structure and evolution of PDs. Reyes-Ruiz & Stepinski (1995) carry out one-dimensional numerical simulations of the viscous ev olution of magnetized PDs. Here we summarize our results. Gas in most parts of a PD is a poor electric conductor, with electrical conductivity lower than 10^4 (cgs units). In order for a PD to contain a magnetic field for a long enough period to produce a significant dynamical effect, the magnetic field must be continuously regenerated by the turbulent $\alpha\Omega$ dynamo to offset the losses due to dissipation. A magnetic field can be maintained only in those parts of the disk where the effective dynamo number, D_{eff} exceeds a certain critical value, $D_{\text{crit}} \simeq 12$. To calculate the physical properties of the disk, we solve the standard, thin disk set of equations modified to incorporate magnetic effects.

---

* Also affiliated with Dept. of Space Physics & Astronomy at Rice University

** Operated by USRA under contract No. NASW-4574 with NASA.

The total pressure is the sum of the gas and magnetic pressures. The standard α-prescription of turbulent viscosity is used; however, both turbulent viscous stress and Maxwell stress are incorporated. Magnetic enhancement of viscosity is not constant either in space, or in time because of the condition $D_{eff} < D_{crit}$ defining the location of the "magnetic gap" changes with time.

2. Results

We have calculated four evolutionary sequences. They are divided into two groups: a high-mass case that starts with a disk of 0.245 $M_\odot$ and an angular momentum of 5.6×10^{52} g cm^2 s^{-1}, and a low-mass case that starts with a disk of 0.11 $M_\odot$ and an angular momentum of 1.8×10^{52} g cm^2 s^{-1}. Within each case we examine disk evolution for two values of α_{ss}: $\alpha_{ss} = 0.01$ and $\alpha_{ss} = 0.002$. High-mass and low-mass scenarios differ in details but are qualitatively similar.

Overall, there is not much qualitative difference between *global* evolutionary properties of magnetized (MPD) and nonmagnetized (NMPD) disk models. MPDs evolve faster. For any given time, the MPD has a smaller mass than the NMPD. At most, the MPD requires up to an order of magnitude shorter time than the equivalent NMPD to lose the same percentage of initial mass. The maximum accretion rate of the MPD can be up to an order of magnitude higher then the maximum accretion rate of the NMPD; however, those differences last only for a short period of time. Throughout most of their evolution, accretion rates of MPD and NMPD are very close to each other. The inclusion of a self-generated magnetic field introduces the existence of a density bulge where density increases by a factor of ≈ 4. This bulge exists at the location of the magnetic gap. The presence of a density bulge can influence the character of the planetary system that presumably emerges from the PD. For example, in order to form the core of Jupiter on a proper timescale, a surface density of solids at ~ 5 AU from the Sun must be about $15 - 30$ g cm^{-2}, or 5-10 times greater than predicted by the minimum mass nebula model. In a MPD, this difficulty may have a natural solution, the density bulge may remain unaltered in the region of Jupiter's formation for a long enough time to permit the core of the proper mass to form.

References

Reyes-Ruiz, M. & Stepinski, T.F.: 1995, 'Evolution of Magnetized Protoplanetary Disk', *to appear in Ap. J.* ,

COAGULATION AND SETTLING OF DUST IN A TURBULENT PROTOPLANETARY DISK

MICHAEL F. STERZIK and GREGOR E. MORFILL

Max-Planck Institut für extraterrestrische Physik, 85740 Garching, Germany

and

BERENGERE DUBRULLE

C.N.R.S, Service d'Astrophysique, 91191 Saclay, France

Abstract. We apply our analytic model for the dust diffusivity to calculate the vertical structure of the dust sub-disk in a turbulent protoplanetary nebula. We present a numerical solution of a vertical dust settling equation *and* a coagulation equation for dust grains covering four orders of magnitude in time and grain size.

Key words: circumstellar disks — turbulence — dust coagulation

1. Model

In a turbulent solar nebula small μm-sized dust grains are able to coagulate efficiently into mm-sized particles. Larger dust particles settle more easily towards the midplane, because they begin to decouple from turbulent gas motion that tends to stir up small grains. The fractionation of solids and gas implies modifications of the relevant transport coefficients (viscosity and diffusivity), which in turn have effects on coagulation and settling rates. The derivation of the dust diffusivity is given in Dubrulle *et al.* (1994). For particle sizes of μm to cm, coagulation and settling time scales are compareable, and a variable mean size of the dust population has to be considered. The coagulation equation describing grain growth is usually solved by tedious numerical integration of the integro-differential equation (Weidenschilling 1988; Mizuno 1988; Nakagawa et al. 1986). We reduce the problem to the solution of two coupled PDEs: we assume coagulation of similar sized particles in a narrow size distribution and use an analytic expression for the relative velocity of two dust grains (Morfill 1985).

$$\partial_t N - \partial_z(z\Omega^2\tau_f N) = -\frac{1}{2}\int dm \int dm' \, n(m)n(m')\theta_c(m, m')$$

$\theta_c(m, m')$	$= \pi(a + a')^2 Q\delta v$	: coagulation parameter
δv	$= c_s\frac{\sqrt{2\alpha\Omega\tau_f}}{1+\Omega\tau_f}$	: relative grain velocity
$\Omega\tau_f$	$= \frac{\rho_s a}{\Sigma}$	: friction parameter $\propto$ particle size
m	$= \frac{4}{3}\pi\rho_s a^3$	: dust grain mass
N	$= \int dm \, n(m)$	: dust number density

Astrophysics and Space Science **224**: 567–568, 1995.
© 1995 *Kluwer Academic Publishers.*

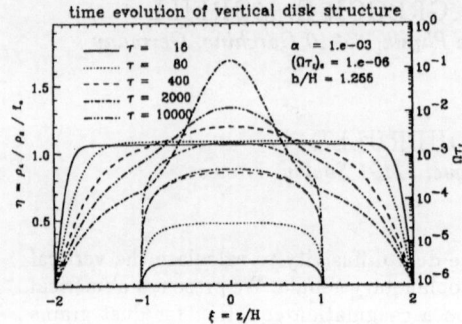

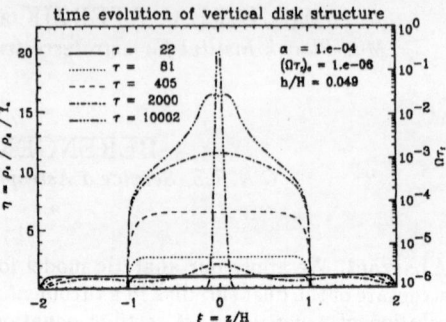

Fig. 1. Relative dust density η and
particle size $\Omega\tau_f$ evolution: $\alpha = 10^{-3}$.

Fig. 2. Relative dust density η and
particle size $\Omega\tau_f$ evolution: $\alpha = 10^{-4}$

2. Results

In Figure 1 and 2 we show the time evolution of the dust to gas density
ratio η (normalized to cosmic abundance $f_c = 0.01$) in the vertical direction
for $\alpha = 10^{-3}$ and 10^{-4} (thick lines). Initally the dust consists of $\Omega\tau_f = 10^{-6}$
($\approx 1\mu m$) sized grains and is mixed homogenously throughout two gas scale
heights H. Thin lines indicate the vertical distribution of the mean particle
size that is confined within one scale height. For both turbulence strengths
rapid grain growth up to mm occures during the first 2000 dynamical time
scales. Whereas in the case of higher α the turbulence is strong enough to
prevent particles from significant settling, a run-away collapse of the dust
sub-disk happens for lower α: dust density enhancement in the midplane in
turn favors grain collisions (and hence growth). The turbulence is no longer
able to sweep up these grains and the combined action of settling and coagulation leads to a highly nonlinear evolution of η near the midplane. Our
integration stops after 10^4 dynamical timescales with a dust/gas enrichment
of 20 around the midplane; note that the sub-disk has not reached its equilibrium state at this time. An analytic analysis indicates stability as long as
$\Omega\tau_f \leq \alpha^{1/3}f_c^{2/3}$, which is only guaranteed for our high α case.

References

Dubrulle, B., G.E. Morfill, and M.F. Sterzik: 1994, *Icarus* , submitted
Mizuno, H.: 1988, *Icarus* **80**, 189
Morfill, G. E., 1985, in R. Lucas (ed), *Birth and infancy of stars*, Amsterdam, 693
Nakagawa, Y., Sekiya, M. and Hayashi, C.: 1986, *Icarus* **67**, 375
Weidenschilling, S.J., 1988, in J. T. Kerridge & M.S. Mathews (eds) *Meteorites and the
    early solar system*, Tucson, 348

ASCA OBSERVATIONS OF COLLIDING STELLAR WINDS
IN GAMMA VELORUM

I. R. STEVENS

School of Physics and Space Research, Birmingham University, UK

S. L. SKINNER and F. NAGASE

Institute of Space & Astronomical Science, Japan.

M. F. CORCORAN

Code 665, Lab. for High Energy Astrophysics, NASA/GSFC, USA.

A. J. WILLIS

Dept. of Physics & Astronomy, University College London, UK.

A. M. T. POLLOCK

Computer & Scientific Co. Ltd, Sheffield, UK.

and

K. KOYAMA

Dept. of Physics, Kyoto University, Japan.

Abstract. We present new high spectral resolution X-ray observations of the colliding wind binary γ Vel taken with the ASCA satellite. We find two spectral components, one of which is post-shock emission from the colliding winds. Spectral variability is also seen, consistent with current notions of colliding wind phenomena.

Key words: stars:Wolf-Rayet – stars:X-rays – stars:colliding winds

1. Introduction

Gamma Velorum is the closest Wolf-Rayet star, and is a binary (components WC8 + O9I) with $P_{orb} = 78.5$ days and $e = 0.4$. It was observed with ROSAT and seen to be a highly variable source (Willis *et al.* 1994). These observations were interpreted as being due to emission from shocked material from colliding winds. Further to this, γ Vel has been observed twice with the Japanese ASCA satellite, at orbital phases $\phi = 0.4$ and 0.5 (when the O-star is in front of the WR star). ASCA has the advantage of higher spectral resolution and a wider X-ray bandpass ($0.5 - 10$ keV), and is well suited to observing colliding wind emission in massive stars.

2. ASCA observations

The ASCA observations (Fig. 1) find evidence of two spectral components; one soft ($kT_x \sim 0.2$ keV), with a constant absorbing column ($N_H = 10^{22}$ cm^{-2}) and one harder ($kT_x \sim 1.2$ keV), with variable absorption ($N_H = 5 \times 10^{22}$ cm^{-2} at $\phi = 0.4$ and $N_H = 3 \times 10^{22}$ cm^{-2} at $\phi = 0.5$).

Astrophysics and Space Science **224**: 569–570, 1995.
© 1995 *Kluwer Academic Publishers.*

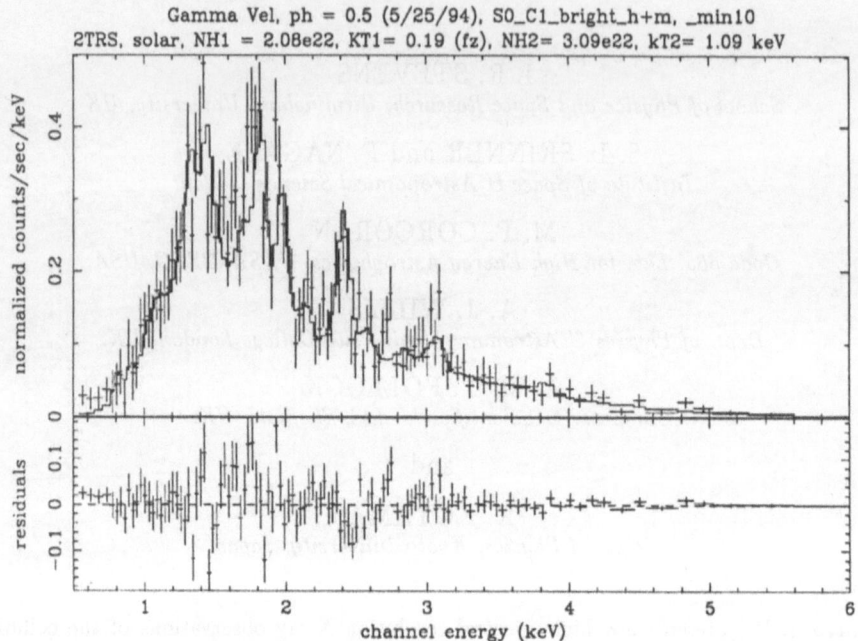

Fig. 1. ASCA Observations of γ Vel, at phase $\phi = 0.5$. The best fit spectral models are shown at the top.

Enhanced line emission at $\phi = 0.5$ seen in Fig. 1 is due to enhanced metal abundances, consistent with the non-solar abundances of the WR star.

These preliminary results agree qualitatively with the previous ROSAT results and the hydrodynamic/spectral models presented in Willis *et al.* (1994). The softer component is probably due to shocks intrinsic to all radiatively driven winds. The hard component is due to X-rays from the wind collision. The lower column at $\phi = 0.5$ is due to viewing the wind collision through the O-star wind rather than through the dense WR wind.

One feature is the low temperature of the hard component. This temperature corresponds to pre-shock wind velocities $\sim 1000 \, \text{km s}^{-1}$, which is lower than anticipated for the component stars. A potential solution is that the opposing radiation fields in close binaries tends to oppose wind acceleration and hence reduce the velocity at which winds collide (Stevens & Pollock 1994).

References

Stevens. I.R. & Pollock, A.M.T., 1994, Monthly Notices of the RAS, 269, 226.
Willis, A.J., Schild, H.E. & Stevens, I.R., 1994, Monthly Notices of the RAS, (submitted).

SEARCHES FOR BRIGHT-RIMMED CLOUDS WITH IRAS POINT SOURCES

K. SUGITANI

College of General Education, Nagoya City University, Mizuho-ku, Nagoya 467, Japan

and

K. OGURA

Kokugakuin University, Higashi, Shibuya-ku, Tokyo 150, Japan

Abstract. We have carried out systematic surveys for small bright-rimmed clouds associated with IRAS point sources in/around HII regions. They are candidate sites for star formation due to radiation-driven implosion.

Key words: Star Formation — IRAS point sources — HII Regions — Radiation-driven Implosion —catalogs

Bright-rimmed clouds are often found in/around relatively old HII regions and have been suspected to be potential sites of star formation due to the compression by converging ionization-shock fronts. They are considered to be originally dense cores of the parental molecular clouds and have emerged after the dispersion of the ambient gas due to UV radiation from OB stars. Their physical conditions match well the theoretical models of radiation-driven implosion (e.g. Bertoldi 1989) and, therefore, induced star formation is generally expected there. We have systematically searched for small bright-rimmed clouds associated with IRAS point sources to locate candidates for such star formation. The POSS atlas has been used in the northern sky, and the ESO(R)/SERC(J) atlases in the southern sky.

Eighty-nine bright-rimmed clouds associated with IRAS point sources have been selected (Table I). At least 19 (20 %) of the 89 clouds are known to be associated with molecular outflows, and six with HH objects. Most of their sizes are <1pc, and the luminosities of the associated IRAS point sources, ~ 10 to $3 \times 10^4 L_\odot$, are much larger than those of the IRAS sources associated with Bok globules or dense cores in dark cloud complexes, both having a similar cloud mass range. This suggests that intermediate-mass stars or multiple star systems are mainly formed in the bright-rimmed clouds. Also, IRAS luminosity to cloud mass ratios are significantly greater than those in Bok globules or dense cores. The details are reported in Sugitani *et al.* (1991) and Sugitani & Ogura (1994)

Astrophysics and Space Science **224**: 571–572, 1995.
© 1995 *Kluwer Academic Publishers.*

TABLE I
Bright-Rimmed Clouds Associated with IRAS Point Sources

#	HII Reg	α(1950)	δ(1950)	IRAS source	HH/ flow	#	HII Reg	α(1950)	δ(1950)	IRAS source	HH/ flow
1	S171	23 56 53.3	67 06 57	23568+6706		45	RCW14	7 16 15.5	-22 00 41	07162-2200	
2	S171	0 01 23.0	68 17 59	00013+6817		46	Gum Neb.	7 17 54.0	-44 29 24	07178-4429	
3	S171	0 02 47.9	67 00 57	00027+6700		47	S306	7 29 36.9	-19 21 07	07296-1921	
4	S185	0 56 00.7	60 37 21	00560+6037		48	Gum Neb.	7 32 56.6	-46 47 34	07329-4647	
5	S190	2 25 14.5	61 20 10	02252+6120	IC1805W	49	S307	7 33 28.0	-18 42 17	07334-1842	
6	S190	2 30 57.7	60 34 41	02309+6034		50	Gum Neb.	7 38 51.4	-42 59 18	07388-4259	
7	S190	2 31 01.7	61 33 40	02310+6133	outflow[1]	51	Gum Neb.	8 07 40.2	-35 56 07	08076-3556	120
8	S190	2 31 48.1	61 06 32	02318+6106		52	Gum Neb.	8 24 16.5	-50 50 44	08242-5050	46/47
9	S190	2 32 37.3	61 10 34	02326+6110		53	Gum Neb.	8 25 03.4	-50 30 34	08250-5030	
10	S199	2 44 23.9	60 12 06	02443+6012		54	NGC2626	8 33 42.6	-40 28 02	08337-4028	135/6
11	S199	2 47 37.4	59 50 54	02476+5950		55	RCW27	8 39 23.4	-40 41 18	08393-4041	
12	S199	2 51 08.3	60 23 35	02511+6023	outflow[1]	56	RCW27	8 41 08.3	-39 49 05	08411-3949	
13	S199	2 57 03.6	60 28 29	02570+6028	outflow[1]	57	RCW32	8 42 18.2	-41 05 19	08423-4105	
14	S199	2 57 35.6	60 17 22	02575+6017	GL4029	58	RCW32	8 43 35.7	-41 05 03	08435-4105	
15	S236	5 20 13.3	33 09 08	05202+3309		59	RCW38	8 56 22.3	-47 11 17	08563-4711	
16	S276	5 17 21.9	-5 55 05	05173-0555	RNO40	60	RCW38	8 58 19.1	-47 19 51	08583-4719	
17	S264	5 28 40.2	12 03 13	05286+1203		61	NGC3503	10 58 09.5	-59 20 02	10581-5920	
18	S264	5 41 45.3	9 07 40	05417+0907	B35	62	NGC3503	10 59 11.6	-59 34 52	10591-5934	
19	S277	5 32 00.4	-3 00 12	05320-0300	outflow[1]	63	NGC3503	11 01 13.5	-59 31 50	11012-5931	
20	S277	5 35 33.2	-1 46 50	05355-0146	Ori-I 2	64	BBW347	11 10 07.4	-58 29 60	11101-5829	
21	S277	5 37 11.8	-3 38 46	05371-0338		65	RCW62	11 30 37.9	-63 11 24	11306-6311	
22	S281	5 35 58.5	-5 15 48	05359-0515		66	RCW62	11 31 31.4	-62 59 45	11315-6259	
23	S249	6 19 56.5	23 11 32	06199+2311		67	RCW62	11 31 42.5	-62 54 43	11317-6254	
24	S275	6 32 16.5	4 27 40	06322+0427		68	RCW62	11 33 12.9	-62 58 15	11332-6258	
25	S273	6 38 17.6	10 17 54	06382+1017	I24	69	RCW62	11 38 49.0	-63 06 34	11388-6306	
26	S296	7 01 26.8	-11 41 17	07014-1141		70	RCW62	11 39 48.3	-62 51 11	11398-6251	
27	S296	7 01 37.9	-11 18 48	07016-1118		71	(Cen R1)	13 05 02.8	-61 54 26	13050-6154	
28	S296	7 02 21.4	-10 17 25	07023-1017		72	RCW75	13 15 51.5	-62 17 58	13158-6217	
29	S296	7 02 32.5	-12 04 51	07025-1204		73	RCW75	13 16 49.3	-62 08 19	13168-6208	
30	S49	18 15 55.0	-13 46 09	18159-1346		74	RCW85	14 15 59.2	-61 11 30	14159-6111	
31	S117	20 48 57.5	44 10 43	20489+4410		75	RCW98	15 51 56.5	-54 30 14	15519-5430	
32	S131	21 30 52.7	57 10 49	21308+5710		76	RCW105	16 06 56.7	-48 58 04	16069-4858	
33	S131	21 31 41.1	57 16 13	21316+5716		77	(σ Sco)	16 16 51.3	-25 26 29	16168-2526	
34	S131	21 32 02.5	57 50 06	21320+5750	outflow[1]	78	(σ Sco)	16 17 50.8	-25 01 01	16178-2501	
35	S131	21 34 35.8	58 18 10	21345+5818		79	RCW108	16 36 14.8	-48 45 54	16362-4845	
36	S131	21 34 40.1	57 14 05	21346+5714		80	RCW108	16 36 31.8	-48 36 35	16365-4836	
37	S131	21 38 53.2	56 22 18	21388+5622	GN21.38.9	81	RCW108	16 37 22.2	-49 11 59	16373-4911	
38	S131	21 39 10.3	58 02 29	21391+5802	IC1396N	82	RCW113/6	16 43 21.0	-41 08 32	16438-4110	
39	S131	21 44 30.8	57 12 29	21445+5712	IC1396E	83	RCW113/6	16 48 43.2	-40 43 09	16487-4043	
40	S131	21 44 38.0	56 55 05	21446+5655		84	RCW113/6	16 50 17.5	-40 02 29	16502-4002	
41	S131	21 44 52.8	57 04 46	21448+5704		85	RCW113/6	16 55 33.3	-42 37 33	16555-4237	
42	S131	21 45 00.1	56 58 30	21450+5658		86	RCW134	17 46 21.5	-31 28 20	17463-3128	
43	S142	22 45 48.5	57 46 59	22458+5746		87	M8	17 59 47.5	-24 22 14	17597-2422	
44	S145	22 27 12.2	63 58 21	22272+6358A	L1206	88	M8	18 01 13.1	-24 07 11	18012-2407	
						89	Simeis 188	18 06 50.7	-24 05 33	18068-2405	

[1]Sugitani & Ogura (1994)

References

Bertoldi 1989, *ApJ*, **346**, 735

Sugitani et al. 1991, *ApJS*, **77**, 59

Sugitani & Ogura 1994, *ApJS*, **92**, 163

SOLID CO SPECTROSCOPY:
EMBEDDED YSOS IN TAURUS–AURIGA

T.C. TEIXEIRA* ** and J.P. EMERSON
*Physics Dept., Queen Mary & Westfield College,
Mile End Road, London E1 4NS, UK*

Abstract. Infrared spectra of the $4.67\mu m$ solid CO feature were obtained for YSOs in Taurus. A first attempt is made to analyse the likely composition of the CO ices. Column densities are derived from the spectra, and compared with the ones predicted for a low mass protostar. The strengths of the CO- and water-ice features towards those objects are compared.

Key words: grain mantles — infrared:ISM:lines and bands — stars: pre-main-sequence

To determine the properties of embedded protostellar sources, it is important to study the structure and composition of their surrounding dusty envelopes. The CO-bearing ices of the dust grains' mantles are H_2O- or CO-dominated, pointing to different stages of grain mantle evolution (Tielens *et al.* 1991).

Spectra of the $4.67\mu m$ solid CO feature towards YSOs in the Taurus region were taken at the UKIRT, using CGS4. Of the 12 observed sources, 5 show clear evidence for the presence of CO ice: L1489, L1524, TMC1A, B5 IRS1, and Re50N IRS1. A first attempt has been made to analyse the likely composition of the ices, taking laboratory data from the literature. A good example of the results is presented in Fig.1. The profile is asymmetric, showing a dominant non-polar component (CO_2:CO), and a weaker polar one (H_2O:CO) appearing as a long wavelength wing. Column densities of solid CO were derived from the spectra, and compared with the total CO column density predicted using the Adams, Lada & Shu (1987) model for a low mass protostar (Fig.2). In Fig.3, the estimated optical depth of the CO-ice feature is compared with the optical depth of the $3.1\mu m$ water-ice feature towards the observed sources.

References

Adams, F.C., Lada, C.J. & Shu, F.H.: 1987, *Astrophys.J.* **312**, 788
Tielens, A.G.G.M., Tokunaga, A.T., Geballe, T.R. & Baas, F.: 1991, *Astrophys.J.* **381**, 181

* also Centro de Astrofísica da Universidade do Porto, Rua do Campo Alegre 823, 4100 Porto, Portugal
** TCT supported by grant BD/2095/92-RM from JNICT– Programa CIENCIA, Portugal

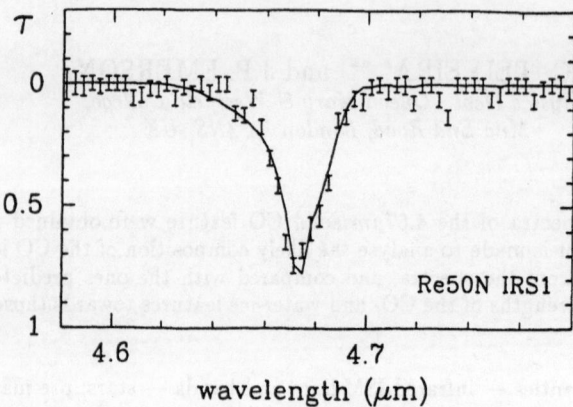

Fig. 1. Spectrum of Re50N IRS1 in optical depth units, and best laboratory data fit (solid line): CO_2:CO (20:1 10K) + H_2O:CO (20:1 40K)

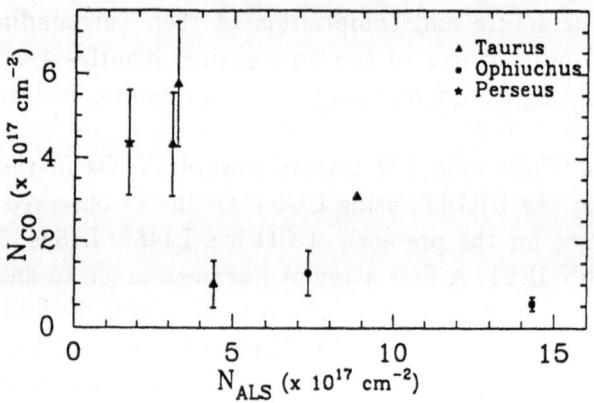

Fig. 2. Observed solid-CO column density *vs.* predicted total (gas + dust) CO column density for a range of temperatures where CO-ices can exist, using the ALS(1987) model for a protostar

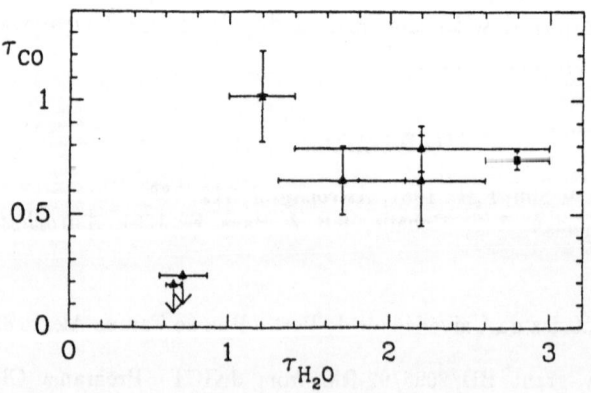

Fig. 3. Optical depth plot: CO-ice *vs.* H_2O- ice. Triangles are sources in Taurus. Re50N IRS1 (square) is in L1641, and B5 IRS1 (star) is in Perseus. Notice the good correlation for Taurus sources.

SI–O STRETCH MODES IN AMORPHOUS SILICATES

S. P. THOMPSON

Department of Physics and Astronomy, Leicester University, University Road, Leicester, LE1 7RH, United Kingdom.

Abstract. The 9.7 μm Trapesium silicate band is modelled as the sum of contributions from tetrahedral Si–O units with varying numbers of non–bridging oxygen atoms. In principle this will allow the silicate grain structure to be determined.

Key words: Dust – IR Bands – Trapesium – Silicate – Laboratory

Circumstellar 9.7μm bands are often modelled by the Trapesium emission function (TEF) attributed to silicate grains (Whittet 1992). Silicates are, however, real materials with measureable properties and laboratory fits to observed profiles are poor (cf. Dorschner *et al.* 1986, Hecht *et al.* 1986). If 9.7μm bands are due to silicate, the poor agreement must be because the right form of silicate has yet to be produced in the laboratory. Here, the TEF is used to place constraints on the silicate required. Writing the composition as xMO$(1 - x)$SiO$_2$, introducing modifying cation oxides (M=Mg, Fe etc.) into a SiO$_2$ network (where the 4 Os coordinating each Si are shared with neighbouring tetrahedra (T)), breaks the T–T bonds, forming non–bridging oxygen (NBO) atoms (Os not shared between T). Crystalline silicates have only one NBO value, but in amorphous silicates a distribution of NBO peaked about the crystalline value is common. NBO's change the Si–O bond's vibrational frequency (Table I). The ranges in wavelength reflect that of the materials/production methods used by various authors. Average λ were taken as initial positions and the TEF (Draine 1985) fitted (cf. Mysen *et al.* 1982, 1985) with Gaussians (Figure 1, Table I) whose λ-fit's may be associated with T–species. The proportions, n_i, of NBO (i =NBO/T) within TEF grains are related to the gaussian intensities, I_i (Mysen *et al.* 1985), $n_i = I_i \sigma_i$. These can not be inferred from Table I since bond scattering cross–sections, σ_i, are not generally available (λ's were found by Raman spectroscopy where σ_i includes polarizability), but could be measured and would allow the structure of the TEF grains to be determined. Figure 1 suggests it has only 4 NBO types. Presumably stars with 9.7μm bands narrower than TEF (Roche & Aitken 1984) have grains with fewer NBO species, possibly due to higher degrees of grain processing as the NBO shift towards single crystalline values. Likewise laboratory samples give poor fits because they do not have suitable n_i. The absence of NBO/T=0 implies SiO$_2$ is a minor component. Studies (Thompson *et al.* 1994) of gell-desication silicates support the modified random network model (Greaves 1985) where regions of SiO$_2$ are bounded by NBO tetrahedra. Between SiO$_2$ regions are the

Astrophysics and Space Science **224**: 575–576, 1995.
© *1995 Kluwer Academic Publishers.*

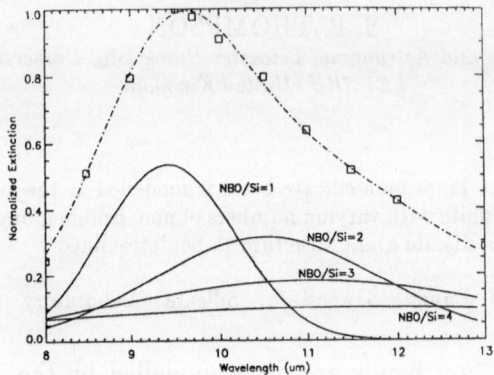

Fig. 1. Gaussian fit to Trapesium emission function

M–atoms. The absence of substantial SiO_2 may suggest a degree of grain processing, allowing extensive modifier distribution, or, formation conditions so far removed from the laboratory as to favour a different structural arrangement.

TABLE I

Si–O wavelengths (Brawer & White 1975, Virgo *et al.* 1980, Kusabiraki & Shiraishi 1981, Kusabiraki 1987) and Gaussian fit parameters.

$\lambda\mu m$	Species	NBO/T	Average λ	λ–fit	Intensity	Sigma
11.75 – 11.49	SiO_4	4	11.62	11.55	0.099(7)	4.0
11.05 – 10.87	Si_2O_7	3	10.96	11.0	0.181	2.21
10.31 – 10.52	SiO_3	2	10.41	10.52	0.314	1.27
9.52 – 9.09	Si_2O_5	1	9.30	9.42	0.534	0.747
9.38 – 8.33	SiO_2	0	8.85	–	–	–

References

Brawer, S.A., White, W.W.: 1975, *J. Chem. Phys.* **63**, 2421.
Dorschner, J., Friedemann, C., Gütler, J., Henning, Th., Wagner, H.:1986,*MNRAS* **218**, 37.
Draine, B.T.: 1985,*Ap. J. Suppl. Ser.* **57**, 587.
Greaves, G.N.: 1985, *J. Non–Cryst. Sol.* **71**, 203.
Hecht, J.H., Russell, R.W., Stephens, J.R., Grieve, P.R.: 1986,*Ap. J.* **309**, 90.
Kusabiraki, K., Shiraishi, Y.: 1981, *J. Non–Cryst. Sol.* **44**, 365.
Kusabiraki, K.: 1987,*J. Non–Cryst. Sol.* **95,96**, 411.
Mysen, B.O., Finger, L.F., Virgo, D., Seifert, F.A.: 1982,*Am. Miner.* **67**, 686.
Mysen, B.O., Virgo, D., Seifert, F.A.: 1985,*Am. Miner.* **70**, 88.
Roche, P.F., Aitken, D.K.: 1984, *MNRAS* **209**, 33P.
Thompson, S.P., Evans, A., Jones, A.: 1994 *in preparation.*
Virgo, D., Mysen, B.O., Kushiro, I.: 1980, *Science* **208**, 1371.
Whittet, D.C.B.: 1992, *Dust in the galactic environment*, I.O.P.

DUST SHELLS OF DISTANT YOUNG STARS

H. M. TOVMASSIAN * and S. G. NAVARRO

INAOE, Tonantzintla, Puebla, Pue., Mexico.

G. H. TOVMASSIAN

OAN, UNAM, Ensenada, B.C., Mexico.

and

L. J. CORRAL

INAOE, Tonantzintla, Puebla, Pue., Mexico.

Abstract. The *IRAS* colours of some B stars with anomalous UV extinction confirm that they have circumstellar dust.

Observations of OB associations with the wide-angle ultraviolet space telescope *Glazar* suggested that some of the young, mainly B type, stars are embedded in small, relatively dense dust clouds, which may well be circumstellar dust envelopes (Tovmassian *et al.* 1990, 1991a, 1991b, 1992, 1993). Ten of the stars belong to a nearby subgroup of the Orion complex at a distance of about 270 pc (Tovmassian *et al.* 1991) and 13 to different OB associations at distances from about 40 pc to 2.4 kpc.

To check the suggestion we inspected the *IRAS* data at the positions of these 23 stars and also on the positions of 14 control stars of the same subgroup of the Orion complex, which were not expected to have circumstellar dust shells. We found that 8 stars of the first group had sources in the *IRAS* Point Source Catalogue (PSC) and two of the control stars — in the Faint Source Catalogue. Some of these stars were observed over a strong background. In order to obtain the intrinsic flux densities of these stars at 12, 25 and 60 microns we subtracted the background contribution on FRES-CO images, facilitated from the Infrared Processing and Analyzing Center (IPAC). The results are presented in the [12]–[25], [25]–[60] colour-colour diagram, Fig. 1. The positions of six stars in Fig. 1 indicate that their emission is due to cold dust: HD 12303 in Perseus, HD 76534 in Vela, HD 37303 and HD 37903 in Orion and HD 36917 and 37806 belonging to the subgroup of the Orion complex with suspected circumstellar dust shells. Three stars (HD 76534 in Vela, HD 35007 and HD 35039 in Orion) are placed in the colour-colour diagram in the area of stars with blackbody or free-free IR radiation of the ionized gas (Cheng *et al.* 1992). Two of these stars in Orion are from the list of control stars of the Orion nearby subgroup, which were not expected to have circumstellar shells.

---

* On leave from Byurakan Astrophysical Observatory, Armenia.

Astrophysics and Space Science **224**: 577–578, 1995.
© 1995 *Kluwer Academic Publishers.*

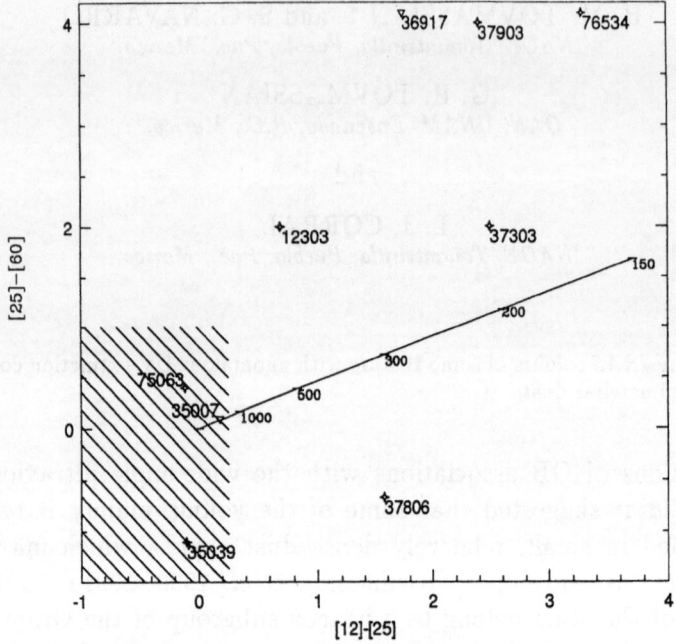

Fig. 1. Colour-colour diagram. The shaded region is the locus of stars without circumstellar dust shells.

We conclude that two of the 10 Orion subgroup stars and four of the 13 stars in different OB associations are indeed embedded in dust shells. Circumstellar dust shells were detected from none of 14 control stars which were not suspected to have them. Stars with detected circumstellar dust shells are at distances from about 40 to 730 pc and should be much more massive than the known ones. The position of the star HD 37806 on the colour-colour diagram shows that its circumstellar shell must be of the same nature as that of the Be star, 51 Oph (Waters *et al.* 1988).

References

Cheng K.-P., Bruhweiler F.C., Kondo Y., Grady C.A.: 1992, *ApJ Lett.* **396**, L83

Tovmassian H.M., Hovhannessian R.Kh., Epremian R.A., Huguenin, D.: 1990, *Astrofisica* **33**, 329

Tovmassian H.M., Hovhannessian R.Kh., Epremian R.A., Huguenin D.: 1991a, *Astrofisika* **34**, 301

Tovmassian H.M., Hovhannessian R.Kh., Epremian R.A., Huguenin D., Victorenko A.S., Serebrov A.A.: 1991b, *AZh* **68**, 942

Tovmassian H.M., Hovhannessian, R. Kh., Epremian, R. A., Huguenin, D., Khodjayants, Yu. M., Krmoyan, M. N., Kashin, A. L., Alexandrov, A. P., Romanenko, Yu. V.: 1992, *Ap. Space Sci.* **188**, 217

Tovmassian H.M., Hovhannessian R.Kh., Epremian R.A., Huguenin D.: 1993, *AJ* **105**, 627

Waters L.B.F.M., Cote J., Geballe T.R.: 1988, *Astron.Astrophys.* **203**, 348

MODELLING AND DISTANCE DETERMINATION OF PLANETARY NEBULAE

P.A.M. VAN HOOF and G.C. VAN DE STEENE

Kapteyn Astronomical Instute, P.O. Box 800, NL-9700 AV Groningen

Abstract. At present the lack of a good, generally accepted distance scale is the major limitation in the study of Planetary Nebula (PN) evolution and their distribution. We describe our fully self-consistent method to derive the physical parameters of a PN, including its distance, from a set of observables. For this we calculate various models and search for a best fit of the model predictions to the observables. An accurate determination of the angular diameter is important for a good distance determination. We think that we shall achieve an accuracy of 30% in distance. We estimate the error in stellar temperature at less than 15% and the uncertainty in the abundances at 0.10 dex.

Key words: planetary nebulae — distance determination — photo-ionization modelling

1. The Method

To model a planetary nebula we use our modified version of the photo-ionization code Cloudy 84.06 (Ferland 1993). For the model we make the following assumptions:

1. The central star has a blackbody spectrum.
2. The nebula is a spherically symmetric, constant density, ionized shell surrounded by a r^{-2}-density, neutral shell
3. The PN-type dust is intermixed with the gas.

Despite its apparent simplicity this model should give reasonable predictions, because our observables are integrated quantities and we are only interested in the mean properties of the nebula.

The above assumptions give the following free parameters: the stellar temperature, the number of hydrogen-ionizing photons, the total hydrogen density in the ionized region, the inner radius of the nebula, the dust to gas ratio, the distance to the nebula and the abundances in the nebula. With given values for these input parameters, it is possible to calculate a model for the nebula with Cloudy, predicting the continuum and line fluxes, photometric magnitudes (with line contamination) and the Strömgren radius.

Our objective is to derive from a given set of observables the physical parameters of the PN. Therefore we assume that there exists a unique set of input parameters, for which the resulting model predictions give the best fit to the observables. These input parameters are then considered the best estimate for the physical properties of the PN. To compare the model predictions with the observed quantities we calculate the goodness-of-fit estimator χ^2. This continuous function is minimized by varying all the input param-

Astrophysics and Space Science **224**: 579–580, 1995.

eters of the model, using the algorithm AMOEBA (Press *et al.*, 1988). The observed quantities we need are:

1. The spectrum of the nebula. Usually this is an optical spectrum, but might also be a UV or IR spectrum. The line ratios make it possible to constrain the stellar temperature, the density and the electron temperature in the nebula. They are also needed to determine the abundances.
2. Since we include dust in the model we also want information on the mid- and far-infrared continuum. For this we use the IRAS fluxes.
3. To constrain the emission measure, we need either a radio continuum measurement (e.g. at 6 cm) or the absolute flux value of some hydrogen recombination line (usually Hβ).
4. We also need a very accurate angular diameter of the nebula.

2.   Testing the Method

We calculated a model PN at a distance of 3.94 kpc to investigate how well the method could retrieve the physical parameters and distance using the computed "observables". The model is typical of a very young PN with low excitation. When we used the correct value for the computed observables, we could retrieve the physical parameters and the distance within 7.5%, and the luminosity within 15%. When we introduced "measurement" errors on the angular diameter, we found a strong correlation between the distance and the angular size. For our model nebula the dependence is at least quadratic.

3.   Conclusions

The method needs further testing to assess the influence of measurement errors on the results. It is already clear that an accurate determination of the angular diameter is important for a good distance determination. We think that we shall achieve an accuracy of 30% in distance and thus an accuracy of 60% in luminosity. We estimate the error in stellar temperature at less than 15% and the uncertainty in the abundances at 0.10 dex.

Acknowledgements

The authors wish to acknowledge support from NFRA Grants 782-372-033 and 782-372-035. For this research the photo-ionization code Cloudy was used, obtained from the University of Kentucky, USA.

References

Ferland, G.J.: 1993, *University of Kentucky, Department of Physics and Astronomy, Internal Report*
Press, W.H. *et al.*: 1988, *Numerical Recipes*, Cambridge University Press. 289–293

BINARITY OF THE RED-RECTANGLE AND OTHER EXTREMELY FE-POOR (POST?)-AGB STARS.

HANS VAN WINCKEL and CHRISTOFFEL WAELKENS

Institute for Astronomy, KU Leuven, Belgium

Abstract. We present radial velocity measurements proving the binary nature of all the known extremely Fe-deficient Post-ABG stars including HD 44179, the central star of the well known but poorly understood Red-Rectangle nebula. With the observed orbital parameters, it is more likely that the stars represent a particular stage in binary evolution, rather than a typical post-AGB evolutionary stage.

Key words: abundances — stars : Post-AGB — individual Objects: HR 4049, HD 52961, HD 44179 (Red Rectangle), BD+39°4926

1. Introduction

In the sample of optically bright Post-AGB candidate stars, there is a distinct subclass of four objects that display extreme photospheric abundance peculiarities: Fe, Ca, Al and other refractory elements are deficient by 3 or even 4 orders of magnitude while the CNO, S and Zn (for HD 52961) contents are solar within an order of a magnitude. The stars are HD 44179, BD+39°4926, HD 52961 and HR 4049 (Van Winckel *et al.* 1992 and references therein).

It became clear that the extreme low Fe contents are **not** primordial (Bond, 1991; Venn & Lambert, 1990), but acquired during the stellar evolution by a process of re-accretion of pure circumstellar gas which was separated from the dust, which absorbed most of the metals (Van Winckel *et al.* 1992). It has been suggested that the most favourable circumstance for this fractionation process occurs when the circumstellar matter is distributed in a disk around a binary system (Waters *et al.* 1992). The observational basis for this suggestion was that HR 4049 and BD+39°4926 are known to be binaries. In this contribution we discuss the binary nature of HD 44179, HR 4049 and HD 52961.

2. Analysis

For the first time radial velocity variations have been measured for HD 44179. We collected enough accurate radial velocity data to perform a full analysis of the velocity curves of HD 44179 and HR 4049 (Table 1.)

Both orbits have fairly large eccentricities (e=0.31 and e=0.45 resp.) and the mass functions are small (0.143 and $0.049 M_\odot$). The estimated mass for the secondary of HR 4049 falls between 0.58 and 0.72 $M_\odot$ and between 0.38 and 0.51 $M_\odot$ for HD 44179, if we adopt a mass of 0.58 $M_\odot$ for the primary.

Astrophysics and Space Science **224**: 581–582, 1995.
© 1995 *Kluwer Academic Publishers.*

TABLE I

Orbital elements of HR 4049 and HD 44179

	HR 4049	HD 44179
Period	429 d.	298 d.
V_o	-32.9 km/s ± .03	21.5 km/s ± .1
K_1	15.5 km/s ± .04	13.0 km/s ± .4
e	0.306 ± .003	0.45 ± .01
ω	248° ± 3°	28° ± 3°
$a_1 sin(i)$	0.582 AU	0.32 AU
rms	1.7 km/s	2 km/s
mass function	0.143 $M_\odot$	0.049 $M_\odot$

Also for HD 52961 important radial velocity variations have been observed. The apparently large timescale together with the measured peak to peak variations of 26 km/s exclude the possibility that the variations are due to pulsations of the star (Waelkens et al., 1991) and indicate that also HD 52961 is a binary star. Our data are too scarce to determine the orbital period.

We can conclude that **ALL** the extremely Fe poor Post-AGB candidate stars are binaries with periods of the order of a year. Clearly binarity is a crucial entry in the discussion and understanding of the special fractionation process.

Interestingly, the orbital elements of HR 4049 and HD 44179 exclude that the stars have been on the AGB with the present luminosities. The radii would have exceeded the Roche Lobes and the orbits would have been circularized, which clearly did not happen. Some AGB evolution must how-ever have occured since the circumstellar environment of HR 4049 and HD 44179 is carbon rich. It is clear then that the stars represent a peculiar stage in binary evolution, rather than a typical Post-AGB evolutionary stage. For a more detailed discussion we refer to Van Winckel et al. 1994.

References

Bond, H.E., 1991, in *Evolution of stars: the photospheric abundance connection*, eds. Michaud, G., Tutukov, A., Kluwer Dordrecht, 341.

Van Winckel, H., Mathis, J.S., Waelkens, C., 1992, *Nature* **356**, 500.

Van Winckel, H., Waelkens, C., Waters, L.B.F.M.: 1994, 'The extremely iron-deficient 'Post-AGB' stars are binaries', *A&A Letters*, submitted.

Venn, K.A., Lambert, D.L., 1990, *ApJ* **363**, 234.

Waelkens, C., Van Winckel, H., Bogaert, E., Trams, N.R., 1991, *A&A* **251**, 495.

Waters, L.B.F.M., Trams, N.R., Waelkens, C.: 1992, *A&A* **262**, L37.

MODELLING INTERSTELLAR WATER MASERS

J. A. YATES, M. D. GRAY and D. FIELD

School of Chemistry, University of Bristol, Cantock's Close, Bristol, BS8 1TS, U.K.

Abstract. Radiative transfer calculations for interstellar H_2O have been performed using accelerated Λ–iteration (ALI) techniques. The results show strong maser action from known maser transitions, as well as predicting new strong maser transitions for $\nu > 1.5$ THz.

Key words: Masers — radiative transfer

Nine interstellar H_2O maser transitions have been detected between rotational levels in the ground vibrational state. These are the 22, 183, 321, 325, 355, 380, 437, 439 and 471 GHz transitions (Yates *et al.* 1994, and references therein). To investigate the maser mechanism theoretically, approximate solutions of the radiative transfer problem have been obtained mainly by using the Sobolev approximation (e.g. Neufeld & Melnick 1991). This yields qualititative interpretation of observations. Operator-perturbation techniques such as ALI have been developed for stellar atmosphere problems which make no assumptions about the variation of physical conditions, and yield exact solutions to the radiative transfer equations. Therefore solutions can be used to interpret quantitatively observed maser data. We present sample results from the application of ALI to the radiative transfer problem of H_2O in warm (200–2000 K), dense gas, $n_{H_2} = 10^8$ to 10^{10} cm^{-3} with $n_{H_2O} = 10^3$ to 10^5 cm^{-3}, in which there is a zero velocity gradient. The radiative transfer code yields the size of the population inversion and the specific intensity of each maser transition (Jones *et al.* 1994).

All the observed maser inversions are calculated by the code, with maser gains comparable to those given by Sobolev models. The maser pump is confirmed as collisional excitation of H_2O by H_2 molecules, followed by radiative decay. Figure 1 shows how different maser transitions are affected by physical conditions. The 22 GHz maser transition is strongly amplified at low temperatures, whereas the 321 GHz maser transition is amplified most at higher temperatures. Our ALI code shows that the 437, 439, and 471 GHz transitions are strongly inverted by strong dust continuum emission, whereas other observed H_2O maser transitions, e.g. 183 GHz, are quenched by it (see Figure 2).

References

Jones, K. N. and Field, D. and Gray, M. D. and Walker R. N. F.: 1994, *ApJ*, in press.
Neufeld, D. A. and Melnick, G. J.: 1991, *ApJ*, **368**, 215.
Yates, J. A. and Cohen, R. J. and Hills, R. E.: 1994, *submitted to MNRAS*.

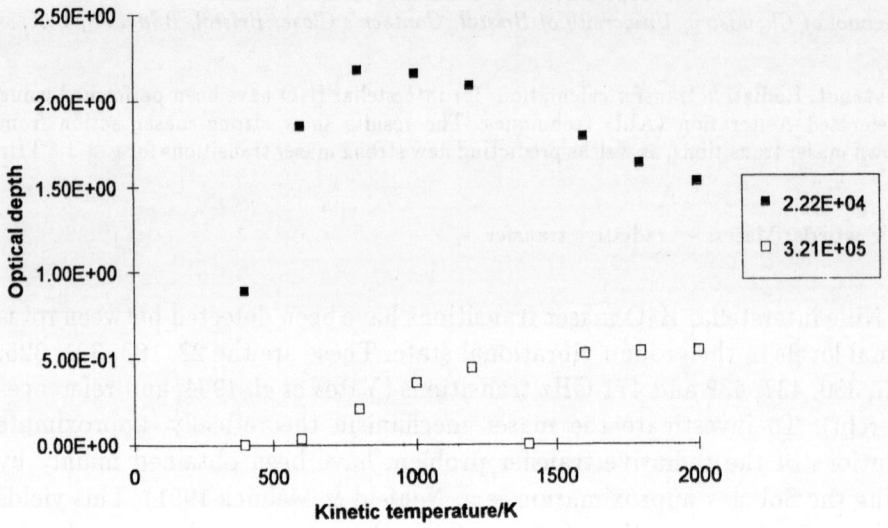

Fig. 1.   The variation of the optical depths of the 22 and 321 GHz maser transitions with kinetic gas temperature. The calculations were performed with $n_{H_2} = 10^9 \mathrm{cm}^{-3}$, $n_{H_2O} = 10^4 \mathrm{cm}^{-3}$ and T_b=3K.

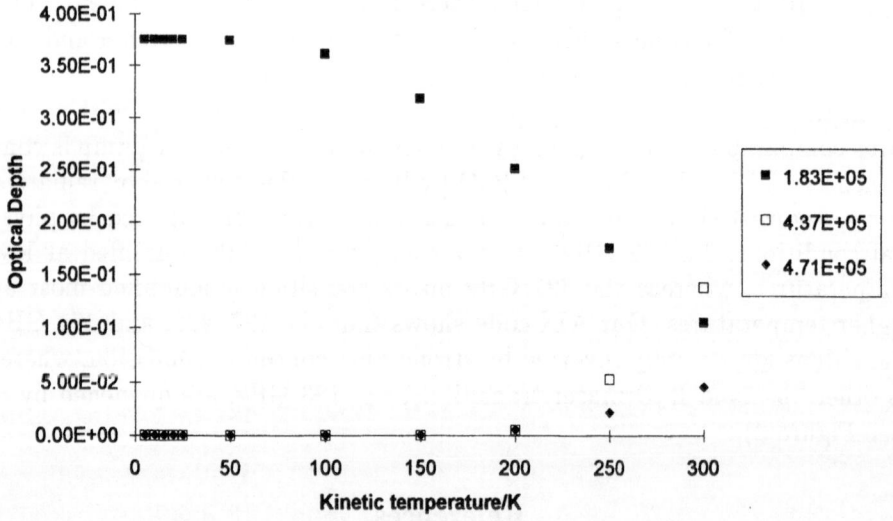

Fig. 2.   The variation of the para 183, 437 and 471 GHz optical depths with dust blackbody temperature. The calculations were performed with $n_{H_2} = 2 \times 10^9 \mathrm{cm}^{-3}$, $n_{H_2O} = 10^5 \mathrm{cm}^{-3}$ and T_k=600 K.

CS STUDIES OF DENSE CORES IN REGIONS OF HIGH

MASS STAR FORMATION:

A SURVEY OF SOUTHERN MOLECULAR MASERS

I. ZINCHENKO and A. LAPINOV

Institute of Applied Physics of the Russian Academy of Sciences 46 Uljanov st., 603600
Nizhny Novgorod, Russia

and

K. MATTILA and M. TORISEVA

Helsinki University Observatory Tähtitorninmäki, FIN-00130 Helsinki, Finland

Abstract. We searched for the CS $J = 2 - 1$ emission towards 29 southern H_2O and H_2O/OH masers and 1 OH maser with the SEST radio telescope. We detected and mapped 24 CS emitting regions probably associated with 27 H_2O masers. The $C^{34}S$ $J = 2 - 1$ and CO $J = 1 - 0$ lines were also observed at the grid positions closest to the CS peaks. Four cores were mapped in the CS $J = 5 - 4$ and $C^{34}S$ $J = 2 - 1$ lines.

The goal of this our project is to extend the list of more thoroughly investigated high-mass star-formation cores by systematic search and observation in the southern sky. It continues the line started in our previous paper on the CS studies of 11 northern cores (Zinchenko et al. 1994)

In the present study we selected H_2O masers as indicators of the regions of massive star formation. They are known to be good pointers to such regions (e.g. Plume et al. 1992) due to their frequent association and relatively accurate positions.

Here we present results of our $J = 2 - 1$ CS and $C^{34}S$ as well as $J = 1 - 0$ CO observations towards 30 H_2O and OH masers in the longitude range $260° - 310°$. We mapped all detected cores in the CS line. The CO and most of the $C^{34}S$ observations are limited to the positions of the CS emission peaks. Four cores have been mapped in the CS $J = 5 - 4$ and $C^{34}S$ $J = 2 - 1$ lines.

The observations were performed in October 1993 and April 1994 with the 15-m SEST telescope on La Silla, Chile. The source list was compiled from the catalogue of non stellar molecular maser sources published by Braz & Epchtein (1983), with additions from Braz et al. (1989) and Scalize et al. (1989).

We detected the CS emission in most of our survey directions. In some cases we found that two masers are located within or very close to a single CS emission region. In total 24 such separate CS sources have been detected associated probably with 27 masers.

The histograms of line area distributions for CS and $C^{34}S$ are presented in Fig. 1. These distributions should not be affected much by the system

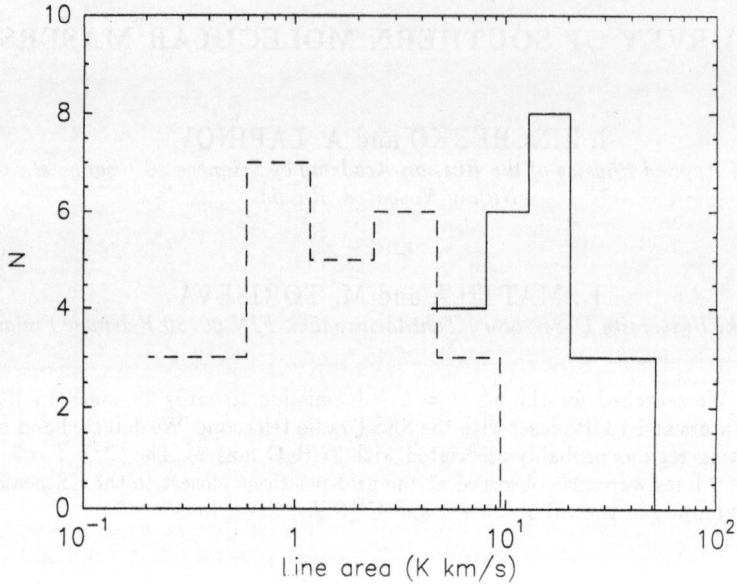

Fig. 1.   Histograms of the CS and C^{34}S line area distributions

sensitivity because the lines were detected in almost all cases. It means that the variations of the CS line area in objects of this kind are not very large. The variations of the CS column density should be of the same order of magnitude.

The examination of the IRAS point source and small scale structure catalogues demonstrates that most of these masers and CS sources are probably associated with rather strong IRAS point sources and two masers are associated with a small scale structure source. No appropriate IRAS association could be found for two masers. However, the CS emission was detected in both cases. Most frequently the nominal IRAS positions are very close to the maser positions.

The CS spectra at the peak positions obtained with higher signal-to-noise ratios demonstrate extended wings in many cases. The differences between the maser and CS velocities reach several tens of km/s and on the average the maser emission is blue-shifted relative to the CS emission.

References

Braz M.A., Epchtein N.: 1983, A&AS 54, 167
Braz M.A., Scalize E. Jr., Gregorio Hetem J.C., Monteiro do Vale J.L., Gaylard M.: 1989, A&AS 77, 465
Plume R., Jaffe D.T., Evans N.J. II: 1992, ApJS 78, 505
Scalize E. Jr., Rodriguez L.F., Mendoza-Torres E.: 1989, A&A 221, 105
Zinchenko I., Forsström V., Lapinov A., Mattila K.: 1994, A&A 288, 601

X-RAY EMISSION AS A TRACER OF CIRCUMSTELLAR MATTER AROUND HERBIG AE/BE STARS

HANS ZINNECKER and THOMAS PREIBISCH

Institut für Astronomie und Astrophysik der Universität Würzburg,
Am Hubland, D-97074 Würzburg, Federal Republic of Germany,
Internet: hans@astro.uni-wuerzburg.de; preib@astro.uni-wuerzburg.de

Abstract. We have detected X-ray emission (1 keV) from young intermediate-mass stars (Herbig Ae/Be stars). Since these stars are not supposed to produce intrinsic X-ray emission (no convection, no coronae), we believe that our results suggest that the X-ray emission actually traces the shock interaction of the Ae/Be star stellar winds with remnant circumstellar matter left over from the star formation process, the presence of which is also indicated by far-infrared (IRAS) and submm/mm continuum data.

Key words: stars: Herbig Ae/Be

We have used ROSAT to search for X-ray emission from 21 Herbig Ae/Be stars. We detected HR 5999, LkHα 349, MWC 1080, AB Aur, BD+46°3471, Z CMa, HD 97048, V892 Tau, TY CrA, V380 Ori, and LkHα 215. We did not detect LkHα 101, AFGL 4029, LkHα 198, V 376 Cas, LkHα 257, GU CMa, LkHα 234, R CrA, T CrA, and ω Ori. X-ray luminosities range from 10^{29} erg/sec for HD 97048 up to 10^{32} erg/sec for MWC1080. The high detection rate is surprising, given that Herbig Ae/Be stars were not expected to exhibit significant X-ray emission, in contrast to their lower-mass analogs, the T Tauri stars. While T Tauri stars have convective envelopes, magnetic dynamos, and hot stellar coronae where their X-ray emission originates, the Herbig Ae/Be stars have radiative envelopes, thus presumably no dynamos and no hot coronae which would give rise to X-ray emission.

Although a *stellar corona* as the source of the X-ray emission seems to be improbable, very recently a model has been proposed (Tout & Pringle 1994) where dynamo activity is sustained by tapping the initial stellar differential rotation. This leads to a magnetically heated corona that could support X-ray luminosities of up to $L_X \approx 10^{31}$ erg/sec for a period of about 10^6 years. If confirmed, the problem of the origin of X-ray from Herbig Ae/Be stars would be solved; nevertheless other possible solutions may exist.

A *disk corona* cannot be excluded as an explanation for the X-ray emission. Since the processes in the generation of a disk corona are not well understood, we can only speculate about this possibility.

Unresolved T Tauri companions may be the source of the X-rays in some cases, but cannot explain the observed X-ray luminosities for the whole sample. If the X-rays originated in the T Tauri companions (cf. Caillault *et al.* 1994), we would not expect to find (as we do) a trend for L_X proportional to L_{bol}. Furthermore, it appears that the average X-ray luminosity of T

Astrophysics and Space Science **224**: 587–588, 1995.

Tauri stars ($L_X \approx 10^{28} - 10^{30}$ erg/sec) is too weak to explain the X-ray luminosities of the Ae/Be stars; not to mention that some Ae/Be stars in all likelihood appear to be single (e.g. AB Aur).

Strong stellar winds seem to be the most probable cause and pre- condition for the X-ray luminosities observed. The kinetic power of the wind is several orders of magnitude higher than the X-ray luminosity. Thus, aside from shocks due to wind instabilities, we expect shocks when the fast wind ($v_\infty \approx 100 - 500$ km/sec) collides with the remnant circumstellar material of the Herbig stars. There is convincing evidence for such material, given the IRAS and submm/mm dust continuum detections (Hillenbrand *et al.* 1992, Mannings 1994; in fact, 16 of our 21 objects have been observed with 12 being detected). If a small fraction ($\lesssim 1\%$) of the wind luminosity ($\lesssim 10^{34}$ erg/sec) is dissipated in the shocks, X-rays will be generated with a luminosity that is just of the order of magnitude that we observe with ROSAT.

There is also an additional possibility for a wind related origin: Recently Zhekov *et al.* (1994) have computed the X-ray emission from colliding winds in young binary systems. They showed that winds typical for those of our sample can produce X-ray luminosities of $10^{30} - 10^{33}$ erg/sec in the ROSAT energy band, as observed. This process could be relevant in a few cases of our sample, but it is unlikely that all our objects are the right kind of binaries for this to be a generally viable mechanism.

Despite some alternatives (especially Tout & Pringle's model), we conclude that at present the collision of high-velocity stellar winds from the Ae/Be stars (which are known to exist) with surrounding circumstellar material (which is also known to exist) provides the most reasonable explanation for the origin of X-ray emission in these young intermediate mass stars. If true, the X-ray emission should cease whenever there is no more circumstellar matter left, i.e. when the stellar wind has swept away the remains of the remnant cloud from which the star has formed.

For more details, including X-ray images, we refer the reader to our paper in Astronomy and Astrophysics (Zinnecker & Preibisch 1994).

References

Caillault, J.P., Gagne, M., & Stauffer, J.R., 1994, Astrophys. J. , 432, 386
Hillenbrand, L.A., Strom, S.E., Vrba, F.J., & Keene, J., 1992, Astrophys. J. , 397, 613
Mannings, V., 1994, Mon. Not. of the RAS , in press
Tout, C.A. & Pringle, J.E., 1994, Mon. Not. of the RAS , in press
Zhekov, S.A., Palla, F., & Myasnikov, A.V., 1994, Mon. Not. of the RAS , in press
Zinnecker, H. & Preibisch, Th., 1994, Astron. & Astrophys. , in press

CONCLUDING

REMARKS

SUMMARY & CONCLUDING REMARKS

DAVID A WILLIAMS

Department of Mathematics, UMIST,
(present address Department of Physics and Astronomy, UCL)

At this conference we have listened to 14 reviews and 56 contributed papers, and we have read what seemed countless posters. We have been here for 5 days, we have discussed 9 separate and distinct areas of study, each of which could easily form the subject of a 5 day meeting, on its own. I congratulate you on surviving. To attempt to summarize in any formal way these enjoyable and enlightening experiences would be a pointless task. Each of us will depart with particular insights and new knowledge, and may re-orient the emphasis of our research as a result of this meeting. All I can do is to highlight a few of the most interesting issues, from my own perspective. I shall also mention some aspects of our subject that – in the excitement of the meeting – did not receive the attention I feel they deserved.

First, I have to say that it is almost impossible to cover in satisfying depth the whole subject of Circumstellar Matter in a five day meeting. What we have had is a successful but rapid and exhausting overview of the most important areas. I'm sure that, like me, you found that just as your interest had been really excited by an excellent review followed by several pertinent contributed papers, the subject of discussion moved swiftly on. So the meeting has been very successful in giving all of us an overview of present problems and achievements in various aspects of Circumstellar Matter, and reviewing that sense of progress and challenge that we all need. Most importantly, it has provided a useful link between the various areas of studies of Circumstellar Matter Areas. The published proceedings of this meeting will certainly provide the main access to these studies for some time to come. Another general aspect of the meeting that has struck me has been the emphasis on describing the morphology rather than the physics of the Circumstellar Matter. At one point, I felt that if I heard the word morphology again, I'd reach for my revolver! Areas of uncertainty or lack of real understanding were in some cases simply delineated, fenced-off and avoided, for the most part. For example, the mechanism by which jets are actually formed and collimated in the star formation process attracted little discussion and was treated as a black box, whereas pictures of the impact of those jets on ambient material were given much prominence. Yet understanding the jet acceleration is surely crucial to our understanding of star formation. As a theoretician, I regretted this preference for the softer descriptive approach, for it meant that the issues were not always as clearly formulated as they might have been. A third general comment was to note the low level

Astrophysics and Space Science **224**: 591–595, 1995.
© *1995 Kluwer Academic Publishers.*

of controversy. Perhaps this was a consequence of attention to description rather than analysis. The younger members of the audience need to know that not all areas of science are conducted in this polite, relaxed and civilized manner. However, the level of activity in our subject is quite astounding, as the oral and poster papers confirm. Progress is rapid, but – fortunately – there is still much to do!

One of the interesting discussions on protostellar and pre-main sequence shells and disks concerned the very real possibility of confusion between disk and envelope material, and the timescales over which disks persist. I think the meeting made progress in emphasizing how these different components were best identified and studied, both in remarks made in this session and also in the later session on chemistry. It was, however, remarked by several people, including I R Stevens in the final session, as surprising that the effects of binarity were largely ignored in this discussion. Perhaps the major uncertainty in interpretation we heard about is in the long wavelength dust opacity. A wide range of values was quoted by various speakers, and most were fairly confident of their views though these were not all consistent. This discussion never really emerged, was only touched upon – but the topic seems fundamental to our understanding of the early stages of star formation.

Substantial progress on protostellar and pre-protostellar cores was described in P André's wide-ranging review and D Ward-Thompson's paper. These struck me as among the most exciting new developments. Observational criteria for the various classes of object, including Class 0, now seem well established. The effort that is being put into these observational studies seems well justified, for as Derek Ward-Thompson put it "If our models can't explain pre-proto-stellar cores, then we haven't a chance of understanding the later and more complicated phases of evolution". This seems sound advice! As he remarked later to me, in the area of star formation, theory had led the observations for some time, but now this situation is reversed and there is much for theoreticians to do. This dominance of observations will be reinforced by the imminent arrival of SCUBA on JCMT, and of ISO. These new techniques should certainly solve questions of the nature of the observed cores: to what extent are they magnetically supported? Are they pressure bounded? We should soon be sure.

The other topic that ran like a thread through that session – and indeed through some other parts of the meeting – was the question of which tracer was best for circumstellar matter. A Wootten admitted that NH_3 was the "wrong" molecule. M Barsony chose H_2CO to good effect, and L Mundy gave a talk in which I agreed with everything he said except his concluding (as I saw it!) heresy "all molecules become unreliable tracers". He also said "avoid molecules, they try to fool you much more than grains. But chemistry drives the morphology". E F van Dishoeck, in her brilliant review later on, argued that this was much too pessimistic a view. Yes, the situation may be complicated, but the potential information content in molecular spectra is enormous. Even the absence of molecules is informative, as in the search for β Pic type objects discussed by H Walker. Personally, I feel that theoretical studies already in the literature of the chemistry and the gas/dust interaction in star forming regions have not yet been fully appreciated by all observers. However, N J Evans developed an impressive model of an ultracompact HII region which incorporates both deposition of molecular ices on dust grains and desorption of those ices back to the gas phase. So the ideas of molecular astrophysics are indeed gaining acceptance. The discussion of hot cores near high mass young stellar objects relating observations of G H Macdonald and of C M Walmsley to the theory as presented by S B Charnley demonstrated one of the great successes of molecular astrophysics. The theory really does work, and remarkably well! We can have confidence in a chemistry which includes gas phase reactions, freeze-out of molecules on dust, surface processing and desorption. We shouldn't minimize this success: it is a major achievement in understanding, and supports our knowledge not only of circumstellar matter, but of cool, dense matter throughout the Galaxy.

This doubt about molecules continued with W R F Dent's presentation which compared gas ($C^{18}O$ emission) with dust emission, and found a poor correlation. All of this assumes that we know how effective dust is as a tracer of material in star forming regions. However, our lack of knowledge on this point was brought out by others in the discussion, notably in a typically penetrating question by C M Walmsley which reminded us that the dust emission at 100 μm is known to underestimate mass. In addition, a further uncertainty is that the dust grains in the circumstellar medium are probably not like those in the interstellar medium; we don't really know their nature, size distribution, structure, and – as remarked above – their long wavelength optical properties. Evidently, further work is needed before we are confident about using the perfect tracer of mass in these regions. The nature of dust in circumstellar regions, as in the interstellar medium, is not immutable, but responds to conditions in the medium in which it is embedded. This important point was emphasized by W W Duley, who presented a convincing picture of carbon dust evolution in various circumstellar environments, using as a basis for his discussion recent laboratory data for carbon materials. This picture will be further developed, but certainly suggests that a combination of laboratory, theory and observations should soon provide a definitive description of the nature of circumstellar dust. Duley startled his audience by cutting up his transparencies to demonstrate the break-up of solid carbon into free-flying PAH and polyyne molecules.

The session on outflows was particularly timely and interesting, and the background provided by C R Masson's review and C J Chandler's talk made this new and fashionable area of study quite accessible. N Hirano's work addressed the Taurus outflows and made the point that outflows begin very early in the evolution of young stellar objects, indeed as soon as they become observable by IRAS. One assumes that this must be a fundamentally important point, though there was no discussion at the meeting to elucidate what it means. The high velocities apparent in star-forming regions would seem to indicate a high probability of the presence of shocks. However, P W J L Brand, in a disarmingly honest appraisal of the situation remarked "none of the existing shock models seem to explain the existing line profiles". He pointed out observations of DR21 as a "terrible warning for theorists not to look at the observations in too much detail". Perhaps we should start dismantling our telescopes at that point when theory and observation agree! Clearly, theoreticians have much to do in this area.

I found it surprising that so many problems associated with late evolutionary stages of high mass stars involved dust and gas in filamentary structures. The conditions that P M Williams was describing, for instance, seemed very hostile to dust formation, yet the process is apparently successful. Yet, I suppose that if dust can be formed in the hostile environment of novae, then the process should also be possible in a binary WR star, even if we don't fully understand either situation. I R Stevens' excellent review of colliding winds model this and gave us some insight into how the appropriate conditions for dust formation might arise.

The review of cool circumstellar environments by P J Huggins made clear that this subject, in which models have had much success, is – in fact – far from fully worked out, and that there is still much to do in terms of understanding the development of asymmetries, clumpiness and some chemical details. He remarked with delightful understatement of IRC+10216 "the archetype seems quite complicated". Nevertheless, the subject of the physical (and especially) chemical circumstellar environments is one that has also been a major success for astronomy and astrophysics, and in a consequence of excellent collaboration between theory, observation, and laboratory work this will, I'm sure, continue into the future.

A common theme in all of circumstellar matter, from early to late stages of evolution, has been the clumpiness and irregularity of the material whether in a cloud, an envelope, or – much to my surprise, as I learned here – even a stellar atmosphere. I was much impressed that a new technical term has been agreed at this meeting to describe this phenomenon – woofly. This term was introduced by several of our US speakers, but rapidly taken up by others. For example, P M Williams said: "stars aren't round but woofly as well". It's a new term to me, but quite expressive. It seems to me that the term could also be applied to some models! As A Zijlstra remarked in his excellent review on stellar evolution "it is not surprising that when astronomers deal with too many variables, their predictions begin to lose precision".

I was told that the standard conclusion of a summary talk was to note the excitement in the field and to plead for more money for telescopes. That scarcely seems necessary. The subject is booming, driven by observations from γ rays (very clever work reported here by N Langer), X rays, UV, optical, IR, sub-mm, mm, long radio. The present achievements and potential of the interferometers described by W Danchi are astounding! The idea of measuring stellar diameters to an accuracy of 1% is incredible. Great progress will clearly be made in the next few years, and if a 5 day meeting cannot be extended then the coverage will need to be more selective.

Finally, it remains to thank Graeme Watt and his colleagues for their magnificent work in putting on this splendid conference and planning all the social events which have been so enjoyable. We are all very pleased that he managed to persuade the authorities in Edinburgh to hold their Festival at the time of our conference, and I'm sure we are particularly impressed that they could be persuaded to put on the Firework Display to celebrate our excellent conference dinner. Thank you, Graeme, from all of us for your hard work.

INDICES

CIRCUMSTELLAR MATTER

1994

LIST OF PARTICIPANTS

Nancy Ageorges:–	MPE, Garching, GERMANY
Geary Albright:–	University of Virginia, USA
Phillipe André:–	Centre d'Etudes de Saclay, FRANCE
Bernhard Aringer:–	University of Vienna, AUSTRIA
Jane Arthur:–	Universidad Nacional Autonoma de Mexico, MEXICO
Lorne Avery:–	Herzberg Institute of Astrophysics, CANADA
Stefano Bagnulo:–	Armagh Observatory, NORTHERN IRELAND
Eric Bakker:–	SRON, Utrecht, NETHERLANDS
Dominique Ballereau:–	Observatoire de Paris-Meudon, FRANCE
Mary Barsony:–	UC Riverside, California, USA
Nicole Berruyer:–	Observatoire de la Cote d'Azur, FRANCE
Nina Beskrovnaya:–	Pulkova Observatory, St. Petersburg, RUSSIA
Peter Brand:–	University of Edinburgh, SCOTLAND
Danielle Briot:–	Observatoire de Paris, FRANCE
Henry Buckley:–	University of Edinburgh, SCOTLAND
Valentin Bujarrabal:–	Centro Astronomico de Yebes, SPAIN
Harold Butner:–	DTM-Carnegie Institute of Washington, USA
John Carr:–	Ohio State University, USA
Mark Casali:–	Royal Observatories, Edinburgh, SCOTLAND
Josephine Chan:–	Max-Planck Research Group, Jena, GERMANY
Claire Chandler:–	NRAO, Socorro, New Mexico, USA
Steve Charnley:–	UC Berkeley/NASA Ames, California, USA
Isabelle Cherchneff:–	Royal Observatories, Edinburgh, SCOTLAND
Ed Churchwell:–	University of Wisconsin, USA
Mark Clampin:–	Space Telescope Science Institute, Maryland, USA
Stuart Clark:–	University of Hertfordshire, ENGLAND
Jim Cohen:–	NRAL Jodrell Bank, ENGLAND
Martin Cohen:–	University of California, Berkeley, USA
Francisco Colomer:–	Centre Astronomico de Yebes, SPAIN
Carlos Correia:–	University of Lisbon, PORTUGAL
Bill Danchi:–	Space Science Laboratory, Berkeley, USA
Chris Davis:–	MPIA, Heidelberg, GERMANY
Francisco de Araujo:–	Observatoire de la Cote D'Azur, FRANCE
Bill Dent:–	JAC, Hawaii, USA
Lynne Deutsch:–	University of Massachussetts, Amherst, USA
Sean Dougherty:–	Liverpool John Moore's University, ENGLAND

Astrophysics and Space Science **224**: 599–603, 1995.

Walt Duley:–   University of Waterloo, CANADA
Wolfgang Duschl:–   ITA, University of Heidelberg, GERMANY
David Emerson:–   University of Edinburgh, SCOTLAND
Sandra Etoka :–   Observatoire de Paris-Meudon, FRANCE
Neal Evans:–   University of Texas, Austin, USA
Stewart Eyres:–   NRAL, Jodrell Bank, ENGLAND
Marco Ferrari-Toniolo:–   Istituto di Astrofisica Spaziale, Frascati, ITALY
Michael Feuchtinger:–   Universitat Wien, AUSTRIA
Sergio Fonti:–   University of Lecce, ITALY
Dan Gezari:–   NASA/Goddard, Maryland, USA
Andy Gibb:–   University of Kent, ENGLAND
Alistair Glasse:–   Royal Observatories, Edinburgh, SCOTLAND
Malcolm Gray:–   University of Bristol, ENGLAND
Lincoln Greenhill:–   Smithsonian Astrophysical Observatory, USA
Martin Groenewegen:–   Institute d'Astrophysique de Paris, FRANCE
Harm Habing:–   Sterrewacht Leiden, NETHERLANDS
Andrew Harrison:–   University of Edinburgh, SCOTLAND
Lee Hartmann:–   Center for Astrophysics, Cambridge, USA
Paul Harvey:–   University of Texas, Austin, USA
Osamu Hashimoto:–   Seikei University, Tokyo, JAPAN
Saeko Hayashi:–   SUBARU Project Office, Tokyo, JAPAN
Will Henney:–   Universidad Nacional Autonoma de Mexico, MEXICO
Naomi Hirano:–   Hitosubashi University, Tokyo, JAPAN
Melvin Hoare:–   MPIA, Heidelberg, GERMANY
Susanne Hofner:–   Universitat Wien, AUSTRIA
Joseph Hora:–   Institute for Astronomy, Honolulu, USA
Patrick Huggins:–   New York University, USA
Vic Hughes:–   Queen's University, Kingston, CANADA
Robert Hurt:–   UC Riverside, California, USA
Busaba Hutawarakorn:–   NRAL, Jodrell Bank, ENGLAND
Alex Il'in:–   Pulkovo Observatory, St. Petersburg, RUSSIA
Syuzo Isobe:–   National Astronomical Observatory, Tokyo, JAPAN
Rob Ivison:–   Royal Observatories, Edinburgh, SCOTLAND
Hideyuki Izumiura:–   Tokyo Gakuger University, JAPAN
Eric Josselin:–   Institute d'Astrophysique de Paris, FRANCE
Kay Justtanont:–   NASA Ames Research Center, California, USA
Jill Knapp:–   Princeton University, New Jersey, USA
Carsten Koempe:–   Uni-Sternwarte Jena, GERMANY
Maria Kun:–   Konkoly Observatory, Budapest, HUNGARY
Jean-Pierre Lafon:–   Observatoire de Paris-Meudon, FRANCE
Jeane-Pierre Lagage:–   Centre d'Etudes de Saclay, FRANCE
Sergei Lamzin:–   Sternberg Astronomical Institute, Moscow, RUSSIA
Norbert Langer:–   MPA, Garching, GERMANY

A M Le Squeren:–   Observatoire de Paris-Meudon, FRANCE
Giuseppe Leto:–   Catania University, ITALY
John Lightfoot:–   Royal Observatories, Edinburgh, SCOTLAND
Michael Lindqvist:–   Sterrewacht Leiden, NETHERLANDS
Silvia Lorenz-Martins:–   Observatoire de la Cote D'Azur, FRANCE
Robert Lucas:–   IRAM, Grenoble, FRANCE
Ed Ludke:–   Universidade Federal de Santa Maria, BRAZIL
Stuart Lumsden:–   Anglo-Australian Observatory, Epping, AUSTRALIA
Geoff Macdonald:–   University of Kent, ENGLAND
Duncan Mackay:–   University of Kent, ENGLAND
Arturo Manchado:–   Instituto de Astrofisica de Canarias, Tenerife, CANARY ISLANDS
Vince Mannings:–   University College London, ENGLAND
Colin Masson:–   Center for Astrophysics, Cambridge, USA
Masafumi Matsumura:–   Kagawa University, Takamatsu, JAPAN
Mike McCartney:–   University of Edinburgh, SCOTLAND
Bruce McCollum:–   Computer Sciences Corporation, Maryland, USA
John McLeod:–   Herzberg Institute of Astrophysics, CANADA
Nigel Minchin:–   Queen Mary & Westfield College, London, ENGLAND
Luis Miranda:–   Universida Autonoma de Madrid, SPAIN
Andre Moitinho de Almeida:–   University of Lisbon, PORTUGAL
Toby Moore:–   University of New South Wales, Canberra, AUSTRALIA
Lee Mundy:–   University of Maryland, USA
Takenori Nakano:–   Nobeyama Radio Observatory, JAPAN
Jim Neff:–   Penn State University, Pennsylvania, USA
Ralph Neuhauser:–   MPE, Garching, GERMANY
Antonella Nota :–   Space Telescope Science Institute, Maryland, USA
Lars Nyman:–   ESO, Santiago, CHILE
Katsuo Ogura:–   Kokugakuin University, Tokyo, JAPAN
Nagayoshi Ohashi:–   Nobeyama Radio Observatory, JAPAN
Masatoshi Ohishi:–   Nobeyama Radio Observatory, JAPAN
Hans Olofsson:–   Stockholm Observatory, SWEDEN
Elena Ortiz:–   Universida Autonoma de Madrid, SPAIN
Rene Oudmaijer:–   Kapteyn Laboratory, Groningen, NETHERLANDS
Isabella Pagano:–   Catania Astrophysical Observatory, ITALY
Elma Parsamian:–   Instituto Nacional de Astrofisica, Optica y Electronica, Pue, MEXICO
Gianni Pascoli:–   Universite de Picardie Jules Verne, Amiens, FRANCE
Guy Perrin:–   DESPA, Observatoire de Paris-Meudon, FRANCE
Paolo Persi:–   Istituto di Astrofisica Spaziale, ITALY
Martin Peters:–   Teikyo University, Tokyo, JAPAN
Tim Phillips:–   Hughes STX Corporation, New York, USA
Mikhail Pogodin:–   Pulkova Observatory, St. Petersburg, RUSSIA

John Porter:–   University of Oxford, ENGLAND
Phil Puxley:–   Royal Observatories, Edinburgh, SCOTLAND
Susanne Ramsay-Howat:–   Royal Observatories, Edinburgh, SCOTLAND
Jonathan Rawlings:–   UMIST, Manchester, ENGLAND
Anita Richards:–   NRAL, Jodrell Bank, ENGLAND
Mercedes Richards:–   University of Virginia, USA
Angels Riera:–   Universitat de Barcelona, SPAIN
Ian Robson:–   JAC, Hawaii, USA
Adrian Russell:–   Royal Observatories, Edinburgh, SCOTLAND
Nicole Saint-Louis:–   Universite de Montreal, CANADA
Eric Schlegel:–   NASA-GSFC/USRA, Maryland, USA
Andreas Schulz:–   MPI, Bonn, GERMANY
Karl Schuster:–   IRAM, St. Martin d'Heres, FRANCE
Chris Skinner:–   Lawrence Livermore National Laboratory, USA
Augustin Skopal:–   Skalnaté Pleso Observatory, SLOVAKIA
Phillipe Stee:–   Observatoire de la Cote d'Azur, FRANCE
Robert Stencel:–   University of Denver Observatories, Colorado, USA
Lars Stenholm:–   Uppsala Astronomical Observatory, SWEDEN
Tomasz Stepinski:–   Lunar and Planetary Institute, Houston, Texas, USA
Michael Sterzik:–   MPE, Garching, GERMANY
Ian Stevens:–   University of Birmingham, ENGLAND
Jim Stone:–   University of Maryland, USA
Koji Sugitani:–   Nagoya City University, JAPAN
Roger Sylvester:–   University College London, ENGLAND
Motohide Tamura:–   National Astronomical Observatory, Tokyo, JAPAN
Jonathan Tedds:–   Royal Observatories, Edinburgh, SCOTLAND
Teresa Teixeira:–   Queen Mary & Westfield College, London, ENGLAND
S P Thompson:–   Leicester University, Leicester, ENGLAND
Hrant Tovmassian:–   Instituto Nacional de Astrofisica, Optica y Electronica, Pue, MEXICO
Corrado Trigilio:–   Istituto di Radioastronomia, CNR, Noto, ITALY
Barry Turner:–   NRAO, Charlottesville, USA
Munetaka Ueno:–   University of Tokyo, JAPAN
Grazia Umana:–   Istituto di Radioastronomia, CNR, Noto, ITALY
Johan van der Walt:–   Potchefstroom University, SOUTH AFRICA
Ewine van Dishoeck:–   Leiden Observatory, NETHERLANDS
Peter van Hoof:–   Kapteyn Institute, Groningen, NETHERLANDS
Hans van Winckel:–   Instituut voor Sterrenkunde, Leuven, BELGIUM
Nikolai Voschchinnikov:–   St. Petersburg University, RUSSIA
Christoffel Waelkens:–   Instituut voor Sterrenkunde, Leuven, BELGIUM
Helen Walker:–   Daresbury Rutherford Appleton Laboratory, ENGLAND
Malcolm Walmsley:–   MPI, Bonn, GERMANY
Derek Ward-Thompson:–   Royal Observatories, Edinburgh, SCOTLAND

Graeme Watt:-   Royal Observatories, Edinburgh, SCOTLAND
Karen Willacy:-   University of Edinburgh, SCOTLAND
David Williams:-   UMIST, Manchester, ENGLAND
Peredur Williams:-   Royal Observatories, Edinburgh, SCOTLAND
J M Winters:-   Technische Universität Berlin, GERMANY
Joachim Wirsick:-   Berlin, GERMANY
Ray Wolstenccroft:-   Royal Observatories, Edinburgh, SCOTLAND
Doug Wood:-   NRAO, Socorro, USA
Al Wootten:-   NRAO, Charlottesville, USA
Gareth Wynn-Williams:-   Institute of Astronomy, Hawaii, USA
Jeremy Yates:-   University of Bristol, ENGLAND
Albert Zijlstra:-   ESO, Garching, GERMANY
Igor Zinchenko:- Russian Academy of Sciences, Nizhny Novgorod, RUSSIA
Hans Zinnecker:-   University of Würzburg, GERMANY
Robert Zylka:-   MPI, Bonn, GERMANY

AUTHOR INDEX

SOURCE INDEX

SUBJECT INDEX

accretion disks   3, 13, 211, 415,
         523, 525, 545, 547, 565
AGB stars   217, 255, 281, 305,
         309, 321, 329, 339, 379,
         383, 419, 421, 433, 439,
         451, 477, 485, 495, 499,
         501, 511, 515
Algol variables   415, 547
ammonia   43
Anglo-Australian Telescope 151, 521
ASCA satellite   367, 553, 569

Beta-Pic like stars   389, 395, 475,
         525
BIMA   81
binary stars   491, 519, 557,
         559, 569, 581
bipolar nebulae   143, 361, 457,
         487, 517, 535, 549, 551

carbon-rich stars & envelopes   217,
         227, 321, 329, 357, 379,
         383, 417, 425, 435, 493,
         495, 501, 503, 511, 527,
         531, 543
Calar Alto Observatory   517, 529
chemistry   81, 177, 181,
         217, 237, 251, 281, 293,
         297, 379, 425, 439, 441,
         507, 517, 543
CO observations   17, 25,
         73, 109, 113, 117, 143,
         305, 321, 433, 467, 497,
         499, 573
colliding winds   367
collision cross-sections   537
compact H II regions   505

Crimean Astr. Obs.   429, 539
CSO   81, 189

distance determination   579
Dodaira Observatory   513
doppler tomography   547
dust composition   217, 267, 297,
         321, 329, 339, 389, 405,
         419, 425, 503, 531, 543,
         563, 567, 573, 575
dust grains   21, 81, 177, 227,
         223, 233, 251, 267, 379,
         441, 507, 521, 573
dust shells   85, 223, 321, 329, 339,
         389, 395, 417, 427, 445,
         447, 449, 459, 463, 469,
         475, 477, 481, 485, 495,
         511, 515, 519, 525, 531,
         535, 539, 543, 545, 549,
         557, 567, 573, 577, 581

Effelsberg   445
episodic mass loss & outflow   151,
         501
ESO   357, 411, 421, 539, 549
evolved stars   261, 281, 293, 297,
         305, 309, 321, 325, 329,
         335, 377, 563
extended red emission   217
extinction curves   223

filamentary structure   533
formaldehyde   51
fluorescence   271

Gamma-ray lines   275